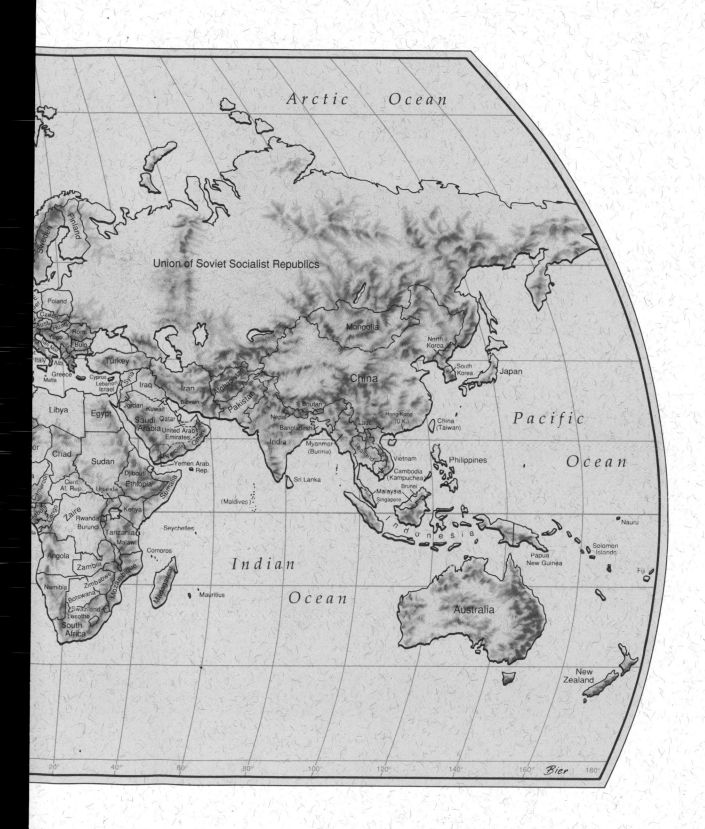

INTRODUCTION TO
GEOGRAPHY

INTRODUCTION TO

GEOGRAPHY

Arthur Getis
University of Illinois, Urbana—Champaign

Judith Getis

Jerome Fellmann
University of Illinois, Urbana—Champaign

Third Edition

WCB **Wm. C. Brown Publishers**

Book Team

Editor *Jeffrey L. Hahn*
Developmental Editor *Lynne M. Meyers*
Production Editor *Ann Fuerste/Connie Balius-Haakinson*
Designer *Mark Elliot Christianson*
Art Editor *Joan M. Soppe*
Photo Editor *Michelle Oberhoffer*
Permissions Editor *Vicki Krug*
Visuals Processor *Joe O'Connell*

 Wm. C. Brown Publishers

President *G. Franklin Lewis*
Vice President, Publisher *George Wm. Bergquist*
Vice President, Publisher *Thomas E. Doran*
Vice President, Operations and Production *Beverly Kolz*
National Sales Manager *Virginia S. Moffat*
Advertising Manager *Ann M. Knepper*
Marketing Manager *John W. Calhoun*
Editor in Chief *Edward G. Jaffe*
Managing Editor, Production *Colleen A. Yonda*
Production Editorial Manager *Julie A. Kennedy*
Production Editorial Manager *Ann Fuerste*
Publishing Services Manager *Karen J. Slaght*
Manager of Visuals and Design *Faye M. Schilling*

Cover © Ken Ross/Viesti Associates, Inc.

The credits section for this book begins on page 454, and is considered an extension of the copyright page.

Library of Congress Catalog Card Number: 89–82465

ISBN 0–697–07919–8 (cloth) / 0–697–13279–X (paper)

Printed in the United States of America by Wm. C. Brown Publishers, 2460 Kerper Boulevard, Dubuque, IA 52001

10 9 8 7 6 5 4 3 2 1

C O N T E N T S

4

Physical Geography: Weather and Climate 81

5

Human Impact on the Environment 121

P A R T 2

The Culture–Environment Tradition 158

6

Population Geography 160

P A R T 3

The Locational Tradition 298

10

Economic Geography 300

11

The Geography of Natural Resources 338

12

Urban Geography 376

P A R T   4

The Area Analysis Tradition 412

13

The Regional Concept 414

P R E F A C E

This third edition of *Introduction to Geography* retains the purpose of its predecessors: to introduce college students to the breadth and excitement of the field of geography. Like previous editions, its content is organized around the major research traditions of the discipline, a structure that adopters of earlier editions have found to be attractive to students and convenient and flexible for instructors.

Although the framework of presentation has been retained in this current edition, we have added and deleted materials to reflect current advances in the several topical areas considered, updated facts, and made every effort to profit by suggestions offered by users and colleagues to improve content. Among these suggestions were helpful ideas for substantial restructuring of the chapters on cultural, behavioral, and economic geography and for recasting the content of the political and urban geography discussions. We are grateful for those suggestions.

We recognize that many students will have only a single college course and textbook in geography. Our purpose for those students is to convey concisely but clearly the nature of the field, its intellectual challenges, and the logical interconnections of its parts. Even if they take no further work in geography, we are satisfied that they will have come into contact with the richness and breadth of our discipline and have at their command new insights and understandings for their present and future role as informed adults. Other students may have the opportunity and interest to pursue further work in geography. For them, we believe, this text will make apparent the content and scope of the subfields of geography, emphasize its unifying themes, and provide the foundation for further work in their areas of interest.

The approach we take is to let the major research traditions of geography dictate the principal themes we explore. As Chapter 1 makes clear to students, the organizing traditions that have emerged through the long history of geographical thought and writing are those of earth-science, culture–environment, location, and area analysis. Each of the four parts of this book centers on one of these geographic perspectives. Within each part, except that on area analysis, are chapters devoted to the subfields of geography, each placed with the tradition to which we think it belongs. Thus, the study of weather and climate is part of the earth-science tradition; population geography is considered under the culture–environment tradition; and urban geography is included with the locational perspective.

Of course, our assignment of a topic may not seem appropriate to all users since each tradition contains many emphases and themes. Some subfields could logically be attached to more than one of the recognized traditions. The rationale for our clustering of chapters is given in the brief introductions to each part of the text. The tradition of area analysis—of regional geography—is presented in a single final chapter that draws upon the preceding traditions and themes and is integrated with them by cross-references.

In this revision we have made every effort to incorporate the most current viewpoints and approaches and to include the data on which they are based. Many of the diagrams, maps, and photographs are new to this edition. Their frequently extensive accompanying captions convey additional information and explanation—that is, are part of the text—not just identification or documentation of the illustration. Chapter introductions are designed to capture students' attention, arousing their interest in the vignette itself and, therefore, in the subject matter that follows. In addition, each chapter contains boxed inserts that further embellish points or ideas included in the text proper. In short, every effort has been made to gain and retain student attention—the essential first step in the learning process.

Chapter 1 prepares the student for the later substantive chapters. It introduces the field of geography as a whole, noting breadth of interests, and the unifying questions, themes, and concepts that structure all geographic inquiry whatever the specialized subfield of interest. It also outlines for the student the organization of the book and explains the several "traditions" forming its framework.

Important to that framework is the terminal single chapter treatment of the area analysis tradition. The case studies and examples that Chapter 13, "The Regional Concept," contains illustrate the regional geographic application of the systematic themes developed by the earlier chapters. Regional understanding has always been an important motivation and justification of geography as a discipline; Chapter 13 is designed to introduce students to the diversity of regional geographic exposition. It may be read either as a separate chapter or in conjunction with the earlier material. That is, each systematic chapter contains a reference to the portion of Chapter 13 where a relevant regional geographic example is to be found. That referenced case study, of course, can be incorporated to demonstrate the relationships of regional and systematic geography, to show the "real world" application of geographic understandings, and to provide a springboard for further case studies as class or instructor interest may dictate. The included regional studies may also serve as models for independent student reports by applying to specific cases the insights and techniques of analysis developed in the separate substantive chapters.

Learning aids at the conclusion of each chapter include a Summary, a list of Key Words together with a page reference to the location where each term is defined and discussed, For Review questions designed to help students check on their understanding of the chapter material, and a Selected References listing of important recent or classic considerations of the subject matter of the chapter. We have tried to include not only relatively widely available recent textbook and other titles, many containing additional extensive bibliographies, but also more specialized articles and monographs useful to students who are motivated to delve more deeply into particular subfields of geography.

At the end of the book we have placed a comprehensive Glossary of terms and, as a special Appendix, a modified version of the *1990 World Population Data Sheet* of the Population Reference Bureau. In addition to basic demographic data and projections for countries, regions, and continents, the *Data Sheet* includes statistics on gross national product, energy consumption, and access to safe water supply. Although inevitably dated and subject to change, the appendix data nonetheless will provide for some years a wealth of useful comparative information for student projects, regional and topical analyses, and study of world patterns.

A useful textbook must be flexible enough in its organization to permit an instructor to adapt it to the time and subject-matter constraints of a particular course. Although designed with a one-quarter or one-semester course in mind, this text may be used in a full-year introduction to geography when employed as a point of departure for special topics and amplifications introduced by the instructor or when supplemented by additional readings or class projects. Moreover, the chapters are reasonably self-contained and need not be assigned in the sequence here presented. The "traditions" structure may be dropped and the chapters rearranged to suit the emphases and sequences preferred by the instructor or found to be of greatest interest to the students. The format of the course should properly reflect the joint contribution of instructor and book rather than be dictated by the book alone.

A number of reviewers, anonymous at the time, greatly improved the content of this book by their critical comments and suggestions. With pleasure we acknowledge the thoughtful assistance rendered by James C. Hughes, Slippery Rock University; Paul Butt, University of Central Arkansas; Edward Babin, University of South Carolina–Spartanburg; and Kavita Pandit, University of Georgia. We are also indebted to W. D. Brooks and C. E. Roberts, Jr., of Indiana State University for the projection used for many of the maps in this book: a modified van der Grinten. We gratefully thank these and unnamed others for their help and contributions and specifically absolve them of responsibility for decisions on content and for any errors of fact or interpretation that users may detect.

Finally, we note with deep appreciation and admiration the efforts of the publisher's "book team" separately named on the reverse of the title page. In particular, we acknowledge the supportive interest of our editor, Jeffrey L. Hahn; the careful and superbly professional guidance of Senior Developmental Editor Lynne M. Meyers; the helpful and dedicated efforts of Senior Production Editor Ann Fuerste; and the design skills of Mark E. Christianson.

Arthur Getis
Judith Getis
Jerome Fellmann

1

Introduction

Vacationers along the Costa Blanca in Spain.

GOVERNMENT inspectors had issued the first of two warnings about the dam nearly a year before, but no repairs had been made. Fortunately, workers spotted the cracks early, and when the dam broke, letting loose a 20-foot (6-m) wave, the villages had been cleared and all train and highway traffic stopped. What poured through the breach was not just water, but a thick deadly brine, the impounded discharge of a major fertilizer plant. On its rush to the river, the brine swept away railroad track, ripped up roads, smashed through a village, inundated 500 acres (200 hectares) of the nation's richest farmland, and spilled into the purest river left in the western part of the country. Worse than the immediate physical destruction were the aftereffects. Two thousand tons of fish were destroyed along with all of the vegetation that could support their replacement. Water supplies to two major and many minor cities were cut off as the formerly pure river became "brinier than the saltiest seawater," and a million tons of salt were deposited in a layer 35 feet (10.5 m) thick at the bottom of a major reservoir 300 miles (480 km) downstream.

The accident happened in 1983 in the western part of the Soviet Union and affected the Dniester River from near the Polish border all the way to the Black Sea. The details of its location are less important than the lessons contained in the event. The wall of brine, the destruction of farmland and fish, the salt layer at the lake bottom, and the scramble for alternate water supplies are dramatic evidence of the pressures humans place upon the environment of which they are a part. Radioactive waste materials and atmospheric discharges through nuclear plant accidents, deadly manufactured chemicals, accelerated erosion through unwise forestry and farming practices, the creation of deserts through overgrazing, the poisoning of soils by salts from faulty irrigation and of groundwater through deep-well injection of the liquid garbage of modern industry—all are evidences of the adverse consequences of human pressures upon natural systems. The social and economic actions of humans occur within the context of the environment and have environmental consequences too dangerous to ignore.

The interaction of human and environmental systems works both ways. People can inflict irreparable damage upon the environment and the environment can exact a frightening toll from the societies that inappropriately exploit it.

November is the cyclone season in the Bay of Bengal. At the northern end of the bay lie the islands and the lowlands of the Ganges Delta, a vast fertile land, mostly below 30 feet (9 m) in elevation, made up of old mud, new mud, and marsh. Densely settled by desperately land-needy people, this delta area is home to the majority of the population of Bangladesh. Early in November of 1970 a low-pressure weather system moved across the Malay Peninsula of Southeast Asia and gained strength in the Bay of Bengal, generating winds of nearly 150 miles (240 km) per hour. As it moved northward, the storm sucked up and drew along with it a high wall of water. On the night of November 12, with a full moon and highest tides, the cyclone and its battering ram of water slammed into the islands and the deltaic mainland. When it had passed, some of the richest rice fields in Asia were gray with the salt that ruined them, islands totally covered with paddies were left as giant sand dunes, and an estimated 500,000 people had perished.

Should the tragedy be called the result of the blind forces of nature, or should it be seen as the logical outcome of a state of overpopulation that forced human encroachment upon lands more wisely left as the realm of river and sea (Figure 1.1)?

As a discipline, geography does not attempt to make value judgments about such questions. Geography does claim, however, to be a valid and revealing approach to contemporary questions of political, economic, social, and ecological concern. Humans and environment in interaction; the patterns of distribution of natural phenomena affecting human use of the earth; the cultural patterns of occupance and exploitation of the physical world—these are the themes of that encompassing discipline called *geography*.

Geographic knowledge is vital to an understanding of the important national and international problems that dominate daily news reports. Acid rain and the greenhouse effect, the decline of manufacturing in the "Rust Belt" and migration to the "Sunbelt," the deterioration of urban areas and the rise of crime associated with the use of illicit drugs, international trade deficits, problems of food supply and population growth in developing countries, turmoil in Central America and the Middle East—all these occur in a geographic context, and geography helps to explain them. To be geographically illiterate is to deny oneself not only the ability to comprehend world problems but also the opportunity to contribute meaningfully to the development of policies to deal with them.

Roots of the Discipline

Geography literally means "description of the earth." It is a description in which people and their activities assume a central position. As a way of thinking about and analyzing the earth's surface, its physical patterns, and human occupance, geography has ancient roots. It grew out of early Greek philosophical concern with "first beginnings"

HURRICANE HUGO
22 SEPTEMBER 1989
1201 AM EST

Figure 1.1

Hurricane Hugo, one of the most devastating storms ever to hit the Caribbean Islands and U.S. mainland, is shown here approaching the South Carolina coast in September 1989. Winds of 135 miles per hour (216 km per hour) and a storm surge of water 20 feet (6 m) high ravaged the islands offshore and did more than $4 billion damage to mainland Charleston and nearby cities and towns. Much of the damage occurred on barrier beaches and along ocean shore resorts that had been heavily developed and built up despite recurring warnings by environmentalists that such a disaster might take place. (See also Figure 9.6.)

and with the nature of the universe. From that broader inquiry, it was refined into a specialized investigation of the physical structure of the earth, including its terrain and its climates, and the nature and character of its contrasting inhabited portions.

To Strabo (*c.* 64 B.C.–A.D. 20), one of the greatest of Greek geographers, the task of the discipline was to "describe the known parts of the inhabited world . . . to write the assessment of the countries of the world [and] to treat the differences between countries." Even earlier, Herodotus (*c.* 484–425 B.C.) had found it necessary to devote much of his book to the lands, peoples, economies, and customs of the various parts of the Persian Empire as necessary background to audience understanding of the causes and course of the Persian wars.

Geographers measured the earth, devised the global grid of latitudes and longitudes, and drew upon this grid surprisingly sophisticated maps of their known world (Figure 1.2). They explored the apparent latitudinal variations in climate and described in numerous works the familiar Mediterranean basin and the more remote, partly rumored lands of northern Europe, Asia, and equatorial Africa. Employing nearly modern concepts, they described river systems, explored cycles of erosion and patterns of deposition, cited the dangers of deforestation, and noted the consequences of environmental abuse. Strabo cautioned against the assumption that the nature and actions of humans were determined by the physical environment they inhabited. He observed that humans were the active elements in a cultural–physical partnership.

So broad and integrated are the concerns of geography and so great the variety of facts, observations, and distributions that guide its analyses, that it has been called the mother of sciences. From its earlier concerns have sprung such specialized, independent disciplines as meteorology, climatology, cultural anthropology, geology, and a host of other fields of inquiry. From such areas of study, geographers draw the background data that contribute to their own broader investigation of human–environment systems and spatial relationships.

Basic Geographic Concepts

Although such investigations may serve broadly to define the field of geography, its nature is best understood through the questions geographers ask and the approaches they employ to answer those questions. Of a physical or cultural phenomenon, they will inquire: What is it? Where is it? How did it come to be what and where it is? Where is it in relation to other physical or cultural realities that affect it or are affected by it? How is it part of a functioning whole? How does its location affect people's lives?

These and similar questions derive from central themes in geography. In answering them, geographers respond by using certain fundamental concepts and terms. Together, the themes, concepts, and terms form the basic structure and vocabulary of geography. They recognize the fundamental truths addressed by geography: that things are rationally organized on the earth's surface and that understanding spatial patterns is an essential starting point for understanding how humans live on the earth.

Geography is about earth places and spaces. The questions that geographers ask originate in basic observations about the nature of places and how places are similar to or different from one another. Those observations, though simply stated, are profoundly important to our comprehension of the world we occupy. They are:

A place may be large or small.
A place has location.
A place has both physical and cultural characteristics.
The characteristics of places develop and change over time.
Places interact with other places.
Places may be generalized into regions of similarities and differences.

Size and Scale

When we say that a place may be large or small, we speak both of the nature of the place itself and of the generalizations that can be made about it. In either instance, geographers are concerned with **scale,** though we may use that term in different ways. We can, for example, study a problem—population, say, or landforms—at the local scale or on a global scale. Here, the reference is purely to the

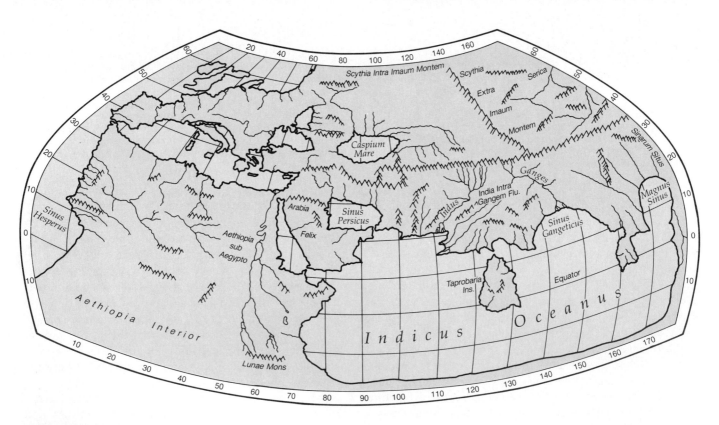

Figure 1.2

Ptolemy's world outline from the Rome edition of 1490. The 2d-century A.D Roman geographer-astronomer Ptolemy (Claudius Ptolemaeus) adopted an already developed map grid of latitude and longitude lines based on the division of the circle into 360° The grid permitted a precise mathematical location for every

recorded place. Unfortunately, errors of assumption and estimation rendered both Ptolemy's map and its accompanying six-volume gazetteer inaccurate. Many variants of Ptolemy's map were published in the 15th and 16th centuries. The version shown here summarizes the extent and content of the original.

size of unit studied. More technically, scale tells us the relationship between the size of an area on a map and the actual size of the mapped area on the surface of the earth. In this sense, as Chapter 2 makes clear, scale is a feature of every map and essential to recognizing what is shown on that map.

In both senses of the word, scale implies the degree of generalization represented (Figure 1.3). Geographic inquiry may be broad or narrow; it occurs at many different size-scales. Climate may be an object of study, but research and generalization focused on climates of the world will differ in degree and kind from study of the microclimates of a city. Awareness of scale is very important. In geographic work, concepts, relationships, and understandings that have meaning at one scale may not be applicable at another scale.

Location

The location of a place may be described in both absolute and relative terms. **Absolute location** records a precise position on the surface of the globe, usually in terms of a mathematically based reference system. The global grid of latitude and longitude is commonly employed for absolute location. In United States property description, a reference to township, section, and range is customary in those parts of the country where that survey system exists. Absolute location is unique to each described place, is independent of any other characteristic or observation about that place, and has obvious usefulness in legal description of places, in measuring the distance separating places, or in finding directions between places on the earth's surface.

When geographers remark that "location matters," however, their reference is usually not to absolute but to **relative location**—the location of a place or thing in relation to that of other places or things. Relative location expresses spatial interconnection and interdependence. It tells us that people, things, and places exist in a world of physical and cultural characteristics that differ from place to place. The attributes and potentialities of a particular locale are understandable only as that locale is seen in its spatial relationship to other places with their own attributes (Figures 1.4 and 12.9). Again, the idea of scale comes into play. Depending upon the place and characteristic studied, spatial relationships may be traced at the local, regional, national, or global scales, with each scale reference to relative location adding to our understanding of place.

New York City, for example, may in absolute terms be described as located at (approximately) latitude 40°43′ N (read as 40 degrees, 43 minutes north) and longitude

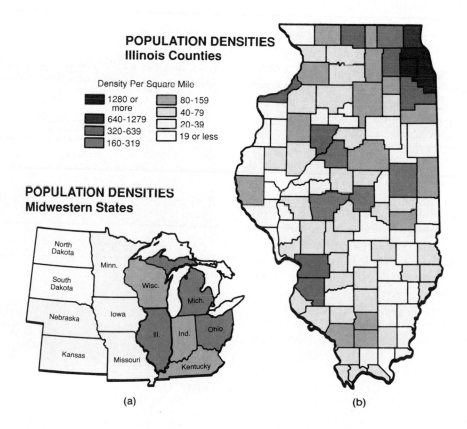

Figure 1.3

"Truth" depends on one's scale of inquiry. Map (a) reveals that the maximum population density of Midwestern states is no more than 319 per square mile (122 per km²). From map (b), however, we see that population densities in two Illinois counties exceed 1280 people per square mile (494 per km²). Were we to reduce our scale of inquiry even further, examining individual city blocks in Chicago, we would find densities as high as 5000 people per square mile (1930 per km²). Scale matters!

73° 58′ W. We have a better understanding of the *meaning* of its location, however, when reference is made to its spatial relationships: to the continental interior through the Hudson-Mohawk lowland corridor; to its position on the eastern seaboard of the United States; to its role as a transportation node and multifunctional economic and administrative center of a major industrial country of the Northern Hemisphere; and to its interconnections, through trade routes, financial ties, and the United Nations, to the international community. Within the city, we gain understanding of the locational significance of, for example, Central Park or the Lower East Side not by reference solely to the street addresses or city blocks they occupy but by their spatial and functional relationships to the total land use, activity, and population patterns of New York City.

Physical and Cultural Attributes

All places have individual physical and cultural attributes distinguishing them from other places and giving them character, potential, and meaning. Geographers are concerned with identifying and analyzing the details of those attributes and, particularly, with recognizing the interrelationship between the physical and cultural components of area: the human–environmental interface.

The physical characteristics of a place refer to such natural aspects of a locale as its climate and soil, the presence or absence of water supplies and mineral resources, its terrain features, and the like. These attributes provide the setting within which human action occurs. They help shape—but do not dictate—how people live. The resource base, for example, is physically determined, though how resources are perceived and utilized is culturally conditioned.

Environmental circumstances directly affect agricultural potential and reliability; indirectly, they may affect such matters as employment patterns, trade flows, population distributions, national diets, and so on. The physical environment simultaneously presents advantages and disadvantages with which humans must deal. Thus, the danger of cyclones in the Bay of Bengal must be balanced against the agricultural bounty derived from the region's favorable terrain, soil, temperature, and moisture conditions. Physical environmental patterns and processes are explored in Chapters 3 and 4 of this book.

At the same time, by occupying a given place, people modify its environmental conditions. The existence of the American Environmental Protection Agency (and its counterparts elsewhere) is reminder that humans are the active and frequently harmful agents in the continuing spatial interplay between the cultural and physical worlds (Figure 1.5). Virtually every human activity leaves its imprint on the earth's soil, water, vegetation, animal life, and other resources, and on the atmosphere common to all earth space, as Chapter 5 makes clear.

Figure 1.4

The reality of *relative location* on the globe may be strikingly different from the impressions we form from flat maps. The position of the USSR with respect to North America when viewed from a polar perspective emphasizes that relative location properly viewed is important to our understanding of spatial relationships and interactions between the two world areas.

Figure 1.5

Sites (and sights) such as this devastation of ruptured barrels and petrochemical contamination near Texas City, Texas, are all-too-frequent reminders of the adverse environmental impacts of humans and their waste products. Many of those impacts are more hidden in the form of soil erosion, water pollution, increased stream sedimentation, plant and animal extinction, deforestation, and the like.

The visible imprint of that human activity is called the **cultural landscape.** It, too, exists at different scales and at different levels of visibility. Differences in agricultural practices and land use between Mexico and southern California are evident in Figure 1.6, while the signs, structures, and people of Los Angeles' Chinatown leave a smaller, more confined imprint within the larger cultural landscape of the metropolitan area itself.

The physical and human characteristics of places are the keys to understanding both the simple and the complex interactions and interconnections between people and the environments they occupy and modify. Those interconnections and modifications are not static or permanent, but are subject to continual change.

Figure 1.6
This Landsat image reveals different land use patterns in California and Mexico. Slowly move your eyes from the Salton Sea (the dark patch near the center of the image) to the agricultural land extending to the southeast and to the edge of the picture. Notice how the regularity of the fields and the bright colors (representing growing vegetation) give way to a marked break, where irregularly shaped fields and less prosperous agriculture are evident. North of the break is the Imperial Valley of California; south of the border is Mexico.

Attributes of Place Are Always Changing

Characteristics of places today are the result of constantly changing past conditions. They are the forerunners of differing human–environmental balances yet to be struck. Geographers are concerned with place at given moments of time. But to understand fully the nature and development of places, to appreciate the significance of their relative locations, and to understand the interplay of their physical and cultural characteristics, geographers must view places as the present result of past operation of distinctive physical and cultural processes (Figure 1.7).

You will recall that one of the questions geographers ask of a place or thing is: How did it come to be what and where it is? This is an inquiry about process and about becoming. The forces and events shaping the physical and explaining the cultural environment of places today are an important focus of geography and are the topics of most of the chapters of this book. To understand them is to appreciate the changing nature of the spatial order of our contemporary world.

Places Interact

The notion of relative location, already introduced, leads directly to another fundamental spatial reality: places interact with other places in structured and comprehensible ways. In describing the processes and patterns of **spatial interaction,** geographers employ ideas of *distance, accessibility,* and *connectivity.*

We may ask how far one place is from another and how that distance affects their interaction. **Distance** is a fundamental consideration in nearly everything people do. At one scale, you make decisions about where to live, eat, and shop based upon the distance you have to travel to carry out your day's activities (see Figure 9.8). At a different scale, distance isolated North America from Europe until ships (and aircraft) were developed that reduced the effective distance between the continents. From this latter example it is evident that distance is more than a linear concept (Figure 1.8). It may be measured in units other than feet or miles, and it may change as technology for bridging space and as attitudes about distance change. Whether place separation is measured in *linear distance, time-distance,* or *psychological distance* (ideas explored further in Chapter 9), interaction falls off as distance increases—a statement of the idea of **distance decay.** Distance decay may be thought of as the result of a basic law of geography: everything is related to everything else, but relationships are stronger when things are near one another and weaken as distance increases.

Consideration of distance implies assessment of **accessibility.** How easy or difficult is it to overcome the friction of distance? That is, how easy or difficult is it to overcome the barrier of the time or the space separation

Figure 1.7

The process of change in a cultural landscape. Before the advent of the freeway, this portion of suburban Long Island, New York, was largely devoted to agriculture (left). The construction of the freeway and cloverleaf interchange ramps altered nearby land use patterns (right) to replace farming with new housing developments and new commercial and light industrial activities.

of places—and how has changing technology of transportation altered former patterns of spatial interaction? In its turn, accessibility suggests **connectivity,** a broader concept implying all the tangible and intangible ways in which places are connected: by physical telephone lines, by street and road systems, by pipelines and sewers, by radio and TV broadcasts beamed across airwaves, and even by natural movements of wind systems and flows of ocean currents.

There is, inevitably, interchange between connected places. **Spatial diffusion** is the process of dispersion of an idea or a thing (a new consumer product or a new song, for example) from a center of origin to more distant points. The rate and extent of that diffusion are affected, again, by the distance separating the origin of the new idea or technology and other places where it is eventually adopted. Diffusion rates are also affected by such factors as population densities, means of communication, obvious advantages of the innovation, and importance or prestige of the originating node. Further discussion of spatial diffusion is found in Chapter 9.

Geography is a study of the dynamics of spatial relationships. Movement, connection, and interaction are part of the social and economic processes that give character to places and regions (Figure 1.9). Geography's study of those relationships recognizes that spatial interaction is not just an awkward necessity but a fundamental organizing principle of the physical and social environment.

Place Similarity and Regions

The distinctive characteristics of places—physical, cultural, locational—immediately suggest to us two geographically important ideas. The first is that no two places on the surface of the earth can be *exactly* the same. Not only do they have different absolute locations, but—as in the features of the human face—the precise mix of physical and cultural characteristics of place is never exactly duplicated. Since geography is a spatial science, the inevitable uniqueness of place would seem to impose impossible problems of generalizing spatial information.

That this is not the case results from the second important idea: the physical and cultural features of an area and the dynamic interconnections of people and places show patterns of spatial similarity. Often the similarities are striking enough for us to conclude that spatial regularities exist. They permit us to recognize and define **regions,** earth areas that display significant elements of uniformity. Places are, therefore, both unlike and like other places, creating patterns of areal differences and coherent spatial similarity.

The problems of the historian and the geographer are similar. Each must generalize about items of study that are essentially unique. The historian creates arbitrary but

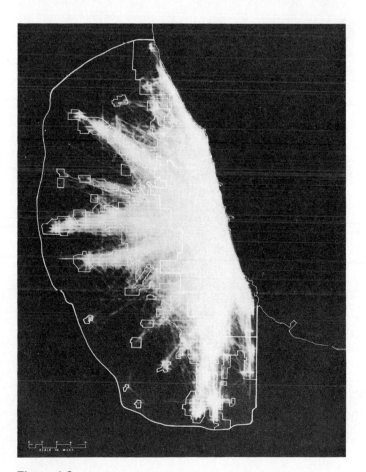

Figure 1.9
The routes of the 5 million automobile trips made each day in Chicago during the late 1950s are recorded on this light-display map. The boundaries of the region of interaction that they create are clearly marked. Those boundaries (and the dynamic region they defined) were subject to change as residential neighborhoods expanded or developed, as population relocations occurred, and as the road pattern was altered over time.

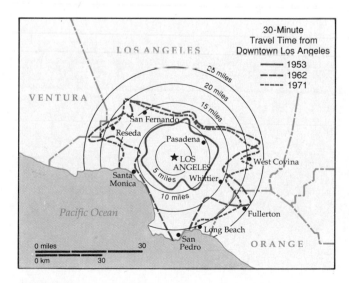

Figure 1.8
Lines of equal travel time (*isochrones*) mark off different linear distances from a given starting point, depending on the condition of the route and terrain and changes in the roads and traffic flows over time. On this map, the areas within 30 minutes' travel time from downtown Los Angeles are recorded for the period 1953 to 1971.

meaningful and useful historical periods for reference and study. The "Roaring Twenties" or the "Victorian Era" are shorthand summary names for specific time spans, internally quite complex and varied, but significantly distinct from what went before or followed after. The region is the geographer's equivalent of the historian's epoch: a device that segregates into component parts the complex reality of the surface of the earth. In both the time and the space need for generalization, attention is focused upon key unifying elements or similarities of the era or area selected for study. In both the historical and geographical cases, the names assigned to those times and places serve to identify the time span or region and to convey between speaker and listener a complex set of interrelated attributes.

Regions are not "given" in nature any more than "eras" are given in the course of human events. Regions are devised; they are spatial summaries designed to bring order to the infinite diversity of the earth's surface. At their root, they are based upon the recognition and mapping of **spatial distributions**—the territorial occurrence of environmental, human, or organizational features selected for study. Although as many spatial distributions exist as there are imaginable physical, cultural, or connectivity elements of area to examine (Figure 1.10), those that are selected for study are those that contribute to the understanding of a specific topic or problem.

Although there are as many individual regions as the objectives of spatial study and understanding demand, two generalized *types* of regions are recognized. A **formal region** is one of essential uniformity in one or a limited number of related physical or cultural features. Your home state is a formal political region within which uniformity of law and administration is found. The Columbia Plateau

or the tropical rain forest are areas of uniform physical characteristics making up formal natural regions (Figure 1.11).

A **functional region,** in contrast, may be visualized as a spatial system. Its parts are interdependent, and throughout its extent the functional region operates as a dynamic, organizational unit. A functional region has unity not in the sense of static content but in the manner of its operational connectivity. It has a node or core area surrounded by the total region defined by the type of control exerted. Trade areas of towns (Figure 1.12), national "spheres of influence," and the territories subordinate to the financial, administrative, wholesaling, or retailing centrality exercised by such regional capitals as Chicago, Atlanta, or Minneapolis are cases in point. Further examples of formal and functional regionalism will be encountered incidentally in following chapters and make up the total content of Chapter 13.

Organization of This Book

The breadth of geographic interest and subject matter, the variety of questions that focus geographic inquiry, and the diversity of concepts and terms employed to analyze likeness and difference from place to place on the earth's surface require a simple, logical organization of topics for presentation to students new to the field. Despite its outward appearance of complex diversity, geography should be seen to have a broad consistency of purpose achieved through the recognition of a limited number of distinct but closely related "traditions." William D. Pattison, who suggested this unifying viewpoint, and J. Lewis Robinson (among others) who accepted and expanded Pattison's reasoning, found that four traditions were ways of clustering geographic inquiry. While not all geographic work

Figure 1.10
Pulmonary tuberculosis mortality rates for white males, 1965–1971. All spatial data may be mapped. As this example of the distribution of pulmonary tuberculosis demonstrates, mapped distributions frequently reveal regional patterns inviting analysis. Areas emphasized are those with above average mortality rates. The question is, Why?

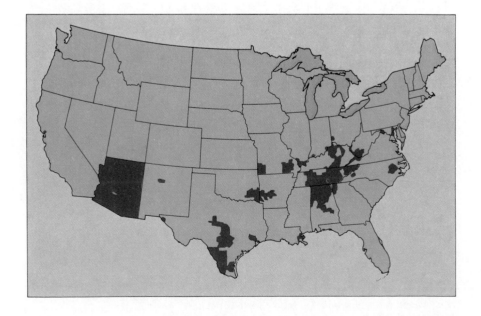

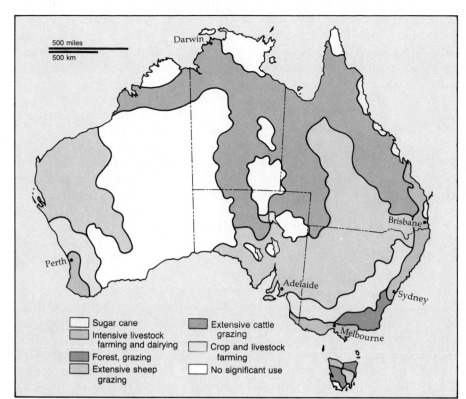

Figure 1.11
This generalized land use map of Australia is composed of *formal regions* whose internal economic characteristics show essential uniformities setting them off from adjacent territories of different condition or use.

Legend:
- Sugar cane
- Intensive livestock farming and dairying
- Forest, grazing
- Extensive sheep grazing
- Extensive cattle grazing
- Crop and livestock farming
- No significant use

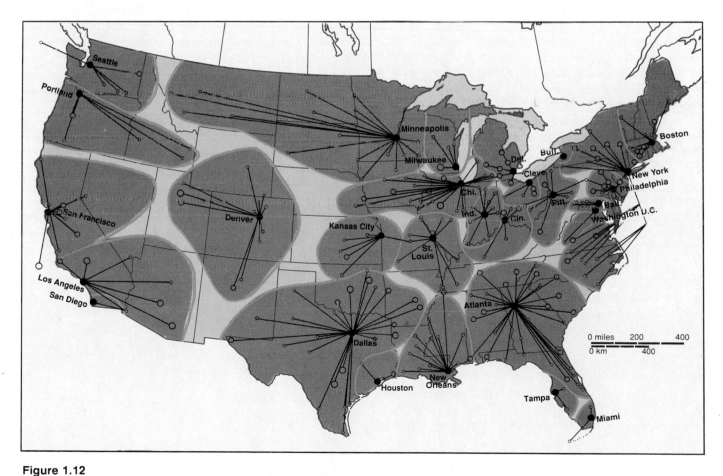

Figure 1.12
The *functional regions* shown on this map are based on linkages between large banks of major central cities and the "correspondent" banks they serve in smaller towns. The regions suggest one form of *connectivity* between principal cities and locales beyond their own immediate metropolitan area.

is confined by the separate themes, one or more of the traditions is implicit in most geographical studies. The unifying themes—the four traditions within which geographers work—are:

the earth-science tradition
the culture–environment tradition
the locational (or spatial) tradition
the area analysis tradition

We have employed these as the device for clustering the chapters of this book, hoping that they will help you to recognize the unitary nature of geography while appreciating the diversity of topics studied by geographers.

The **earth-science tradition** is that branch of the discipline that addresses itself to the earth as the habitat of humans. It is the tradition that in ancient Greece represented the roots of geography, the description of the physical structure of the earth and of the natural processes that give it detailed form. In modern terms, it is the vital environmental half of the study of human–environment systems, which together constitute geography's subject matter. The earth-science tradition prepares the physical geographer to understand the earth as the common heritage of humankind and to find solutions to the increasingly complex web of pressures placed on the earth by its expanding, demanding human occupants. By its nature and content, the earth-science tradition forms the unifying background against which all geographic inquiry is ultimately conducted. Consideration of the elements of the earth-science tradition constitutes Part I (Chapters 2 to 5) of this text.

Part II (Chapters 6 to 9) concerns itself with some of the content of the **culture–environment tradition.** Within this theme of geography, consideration of the earth as a purely physical abstraction gives way to a primary concern with how people *perceive* the environments they occupy. Its focus is upon culture. The landscapes that are explored and the spatial patterns that are central are those that are cultural in origin and expression. People in their numbers, distributions, and diversity; in their patterns of social and political organization; and in their spatial perceptions and behaviors are the orienting concepts of the culture–environment tradition. The theme is distinctive in its thrust but tied to the earth-science tradition because populations exist, cultures emerge, and behaviors occur within the context of the physical realities and patternings of the earth's surface.

The **locational tradition**—or as it is sometimes called, the spatial tradition—is the subject of the chapters of Part III (Chapters 10 to 12). It is a tradition that underlies all of geographic inquiry. As Robinson suggested, if we can agree that geology is rocks, that history is time, and that sociology is people, then we can assert that geography is earth space. The locational tradition is primarily concerned with the distribution of cultural phenomena or physical items of significance to human occupance of the earth.

Part, but by no means all, of the locational tradition is concerned with distributional patterns. More central are scale, movement, and areal relationships. Map, statistical, geometrical, and systems analysis and modeling are among the research tools employed by geographers working within the locational tradition. Irrespective, however, of the analytical tools employed or the sets of phenomena studied—economic activities, resource distributions, city systems, or others—the underlying theme is the geometry, or the distribution, of the phenomenon discussed and the flows and interconnections that unite it to related physical and cultural occurrences.

A single chapter (Chapter 13) consideration of the **area analysis tradition** constitutes Part IV and completes our survey of geography. Again, the roots of this theme may be traced to antiquity. Strabo's *Geography* was addressed to the leaders of Augustan Rome as a summary of the nature of places in their separate characters and conditions—knowledge deemed vital to the guardians of an empire. Imperial concerns may long since have vanished, but the study of regions and the recognition of their spatial uniformities and differences remain. Such uniformities and differences, of course, grow out of the structure of human–environment systems and interrelations that are the study of geography.

The identification of the four traditions of geography is not only an organizational convenience. It is also a recognition that within that diversity of subject matter called geography, unity of interest is ever preserved. The traditions, though recognizably distinctive, are intertwined and overlapping. We hope that their use as organizing themes—and their further identification in short introductions to the separate sections of this book—will help you to grasp the unity in diversity that is the essence of geographic study.

Key Words

absolute location	functional region
accessibility	locational tradition
area analysis tradition	region
connectivity	relative location
cultural landscape	scale
culture–environment tradition	spatial diffusion
distance decay	spatial distribution
earth-science tradition	spatial interaction
formal region	

For Review

1. In what two meanings and for what different purposes do we refer to *location?* When geographers say "location matters," what aspect of location commands their interest?

2. What does the term *cultural landscape* imply? Is the nature of the cultural landscape dictated by the physical environment?

3. How are the ideas of *distance, accessibility,* and *connectivity* related to processes of *spatial interaction?*

4. Why do geographers concern themselves with *regions?* How are *formal* and *functional* regions different in concept and definition?

5. What are the *four traditions* of geography? Do they represent unifying or divided approaches to geographic understanding?

Selected References

Abler, Ronald, John S. Adams, and Peter Gould. *Spatial Organization: The Geographer's View of the World.* Englewood Cliffs, N.J.: Prentice-Hall, 1971.

Broek, J. O. M. *Geography: Its Scope and Spirit.* Columbus, Ohio: Charles E. Merrill, 1965.

Buttimer, Anne, and David Seamon, eds. *The Human Experience of Space and Place.* New York: St. Martin's Press, 1980.

Dickinson, Robert E., and Osbert J. R. Howarth. *The Making of Geography.* Oxford: Clarendon Press, 1933.

Goodall, Brian. *The Facts on File Dictionary of Human Geography.* New York: Facts on File Publications, 1987.

Goudie, Andrew, ed. *The Encyclopaedic Dictionary of Physical Geography.* Oxford, England: Basil Blackwell Ltd., 1985; 1988.

Gould, Peter. *The Geographer at Work.* London: Routledge & Kegan Paul, 1985.

Gregory, K. J. *The Nature of Physical Geography.* London: Edward Arnold, 1985.

Holt-Jensen, Arild. *Geography: Its History & Concepts.* 2d ed. Totowa, N.J.: Barnes and Noble, 1988.

James, Preston E., and Geoffrey J. Martin. *All Possible Worlds: A History of Geographical Ideas.* 2d ed. New York: Wiley, 1981.

Johnston, R. J. *Geography and Geographers.* 2d ed. London: Methuen, 1983.

Lanegran, David A., and Risa Palm. *An Invitation to Geography.* 2d ed. New York: McGraw-Hill, 1978.

McDonald, James R. "The Region: Its Conception, Design and Limitations." *Annals of the Association of American Geographers* 56 (1966): 516–28.

Morrill, Richard L. *The Spatial Organization of Society.* 2d ed. North Scituate, Mass.: Duxbury Press, 1974.

Murphey, Rhoads. *The Scope of Geography.* 3d ed. New York: Methuen, 1982.

Pattison, William D. "The Four Traditions of Geography." *Journal of Geography* 63 (1964): 211–16.

Robinson, J. Lewis. "A New Look at the Four Traditions of Geography." *Journal of Geography* 75 (1976): 520–30.

Thomson, J(ames) Oliver. *History of Ancient Geography.* New York: Biblo and Tannen, 1965.

Warntz, William, and Peter Wolff. *Breakthroughs in Geography.* New York: New American Library, 1971.

Wheeler, James O. "Notes on the Rise of the Area Studies Tradition in U.S. Geography, 1910–1929." *Professional Geographer* 38 (1986): 53–61.

1

The Earth-Science Tradition

For nearly a month the mountain had rumbled, emitting puffs of steam and flashes of fire. Within the past week, on its slopes and near its base, deaths had been recorded from floods, mud slides, and falling rock. A little after 8:00 on the morning of May 8, 1902, the climax came for volcanic Mount Pelée and the thriving port of Saint Pierre on the island of Martinique. To the roar of one of the biggest explosions the world has ever known and to the clanging of church bells aroused in swaying steeples, a fireball of gargantuan size burst forth from the upper slope of the volcano. Lava, ash, steam, and superheated air engulfed the town, and 29,933 people met their death.

More selective but for those victimized just as deadly was the sudden "change in weather" that struck central Illinois on December 20, 1836. Within an hour, preceded by winds gusting to 70 miles per hour, the temperature plunged from 40° F to −30° F. On a walk to the post office through slushy snow, Mr. Lathrop of Jacksonville, Illinois, found, just as he passed the Female Academy, that "the cold wave struck me, and as I drew my feet up the ice would form on my boots until I made a track . . . more like that of [an elephant] than a No. 7 boot." Two young salesmen were found frozen to death along with their horses; one "was partly in a kneeling position, with a tinderbox in one hand, a flint in the other, with both eyes open as though attempting to light the tinder in the box." Others died, too—inside horses that had been disemboweled and used as makeshift shelters; in fields, woods, and on roads both a short distance and an eternity from the travelers' destinations.

Fortunately, few human–environmental encounters are so tragic as these. Rather, the physical world in all its spatial variation provides the constant background against which the human drama is played. It is to that background that physical geographers, acting within the earth-science tradition of the discipline, direct their attention. Their primary concern is with the natural rather than the cultural landscape and with the interplay between the encompassing physical world and the activities of humans.

The interest of those working within the earth-science tradition of geography, therefore, is not in the physical sciences as an end in themselves but in physical processes as they create landscapes and environments of significance to humankind. The objective is not solely to trace the physical and chemical reactions that have produced a piece of igneous rock or to describe how a glacier has scoured the rock; it is the relationship of the rock to people. Geographers are interested in what the rock can tell us about the

evolution of the earth as the home of human beings or in the significance of certain types of rock for the distribution of mineral resources or of fertile soil.

The questions that physical geographers ask do not usually deal with the catastrophic, though catastrophe is an occasional element of human–earth-surface relationships. Rather, they ask questions that go to the root of understanding the earth as the home and the workplace of humankind. What does the earth as our home look like? How have its components been formed? How are they changing, and what changes are likely to occur in the future through natural or human causes? How are environmental features, in their spatially distinct combinations, related to past, present, and prospective human use of the earth? These are some of the basic questions that geographers who follow the earth-science tradition attempt to answer.

The four chapters in Part I of our review of geography deal, in turn, with maps, landforms, weather and climate, and human impact on the environment. It is easy to see why the earth's landforms, weather, and climate come within the earth-science tradition, but it may be less clear why we begin this part with a discussion of maps and end with a consideration of human impact on the physical world.

Maps are our starting point, for it is impossible to talk of geography or to appreciate its content and lessons without understanding maps. Geography is a spatial science, and since ancient times maps have been used to represent, characterize, and interpret the nature of space. All geographers use maps as basic analytical tools, whether the maps be simple sketches made during field investigations or computer-drawn maps based on data from earth satellites. Mapmakers traditionally sought ways to represent the distribution of land and water features of the earth and to portray the character of the physical landscape. Since the 1700s, they have turned increasing attention to map rendition of cultural and political distributions and to the preparation of thematic maps. In the last few decades,

geographers have begun to draw maps based on projections that distort time or space or that are intended to portray "psychological" or "perceived" reality. Whatever the state of their art or the thrust of their attention, mapmakers have been motivated to express the distribution of things central to their interests. As a spatial, distributional science, geography properly begins with a discussion of the nature of maps.

The treatment of so vast a field as landforms (Chapter 3) must be highly selective within an introductory text. The aim is to summarize the great processes by which landforms are created and to depict the general classes of features resulting from those processes without becoming overly involved in scientific reasoning and technical terms.

In Chapter 4, "Physical Geography: Weather and Climate," the major elements of the atmosphere—temperature, precipitation, and air pressure—are discussed, and their regional generalities and associations in patterns of world climates are presented. It is the study of weather and climate that adds coherence to our understanding of the earth as the home of people. Frequent reference will be made to the climatic background of human activities in later sections of our study.

Part I concludes with a look at the impact humans have had upon the earth's surface and resources. Chapter 5, "Human Impact on the Environment," explores some of the implications of a basic geographic observation: that in the interaction between humans and the physical environment, humans are the active and frequently destructive agents of change. Human exploitation of the earth results in both intended and unanticipated changes to the natural environment.

The focus of Part I, therefore, is on the earth as the environment and habitat of humans. Both the diversity and reciprocal nature of human–environmental interactions are recurring themes not only of the following four chapters but also of geography as a field of inquiry.

Mt. McKinley as seen from Thorofare Pass, Devali National Park, Alaska.

C H A P T E R

Maps

Landsat image of the United Kingdom.

Imagine that all of the maps in the world were destroyed at this moment. Traffic on highways would slow to a crawl. Control towers would be overwhelmed as pilots radioed airports for landing instructions. Ships at sea would be forced to navigate without accurately charting their positions. Rangers would be sent to rescue hikers stranded in national parks and forests. Disputes would erupt over political boundaries. Military planners would not know how to channel troop movements or where to aim artillery. People would have to rely on verbal descriptions to convey information, but no number of words could do justice to the detailed information contained on a map of even moderate complexity. Maps are tersely efficient at indicating the location of things relative to one another, and the information void created by their disappearance would have to be filled by volumes of description.

Maps are as fundamental a means of communication as the printed or spoken word or as photographs. In every age, people have produced maps, whether made out of sticks or shells, scratched on clay tablets, drawn on parchment, or printed on paper. The emperor Charlemagne even had maps made on solid plates of silver. The map is the most efficient way to portray parts of the earth's surface, to record political boundaries, and to indicate directions to travelers. In addition to their usefulness to the general populace, maps have a special significance to the geographer. In the process of studying the surface of the earth, geographers depend on maps to record, present, and aid them in analyzing the location of points, lines, or objects.

Modern scientific mapping has its roots in the 17th century, although the earth scientists of ancient Greece are justly famous for their contributions. They recognized the spherical form of the earth and developed map projections and the grid system. Unfortunately, much of the cartographic tradition of Greece was lost to Europe during the Middle Ages and essentially had to be rediscovered. Several developments during the Renaissance gave impetus to accurate cartography. Among these were the development of printing, the rediscovery of the work of Ptolemy and other Greeks, and the great voyages of discovery.

In addition, the rise of nationalism in many European countries made it imperative to determine and accurately portray boundaries and coastlines as well as to depict the kinds of landforms contained within the borders of a country. During the 17th century, important national surveys were undertaken in France and England. Many conventions in the way data are presented on maps had their origin in these surveys. This early concern about *physical* maps—that is, maps portraying the earth's physical features—allows us to include our discussion of maps within the part of this book exploring topics in the earth-science tradition of geography.

Knowledge of the way information is recorded on maps enables us to read and interpret them correctly. To be on guard against drawing inaccurate conclusions or to avoid being swayed by distorted or biased presentations, we must be able to understand and assess the ways in which facts are represented. Of course, all maps are necessarily distorted because of the need to portray the round earth on a flat surface, to use symbols to represent objects, to generalize, and to record features at a different size than they actually are. This distortion of reality is necessary because the map is smaller than the things it depicts and because its effective communication depends upon selective emphasis of only a portion of reality. As long as map readers know the limitations of the commonly used types of maps and understand what relationships are distorted, they may easily interpret maps correctly.

Locating Points on a Sphere: The Grid System

In order to visualize the basic system for locating points on the earth, think of the world as a sphere with no markings whatever on it. There would, of course, be no way of describing the exact location of a particular point on the sphere without establishing some system of reference. We use the **grid system**, which consists of a set of imaginary lines drawn across the face of the earth. The key reference points in that system are the North and South Poles and the equator, which are given in nature (Figure 2.1), and the prime meridian, which is agreed upon by cartographers.

The North and South Poles are the end points of the axis about which the earth spins. The line that encircles the globe halfway between the poles, and that consequently is perpendicular to the axis, is the *equator*. We

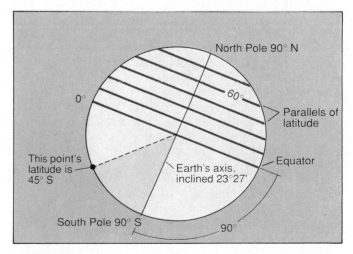

Figure 2.1

The grid system: parallels of latitude. Note that the parallels become increasingly shorter closer to the poles. On the globe, the 60th parallel is only one-half as long as the equator.

can describe the location of a point in terms of its distance north or south of the equator measured as an angle at the earth's center. Because a circle contains 360 degrees, the distance between the two poles is 180 degrees and between the equator and each pole, 90 degrees. **Latitude** is the measure of distance north and south of the equator, which is itself, of course, zero degrees. As is evident in Figure 2.1, the lines of latitude, which are parallel to each other and to the equator, run east-west.

The polar circumference of the earth is 24,899 miles; thus, the distance between each degree of latitude equals 24,899 ÷ 360, or about 69 miles (111 km). If the earth were a perfect sphere, all degrees of latitude would be equally long. Due to the slight flattening of the earth in polar regions, degrees of latitude are slightly longer near the poles (69.41 mi; 111.70 km) than near the equator (68.70 mi; 110.56 km).

To record the latitude of a place in a more precise way, degrees are divided into 60 *minutes* (') and each minute into 60 *seconds* ("), exactly like an hour of time. One minute of latitude is about 1.15 miles (1.85 km), and one second of latitude about 101 feet (31 m). The latitude of the center of Chicago, Illinois, is written 41°51'50" North. This system of subdivision of the circle is derived from the sexagesimal (base 60) numerical system of the ancient Babylonians.

Because the distance north or south of the equator is not by itself enough to locate a point in space, we need to specify a second coordinate to indicate distance east or west from an agreed-upon reference line. As a starting point for east-west measurement, cartographers in most countries use as the **prime meridian** an imaginary line passing through the Royal Observatory at Greenwich, England. That prime meridian was selected as the zero-degree longitude by an international conference in 1884. Like all *meridians*, it is a true north-south line connecting the poles of the earth (Figure 2.2). ("True" north and south vary from magnetic north and south, the direction of the earth's magnetic poles, to which a compass needle points.) Meridians are farthest apart at the equator, come closer and closer together as latitude increases, and converge at the North and South Poles. Unlike parallels of latitude, all meridians are the same length.

Agreement on a Prime Meridian

Latitudinal location relative to the equator, a line given in nature, can be determined by measurement of the angular height of the sun above the horizon at noon on a known date. Measuring devices for this purpose (*astrolabes*) appear to have been in use by the Babylonians and the Egyptians several thousand years before the Greeks addressed the problem of constructing the globe grid and developing world maps.

The determination of longitude—the other necessity for accurate maps—was more difficult. No reference line equivalent to the equator existed, and no convenient motion of sun or stars indicated the location east or west of a naturally given baseline. This lack was merely annoying when one was traveling on land, where surface distances could be measured, or on coasting vessels from whose decks landfalls could be sighted. It was potentially disastrous when long sea voyages out of sight of land became common in the 16th and 17th centuries. Even skilled navigators using dead reckoning* could, and did, miss destinations by scores and sometimes hundreds of miles.

Innumerable proposals were made for using water wheels attached to ships' hulls and for other mechanical devices for measuring distance traveled at sea. All were rendered useless by the currents and winds acting on sailing vessels. The only way longitude could be determined was by the measurement at a given instant of the difference between the time at a known point and the time on a reference line. Since the earth rotates through 360 degrees in 24 hours, each hour of time is the equivalent of 15 degrees of longitude, and each degree of longitude equals 4 minutes of time. The answer to the longitude problem had to be a small but extremely reliable *chronometer* (clock) dependent upon a spring-driven movement.

In 1714 an act of the English Parliament established a prize of 20,000 pounds for any device that could determine longitude within 30 minutes of the globe grid, that is, within 2 minutes of time or 34 miles. John Harrison, to whom the prize was eventually awarded, put the chronometer of his invention to an official sea trial on a trip from Portsmouth, England, to Jamaica in the winter of 1761–1762. It proved to be in error by only 1 1/4 nautical miles.

The problem of the accurate determination of the longitude of any point at sea or on land was solved, but only with reference to a prime meridian that could be established wherever one wished. For patriotic reasons, most countries located the prime meridian within their own borders, and by the 19th century over a score of different ones were in common use. Recognizing the desirability of establishing a universal prime meridian on which acceptable world maps could be based, scientists held a series of conferences as first suggested by the third International Geographical Congress meeting in Venice in 1881. Delicate negotiations finally produced agreement among 25 countries that 0° longitude would refer to the meridian passing through the Royal Observatory at Greenwich, England. This convention is now almost universally accepted, though one occasionally sees maps based on a different prime meridian.

*Dead reckoning: inferring the vessel's position without astronomical observation, for example, by estimating the course and distance traveled from a previously determined position.

Longitude is the angular distance east or west of the prime meridian. Directly opposite the prime meridian is the 180th meridian, located in the Pacific Ocean. East-west measurements range from 0° to 180°, that is, from the prime meridian to the 180th meridian in each direction. Like parallels of latitude, degrees of longitude can

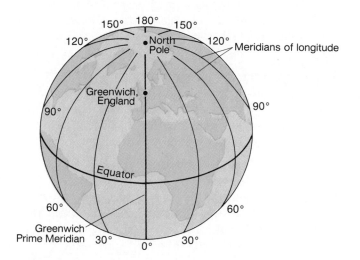

Figure 2.2

The grid system: meridians of longitude. These arbitrary but conventional lines, together with the parallels based upon the naturally given equator, constitute the globe grid. Since the meridians converge at the poles, the distance between degrees of longitude becomes shorter as one moves away from the equator.

be subdivided into minutes and seconds. However, distance between adjacent degrees of longitude decreases away from the equator because the meridians converge at the poles. With the exception of a few Alaskan islands, all places in North and South America are in the area of west longitude; and with the exception of a portion of the Chukchi Peninsula of the USSR, all places in Asia and Australia have east longitude.

Time depends on longitude. The earth, which makes a complete 360-degree rotation once every 24 hours, is divided into 24 time zones centered on meridians at 15-degree intervals. *Greenwich mean time* (GMT) is the time at the prime meridian. The **International Date Line**, where each new day begins, generally follows the 180th meridian. As Figure 2.3 indicates, however, the date line deviates from the meridian in some places in order to avoid having two different dates within a country or island group.

By citing the degrees, minutes, and, if necessary, seconds of longitude and latitude, we can describe the location of any place on the earth's surface. To conclude our earlier example, Chicago is located at 41°52′N, 87°40′W. Singapore is at 1°20′N, 103°57′E (Figure 2.4).

Map Projections

The earth can be represented with reasonable accuracy only on a globe, but globes are not as convenient as flat maps to store or to use, and they cannot depict much detail.

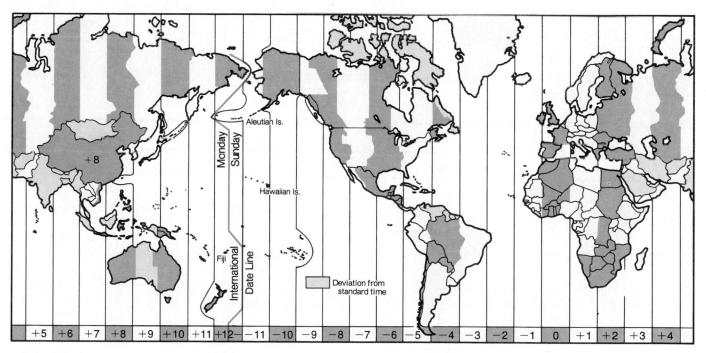

Figure 2.3

World time zones. Each time zone is about 15° wide, but variations occur to accommodate political boundaries. Thus, the International Date Line zigzags so that Siberia has the same date as the rest of the USSR and so as not to split the Aleutian Island and Fiji Island groups. New days begin at the date line and proceed westward, so that west of the line is always one day later than east of the line. The figures at the bottom indicate the number of hours to be added or subtracted from Greenwich mean time to get local time.

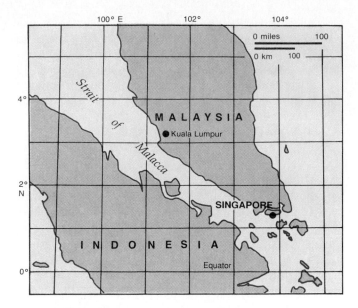

Figure 2.4
The latitude and longitude of Singapore are 1°20'N, 103°57'E. What are the coordinates of Kuala Lumpur?

For example, if we had a large globe with a diameter of, say, 3 feet, we would have to fit the details of nearly 50,000 square miles of earth surface in an area 1 inch on a side. Obviously, a globe of reasonable size cannot show the transportation system of a city or the location of very small towns and villages. In transforming a globe into a map, we cannot flatten the curved surface and keep all the properties of the original intact. So, in drawing a map, the relationships between points are inevitably distorted in some way. To the geometer, a sphere is a *nondevelopable* surface. It can only be *projected* upon a plane surface; the term **projection** designates the method of representing a curved surface on a flat map.

Properties of Map Projections

Because no projection can be entirely accurate, the serious map user needs to know in what respects a particular map correctly reproduces, and in what respects it distorts, earth features. The four main properties of maps—area, shape, distance, and direction—are distorted in different ways and to different degrees by various projections.

Area
Some projections enable the cartographer to represent the areas of regions in correct or constant proportion to earth reality. That means that any square inch on the map represents an identical number of square miles (or of similar units) anywhere else on the map. As a result, the shape of the portrayed area is inevitably distorted. A square on the earth, for example, may become a rectangle on the map, but that rectangle has the correct area. Such projections are called **equal-area** or **equivalent**.

Shape
Although no projection can provide correct shapes for large areas, some accurately portray the shapes of small areas by preserving correct angular relationships. That is, an angle on the globe is rendered correctly on the map. Maps that have true shapes for small areas are called **conformal**. Parallels and meridians always intersect at right angles on such maps, as they do on a globe. *A map cannot be both equivalent and conformal.*

Distance
Distance relationships are nearly always distorted on a map, but some projections do maintain true distances in one direction or along certain selected lines. Others, called **equidistant projections,** show true distance in all directions, but only from one or two central points. (See, for example, Figure 2.9.)

Direction
As is true of distances, directions between all points cannot be shown without distortion. Projections do exist, however, that enable the map user to measure correctly the directions from a single point to any other point.

Types of Projections

While all projections can be described mathematically, some can be thought of as being constructed by geometrical techniques rather than by mathematical formulas. In geometrical projections, the grid system is transferred from the globe to a geometrical figure, such as a cylinder or a cone, which, in turn, can be cut and then spread out flat without any stretching or tearing. The surfaces of cylinders, cones, and planes are said to be **developable surfaces**: the first two because they can be cut and laid flat without distortion, the third because it is flat at the outset.

The selection of the surface to be developed or of the specific mathematical formula to be employed is determined by the properties of the globe grid that one elects to retain. **Globe properties** are as follows:

1. All meridians are of equal length; each is one-half the length of the equator.
2. All meridians converge at the poles and are true north-south lines.
3. All lines of latitude (parallels) are parallel to the equator and to each other.
4. Parallels decrease in length as one nears the poles.
5. Meridians and parallels intersect at right angles.
6. The scale on the surface of the globe is everywhere the same in every direction.

Cylindrical Projections
The **Mercator projection** is one of the most commonly used (and misused) **cylindrical projections**. Named for its inventor, it has been mathematically adjusted to serve navigational purposes. Suppose we roll a piece of paper around

a globe so that it touches the sphere at the equator. Instead of the paper being the same height as the globe, however, it extends far beyond the poles. If we place a light source at the center of the globe, the light will project a shadow map upon the cylinder of paper. When we unroll the cylinder, the map will appear as shown in Figure 2.5.

Note the variance between the grid we have just projected and the true properties of the globe grid. The grid lines cross each other at right angles, as they do on the globe, and they are all straight north-south or east-west lines. But the meridians do not converge at the poles as they do on a globe. Instead they are equally spaced, parallel, vertical lines. Because the meridians are everywhere equally far apart, the parallels of latitude have all become the same length. The Mercator map balances the spreading of meridians toward the poles by spacing the latitude lines farther and farther apart. Mathematical tables show the cartographer how to space the parallels to balance the spreading of meridians so that the result is a conformal projection. The shapes of small areas are true, and even large regions are fairly accurately represented.

Note that the Mercator map in Figure 2.5 stops at 75°N and 60°S. A very large map would be needed to show the polar regions, and the sizes of those regions would be enormously exaggerated. The poles themselves can never be shown on a Mercator projection tangent at the equator. In fact, the enlargement of areas with increasing latitude is so great that a Mercator map should not be published without a scale of miles like that shown.

The Mercator projection has often been misused in atlases and classrooms as a general-purpose world map. Whole generations of schoolchildren have grown up convinced that the state of Alaska is nearly the size of all the lower 48 states put together. To check the distortion of this projection, compare the relative sizes of Greenland and Canada on Figure 2.5 with those on Figure 2.6, an equal-area projection.

The proper use of the Mercator is for navigation. In fact, it is the standard projection used by navigators because of a peculiarly useful property: a straight line drawn anywhere on the map is a line of constant compass bearing. If such a line, called a *rhumb line*, is followed, a ship's or plane's compass will show that the course is always at a constant angle with respect to geographic north. On no other projection is a rhumb line both straight and true as a direction.

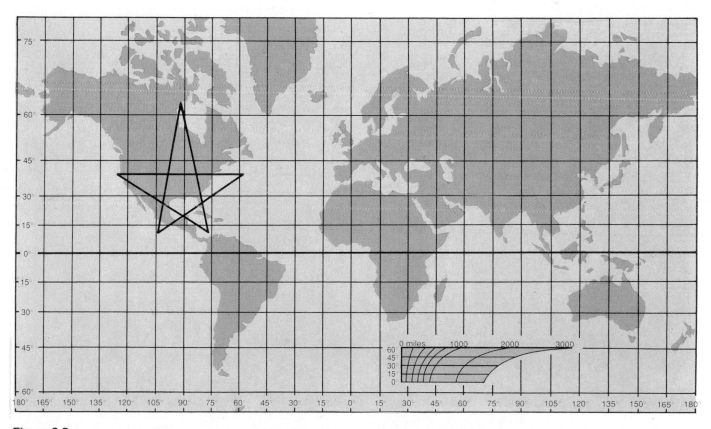

Figure 2.5

Distortion on the Mercator projection. A perfect five-pointed star was drawn on a globe, and the latitude and longitude of the points of the star were transferred to the Mercator map shown here. The manner in which the star is distorted reflects the way the projection distorts land areas. Mercator maps are usually accompanied, as here, by a diagram showing the varying scale of distance at different latitudes.

Figure 2.6
The Lambert planar equal-area projection is mathematically derived to display the property of equivalence.

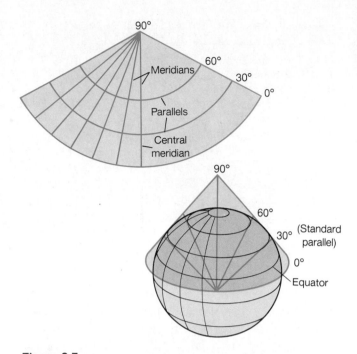

Figure 2.7
A simple conic projection with one standard parallel. Most conics are adjusted so that the parallels are spaced evenly along the central meridian.

Conic Projections

Of the three developable geometric forms—cylinder, cone, and plane—the cone is the closest in form to one-half of a globe. **Conic projections**, therefore, are widely used to depict hemispheres or smaller parts of the earth.

A useful projection in this category, and the easiest to understand, is the *simple conic* projection. Imagine that a cone is laid over half the globe, as in Figure 2.7, tangent to the globe at the 30th parallel. Distances are true only along the tangent circle, which is called the *standard parallel*. When the cone is developed, of course, the standard parallel becomes an arc of a circle, and all other parallels become arcs of concentric circles. With a central light source, the parallels become increasingly farther apart as they approach the pole, and distortion is accordingly exaggerated.

One can lessen the amount of distortion by shortening the length of the central meridian, spacing the parallels of latitude at equal distances on that meridian, and making the 90th parallel (the pole) an arc rather than a point. Most of the conic projections that are in general use employ such mathematical adjustments. When more than one standard parallel is used, a *polyconic* projection results.

Conic projections can be adjusted to achieve desired qualities; hence, they are widely used. They are particularly suited for showing areas in the midlatitudes that have a greater east-west than north-south extent. Both area and shape can be represented by conic projections without serious distortion (Figure 2.8). Many official map series, such

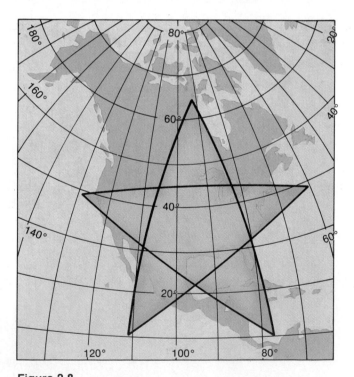

Figure 2.8
The polyconic projection. The map is produced by bringing together east-west strips from a series of cones, each tangent at a different parallel. This projection differs from the simple conic in that the parallels of latitude are not arcs of concentric circles and the meridians are curved rather than straight lines. While neither equivalent nor conformal, the projection portrays shape well. Note how closely the star resembles a perfect five-pointed star.

The Peters Projection

Although map projections might appear to be a rather sterile, uninteresting topic, the projection shown here is creating significant controversy. Developed and promoted by a German named Arno Peters, the Peters projection purports to reflect concern for the problems of the Third World by providing a less European-centered representation of the world. In contrast to the Mercator projection, which distorts areas greatly, the Peters is an equal-area map; therefore, it is better suited than the Mercator for comparing distributions. Peters claims that his map shows the densely populated parts of the earth, the countries of the Third World, in proper proportion to one another. He has persuaded a number of agencies with a special interest in the Third World, including the World Council of Churches and several United Nations organizations, to adopt the map.

But the Peters projection is as inappropriate as the Mercator, argue two geographers (P. Porter and P. Voxland, "Distortion in Maps: The Peters Projection and Other Devilments," *Focus*, Summer 1986). Characterizing Peters's cartography as "perverse and

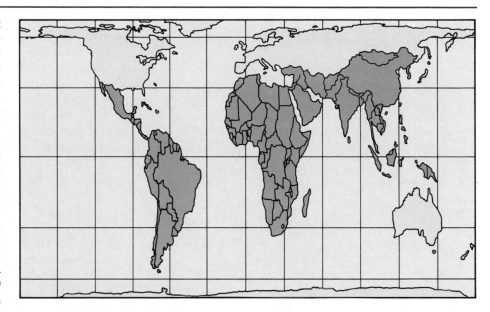

wrongheaded," they criticize his projection on several grounds.

First, it badly distorts shapes in the tropics and at high latitudes. Although Peters claims to support the Third World, his map preserves shapes more accurately for the developed world. Second, distances and directions cannot be measured except under very limited conditions. Third, Peters claims

his projection will meet every cartographic need, which is simply incorrect. Fourth, the projection is not in fact new but identical to an equal-area projection developed by James Gall in 1885. Finally, the authors point out that many projections yield an equal-area world map with less distortion of shapes than does the Peters projection.

---

as the topographic sheets and the world aeronautical charts of the U.S. Geological Survey (USGS) and the Aerospace Center of the Defense Mapping Agency, use types of conic projections. They are also used in many atlases.

Planar Projections

Planar (also called azimuthal) **projections** are constructed by placing a plane surface tangent to the globe at a single point. Although the plane may touch the globe anywhere the cartographer wishes, the polar case with the plane centered on either the North or the South Pole is easiest to visualize (Figure 2.9a).

This *equidistant* projection is useful because it can be centered anywhere, facilitating the correct measurement of distances from that point to all others. For this reason it is often used to show air navigation routes that originate from a single place. When the plane is centered on places other than the poles, the meridians and the parallels become curiously curved, as is evident in Figure 2.9b.

Because they are particularly well suited for showing the arrangement of polar landmasses, planar maps are commonly used in atlases. Depending on the particular

projection used, true shape, equal area, or some compromise between them can be depicted. In addition, one of the planar projections is widely used for navigation and telecommunications. The *gnomonic* projection, shown in Figure 2.10, is the only one on which all *great circles* (or parts thereof) appear as straight lines. Because great circles are the shortest distances between two points, navigators need only connect the points with a straight line to find the shortest route.

Other Projections

Projections can be developed mathematically to show the world or a portion of it in any shape that is desired: ovals, hearts, trapezoids, stars, and so on. One often-used projection is Goode's Homolosine. Usually shown in its interrupted form, as in Figure 2.11, it is actually a combination of two different projections. This equal-area projection also represents shapes well.

Not all maps are equal-area, conformal, or equidistant; many, such as the widely used polyconic (Figure 2.8), are compromises. In fact, some very effective projections

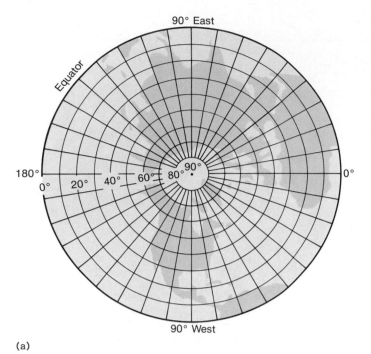

(a)

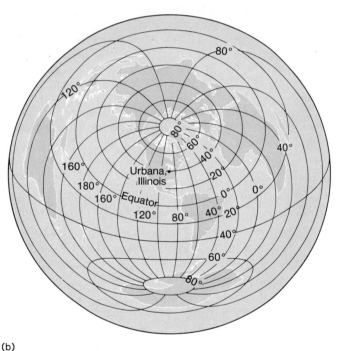

(b)

Figure 2.9

(a) The planar equidistant projection. Parallels of latitude are circles equally spaced on the meridians, which are straight lines. This projection is particularly useful because distances from the center to any other point are true. If the grid is extended to show the Southern Hemisphere, the South Pole is represented as a circle instead of a point. **(b) A planar equidistant projection centered on Urbana, Illinois.** The scale of miles applies only to distances from Urbana or on a line through it. The scale on the rim of the map, representing the antipode of Urbana, is infinitely stretched.

are non-Euclidian in origin, transforming space in unconventional ways. Distances may be measured in nonlinear fashion (in terms of time, cost, number of people, or even perception), and maps that show relative space may be constructed from these data. Two examples of such transformations are shown in Figure 2.12.

Mapmakers must be conscious of the properties of the projections they use, selecting the one that best suits their purposes. If the map shows only a small area, the choice of a projection is not critical—virtually any can be used. The choice becomes more important when the area to be shown extends over a considerable longitude and latitude; then the selection of a projection depends on the purpose of the map. As we have seen, Mercator or gnomonic projections are useful for navigation. If numerical data are being mapped, the relative sizes of the areas involved should be correct, so that one of the many equal-area projections is likely to be employed. Most atlases indicate which projection has been used for each map, thus informing the map reader of the properties of the maps and their distortions.

Selection of the map grid, determined by the projection, is the first task of the mapmaker. A second decision involves the scale at which the map is to be drawn.

Scale

The **scale** of a map is the ratio between the measurement of something on the map and the corresponding measurement on the earth. Scale is typically represented in one of three ways: verbally, graphically, or numerically as a representative fraction. As the name implies, a *verbal* scale is given in words, such as "1 inch to 1 mile." A *graphic* scale of 1 inch to 1 mile is shown below.

miles

A *representative fraction* (RF) scale gives two numbers, the first representing the map distance and the second the ground distance. The fraction may be written in a number of ways. There are 5280 feet in 1 mile and 12 inches in 1 foot; 5280 times 12 equals 63,360, the number of inches in 1 mile. The fractional scale of a map at 1 inch to 1 mile can be written as 1:63,360 or 1/63,360. On the simpler metric scale, 1 centimeter to 1 kilometer is 1:100,000. The units used in each part of the fractional scale are the same, thus 1:63,360 could also mean that 1 foot on the map represents 63,360 feet on the ground, or 12 miles—which is, of course, the same as 1 inch represents 1 mile. Numerical scales are the most accurate of all scale statements and can be understood in any language. Figure 2.14 is based on a fractional scale of 1:24,000.

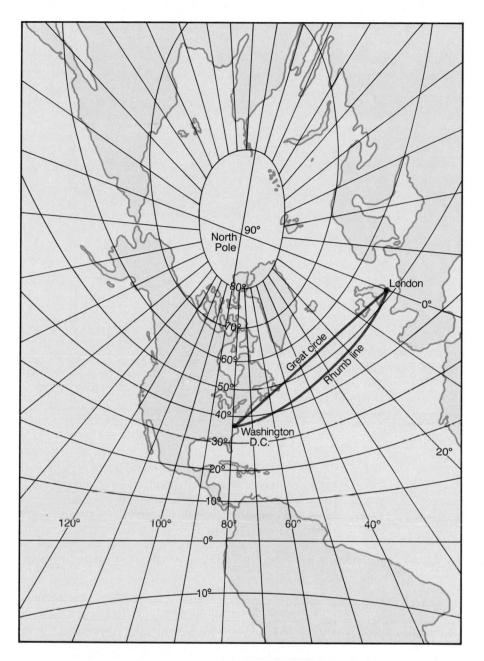

Figure 2.10
The gnomonic projection is the only one on which all great circles appear as straight lines. Rhumb lines are curved. In this sense, it is the opposite of the Mercator projection, on which rhumb lines are straight and great circles are curved.

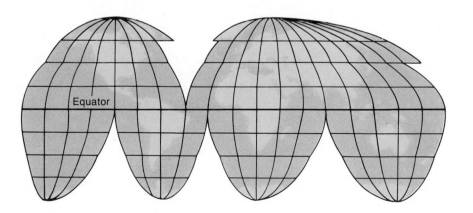

Figure 2.11
Goode's Homolosine projection, a combination of two different projections. This projection can also interrupt the continents to display the ocean areas intact.

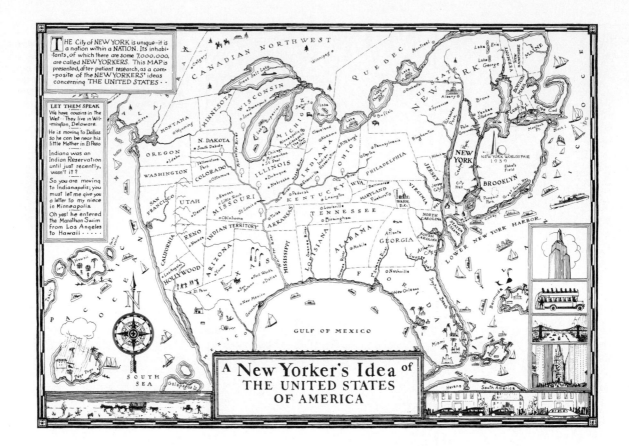

(a)

$160
$140
$120
$100
$80
$60
$40
$20

Boston
Newark
New York
Philadelphia
San Francisco
Dayton
Champaign
Washington, D.C.
Los Angeles
Charlotte
Dallas/Ft. Worth
Tampa
Orlando
Miami

(b)

Figure 2.12

(a) A New Yorker's idea of the United States of America portrays the country not as it actually is but as the artist conceives it. Think of it as a map based on an unknown projection. Distance, area, shape, and direction are all distorted, and a "psychological" rather than a literally true map is achieved. Such mental maps are discussed in Chapter 9.
(b) Airline cost distance from Champaign, Illinois. A distance-transformation map based upon one-way coach airfares, November 1986.

The terms *large-scale* and *small-scale* are often applied to maps. A large-scale map, such as a plan of a city, shows a restricted area in considerable detail, and the ratio of map to ground distance is relatively large, for example 1:600 (1 inch on the map represents 600 inches or 50 feet on the ground). At this scale, features, such as buildings and highways, that are usually represented by symbols on smaller scale maps can be drawn to scale. Small-scale maps, such as those of countries or continents, have a much smaller ratio. The scale may be 1 inch to 100 or even 1000 miles (1:6,336,000 or 1:63,360,000).

Each of the four maps in Figure 2.13 is drawn at a different scale. Although each is centered on Boston, notice how scale affects both the area that can be shown in a square that is 2 inches on a side and the amount of detail that can be depicted. On Map (a), at a scale of 1:25,000, about 2.6 inches represent 1 mile, so that the 2-inch square shows less than 1 square mile. At this scale, one can identify individual buildings, highways, rivers, and other landscape features. Map (d), drawn to a scale of 1 to 1 million (1:1,000,000 or 1 inch represents almost 16 miles), shows an area of almost 1000 square miles. Now only major features, such as main highways and the location of cities, can be shown, and even the symbols used for that purpose occupy more space on the map than the features depicted would if they were drawn true to scale.

Small-scale maps like (c) and (d) in Figure 2.13 are said to be very *generalized*. They give a general idea of the relative locations of major features but do not permit

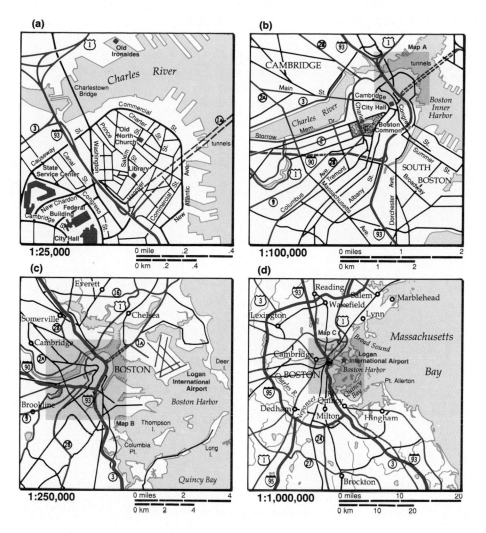

(a) 1:25,000

0 mile .2 .4
0 km .2 .4

(b) 1:100,000

0 miles 1 2
0 km 1 2

(c) 1:250,000

0 miles 2 4
0 km 2 4

(d) 1:1,000,000

0 miles 10 20
0 km 10 20

Figure 2.13
The effect of scale on area and detail.
The larger the scale, the greater the number and kinds of features that can be included because the map shows a more restricted area than does a map at a smaller scale.

accurate measurement. They show significantly less detail than do large-scale maps and typically smooth out such features as coastlines, rivers, and highways.

Topographic Maps and Terrain Representation

When we speak of the topography of an area, we refer to its terrain. **Topographic maps** portray the surface features of relatively small areas, often with great accuracy (Figure 2.14). They not only show the elevations of landforms, streams, and other water bodies but may also display features that people have added to the natural landscape. These might include transportation routes, buildings, and such land uses as orchards, vineyards, and cemeteries. Boundaries of all kinds, from state boundaries to field or airport limits, may also be depicted on topographic maps.

The U.S. Geological Survey (USGS), the chief federal agency for topographic mapping in this country, produces several map series, each on a standard scale. A small-scale topographic series at 1:250,000, or 1 centimeter to 2.5 kilometers, is complete for the United States. Other series are at scales of 1:125,000, 1:62,500, and 1:24,000.

A single map in one of these series is called a *quadrangle*. Topographic quadrangles at the scale of 1:24,000 exist for two-thirds of the United States excluding Alaska. Each map covers a rectangular area 7.5 minutes of latitude by 7.5 minutes of longitude. As is evident from Figure 2.14, these quadrangle maps can provide detailed information about the natural and cultural features of an area.

In Canada, the responsibility for national mapping lies with the Survey and Mapping Branch of the Department of Energy, Mines, and Resources. Maps at a scale of 1:250,000 are available for the entire country; the more heavily populated southern part of Canada is covered by 1:50,000-scale maps. Provincial mapping agencies produce detailed maps at even larger scales.

The USGS produces a sheet listing the symbols it employs on topographic maps (Figure 2.15), and some older maps provide legends on the reverse side. Note that in the case of running water, separate symbols are used in the legend and on the map to depict perennial (permanent) streams, intermittent streams, and springs; the location and size of rapids and falls are also indicated. There are three different symbols for dams and two more for

Figure 2.14

A portion of the Santa Barbara, California, 7.5-minute series of U.S. Geological Survey topographic maps. The fractional scale is 1:24,000. The shoreline shown represents the approximate line of mean high water. The pink tint indicates built-up areas, in which only landmark buildings are shown. The light purple tint is used to indicate extensions of urban areas.

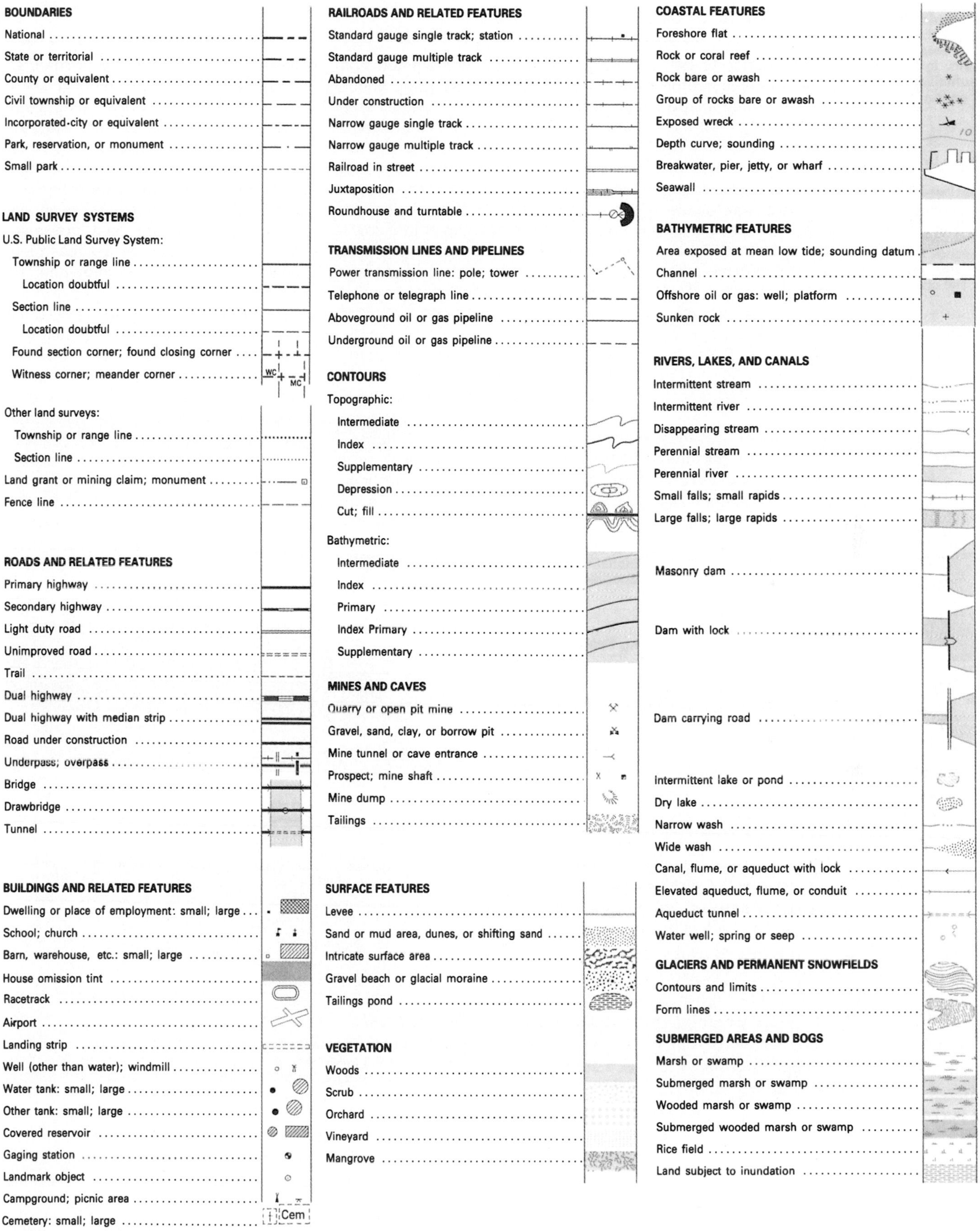

BOUNDARIES

National ....

State or territorial ....

County or equivalent ....

Civil township or equivalent ....

Incorporated-city or equivalent ....

Park, reservation, or monument ....

Small park ....

LAND SURVEY SYSTEMS

U.S. Public Land Survey System:

Township or range line ....

Location doubtful ....

Section line ....

Location doubtful ....

Found section corner; found closing corner ....

Witness corner; meander corner ....

Other land surveys:

Township or range line ....

Section line ....

Land grant or mining claim; monument ....

Fence line ....

ROADS AND RELATED FEATURES

Primary highway ....

Secondary highway ....

Light duty road ....

Unimproved road ....

Trail ....

Dual highway ....

Dual highway with median strip ....

Road under construction ....

Underpass; overpass ....

Bridge ....

Drawbridge ....

Tunnel ....

BUILDINGS AND RELATED FEATURES

Dwelling or place of employment: small; large ...

School; church ....

Barn, warehouse, etc.: small; large ....

House omission tint ....

Racetrack ....

Airport ....

Landing strip ....

Well (other than water); windmill ....

Water tank: small; large ....

Other tank: small; large ....

Covered reservoir ....

Gaging station ....

Landmark object ....

Campground; picnic area ....

Cemetery: small; large ....

RAILROADS AND RELATED FEATURES

Standard gauge single track; station ....

Standard gauge multiple track ....

Abandoned ....

Under construction ....

Narrow gauge single track ....

Narrow gauge multiple track ....

Railroad in street ....

Juxtaposition ....

Roundhouse and turntable ....

TRANSMISSION LINES AND PIPELINES

Power transmission line: pole; tower ....

Telephone or telegraph line ....

Aboveground oil or gas pipeline ....

Underground oil or gas pipeline ....

CONTOURS

Topographic:

Intermediate ....

Index ....

Supplementary ....

Depression ....

Cut; fill ....

Bathymetric:

Intermediate ....

Index ....

Primary ....

Index Primary ....

Supplementary ....

MINES AND CAVES

Quarry or open pit mine ....

Gravel, sand, clay, or borrow pit ....

Mine tunnel or cave entrance ....

Prospect; mine shaft ....

Mine dump ....

Tailings ....

SURFACE FEATURES

Levee ....

Sand or mud area, dunes, or shifting sand ......

Intricate surface area ....

Gravel beach or glacial moraine ....

Tailings pond ....

VEGETATION

Woods ....

Scrub ....

Orchard ....

Vineyard ....

Mangrove ....

COASTAL FEATURES

Foreshore flat ....

Rock or coral reef ....

Rock bare or awash ....

Group of rocks bare or awash ....

Exposed wreck ....

Depth curve; sounding ....

Breakwater, pier, jetty, or wharf ....

Seawall ....

BATHYMETRIC FEATURES

Area exposed at mean low tide; sounding datum .

Channel ....

Offshore oil or gas: well; platform ....

Sunken rock ....

RIVERS, LAKES, AND CANALS

Intermittent stream ....

Intermittent river ....

Disappearing stream ....

Perennial stream ....

Perennial river ....

Small falls; small rapids ....

Large falls; large rapids ....

Masonry dam ....

Dam with lock ....

Dam carrying road ....

Intermittent lake or pond ....

Dry lake ....

Narrow wash ....

Wide wash ....

Canal, flume, or aqueduct with lock ....

Elevated aqueduct, flume, or conduit ....

Aqueduct tunnel ....

Water well; spring or seep ....

GLACIERS AND PERMANENT SNOWFIELDS

Contours and limits ....

Form lines ....

SUBMERGED AREAS AND BOGS

Marsh or swamp ....

Submerged marsh or swamp ....

Wooded marsh or swamp ....

Submerged wooded marsh or swamp ....

Rice field ....

Land subject to inundation ....

Figure 2.15
Standard U.S. Geological Survey topographic map symbols.

Cartographic Falsification: Maps as Lies

Most people have a tendency to believe what they see in print. It is useful to be reminded that the printed image doesn't always mirror reality, that the truth can be distorted. For a variety of reasons, maps have often lied. Sometimes the cause has been ignorance, as when cartographers of the Middle Ages filled the unknown interiors of continents with mythical beasts. Other times the motivation for cartographic distortion has been propaganda; the maps of Nazi Germany fall into this category. In the most recent case to come to public attention, that of Soviet maps, the apparent purpose is to thwart foreign military and intelligence operations.

The chief cartographer of the USSR acknowledged in 1988 that for 50 years the policy of the Soviet Union was to deliberately falsify almost all publicly available maps of the country.

The policy apparently stemmed from the belief that cartography should serve military needs and began when the government mapmaking agency was put under the control of the security police. The announcement came as no surprise to westerners, who had known for years, for example, that the most reliable street map of Moscow is the one that is produced by the U.S. Central Intelligence Agency and that is a prized item on the city's black market.

The types of cartographic distortions on Soviet maps include the displacement and omission of features and the use of incorrect grid coordinates. The routes of highways, rivers, and railroads are sometimes altered by as much as 6 miles; the locations of rivers and dams can be off by a similar amount. A city or town might be shown on the east bank of a river when in fact it is on the west bank. Peninsulas, headlands, and other coastal features might be displaced by several miles or be absent entirely from the map. Even when features are shown correctly, the latitude and longitude grid can be misplaced. The distortion is not consistent over time. Thus, a town shown on one side of a river on one map might appear on the opposite bank on a later map, move several miles on a subsequent edition, and disappear altogether on yet another.

Two factors may have led to the decision to admit to the fabrication of Soviet maps. First, recent developments in remote sensing have enabled foreign countries to make their own accurate maps from satellite data. Second, Mikhail Gorbachev's policy of "openness" has widened public access to information long restricted for reasons of security or ideology.

---

types of bridges. On maps of cities, where it would be impossible to locate every building separately, the built-up area is denoted by special tints, and only streets and public buildings are shown.

The principal device used to show elevation on topographic maps is the **contour line**, along which all points are of equal elevation above a datum plane, usually mean sea level. Contours are imaginary lines, perhaps best thought of as the outlines that would occur if a series of progressively higher horizontal slices were made through a vertical feature. Figure 2.16 shows the relationship of contour lines to elevation for an imaginary island.

The *contour interval* is the vertical spacing between contour lines, and it is normally stated on the map. Contour intervals of 10 and 20 feet are often used, though in relatively flat areas the interval may be only 5 feet. In mountainous areas, the spacing between contours is greater for graphic convenience. On Figure 2.14, the contour interval is 50 feet. The more irregular the surface, usually, the greater is the number of contour lines that will need to be drawn; the steeper the slope, the closer are the contour lines rendering that slope.

Contour lines are the most accurate method of representing terrain, giving the map reader information about the elevation of any place on the map and the size, shape, and slope of all relief features. They are not truly pictorial, however. To heighten the graphic effect of a topographic map, contours are sometimes supplemented by the use of **shaded relief**. This method of representing the three-dimensional quality of an area is illustrated by Figure 2.17.

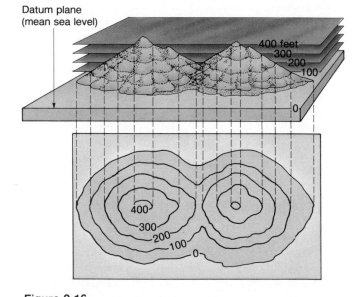

Figure 2.16

Contours drawn for an imaginary island. The intersection of the landform by a plane held parallel to sea level is a contour representing the height of the plane above sea level.

An imaginary light source, usually in the northwest, can be thought of as illuminating a model of the area, simulating the appearance of sunlight and shadows. Portions that are in the shadow are darkened on the map.

The tremendous amount of information on topographic maps makes them useful to engineers, regional planners, land use analysts, and developers as well as to

Figure 2.17
A shaded-relief map of a portion of the Swiss Alps near Interlaken. The scale is 1:50,000. Notice how effectively shading portrays a three-dimensional surface.

hikers and casual users. Given such a wealth of information, the experienced map reader can make deductions about both the physical character of the area and the economic and cultural use of the land.

Patterns and Quantities on Maps

The study of the spatial pattern of things, whether people, cows, or traffic flows, is the essence of the locational tradition in geography, a subject explored in Part III of this book. Maps are used to record the location of these phenomena, and different kinds of techniques are employed to depict their presence or numbers at specific points, in given areas, or along lines.

Quantities at Points

A topographic map, using symbols, shows the location at points of many kinds of things, such as churches, schools, and cemeteries. Each symbol counts as one occurrence. Sometimes, however, our interest is in showing the variation in the number of things that exist at several points, for example, the population of selected cities, the tonnage handled at certain terminals, or the number of passengers at given airports.

There are two chief means of symbolizing such distributions, as Figure 2.18 indicates. One method is to choose a symbol, usually a dot, to represent a given quantity of the mapped item (such as 50 people) and to repeat

Figure 2.18
(a) The distribution of country music radio stations in the late 1970s. The dot map provides a visual impression of specific locations and distributions.
(b) Metropolitan areas with more than 100,000 Mexican-Americans in 1980. On this map, the area of the circle is proportional to the total Hispanic population. The scale in the bottom left corner aids the reader in interpreting the map.

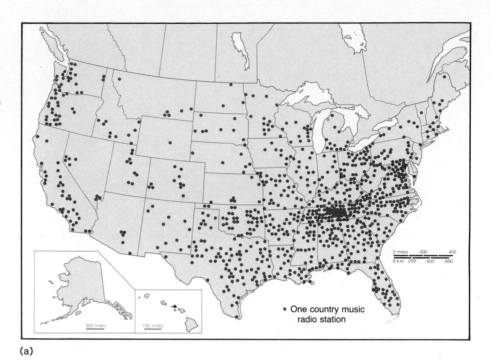

• One country music radio station

(a)

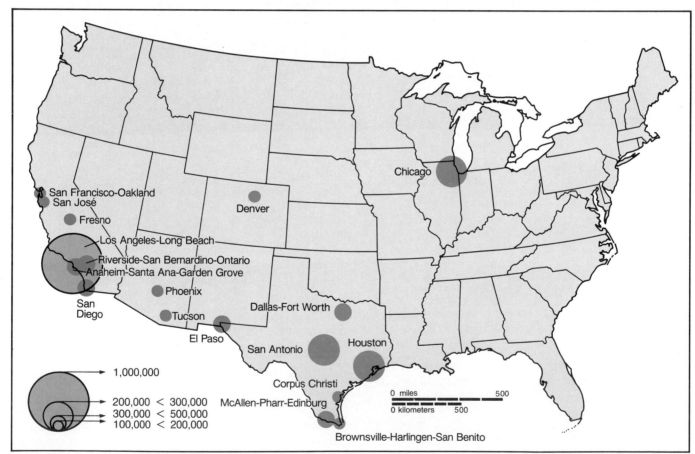

San Francisco-Oakland
San José
Fresno
Denver
Chicago
Los Angeles-Long Beach
Riverside-San Bernardino-Ontario
Anaheim-Santa Ana-Garden Grove
San Diego
Phoenix
Tucson
El Paso
Dallas-Fort Worth
San Antonio
Houston
Corpus Christi
McAllen-Pharr-Edinburg
Brownsville-Harlingen-San Benito

1,000,000
200,000 < 300,000
300,000 < 500,000
100,000 < 200,000

(b)

that symbol as many times as necessary. Such a map is easily understood because the dots give the map reader a visual impression of the pattern. Sometimes pictorial symbols, for example, human figures or oil barrels, are used instead of dots.

If the range of the data is great, the cartographer may find it inconvenient to use a repeated symbol. For example, if one port handles 50 or 100 times as much tonnage as another, that many more dots would have to be placed on the map. To circumvent this problem, the cartographer can choose a second method and use proportional symbols. The size of the symbol is varied according to the quantities represented. Thus, if bars are used, they can be shorter or longer as necessary. If squares or circles are used, the *area* of the symbol ordinarily is proportional to the quantity shown (Figure 2.18b). There are occasions, however, when the range of the data is so great that even circles or squares would take up too much room on the map. In such cases, three-dimensional symbols, usually spheres or cubes, are used, and their *volume* is proportional to the data. Unfortunately, many people fail to perceive the added dimension implicit in volume, and most cartographers do not recommend the use of such symbols (Figure 2.19).

Quantities in Area

Maps showing quantities in area fall into two general categories: those portraying the areas within which a kind of phenomenon occurs and those showing amounts by area. Atlases contain numerous examples of the first category, such as patterns of religions, languages, political entities,

vegetation, or types of rock. Normally, different colors or patterns are used for different areas, as in Figure 2.20. There are three main problems involved in using such maps: (1) they give the impression of uniformity to areas that may actually contain significant variations; (2) boundaries attain unrealistic precision and significance, implying abrupt changes between areas when, in reality, the changes may be gradual; and (3) unless colors are chosen wisely, some areas may look more important than others.

Variation in the *amount* of a phenomenon from area to area may be shown by covering subareas with different shades, colors, or patterns. Because the use of different colors or patterns may suggest qualitative rather than quantitative differences—that is, differences in kind rather than degree—the best cartographic practice is to use a single hue and vary its value or intensity. Figure 2.21 is an example of such a map.

An interesting variation of these maps is the **cartogram**, a map simplified to present a single idea in a diagrammatic way. Railroad and subway route maps commonly take this form. Depending on the idea that the

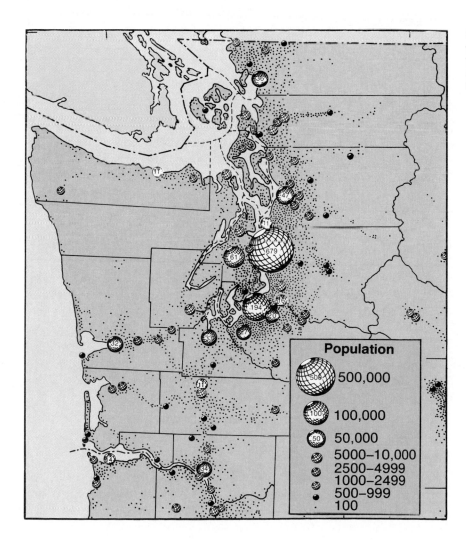

Figure 2.19

The distribution of population in western Washington in 1950. The cartographer used proportional spheres to represent large urban population concentrations.

Population

500,000
100,000
50,000
5000–10,000
2500–4999
1000–2499
500–999
100

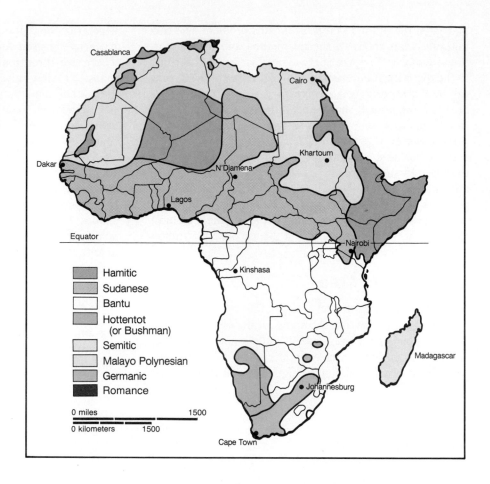

Figure 2.20

Language regions of Africa. Maps such as this one may give the false impression of uniformity within a given area, for example, that Bantu and no other languages are spoken over much of Africa. Such maps are intended to represent only the predominant language in an area.

Legend:
- Hamitic
- Sudanese
- Bantu
- Hottentot (or Bushman)
- Semitic
- Malayo Polynesian
- Germanic
- Romance

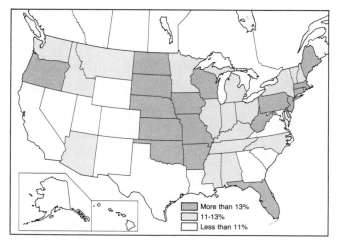

Figure 2.21

Percentage of population 65 years of age and over by state, 1980. Quantitative variation by area is more easily visualized in map than in tabular form.

Legend:
- More than 13%
- 11-13%
- Less than 11%

cartographer wishes to convey, the sizes and shapes of areas may be altered, distances and directions may be awry, and contiguity may or may not be preserved. In Figure 2.22, the sizes of the areas represented have been purposely distorted so that they are proportional to the population, and the map has been oriented with south at the top. Two other cartograms appear as Figures 11.7 and 11.11.

Quantities along Lines

Some lines on maps do not have numerical significance. The lines representing rivers, political boundaries, roads, and railroads, for example, are not quantitative. They are indicated on maps by such standardized symbols as the ones shown here and in Figure 2.15.

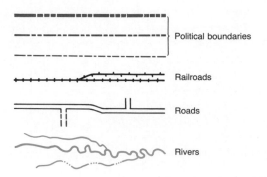

Political boundaries

Railroads

Roads

Rivers

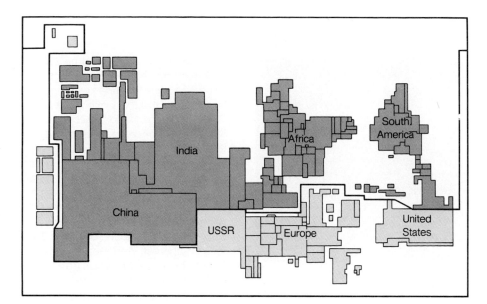

Figure 2.22
On this **cartogram,** the areas of countries have been made proportional to their populations. The map has been oriented with south at the top to make the point that northern countries contain only a small minority of the world's people.

Often, however, lines on maps do denote specific numerical values. Contour lines that connect points of equal elevation above mean sea level are a kind of **isoline,** or line of constant value. Other examples of isolines are isohyets (equal rainfall), isotherms (equal temperature), and isobars (equal barometric pressure). The implications of isolines as regional boundaries are discussed in Chapter 13.

Flow line maps are used to portray traffic or commodity flows along a given route, usually a waterway, a highway, or a railway. The location of the route taken, the direction of movement, and the amount of traffic can all be depicted. The amount shown may be either the total or a per mile figure. In Figure 2.23, the width of the flow line is proportional to the number of vehicles. Another flow line map appears in Figure 11.6.

Remote Sensing

When topographic maps were first developed, it was necessary to obtain the data for them through fieldwork, a slow and tedious process, which involved relating a given point on the earth's surface to other points by measuring its distance, direction, and altitude. The technological developments that have taken place in aerial photography since the 1930s have made it possible to speed up production and greatly increase the land area represented on topographic maps. Aerial photography is only one of a number of remote sensing techniques now employed.

Remote sensing is a relatively new term, but the process it describes—detecting the nature of an object from a distance—has been going on for well over a century. Soon after the development of the camera, photographs were made from balloons and kites. Even carrier pigeons wearing miniature cameras that took exposures automatically at set intervals were used to take aerial photographs

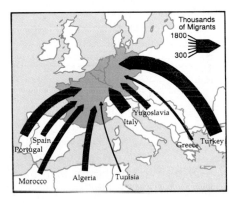

Figure 2.23
A flow line map showing the number of "guest workers" immigrating to Western Europe in the early 1980s.

of Paris. The airplane, first used for mapping in the 1930s, provided a platform for the camera and the photographer so that it was possible to take photographs from planned positions.

Aerial Photography

Although there is now a variety of sensing devices, aerial photography employing cameras with returned film is perhaps the most widely used remote sensing technique. Mapping from the air has certain obvious advantages over surveying from the ground, the most evident being the bird's-eye view that the cartographer obtains. Using stereoscopic devices, the cartographer can determine the exact slope and size of features, such as mountains, rivers, and coastlines. Areas that are otherwise hard to survey, such as mountains and deserts, can be mapped easily from the air. Furthermore, millions of square miles can be surveyed in a very short time. Aerial photographs must, of

Geodetic Control Data

The horizontal position of a place, specified in terms of latitude and longitude, constitutes only two-thirds of the information needed to locate it in three-dimensional space. Also needed is a vertical control point defining elevation, usually specified in terms of altitude above sea level. Together, the horizontal and vertical positions comprise *geodetic control data*. A network of more than one million of these precisely known points covers the entire United States.

Each point is indicated by a bronze marker fixed in the ground. You may have seen some of the vertical markers, called *bench marks*, on mountaintops, hilltops, or even on city sidewalks. Every U.S. Geological Survey (USGS) map shows the markers in the area covered by the map, and the USGS maintains Geo-

detic Control Lists containing the description, location, and elevation of each marker.

These lists were revised in 1987 when, after 12 years' effort, federal scientists completed the recalculation of the precise location of some 250,000 bench marks across the country. In using a satellite locating system for the first national resurvey of control points since 1927, the National Oceanic and Atmospheric Administration (NOAA) found, for example, that New York's Empire State Building is 120.5 feet northeast of where it formerly officially stood. The Washington Monument has been moved northeast by 94.5 feet, the dome of California's state capitol in Sacramento has been relocated 300.9 feet southwest, and Seattle's Space Needle now has a position 305.26 feet west and 65.53 feet south of where maps now show it. The satellite survey provides much more accurate locations than did the old system of land measurement of distances and angles. The result is more accurate maps and more precise navigation.

---

course, be interpreted by using such clues as the size, shape, tone, and color of the recorded objects before maps can be made from them. Maps based on aerial photographs can be made quickly and revised easily so that they are kept up-to-date. With aerial photography, the earth can be mapped more accurately, more completely, and more rapidly than ever before.

In 1975 the U.S. Department of the Interior instituted the National Mapping Program to improve the collection and analysis of cartographic data and to prepare maps that would assist decision makers who deal with resource and environmental problems. One of the first goals of the program was to achieve complete coverage of the country with orthophotographic imagery for all areas not already mapped at the scale of 1:24,000.

An **orthophotomap** is a topographic map on which natural and cultural features are portrayed by color-enhanced photographic images with certain map symbols (contours, boundaries, names) added as needed for interpretation. Note in Figure 2.24 that in contrast to the conventional topographic map, the photograph is the chief means of representing information. Orthophotomaps have a variety of uses, aiding forest management, soil surveys, geological investigations, flood hazard and pollution studies, and city planning.

Standard photographic film detects reflected energy within the visible portion of the electromagnetic spectrum (Figure 2.25). Although they are invisible, *near-infrared*

wavelengths can be recorded on special sensitized infrared film. Discerning and recording objects that are not visible to the human eye, infrared film has proved particularly useful for the classification of vegetation and hydrographic features. Color-infrared photography yields what are called **false-color images,** "false" because the film does not produce an image that appears natural. For example, leaves of healthy vegetation have a high infrared reflectance and are recorded as red on color-infrared film, while unhealthy or dormant vegetation appears as blue, green, or gray.

Nonphotographic Imagery

For wavelengths longer than 1.2 micrometers (a micrometer is 1 one-millionth of a meter) on the electromagnetic spectrum, sensing devices other than photographic film must be used. **Thermal scanners**, which sense the energy emitted by objects on earth, are used to produce images of thermal radiation (Figure 2.26). That is, they record the longwave radiation (which is proportional to surface temperature) emitted by water bodies, clouds, and vegetation as well as by buildings and other structures. Unlike conventional photography, thermal sensing can be employed during nighttime as well as daytime, giving it military applications. It is widely used for studying various aspects of water resources, such as ocean currents, water pollution, surface energy budgets, and irrigation scheduling.

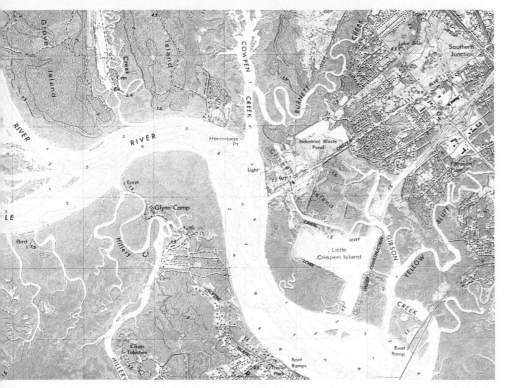

Figure 2.24
Topographic map (top) and orthophotomap (bottom) of a portion of the Brunswick West quadrangle in southeastern Georgia. Orthophotomaps are well suited for portraying swamp and marshlands; flat, sandy terrain; and urban areas.

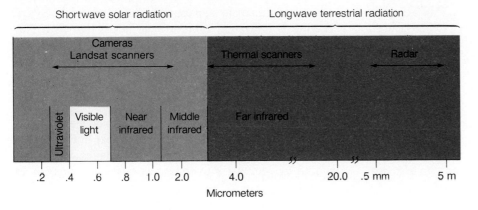

Figure 2.25
Wavelengths of the electromagnetic spectrum in micrometers. One micrometer equals 1 one-millionth of a meter. The human eye can detect light in only the small portion of the spectrum that is colored yellow. Radiation from the sun has short wavelengths. Wavelengths longer than 4 micrometers characterize terrestrial radiation.

Operating in a different band of the electromagnetic spectrum, **radar** systems can also be used during the day or night. They transmit pulses of energy toward objects and sense the energy reflected back. The data are used to create images such as that shown in Figure 2.27, which was produced by radar equipment mounted on an airplane. Because radar can penetrate clouds and vegetation as well as darkness, it is particularly useful for monitoring the locations of airplanes, ships, and storm systems and for detecting variation in soil moisture conditions.

Satellite Imagery

In the last 30 years, both manned and unmanned spacecraft have supplemented the airplane as the vehicle for imaging the terrain. Concurrently, many steps have been taken to automate mapping, including the use of electronic mapping techniques, automatic plotting devices, and automatic data processing. Many images are now taken either from continuously orbiting satellites, such as those in the U.S. Landsat series and the French SPOT series, or from manned spacecraft flights, such as those of the Apollo and Gemini missions. Among the advantages of satellites are the speed of coverage and the fact that views of large regions can be obtained.

In addition, because they are equipped to record and report back to the earth information in the infrared portion of the electromagnetic spectrum, these satellites enable us to map the invisible. A number of federal agencies in the United States, Japan, and the USSR have launched satellites specifically to monitor the weather. Data obtained by satellites have greatly improved weather forecasting and the tracking of major storm systems and in the process have saved countless lives. The satellites are the source of the weather maps shown daily on television and in newspapers.

The **Landsat satellites**, first launched in 1972, take about 1 hour and 40 minutes to orbit the globe and can provide repetitive coverage of almost the entire globe every 18 days. Rather than recording data photographically, the Landsat satellite relays electronic signals to receiving stations, where computers convert them into photolike images that can be adjusted to fit special map projections. Composite images can be made by combining information from different wavelengths of light energy.

Landsat carries scanning instruments that pick up sunlight reflected by foliage, water, rocks, and other objects. One sensor, the multispectral scanner (MSS), covers the visible and near-infrared range, from 0.4 to 1.1 micrometers. The other sensor, the higher resolution thematic mapper (TM), has seven wavebands, several in the thermal ranges up to 11.7 micrometers.

Landsat's cameras are capable of resolving objects about 100 feet apart. Even sharper images are yielded by the French *SPOT* (Satellite Probatoire d'Observation de la Terre) *satellite* launched in 1986. Its sensors can show objects that are less than 35 feet apart and can also pro-

Figure 2.26

A thermal radiation image of an area south of Manhattan, Kansas, June 6, 1987. The warm (red, orange) colors represent areas with the highest surface temperatures—the bare fields in the upper right of the image and the road. The higher elevations in the sandstone region on the left-hand side of the image are warmer than the lowlands and woodlands at lower elevations. (From the CESAR Laboratory of the San Diego State University. Courtesy of Allen Hope.)

duce three-dimensional pictures. Like Landsat, the SPOT satellite is in a polar orbit, which means that as it flies from south to north, the earth turns below it so that each orbit covers a strip of surface adjacent to the previous one. SPOT images the earth at the same local time on consecutive passes and repeats its pattern of successive ground tracks at 26-day intervals.

A Landsat image of the Los Angeles metropolitan area appears in Figure 2.28. Others appear in Figures 1.6 and 3.22. As they indicate, Landsat images can be enhanced by computers to appear as false-color images. In

Figure 2.27
Radar image of the San Francisco peninsula. The San Andreas
fault is clearly visible.

Figure 2.28
Landsat image of southern California.
The Santa Barbara channel appears black
and the Los Angeles metropolitan region
gray. The fan-shaped stain spreading into
the Mojave Desert near the right-center
margin is silt and other material carried
down the mountains by water runoff.
Analyses of Landsat images have
practical applications in agriculture and
forest inventory, land use classification,
identification of geologic structures and
associated mineral deposits, and
monitoring of natural disasters.

Figure 2.28, growing healthy vegetation appears as red, clear water as black, and urban areas as gray. The images have been used to produce small-scale maps at scales ranging from 1:250,000 to 1:1,000,000.

Mapping is only one of the applications of remote sensing, which has also proven to be an effective method of conducting resource surveys and monitoring the natural environment. Geologists have found remote sensing to be particularly useful in conducting resource surveys in desert and remote areas. For example, information about vegetation or folding patterns of rocks can be used to help identify likely sites for mineral or oil prospecting. Remote sensing imagery has been used to monitor a variety of environmental phenomena, including water pollution, the effects of acid rain, and rain forest destruction. As noted earlier, weather satellites can monitor frontal systems and are a valuable contribution to worldwide weather forecasting. Because remotely sensed images can be used to calculate such factors as biomass production and rates of transpiration and photosynthesis, they are invaluable for modeling relationships between the atmosphere and the earth's surface.

Automated Cartography and Geographic Information Systems

A recent development in cartography is the use of computers to assist in making maps. Within the last two decades, computers have become an integral part of almost every stage of the cartographic process, from the collection and recording of data to the production and reproduction of maps. Although the initial cost of equipment is high, the investment is repaid in the more efficient and accurate production and revision of maps.

Computers are at the heart of what have come to be known as a **geographic information system (GIS)**. A GIS is a tool for storing and manipulating geographic information in a computer. The three major components of a GIS are (1) the digital map data, (2) the hardware used to enter, store, retrieve, process, and display these data, and (3) all the computer software used to perform GIS operations.

The first step in developing a GIS is to create the **geographic database** (a digital record of geographic information) from such sources as maps, field surveys, aerial photographs, satellite imagery, and so on. Geographic information is of two types: locational data and nonlocational, or attribute, data. *Locational* data describe the location of and connections between point, line, and area features. These data can be stored in a GIS in several different formats, and they are commonly represented as a series of x,y coordinates. That is, each point is recorded as a single x,y location, and lines are indicated by a series of x,y coordinates (Figure 2.29). Areas (usually called polygons) are represented as a series of x,y coordinates defining line segments that enclose them. All locational map features are stored in a computer as lists of x,y digits (hence the term "digital" map database).

The second type of geographic information is descriptive. It gives the characteristics of the point, line, and area features in terms of certain *attributes*, which may be either qualitative (e.g., the types and names of roads in a given area) or quantitative (e.g., the widths of the roads). Attributes are stored in the computer in tabular form as sets of numbers and characteristics to be accessed when they are needed to create maps or perform analyses.

Although there are a number of devices that can be used to digitize an existing map, such as a mouse connected to a personal computer, the fastest is the *raster scanner*, consisting of a scanning head and a large cylindrical drum on which the map is placed (Figure 2.30). The scanning head takes readings as the drum rotates. When one rotation is completed, the head moves a fraction of a millimeter and repeats the process. The scanner optics subdivide the area under view into small units called *pixels* (picture elements), each representing an area considerably smaller than one-tenth the size of the period at the end of this sentence. Each pixel has a value between 0 and 255. The brighter the pixel, the higher the numerical value. Pixels of the same value are converted to a series of x,y coordinates.

Once geographic information is in the computer in digital form, the data can be manipulated, analyzed, and displayed with a speed and precision not otherwise possible. Because computers can process millions of facts in seconds, they are particularly useful for researchers who need to analyze many variables simultaneously. The development of geographic information systems has deemphasized the use of the map to store information and enabled researchers to concentrate on using the map for analyzing and communicating spatial information. With the appropriate software, a computer operator can display any combination of data, showing the relationships among variables almost instantly. In this sense, computers make possible the production of maps that were virtually impossible to generate 20 years ago.

Figure 2.31 illustrates how new map features can be created by merging information from several different variables. In this example, planners are trying to find the best sites for a new well. The first four images depict different geological and topographical features bearing on the decision. The first shows kinds of land use and cover. Eastern brushlands, forested wetlands, and deciduous, evergreen, and mixed forests are considered acceptable for well sites. The second image indicates sources of pollution. Areas within 100 meters of polluted streams and within 500 meters of pollution point sources should be excluded from consideration as well sites. Surface materials are shown in the third image. A well site should have

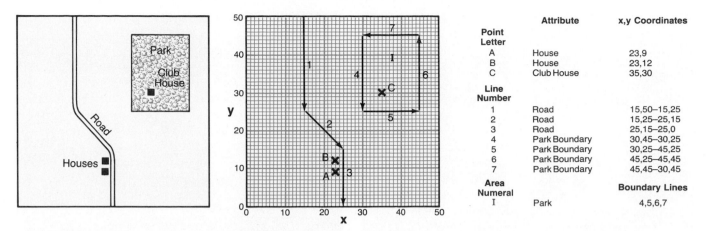

Point Letter	Attribute	x,y Coordinates
A	House	23,9
B	House	23,12
C	Club House	35,30
Line Number		
1	Road	15,50–15,25
2	Road	15,25–25,15
3	Road	25,15–25,0
4	Park Boundary	30,45–30,25
5	Park Boundary	30,25–45,25
6	Park Boundary	45,25–45,45
7	Park Boundary	45,45–30,45
Area Numeral		**Boundary Lines**
I	Park	4,5,6,7

Figure 2.29

Digitizing a map. Digitizing treats a map as a graph. Coordinates x, y define the positions of the points, lines, and areas that constitute geographic information. A complete digital description of the map's contents would include precise plotting of the route of the road.

Figure 2.30

A raster scanner, which converts maps into digital form.

coarse-grained material. The depth of water saturation, which should be at least 41 feet, is indicated in the fourth image. Combining data from the first four images yields the potential well sites, shown in green in the final image.

To give another example, suppose that a state needed to select a site for a radioactive waste disposal facility. With a database drawn from maps of earthquake intensity, oil and gas field locations, sand and gravel resources, aquifers, and other distributions, the suitability of sites could be evaluated. Areas prone to earthquakes or subsidence could be immediately excluded from consideration, while those with such characteristics as simple geologic structure, surficial materials of low permeability, and deep aquifers would be more favorably assessed.

GIS operations can produce several types of output: displays on a computer monitor, listings of statistics, or hard copy. When a map is to be produced, the operator can quickly call up the desired data. Geographic information systems are particularly useful for revising existing maps because outdated data—for example, population sizes—can be erased from the file and replaced. Several software packages are now available for map production, particularly for commonly used projections. They enable the cartographer to produce plots of boundaries, coastlines, and grid systems at any desired scale. Some software is intended for use only on large-capacity computers, but software packages for personal computers are also available.

A number of bureaus and agencies in the federal government use geographic information systems. Currently, the USGS is digitizing categories of data from its 7.5-minute topographic quadrangles and hopes to complete the National Digital Cartographic Data Base by the year 2000. It will include all the information that now appears on the agency's maps. The USGS is also helping the Census Bureau produce a GIS that will incorporate census data as well as various physical and cultural landscape features. At the state and local levels, bureaus and agencies concerned with resource use and urban and regional planning are especially attracted by the usefulness of GIS.

A growing number of companies in the private sector are also using computerized mapmaking systems. Among others, oil and gas companies, restaurant chains, soft drink bottlers, and car rental companies rely on GIS to perform such diverse tasks as identifying drilling sites, picking locations for new franchises, analyzing sales territories, and calculating optimal driving routes.

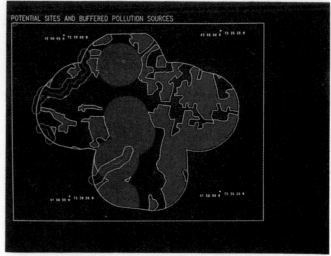

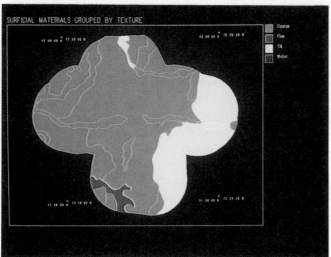

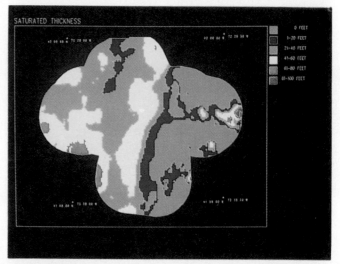

Figure 2.31
Geographic information systems enable researchers to create new maps by merging information from different variables. The first four images show different geological and topographical features that would affect the decision of where to site a new water well. Combining data from them yields the potential well sites, shown in green in the final image. The existing water service area is shown in yellow.

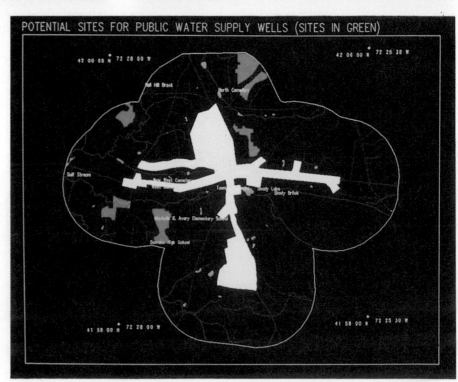

Summary

In this chapter, we have not attempted to discuss all aspects of the field of cartography. Mapmaking and map design, systems of land survey, map compilation, and techniques of map reproduction are among the topics that have been deliberately omitted. Our intent has been to introduce those aspects of map study that will aid in map reading and interpretation and to hint at new techniques of map creation and of the design and purpose of geographic information systems.

Maps are among the oldest and most basic means of communication. They are as indispensable to the geographer as words, photographs, or quantitative techniques of analysis. Geographers are not unique in their dependence on maps. People involved in the analysis and solution of many of the problems facing the world also rely on them. Environmental protection, control of pollution, conservation of natural resources, and land use planning are just a few of the issues that call for the accurate representation of elements on the earth's surface.

The roots of modern mapping lie in the 17th century. Three key developments that made it both desirable and possible to map the surface features of the earth were the rediscovery of the works of earth scientists of ancient Greece, transoceanic voyages, and the invention of the printing press.

The grid system of longitude and latitude is used to locate points on the earth's surface. Latitude is the measure of distance north and south of the equator, while longitude is the angular distance east or west of the prime meridian. Both latitude and longitude are given in degrees, which for greater precision can be subdivided into minutes and seconds.

All systems of representing the curved earth on a flat map distort one or more earth features. Any given projection will distort area, shape, distance, and/or direction. Cartographers select the projection that best suits their purpose and may elect to use an equal-area, conformal, or equidistant projection or one that shows correctly the directions from a single point to all others. Many useful projections have none of these qualities, however.

Among the most accurate and most useful large-scale maps are the topographic quadrangles produced by a country's chief mapping agency. They contain a wealth of information about both the physical and cultural landscape and are used for a variety of purposes.

In recent years remote sensing techniques have resulted in the faster and more accurate mapping of the earth. Both aerial photography and satellite imagery can discern reflected energy in the visible and near-infrared portions of the electromagnetic spectrum. Applications of remote sensing include mapping, monitoring the environment, and conducting resource surveys. The need to store, process, and retrieve the vast amounts of data generated by remote sensing has spurred the development of geographic information systems.

As you read the remainder of this book, note the many different uses of maps. For example, notice in Chapter 3 how important maps are to your understanding of the theory of continental drift, in Chapter 7 how maps aid geographers in identifying cultural regions, and in Chapter 9 how behavioral geographers use maps to record people's perceptions of space.

Key Words

cartogram	isoline
conformal projection	Landsat satellite
conic projection	latitude
contour line	longitude
cylindrical projection	map scale
developable surface	Mercator projection
equal-area projection	orthophotomap
equidistant projection	planar projection
equivalent projection	prime meridian
false-color image	projection
geographic database	radar
geographic information	remote sensing
system	shaded relief
globe properties	thermal scanner
grid system	topographic map
International Date Line	

For Review

1. What important map and globe reference purpose does the *prime meridian* serve? Is the prime or any other meridian determined in nature or devised by humans? How is the prime meridian designated or recognized?

2. What happens to the length of a degree of longitude as one approaches the poles? What happens to a degree of latitude between the equator and the poles?

3. From a world atlas, determine in degrees and minutes the locations of New York City; Moscow, USSR; Sydney, Australia; and your hometown.

4. List at least five properties of the globe grid. Examine the projections used in Figures 2.5, 2.8, and 2.9a. In what ways do each of these projections adhere to or deviate from globe grid properties?

5. Briefly make clear the differences in properties and purposes of *conformal, equivalent*, and *equidistant* projections. Give one or two examples of the kinds of map information that would best be presented on each type of projection. Give one

or two examples of how misunderstandings might result from data presented on an inappropriate projection.

6. In what different ways may *map scale* be presented? Convert the following map scales into their verbal equivalents:

    1:1,000,000   1:63,360   1:12,000

7. What is the purpose of a *contour line*? What is the *contour interval* on Figure 2.14? What landscape feature is implied by closely spaced contours?

8. What kinds of data acquisition are suggested by the term *remote sensing*? Describe some ways in which different portions of the spectrum can be sensed. To what uses are remotely sensed images put?

9. What are the basic components of a *geographic information system*? How is spatial information recorded in a *geographic database*? What are some of the uses of computerized mapmaking systems?

10. The table below gives the gross national product per capita of selected European countries (in U.S. dollars, 1988). On outline maps or as cartograms, represent the data in two different ways.

Austria	15,560	Luxembourg	22,600
Belgium	14,550	Netherlands	14,530
Denmark	18,470	Norway	20,020
France	16,080	Portugal	3,670
W. Germany	18,530	Spain	7,740
Greece	4,790	Sweden	19,150
Ireland	7,480	Switzerland	27,260
Italy	13,320	United Kingdom	12,800

Selected References

American Cartographer 14, no. 3 (July 1987). Special issue.

American Cartographic Association, Committee on Map Projections. *Choosing a World Map: Attributes, Distortions, Classes, Aspects*. Special Publication No. 2. Falls Church, Va.: American Congress on Surveying and Mapping, 1988.

————. *Which Map Is Best? Projections for World Maps*. Special Publication No. 1. Falls Church, Va.: American Congress on Surveying and Mapping.

Brown, Lloyd A. *The Story of Maps*. Boston: Little, Brown, 1949; reprint ed., New York: Dover 1977.

Burnside, Clifford D. *Mapping from Aerial Photographs*. 2d ed. London: Collins, 1985.

Burrough, P. A. *Principles of Geographical Information Systems for Land Resources Assessment*. Oxford, England: Clarendon Press, 1986.

Campbell, John. *Introductory Cartography*. Englewood Cliffs, N.J.: Prentice-Hall, 1984.

Clarke, Keith C. *Analytical and Computer Cartography*. Englewood Cliffs, N.J.: Prentice-Hall, 1990.

Crone, G. R. *Maps and Their Makers*. 5th ed. Folkestone, Kent: Dawson & Son, 1978.

Dent, Borden D. *Cartography: Thematic Map Design*. 2d ed. Dubuque, Ia.: Wm. C. Brown, 1990.

Gould, Peter, and Rodney White. *Mental Maps*. 2d ed. Boston: Allen & Unwin, 1986.

Harley, J. B., and D. Woodward, eds. *The History of Cartography*. Vol. 1. Chicago: University of Chicago Press, 1987.

Lillesand, Thomas M., and Ralph W. Kiefer. *Remote Sensing and Image Interpretation*. 2d ed. New York: Wiley, 1987.

Miller, Victor, and Mary Westerbrook. *Interpretation of Topographic Maps*. Columbus, Ohio: Charles E. Merrill, 1989.

Monmonier, Mark S. *Technological Transition in Cartography*. Madison: University of Wisconsin Press, 1985.

Muehrcke, Phillip C. *Map Use: Reading, Analysis, Interpretation*. 2d ed. Madison, Wis.: JP Publications, 1986.

————. *Thematic Cartography*. Association of American Geographers, Commission on College Geography Resource Paper No. 19. Washington, D.C.: Association of American Geographers, 1972.

Robinson, Arthur H., Randall D. Sale, Joel Morrison, and Phillip C. Muehrcke. *Elements of Cartography*. 5th ed. New York: Wiley, 1984.

Sabins, Floyd F., Jr. *Remote Sensing: Principles and Interpretation*. 2d ed. New York: Freeman & Co., 1987.

Thomas, Ian L. *Classification of Remotely Sensed Images*. Bristol, England: Adam Hilger, 1987.

Thompson, Morris M. *Maps for America: Cartographic Products of the U.S. Geological Survey and Others*. Washington, D.C.: U.S. Government Printing Office, 1982.

Thrower, Norman J. W. *Maps and Man: An Examination of Cartography in Relation to Culture and Civilization*. Englewood Cliffs, N.J.: Prentice-Hall, 1972.

Wilford, J. N. *The Mapmakers*. New York: Knopf, 1981.

C H A P T E R

3

Physical Geography: Landforms

The Himalaya mountains, India.

BEFORE the morning of November 14, 1963, cartographers thought they had finished their work in Iceland. They certainly had been at it long enough—since the island's first appearance on a world map by Eratosthenes some two centuries before Christ. It was a well-mapped area early in the Middle Ages and could be set aside as one of the certainties of the North Atlantic. Then with a roar, a pillar of steam and fire, and a hail of ash, nature demanded that the cartographers go back to work. Surtsey, the Dark One—god of fire and destruction—rose from the sea 20 miles off the Icelandic coast to become new land—1 mile long, 600 feet high, 670 acres of rock, ash, and lava.

Cartographers had another job, as would have been their lot had they been around for, say, the last 100 million years. Things just won't stand still. And not only little things, like new islands, or big ones, like mountains rising and being worn low to swampy plains, but also monstrous things: continents that wander about like nomads and ocean basins that expand, contract, and split up the middle like worn-out coats. It is a fascinating story, this formation and alteration of the home of humans, which at first glance seems so eternal and unchanging.

Geologic time is long, but the forces that give shape to the land are timeless and constant. Processes of creation and destruction are continually at work to fashion the seemingly eternal structure upon which humankind lives and works. Two types of forces interact to produce those infinite local variations in the surface of the earth called *landforms*: (1) forces that push, move, and raise the earth's surface; and (2) forces that scour, wash, and wear down the surface. Mountains rise and then are worn away. The eroded material—soil, sand, pebbles, rocks—is transported to new locations and helps to create new landforms. How long these processes have worked, how they work, and their effects are the subject of this chapter.

Much of the research needed to create the story of landforms results from the work of geomorphologists. *Geomorphology*, a branch of the fields of geology and physical geography, is the study of the origin, characteristics, and development of landforms. Geomorphology emphasizes the study of the various processes that influence the erosion, transportation, and deposition of materials. A modern thrust is in the area of the interrelationships between plant and animal life and landforms. In a single chapter we can but begin to explore the many and varied contributions of geomorphologists. After discussing the context within which landform change takes place, we consider the forces that are building up the earth's surface and then review the forces wearing it down. Since most earth surface-changing processes occur over long periods of time, amounting to millions of years, perspective can be gained by developing a sense of what is meant by the time span within which the drama of earth change takes place.

Geologic Time

The earth was formed about 4.7 billion years ago. When we think of a person who lives to be 100 years old as having had a long life, it becomes clear that the earth is incredibly old indeed. Because our usual concept of time is dwarfed when we speak of billions of years, it is useful to compare the age of the earth with something that is more familiar.

Imagine that the height of the World Trade Center in New York City represents the age of the earth. The twin towers are 110 stories, or 1353 feet (412 m), tall. In relative terms, even the thickness of an average piece of paper laid on top of the roof would be too great to represent an average person's lifetime. Of the total building height, 4.7 stories represent the 200 million years that have elapsed since the present ocean basins began to form. The first hominids, or humanlike creatures, made their appearance on earth about 15 million years ago, or the equivalent of the height of one-third of a story. Earth history is so long and involves so many major geologic events that scientists have divided it into a series of recognizable, distinctive stages. These are depicted in Figure 3.1.

At this moment, the landforms on which we live are ever so slightly being created and destroyed. The processes involved have been in operation for so long that any given location most likely was the site of ocean and land at a number of different times in its past. Many of the landscape features on earth today can be traced back several hundred million years. The processes responsible for building up and for tearing down those features are occurring simultaneously but usually at different rates.

In the last 40 years scientists have developed a useful framework within which one can best study our constantly changing physical environment. Their work is based on the early 20th century geological studies of Alfred Wegener, one of the pioneers of the rapidly evolving scientific theory of **plate tectonics**. He believed that the present continents were once united in one supercontinent and that over many millions of years the continents broke away from each other and slowly drifted to their current positions. New evidence and new ways of rethinking old knowledge have led to wide acceptance in recent years by earth scientists of the idea of moving continents.

Movements of the Continents

The landforms mapped by cartographers are only the surface features of a thin cover of rock, the earth's *crust*. Above the core and the lower mantle of the earth, there is a partially molten plastic layer called the **asthenosphere**

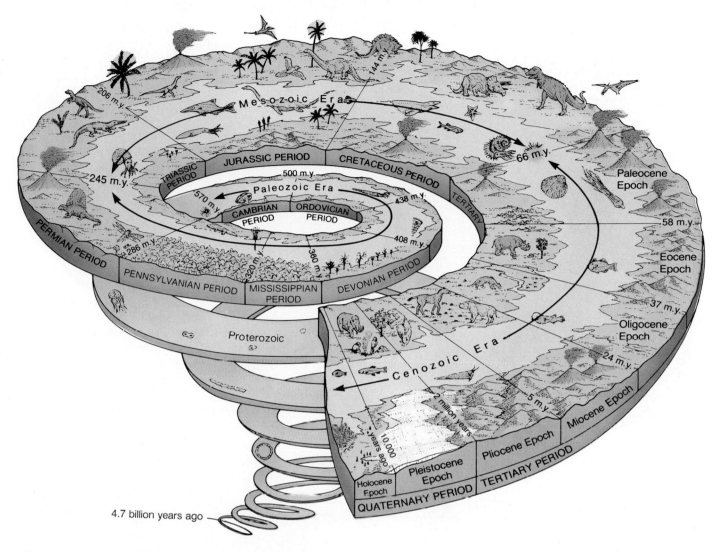

Figure 3.1
A diagrammatic history of the earth. The sketch depicts some of the known characteristics of the named geologic periods.

(Figure 3.2). The asthenosphere supports a thin but strong solid shell of rocks called the **lithosphere**, of which the outer, lighter portion is the earth's crust. The crust consists of one set of rocks found below the oceans and another set that makes up the continents.

The lithosphere is broken into about ten large, rigid plates, each of which, according to the theory of plate tectonics, slides or drifts very slowly over the heavy semimolten asthenosphere. A single plate may contain both oceanic and continental crust. Figure 3.3 shows that the North American plate, for example, contains the northwest Atlantic Ocean and most, but not all, of North America. The peninsula of Mexico (Baja California) and part of California are on the Pacific plate.

Scientists are not certain why the lithospheric plates move. One reasonable theory suggests that heat and heated material from the earth's interior rise by convection into particular crustal zones of weakness. These zones are sources for the divergence of the plates. The cooled materials then sink downward in subduction zones (discussed below). In this way, the plates are thought to be set in motion. Strong evidence indicates that about 200 million years ago the entire continental crust was connected in one super continent, to which Wegener gave the name **Pangaea** ("all earth"). Pangaea was broken into plates as the seafloor began to spread, the major force coming from the widening of what is now the Atlantic Ocean (Figure 3.4).

Materials from the asthenosphere have been rising along the Atlantic Ocean fracture and, as a result, the seafloor has continued to spread. The Atlantic Ocean is now 4300 miles (6920 km) wide at the equator. If it widened by a bit less than 1 inch (2.5 cm) per year, as scientists have estimated, one could calculate that the separation of the continents did in fact begin about 200 million years

Figure 3.2

(a) **The outer zones of the earth** (not to scale). The *lithosphere* includes the crust and the uppermost mantle. The *asthenosphere* lies entirely within the upper mantle. (b) The very thin crust of the earth overlies a layered planetary interior. Zonation of the earth occurred early in its history. Radioactive heating melted the original homogeneous planet. A dense iron core settled to the center; a surface and the remnant lower mantle, overlain by a transition zone and the asthenosphere, formed between them. Escaping gases eventually created the atmosphere and the oceans.

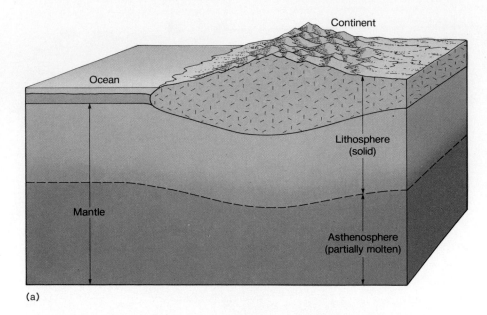

(a)

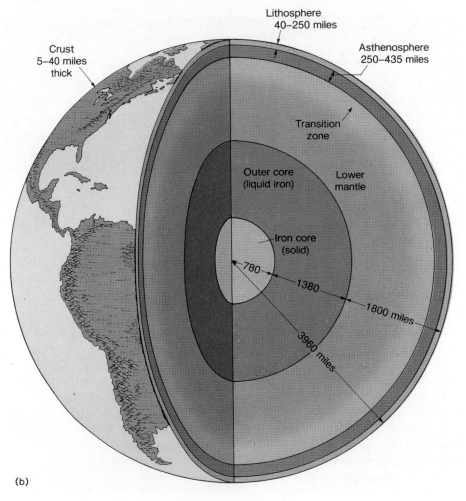

(b)

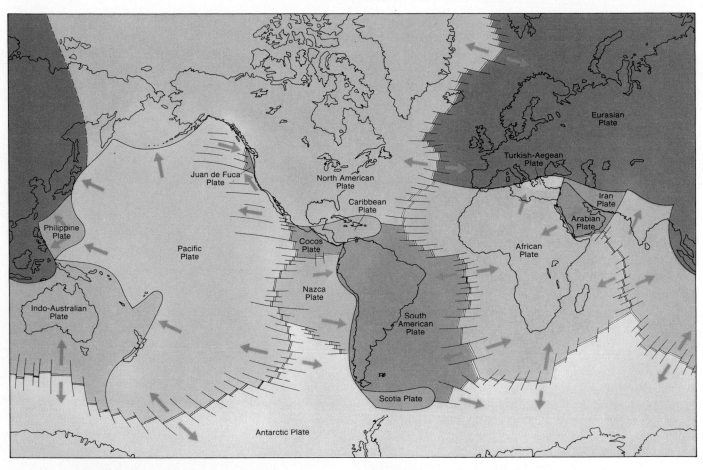

Figure 3.3

The large lithospheric plates move as separate entities and collide. Assuming the African plate to be stationary, relative plate movements are shown by arrows. Seafloor spreading, the triggering mechanism, takes place along the axes of the ridges.

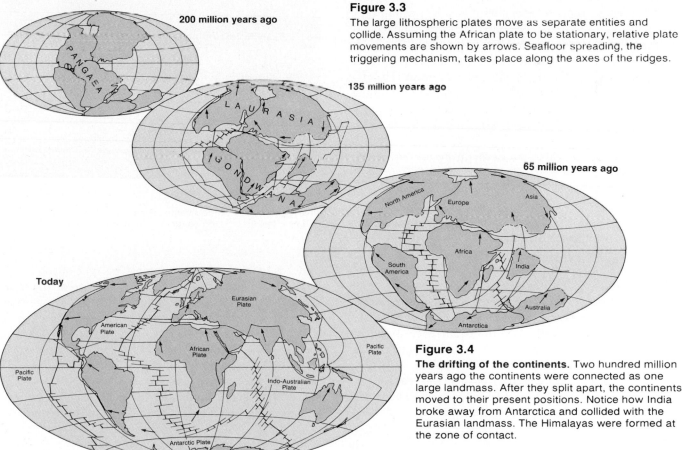

Figure 3.4

The drifting of the continents. Two hundred million years ago the continents were connected as one large landmass. After they split apart, the continents moved to their present positions. Notice how India broke away from Antarctica and collided with the Eurasian landmass. The Himalayas were formed at the zone of contact.

ago. Notice on Figure 3.5 how the ridge line that makes up the axis of the ocean runs parallel to the eastern coast of North and South America and the western coast of Europe and Africa. Scientists were led to the theory of the **continental drift** of lithospheric plates by the amazing fit of the continents.

According to this theory, collisions occurred as the lithospheric plates moved. The pressure exerted at the intersections of plates resulted in earthquakes, which over periods of many years combined to change the shape and the features of the landforms. Figure 3.6 shows the location of near-surface earthquakes for a recent time period. Comparison with Figure 3.3 illustrates that the areas of greatest earthquake activity are at plate boundaries.

The famous San Andreas fault of California is part of a long fracture separating two lithospheric plates, the North American and the Pacific. Earthquakes occur along **faults** (sharp breaks in rocks along which there is slippage) when the tension and the compression at the junction become so great that only an earth movement can release the pressure. The San Andreas case is called a *transform* fault, which occurs when one plate slips past another in a horizontal motion. Because the Atlantic Ocean is still widening at the rate of about 1 inch (2.5 cm) per year, earthquakes must occur from time to time to relieve the stress along the tension zone in the mid-Atlantic and along other fracture lines, such as the San Andreas fault.

Despite the availability of scientific knowledge about earthquake zones, the general disregard for this danger is a difficult cultural phenomenon with which to deal. Every year there are hundreds and sometimes thousands of casualties resulting from inadequate preparation for earthquakes. In some well-populated areas, the chances that damaging earthquakes will occur are very great. The distribution of earthquakes shown in Figure 3.6 implies the potential dangers to densely settled areas of Japan, the Philippines, parts of Southeast Asia, and the western rim of the Americas.

In the aftermath of the devastating earthquake in the Armenian region of the Soviet Union on December 7, 1988, there has been a hastening of activity to develop worldwide networks of highly sensitive seismic stations. There are at least seven major national and international networks now under construction. The purpose of the networks is to provide warnings of future earthquakes and to learn more about the forces at work in the earth's interior that keep the continental plates in motion.

Movement of the lithospheric plates results in the formation of deep-sea trenches and continental-scale mountain ranges as well as in the occurrence of earthquakes. The continental crust is made up of lighter rocks than is the oceanic crust. Thus, where plates with different types of crust at their edges push against each other, there is a tendency for the denser oceanic crust to be forced down into the asthenosphere, causing long and deep

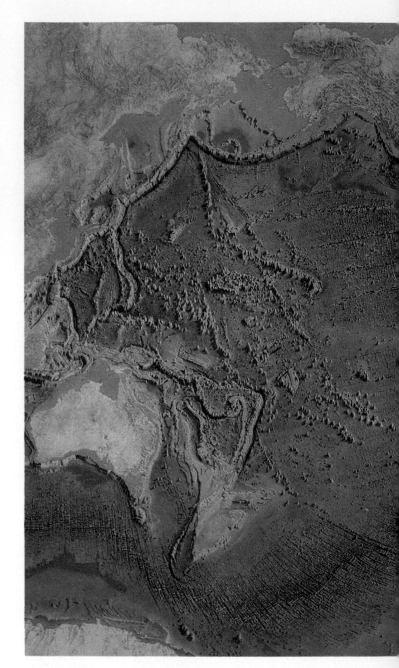

Figure 3.5
The configuration of the Atlantic Ocean floor is evidence of the dynamic forces shaping continents and ocean basins.

trenches to form below the ocean. This type of collision is termed **subduction** (Figure 3.7a). The edge of the overriding continental plate is uplifted to form a mountain chain that runs close to and parallel with the offshore trench. The subduction zones of the world are shown in Figure 3.7b.

Most of the Pacific Ocean is underlain by a plate that, like the others, is constantly pushing and being pushed. The continental crust on adjacent plates is being forced to rise and fracture, making an active earthquake

The map on pp. 50–51 was incorrectly printed.
Please refer to this map for figure 3.5. The
configuration of the Atlantic Ocean floor is evidence
of the dynamic forces shaping continents and
ocean basins.

World Ocean Floor by Bruce C. Heezen and Marie Tharp, 1977 and copyright by Marie Tharp 1977. Reproduced by permission of Marie Tharp, 1 Washington Ave. South Nyack, NY 10960.

and volcano zone of the rim of the Pacific Ocean (sometimes called the "ring of fire"). In recent years, major earthquakes and volcanic activity have occurred in Colombia, Mexico, Central America, the Pacific Northwest of the United States, Alaska, the Armenian region of the USSR, China, and the Philippines. Figure 3.8 shows the location of volcanoes that are known to have erupted at some time in the past. The tremendous explosion that rocked Mount St. Helens in the state of Washington in 1980 is an example of continuing volcanic activity along the Pacific rim. The many scientists who believed that a

damaging earthquake would occur along the San Andreas fault were not surprised by the October 17, 1989, earthquake whose epicenter on the fault was 40 miles south of San Francisco (see box "Danger near the San Andreas fault").

Plate intersections are not the only locations that are susceptible to readjustments in the lithosphere. In the process of continental drift, the earth's crust has been cracked or broken in virtually thousands of places. Some of the breaks are weakened to the point that they allow molten material from the asthenosphere to find its way to

Danger Near the San Andreas Fault

The earthquakes of June 12, 1989, in Los Angeles and October 17, 1989, near Santa Cruz south of San Francisco reminded many that the vast majority of people in California live close to the San Andreas fault and its related faults. Since October 1987, in the Los Angeles basin there have been ten earthquakes registering between 4.3 and 5.9, on the Richter scale (see box on the Richter scale), and in the San Francisco area a major earthquake occurred measuring 7.1. Experts who have been monitoring seismic patterns in that area predict that more major earthquakes are likely to occur along the fault before the year 2010. Great earthquakes have occurred to the east of Los Angeles about every 145 years, plus or minus several decades. The last major quake took place there in 1857, well before the city became the huge metropolis it is today. The San Francisco earthquake of 1906 caused fires to break out, which destroyed the city. In the 1989 earthquake, which caused billions of dollars of damage, a freeway collapsed during rush hour, killing many people.

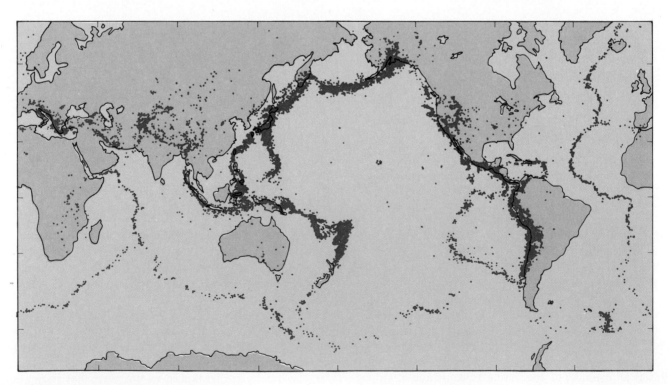

Figure 3.6
World earthquake epicenters, 1961–1967.

the surface. The molten material may explode out of a volcano or ooze out of cracks. Later in this chapter, when we discuss the earth building forces, we will return to the discussion of volcanic activity. First, however, it is necessary to describe the materials that make up the earth's surface.

Earth Materials

The rocks of the earth's crust vary according to mineral composition. Rocks are made up of particles that contain various combinations of such common elements as oxygen,

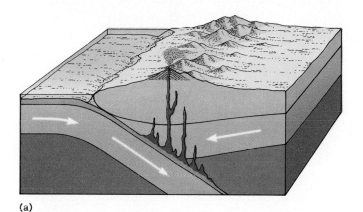

(a)

silicon, aluminum, iron, and calcium, together with less abundant elements. A particular chemical combination that has a hardness, density, and definite crystal structure of its own is called a **mineral**. Some well-known minerals are quartz, feldspar, and silica. Depending on the nature of the minerals that form them, rocks may be hard or soft, dense or open, one color or another, chemically stable or not. While some rocks resist decomposition, others are very easily broken down. Among the more common varieties of rock are granites, basalts, limestones, sandstones, and slates.

Although one can classify rocks according to their physical properties, the more common approach is to classify them by the way they evolved. The three main groups of rocks are igneous, sedimentary, and metamorphic.

Igneous Rocks

Igneous rocks are formed by the cooling and hardening of earth material. Weaknesses in the crust give molten material from the asthenosphere an opportunity to find its way into or onto the crust. When the molten material cools, it hardens and becomes rock. The name for underground

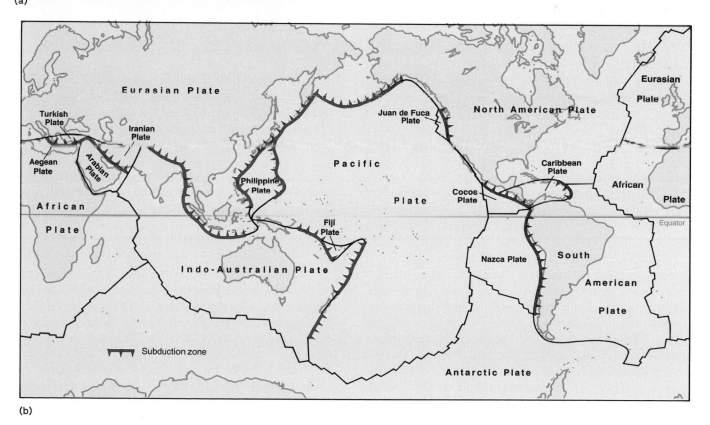

(b)

Figure 3.7

The process of *subduction*. (a) When lithospheric plates collide, the heavier oceanic crust is usually forced beneath the lighter continental material. Deep-sea trenches, mountain ranges, volcanoes, and earthquakes occur along the plate collision lines. (b) The subduction zones of the world.

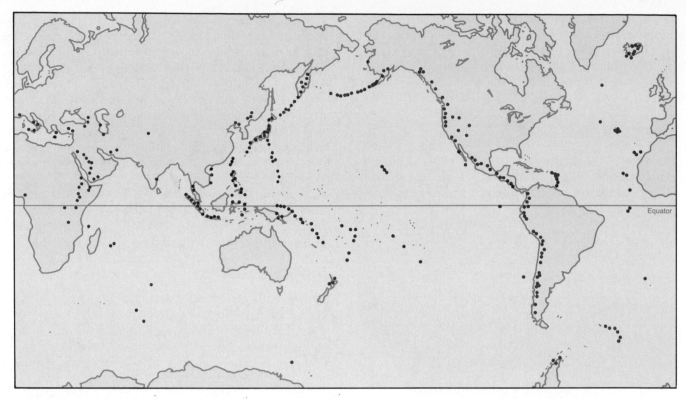

Figure 3.8
The distribution of volcanoes. Note the association of volcanic activity with plate boundaries as shown on Figure 3.7.

molten material is *magma*; above ground it is *lava*. *Intrusive* igneous rocks were formed below ground level by the hardening of magma, while *extrusive* igneous rocks were created above nearby ground level by the hardening of lava (Figure 3.9).

The chemicals making up lava and magma are fairly uniform, but depending on the speed of cooling, different minerals form. Because it is not exposed to the coolness of the air, magma hardens slowly, allowing silicon and oxygen to unite and form quartz, a hard, dense mineral. With other components, grains of quartz combine to form the rock called *granite*.

The lava that oozes out onto the earth's surface and makes up a large part of the ocean basins contains a considerable amount of sodium or calcium aluminosilicates. These form the mineral called *feldspar* and together with several other minerals make up *basalt*, the most common rock on earth. If instead of oozing the lava erupts from a volcano crater, it may cool very rapidly. Some of the rocks formed in this manner contain air spaces and are light and angular, such as *pumice*. Some may be dense, even glassy, as is *obsidian*. The glassiness occurs when lava meets standing water and suddenly cools.

Sedimentary Rocks

Sedimentary rocks are composed of particles of gravel, sand, silt, and clay that were eroded from already existing rocks. Surface waters carry the sediment to oceans, marshes, lakes, or tidal basins. Compression of these materials by the weight of additional deposits on top of them, and a cementing process brought on by the chemical action of water and certain minerals, causes sedimentary rock to form.

Sedimentary rocks evolve under water in horizontal beds called *strata*. Usually one type of sediment collects in a given area. If the particles are large—for instance, the size of gravel—a gravelly rock called *conglomerate* forms. Sand particles are the ingredient for *sandstone*, while silt and clay form *shale* or *siltstone*.

Sedimentary rocks also derive from organic material, such as coral, shells, and marine skeletons. These materials settle into beds in shallow seas and congeal, forming *limestone*. If the organic material is mainly vegetation, it can develop into a sedimentary rock called *coal*. *Petroleum* is also a biological product, formed during the millions of years of burial by chemical reactions that transform some of the organic material into liquid and gaseous compounds. The oil and gas are light, therefore they ooze through the pores of the surrounding rock to places where dense rocks block their upward movement.

Sedimentary rocks vary considerably in color (from coal black to chalk white), hardness, density, and resistance to chemical decomposition. Large parts of the continents contain sedimentary rocks. Nearly the entire

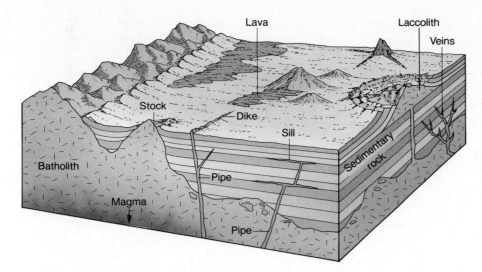

Figure 3.9
Extrusive and intrusive forms of volcanism. Lava and *ejecta* (ash and cinders) are extrusions of rock material onto the earth's surface in the form of cones or horizontal flows. *Batholiths* and *laccoliths* are irregular masses of crystalline rock that have cooled slowly below the earth's surface (intrusions), and in this diagram they have become surface features because of the erosion of overlying material.

eastern half of the United States is overlain with these rocks, for example. Such formations indicate that in the geologic past, seas covered even larger proportions of the earth than they do today.

Metamorphic Rocks

Metamorphic rocks are formed from igneous and sedimentary rocks by earth forces that generate heat, pressure, or chemical reaction. The word *metamorphic* means "changed shape." The internal earth forces that cause the movement and collision of lithospheric plates may be so great that by heat and pressure, the mineral structure of a rock changes, forming new rocks. For example, under great pressure, shale, a sedimentary rock, becomes *slate*, a rock with different properties. Limestone under certain conditions may become *marble*, and granite may become *gneiss* (pronounced "nice"). Materials metamorphosed at great depth and exposed only after overlying surfaces have been slowly eroded away are among the oldest rocks known on earth. Like igneous and sedimentary rocks, however, their formation is a continuing process.

Rocks are the constituent ingredients of most landforms. Their strength or weakness, their permeability, and their chemical content control the way they respond to the forces that shape and reshape them. Two principal processes are at work altering rocks: the tectonic forces that tend to build landforms up and the gradational processes that wear landforms down. All rocks are part of the *rock cycle* through which old rocks are continually being transformed into new ones by these processes. No rocks have been preserved unaltered throughout the earth's history.

Tectonic Forces

The earth's crust is altered by the constant forces resulting from plate movement. *Tectonic* (generated from within the earth) processes shaping and reshaping the

earth's crust are of two types: diastrophic and volcanic. **Diastrophism** is the great pressure acting on the plates that deforms the surface by folding, twisting, warping, breaking, or compressing rock. **Volcanism** is the force that transports heated material to or toward the surface of the earth. When particular places on the continents are under pressure, the changes that take place can be as simple as the bowing or cracking of rock or as dramatic as lava exploding from the crater and sides of a Mount St. Helens.

Diastrophism

In the process of continental drift, pressures build in various parts of the earth's crust, and slowly, over thousands of years, the crust is transformed. By studying rock formations, geologists are able to trace the history of the development of a region. Over geologic time, most continental areas have been subjected to both tectonic and gradational activity—to building up and tearing down. They usually have a complex history of broad warping, folding, faulting, and leveling. Some flat plains in existence today may hide a history of great mountain development in the past.

Broad Warping

Great forces resulting perhaps from the movement of continents may bow an entire continent. Also, the changing weight of a large region, perhaps due to melting continental glaciers, may result in the **warping** of the surface. For example, the down-warping of the eastern United States is evident in the many irregularly shaped stream estuaries. As the coastal area was warped downward, the sea has advanced, forming estuaries and underwater canyons.

Folding

When the pressure caused by moving continents is great, layers of rock are forced to buckle. The result may be a warping or bending effect, and a ridge or a series of parallel ridges or **folds** may develop. If the stress is pronounced, great wavelike folds form (Figure 3.10). The folds can be thrust upward many thousands of feet and laterally for many miles. The folded ridges of the eastern United States are at present low parallel mountains (1000–3000 feet—300 to 900 m—above sea level), but the rock evidence suggests that the tops of the present mountains were once the valleys between 30,000-foot (9100-m) crests (Figure 3.11).

Faulting

A fault is a break or fracture in rock. The stress causing a fault results in displacement of the earth's crust along the fracture zone. Figure 3.12 depicts examples of fault types. There may be uplift on one side of the fault or downthrust on the other. In some cases, a steep slope known as a fault *escarpment*, which may be several hundred miles long, is formed. The stress can push one side up over the other side, or a separation away from the fault may cause sinking of the land and create a *rift valley* (Figure 3.13).

Many faults are merely cracks (called *joints*) with little noticeable movement along them, but in other cases, mountains, such as the Sierra Nevada of California, have risen as the result of faulting. In some instances, the movement has been horizontal along the surface rather than up or down. The San Andreas transform fault mentioned earlier and pictured in Figure 3.14 is such a case.

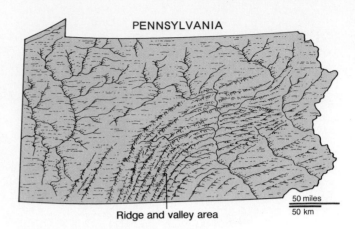

PENNSYLVANIA

Ridge and valley area

50 miles
50 km

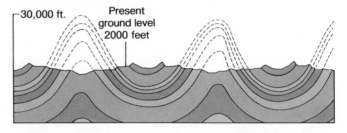

Figure 3.11
The ridge and valley region of Pennsylvania, now eroded to hill lands, is the relic of 30,000-foot (9100-m) folds that were reduced to form *synclinal* (downarched) hills and *anticlinal* (uparched) valleys. The rock in the original troughs, having been compressed, was less susceptible to erosion.

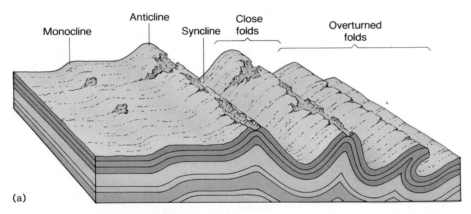

(a)

Monocline | Anticline | Syncline | Close folds | Overturned folds

Figure 3.10
Degrees of folding vary from slight undulations of strata with little departure from the horizontal to highly compressed or overturned beds. (a) Diagram of stylized forms of folding. (b) An overturned fold in the Appalachian Mountains.

(b)

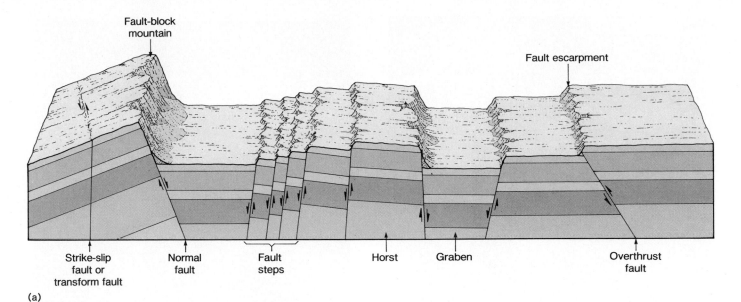

Figure 3.12
Faults, in their great variation, are common features of mountain belts where deformation is great. (a) The different forms of faulting are categorized by the direction of movement along the plane of fracture. (b) The major faults in California with the epicenters of two strong 1986 and one strong 1989 quakes noted.

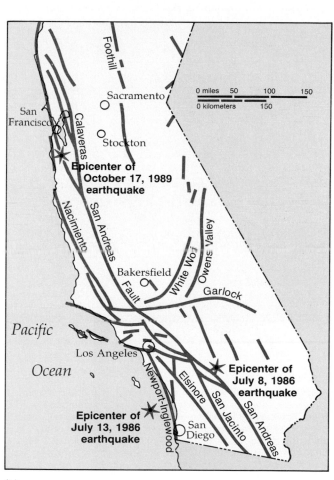

(a)

Fault-block mountain

Fault escarpment

Strike-slip fault or transform fault

Normal fault

Fault steps

Horst

Graben

Overthrust fault

(b)

Foothill

Sacramento

San Francisco

Calaveras

Stockton

Epicenter of October 17, 1989 earthquake

Nacimiento

San Andreas

Owens Valley

Bakersfield

White Wolf

Fault

Garlock

Pacific

Los Angeles

Ocean

Newport-Inglewood

Elsinore

San Jacinto

San Andreas

Epicenter of July 8, 1986 earthquake

Epicenter of July 13, 1986 earthquake

San Diego

0 miles 50 100 150
0 kilometers 150

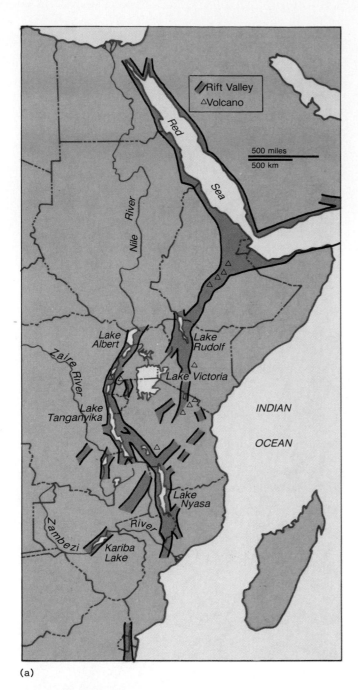

(a)

Figure 3.13

(a) Great fractures in the earth's crust resulted in the creation, through subsidence, of an extensive *rift valley* system in East Africa. The parallel faults, some reaching more than 2000 feet (610 m) below sea level, are bordered by steep walls of the adjacent plateau, which rises to 5000 feet (1500 m) above sea level and from which the structure dropped. (b) In Tanzania, the same tectonic forces have created a rift valley, one edge of which is shown here.

(b)

Earthquakes

Whenever movement occurs along a fault or at some other point of weakness, an earthquake results. The greater the movement, the greater the magnitude of the earthquake. Tension builds in rock as tectonic forces are applied, and when finally a critical point is reached, the earthquake occurs and tension is reduced. The earthquake that occurred in Alaska on Good Friday in 1964 was one of the strongest known. Although the stress point of that earthquake was below ground 75 miles from Anchorage, vibrations called *seismic waves* caused earth movement in the weak clays under the city. Sections of Anchorage literally slid downhill, and part of the business district dropped 10 feet (3 m). Table 3.1 indicates the kinds of effects associated with earthquakes of different magnitudes, and Figure 3.15 shows various types of earthquake-induced damage.

If an earthquake occurs below an ocean, the movement can cause a **tsunami**, a large destructive sea wave. Though not noticeable on the open sea, a tsunami may become 30 or more feet high as it approaches land sometimes thousands of miles from the earthquake site. The islands of Hawaii now have a tsunami warning system that was developed following the devastation at Hilo in 1946.

Figure 3.14

The San Andreas fault in southern California. A *transform fault*, the San Andreas marks a part of the slipping boundary between the Pacific and the North American plates. The inset map shows the relative southward movement of the American plate; dislocation averages about 0.4 inches (1 cm) per year.

Figure 3.15

Two views of earthquake damage. (a) A portion of the top deck of Interstate 880 collapsed when a major earthquake struck the San Francisco Bay area on October 17, 1989. Cars on the lower deck were crushed. (b) The Mexican earthquake of September 1985 destroyed hundreds of buildings in Mexico City and took thousands of lives.

(a)

(b)

The Richter Scale

In 1935, C. F. Richter devised a scale of earthquake *magnitude*. An earthquake is really a form of energy expressed as wave motion passing through the surface layer of the earth. Radiating in all directions from the earthquake focus, seismic waves gradually dissipate their energy at increasing distances from the *epicenter* (the point on the earth's surface directly above the focus). On the Richter scale, the amount of energy released during an earthquake is estimated by measurement of the ground motion that occurs. Seismographs record earthquake waves, and by comparison of wave heights, the relative strength of quakes can be determined. Although Richter scale numbers run from 0 to 9, there is no absolute upper limit to earthquake severity. Presumably, nature could outdo the magnitude of the most intense earthquakes so far recorded, which reached 8.5–8.6.

Because magnitude, as opposed to intensity, can be measured accurately, the Richter scale has been widely adopted. Nevertheless, it is still only an approximation of the amount of energy released in an earthquake. In addition, the height of the seismic waves can be affected by the rock materials under the seismographic station, and some seismologists believe that the Richter scale underestimates the magnitude of major tremors.

Table 3.1

Richter Scale of Earthquake Magnitude

Magnitude[a]	Characteristic Effects of Earthquakes Occurring Near the Earth's Surface[b]
0	not felt
1	not felt
2	not felt
3	felt by some
4	windows rattle
5	windows break
6	poorly constructed buildings destroyed; others damaged
7	widespread damage; steel bends
8	nearly total damage
9	total destruction

[a]Since the Richter scale is logarithmic, each increment of a whole number signifies a 10-fold increase in magnitude. Thus a magnitude 4 earthquake produces a registered effect upon the seismograph 10 times greater than a magnitude 3 earthquake.

[b]The damage levels of earthquakes are presented in terms of the consequences that are felt or seen in populated areas; the recorded seismic wave heights remain the same whether or not there are structures on the surface to be damaged. The actual impact of earthquakes upon humans varies not only with the severity of the quake and such secondary effects as tsunamis or landslides but also with the density of population in the area affected.

The Tsunami

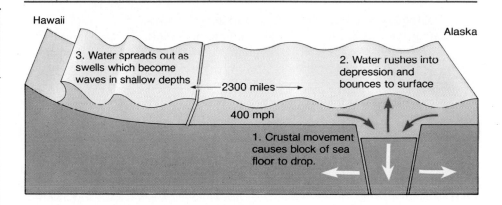

A tsunami follows any submarine earthquake that causes fissures or cracks in the earth's surface. Water rushes in to fill the depression caused by the falling away of the ocean bottom. The water then moves outward, building in momentum and rhythm, as swells of tremendous power. The waves that hit Hawaii following the April 1, 1946, earthquake off Dutch Harbor, Alaska, were moving at approximately 400 miles (640 km) an hour with a crest-to-crest spacing of some 80 miles (130 km).

The long swells of a tsunami are largely unnoticed in the open ocean. Only when the wave trough scrapes sea bottom in shallow coastal areas does the water pile up into precipitous peaks. The seismic sea waves at Hilo on the exposed northeast part of the island of Hawaii were estimated at between 45 and 100 (14 and 30 m) feet in height. The water smashed into the city, deposited 14 feet (4.25 m) of silt in its harbor, left fish stranded in palm trees, caused many millions of dollars in damage, and resulted in 173 deaths; many were people who had gone to the shore to see the giant waves arrive.

The Mexican Earthquake

An earthquake registering 8.1 on the Richter scale struck at the edge of the North American plate along Mexico's Pacific Coast about 220 miles (350 km) from Mexico City on September 16, 1985. Pushed by the huge Nazca plate to the south, the Cocos plate slammed into and under the North American plate (Figure 3.3). About one thousand times as much energy as was released at Hiroshima was expended over western Mexico. Destruction was heaviest in the teeming capital of Mexico City. Wave after wave of violent earth shaking continued for about three minutes. The next day a violent aftershock registering 7.6 on the Richter scale added more fear and grief to the people of Mexico.

A large part of Mexico City sits on an old lake basin made up of silt, clay, sand, and gravel—a very unstable foundation for a city with tall buildings. The rock between the epicenter and Mexico City absorbed most of the quake's force, but, nonetheless, the buildings in the downtown swayed. Fortunately, relatively few collapsed. Most were built to high standards specifically designed to resist earthquakes, and they held. Unfortunately, the 7000 buildings that did collapse killed 9000 people, caused 30,000 injuries, and left 95,000 people homeless.

A similar quake at the edge of the Cocos plate caused additional widespread death and destruction in October, 1986, in San Salvador, the capital of El Salvador. Earthquakes do their greatest damage not directly but indirectly, by putting people living in dense settlements in jeopardy of being destroyed by buildings that fall, by landslides and mudflows that descend on them, and by tsunamis that ravage coastal areas.

Earthquakes occur daily in a number of places throughout the world. Most are slight and are noticeable only on *seismographs*, the instruments that record seismic waves, but from time to time there are large-scale earthquakes, such as those in Guatemala and China in 1976 or the 1986 El Salvador quake that killed more than 1000 people and left 200,000 homeless. Most earthquakes take place on the rim of the Pacific (Figure 3.7), where stress from the outward-moving lithospheric plates is greatest. The Aleutian Islands of Alaska, Japan, Central America, and Indonesia experience many moderately severe earthquakes each year.

Volcanism

The second tectonic force is volcanism. The most likely places through which molten materials can move toward the surface are at the intersections of plates, but other fault-weakened zones are also subject to volcanic activity (compare Figures 3.3 and 3.8).

If sufficient internal pressure forces the magma upward, weaknesses in the crust, or faults, enable molten materials to reach the surface. The material ejected onto the earth's surface may arrive as a great explosion forming a steep-sided cone termed a *strato* or *composite volcano* (Figure 3.16), or the eruption may be without explosions, forming a gently sloping *shield volcano*.

The major volcanic belt of the world coincides with the major earthquake and fault zones. This is the zone of convergence between two plates. A second zone of volcanic activity is at diverging plate boundaries, such as in the center of the Atlantic Ocean. Molten material can either flow smoothly out of a crater or be shot into the air with great force. Some relatively quiet volcanoes have long gentle slopes indicative of smooth flow, while explosive volcanoes have steep sides. Steam and gases are constantly escaping from the nearly 300 volcanoes active in the world today.

When pressure builds, a crater can become a boiling cauldron with steam, gas, lava, and ash all billowing out (Figure 3.17). In the case of Mount St. Helens in 1980, a large bulge had formed on the north slope of the mountain. An earthquake preceded an explosion in the bulging area shooting debris into the air, completely devastating an area of about 150 square miles, causing about 4 inches

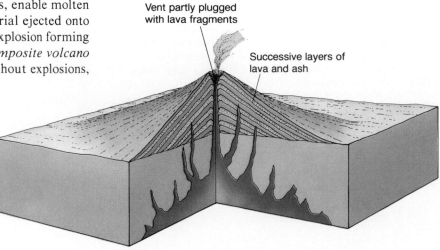

Vent partly plugged with lava fragments

Successive layers of lava and ash

Figure 3.16

Sudden decompression of gases contained within lavas results in explosions of rock material to form ashes and cinders. Composite volcanoes, such as the one diagrammed, are composed of alternate layers of solidified lava and of ash and cinders.

(a)

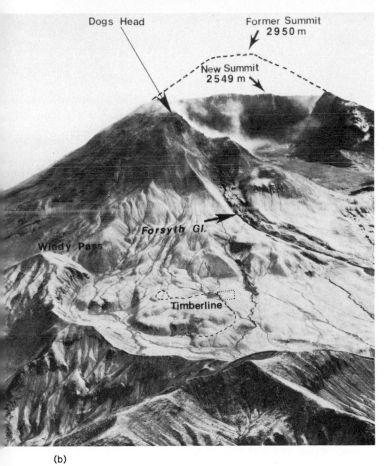

(b)

(c)

Figure 3.17

(a) Mount St. Helens, Washington, before it erupted. (Mount Rainier is in the background.) (b) Aftermath of the May 18, 1980, eruption of Mount St. Helens. (c) Lava exploding from a volcano in Hawaii.

of ash to rain down on most of Washington and parts of Idaho and Montana, and reducing the elevation of the mountain by over 1000 feet.

In many cases the forces beneath the crust are not great enough to allow the magma to reach the surface. In these instances the magma hardens into a variety of underground formations of igneous rock that do little to affect surface landform features. However, gradational forces may erode overlying rock, so that the igneous rock, which usually is hard and resists erosion, becomes a surface feature. The Palisades, a rocky ridge facing New York City from the west, is such a landform. On other occasions, a weakness below the earth's surface may allow the growth of a mass of magma that is denied exit to the surface because of firm overlying rock. Through the pressure it exerts, however, the magmatic intrusion may still buckle, bubble, or break the surface rocks, and domes of considerable size may develop, such as the Black Hills of South Dakota (Figure 13.5).

Evidence from the past shows that sometimes lava has flowed through fissures or fractures without volcanoes forming. These oozing lava flows have covered large areas to great depth. The Deccan Plateau of India and the Columbia Plateau of the U.S. Pacific Northwest are examples of this type of process (Figure 3.18).

Gradational Processes

Gradational processes are responsible for the reduction of land surface. If a land surface where a mountain once stood is now a low, flat plain, gradational processes have been at work. The material that has been worn, scraped, or blown away is deposited in new places, and as a result, new landforms are created. In terms of geologic time, the Himalayas are a recent phenomenon; gradational processes, although active there just as they are active on all land surfaces, have not yet had time to reduce the huge mountains.

There are three kinds of gradational processes: *weathering*, *gravity transfer*, and *erosion*. Weathering processes, both mechanical and chemical, prepare bits of rock for their role in the creation of soils and for their movement to new sites by means of gravity or erosion. The force of gravity acts to transfer any loosened, higher lying material, and the agents of running water, moving ice, wind, waves, and currents erode and carry the loose materials to other areas, where landforms are created or changed.

Mechanical Weathering

Mechanical weathering is the physical disintegration of earth materials at or near the surface. A number of processes cause mechanical weathering, the three most important being frost action, the development of salt crystals, and root action.

Figure 3.18
Fluid basaltic lavas created the Columbia Plateau, covering an area of 50,000 square miles (130,000 km²). Some individual flows were more than 300 feet (100 m) thick and spread up to 40 miles (60 km) from their original fissures.

If the water that soaks into a rock (between particles or along joints) freezes, ice crystals grow and exert pressure on the rock. If the process is repeated—freezing, thawing, freezing, thawing, and so on—there is a tendency for the rock to begin to disintegrate. Salt crystals act similarly in dry climates, where groundwater is drawn to the surface by *capillary* action (water rising because of surface tension). This action is similar to the process in plants whereby liquid plant nutrients move upward through the stem and leaf system. Evaporation leaves behind salt crystals, which help disintegrate rocks. Roots of trees and other plants may also find their way into rock joints and, as they grow, break and disintegrate rock. These are all mechanical processes because they are physical in nature and do not alter the chemical composition of the material upon which they act.

Chemical Weathering

A number of **chemical weathering** processes cause rock to decompose rather than to disintegrate, that is, to separate into component parts by chemical reaction rather than to

fragment. The three most important are oxidation, hydrolysis, and carbonation. Because each of these processes depends on the availability of water, there is less chemical weathering in dry and cold areas than in moist and warm ones. Chemical reactions are speeded in the presence of moisture and heat.

Oxidation occurs when oxygen combines with rock minerals such as iron to form oxides; as a result, some rock areas in contact with the oxygen begin to decompose. Decomposition also results when water comes into contact with certain rock minerals, such as aluminosilicates. The chemical change that occurs is called hydrolysis. When carbon dioxide gas from the atmosphere dissolves in water, a weak carbonic acid forms. The action of the acid, called carbonation, is particularly evident on limestone; the calcium bicarbonate salt that is created is readily dissolved and removed by ground and surface water.

Weathering, either mechanical or chemical, does not itself create distinctive landforms. Nevertheless, it acts to prepare rock particles for erosion and for the creation of soil. After the weathering process decomposes rock, the force of gravity and the erosional forces of running water, wind, and moving ice are able to carry the weathered material to new locations.

Soils

Mechanical and chemical weathering create soil. Soil is the thin layer of fine material resting on bedrock. When digging down into the soil one can recognize a series of layers of different colors. Soil scientists identify three sections. The top, or A-horizon, is usually the darkest; it may contain organic matter, such as decayed plant leaves, twigs, and animal remains, and clays and sand grains. The second level, the B-horizon, is much lighter in color. It is made up largely of clay with small and large bits of minerals and only small amounts of organic matter. The lowest level, the C-horizon, is the upper, broken surface of bedrock mixed with clay. The older the soil and the warmer and wetter the climate, the deeper the C-horizon will be and the more discernible will be the three horizons (Figure 3.19).

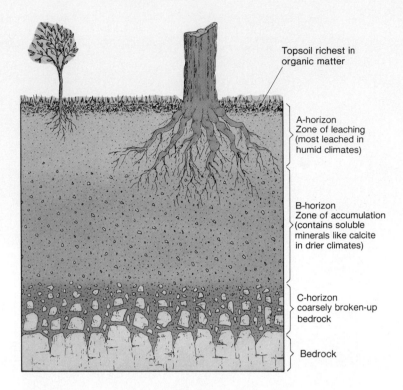

Figure 3.19
A representative soil profile showing the various soil horizons.

Topsoil richest in organic matter

A-horizon
Zone of leaching (most leached in humid climates)

B-horizon
Zone of accumulation (contains soluble minerals like calcite in drier climates)

C-horizon
coarsely broken-up bedrock

Bedrock

Soil type is much more a function of the climate of the region in which it occurs than it is of the kind of bedrock below it. Temperature and rainfall act on the minerals in conjunction with the decay of the overriding vegetation to form soils. The process of soil formation takes many hundreds and sometimes thousands of years, but dust storms and poor conservation practices by farmers and land managers can deplete an area of its crucial A-horizon in just a few years. We will return to the topic of soils in the next chapter.

Gravity Transfer

The force of **gravity transfer**—that is, the attraction of the earth's mass for bodies at or near its surface—is constantly pulling on all materials. Small particles or huge boulders, if not held back by bedrock or other stable material, will fall down slopes. Spectacular acts of gravity include avalanches and landslides. More widespread but less noticeable are movements such as soil-creep and the flow of mud down hillsides.

Especially in dry areas, a common but very dramatic landform created by the accumulation of rock particles at the base of hills and mountains is the *talus slope*, pictured in Figure 3.20. As pebbles, particles of rock, or even larger stones break away from the exposed bedrock on a mountainside because of weathering, they fall and accumulate, producing large conelike landforms. The larger rocks travel farther than the fine-grained sand particles, which remain near the top of the slope.

Erosional Agents and Deposition

Erosional agents, such as wind and water, carve already existing landforms into new shapes. The material that has been worn, scraped, or blown away is deposited in new places, and new landforms are created. Each erosional agent is associated with a distinctive set of landforms.

Running Water

There is no more important erosional agent than running water. Water, whether flowing across land surfaces or in stream channels, plays an enormous role in wearing down and building up landforms.

The ability of running water to erode depends on the amount of precipitation, on the length and steepness of the slope, and on the kind of rock and vegetative cover. The steeper the slope, the faster the flow and, of course, the more rapid the erosion. If the vegetative cover that slows the flow of water is reduced, perhaps because of farming or livestock grazing, erosion can be severe, as shown in Figure 3.21.

Even the impact force of precipitation—heavy rain or hail—can cause erosion. After hard rain dislodges soil, the surface becomes more compact, and therefore further precipitation fails to penetrate the soil. The result is that more water, prevented from seeping into the ground, becomes available for surface erosion. Soil and rock particles in the water are carried to streams, leaving behind gullies and small stream channels.

(a)

(b)

Figure 3.20
Examples of mechanical weathering. (a) Rockfall from this butte in the Grand Canyon has created a pronounced talus slope. (b) Creeping soil has caused trees to tilt.

Both the force of water and the particles contained in the stream are agents of erosion. The particles act as abrasives, scouring the surface over which they move. Abrasion, or wearing away, takes place when the particles strike against stream channel walls and along the streambed. Large particles, such as gravel, slide along the streambed because of the force of the current, grinding rock on the way. Floods and rapidly moving water are responsible for dramatic changes in channel size and configuration, sometimes forming new channels. In cities, where paved surfaces cover soil that would otherwise have absorbed or held water, runoff is accentuated so that nearby rivers and streams rapidly increase in size and velocity after heavy rains, and oftentimes flash floods and severe erosion result.

Small particles, such as clay and silt, are suspended in water and constitute (together with material dissolved in the water or dragged along the bottom) the *load* of a stream. Rapidly moving floodwaters carry huge loads. As high water or floodwater recedes and stream velocity decreases, sediment contained within the stream no longer remains suspended, and particles begin to settle. Heavy, coarse materials drop most quickly; fine particles are carried farther. The decline in velocity and the resulting deposition are especially pronounced and abrupt when streams meet slowly moving water in bays, oceans, and lakes. Silt and sand accumulate at the intersections, creating *deltas*, as pictured in Figure 3.22.

A great river, such as the Chang Jiang (Yangtze) in China, has a large, growing delta, but less prominent deltas are found at the mouths of many streams. Until the recent completion of the Aswan Dam, the huge delta of the Nile River had been growing, but now the silt is being dropped in Lake Nasser behind the dam.

In plains adjacent to streams, land is sometimes built up by the deposition of stream load. If the deposited material is rich, it may be a welcome and necessary part of farming activities, as historically it was in Egypt along the

Figure 3.21
Gullying can result from poor farming techniques, including overgrazing by livestock or many years of continuous row crops. Surface runoff removes topsoil easily when vegetation is too thin to protect it.

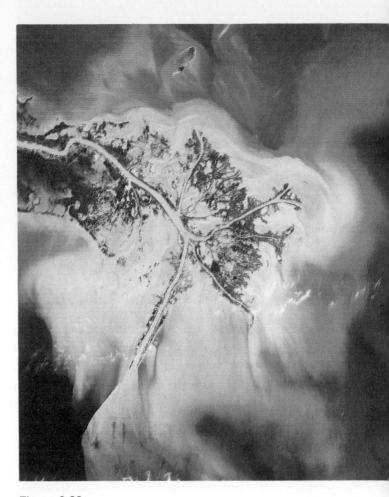

Figure 3.22
Landsat image of the delta of the Mississippi River. Notice the ongoing deposition of silt and the effect that both river and gulf currents have on the movement of the silt.

Nile. Should the deposition be composed of sterile sands and boulders, however, formerly fertile bottomland may be destroyed. By drowning crops or inundating inhabited areas, the floods themselves, of course, may cause great human and financial loss. More than 900,000 lives were lost in the floods of the Huang He (Yellow River) of China in 1887.

Stream Landscapes

A somewhat misleading concept of landscape evolution conveniently places a given stream in a particular stage in its geologic history. Streams seem to have recognizable stages from *early youth* to *maturity* to *old age*. The concept is misleading because each stream and its surrounding landscape has its own history of change that does not necessarily follow a path from youth to old age. A landscape is in a particular state of balance between the uplift of land and its erosion. It does not follow that rapid uplift will be followed by nicely ordered stages of erosion. Recall that uplift and erosion take place simultaneously.

At a given location one force may be greater than the other at a given time, but as yet there is no way to predict the "next" stage of landscape evolution.

Perhaps the most important factor in differentiating the effect of streams on landforms is whether the recent (say, the last several million years) climate has tended to be humid or arid.

Stream Landscapes in Humid Areas

Perhaps weak surface material or a depression in rock allows the development of a stream channel. In its downhill run in mountainous regions, a stream may flow over precipices, forming *falls* in the process. The steep downhill gradient allows streams to flow rapidly, cutting narrow V-shaped channels in the rock (Figure 3.23a). Under these conditions, the erosional process is greatly accelerated. Over time, the stream may have worn away sufficient rock for the falls to become rapids, and the stream channel becomes incised below the height of the surrounding landforms, as can be seen in the upper reaches of the Delaware, Connecticut, and Tennessee rivers.

(a)

(b)

Figure 3.23

(a) V-shaped valley of a rapidly downcutting stream, the Salmon River in Idaho. (b) Valley of a meandering Wyoming stream, the Snake River, which is in the process of widening its floodplain.

In humid areas, the effect of stream erosion is to round landforms (Figure 3.23b). Streams flowing down moderate gradients tend to carve valleys that are wider than those in mountainous areas. Surrounding hills become rounded, and valleys, called **floodplains**, flatten. Streams work to widen the floodplain. Their courses meander, constantly carving out new erosional channels. The channels left behind as new ones are cut become *oxbow*-shaped lakes, hundreds of which are found in the Mississippi River floodplain. In the nearly flat floodplains, the highest elevations may be the banks of the rivers, where *natural levees* are formed by the filtering of silt at river edges during floods. The lower Mississippi River is blocked from view from nearby roads by natural levees. Flood waters that breach the levees are particularly disastrous. Damaging everything in their path, they fill the floodplain as they equalize its elevation with that of the swollen river.

Stream Landscapes in Arid Areas

A distinction must be made between the results of stream erosion in humid as opposed to arid areas. The lack of vegetation in arid regions greatly increases the erosional force of running water. Water originating in mountainous areas sometimes never reaches the sea if the channel runs through a desert. In fact, stream channels may be empty except during rainy periods, when water rushes down the hillsides to collect and form temporary lakes called *playas*. In the process, **alluvium** (sand and mud) builds up in the lakes and at lower elevations, and *alluvial fans* are formed along hillsides (Figure 3.24). The fan is produced by the scattering of silt, sand, and gravel outward as the stream

Figure 3.24

Alluvial fans, such as this one in Death Valley, California, are built where the velocity of streams is reduced as they flow out on the more level land at the base of the mountain slope. The abrupt change in slope and velocity greatly reduces the stream's capacity to carry its load of coarse material. Deposition occurs, choking the stream channel and diverting the flow of water. With the canyon mouth fixing the head of the alluvial fan, the stream sweeps back and forth, building and extending a broad area of deposition.

reaches the lowlands at the base of the slope it traverses. If the process has been particularly long-standing, alluvial deposits may bury the eroded mountain masses. In desert regions in Nevada, Arizona, and California, it is not unusual to observe partially buried mountains poking through alluvium.

Because the streams in arid areas have only a temporary existence, their erosional power is less certain than that of the freely flowing streams of humid areas. In some instances, they may barely mark the landscape; in other cases, swiftly moving water may carve deep, straight-sided *arroyos*. Often, water may rush onto an alluvial plain in a complicated pattern resembling a multistrand braid, leaving in its wake an alluvial fan. The channels resulting from this rush of water are called *washes*. The erosional power of unrestricted running water in arid regions is dramatically illustrated by the steep-walled configuration of *buttes* and *mesas* (large buttes), such as those in Mitchell Mesa, Texas, shown in Figure 3.25.

Groundwater

Some of the water supplied by rain and snow sinks underground into the pores and cracks in rocks and soil, not in the form of an underground pond or lake but simply as very wet subsurface material. When it accumulates, a zone of saturation forms. As indicated in Figure 3.26, the upper level of this zone is the **water table**; below it, the soils and rocks are saturated with water. Groundwater moves constantly but very slowly (only a few feet or inches a day). Most remains underground, seeking the lowest level. However, when the surface of the land dips below the water table, ponds, lakes, and marshes form. Some water finds its way to the surface by capillary action in the ground or in vegetation. Groundwater, particularly when charged with carbon dioxide, dissolves soluble materials by a chemical process called *solution*.

Although groundwater tends to decompose many types of rocks, its effect on limestone is most spectacular. Many of the great caves of the world have been created by the underground movement of water through limestone regions. Water sinking through the overlying rock leaves carbonate deposits as it drips. The deposits hang from the cave roofs (*stalactites*) and build upward from the cave floors (*stalagmites*). In some areas, the uneven effect of groundwater erosion of limestone leaves a landscape pockmarked by a series of *sinkholes*, surface depressions in an area of plentiful caverns. **Karst topography** refers to a large limestone region marked by sinkholes, caverns, and underground streams, as shown in Figure 3.27. Central Florida, a karst area, has suffered

Figure 3.25

The resistant caprock of the mesa protects softer, underlying strata from downward erosion. Where the caprock is removed, lateral erosion lowers the surface, leaving the mesa as an extensive and pronounced relic of the former higher-lying landscape. Here the mesa walls have a sloping accumulation of talus.

considerable damage from the creation and widening of sinkholes. This type of topography gets its name from a region on the Adriatic Sea at the Italy-Yugoslavia border. The Mammoth Cave region of Kentucky, another karst area, has miles of interconnected limestone caves.

Glaciers

Another way erosion and deposition occur is through the effect of glaciers. Although they are much less extensive today, glaciers covered a large part of the earth's land area as recently as 8000 to 15,000 years ago during the *Pleistocene* geologic epoch (see Figure 3.1). Many landforms were created by the erosional or depositional effects of glaciers.

Glaciers form only in very cold places with short or nonexistent summers, where annual snowfall exceeds annual snowmelt and evaporation. The weight of the snow causes it to compact at the base and form ice. When the snowfall reaches a thickness of several hundred feet, the

Figure 3.26

The groundwater table generally follows surface contours but in subdued fashion. Water flows slowly through the saturated rock, emerging at earth depressions that are lower than the level of the water table. During a drought, the table is lowered and the stream channel becomes dry.

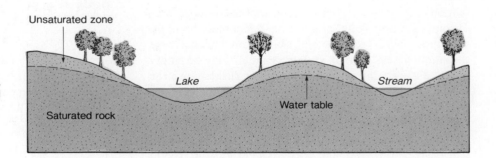

Unsaturated zone

Lake

Stream

Water table

Saturated rock

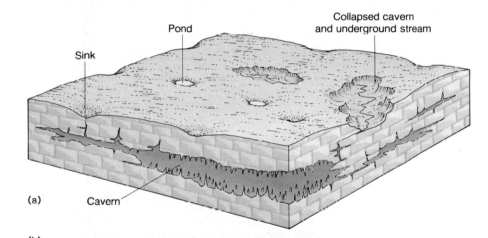

Figure 3.27

Limestone erodes easily in the presence of water. (a) *Karst* topography, such as that shown here, occurs in humid areas where limestone in flat beds is at the surface. (b) This satellite photo of central Florida shows the many round lakes formed in the sinkholes of a karst landscape.

(a)

(b)

ice at the bottom becomes plasticlike and begins to move slowly. A **glacier**, then, is a large body of ice moving slowly down a slope or spreading outward on a land surface (Figure 3.28). Some glaciers appear to be stationary simply because the melting and evaporation at the glacier's edge equal the speed of the ice advance. Glaciers can, however, move as much as several feet per day.

Most theories of glacial formation concern earth climatic cooling. Perhaps some combination of the following theories explains the evolution of glaciers. The first theory attributes the ice ages to periods when there may have been excessive amounts of volcanic dust in the atmosphere. The argument is that the dust, by reducing the amount of solar energy reaching the earth, effectively lowered temperatures at the surface. A second theory attributes the ice ages to known changes in the shape, the tilt, and the seasonal positions of the earth's orbit around the sun over the last half-million years. Such changes alter the amount of solar radiation received by the earth and its distribution over the earth. A recent theory suggests that when large continental plates drift over polar regions, temperatures on earth become more extreme and as a result induce the development of glaciers. This theory, of course, cannot explain the most recent ice ages.

Today, continental-size glaciers exist only on Antarctica and Greenland, but mountain glaciers are found in many parts of the world. About 10% of the earth's land area is under ice. During the most recent advance of ice, the continental ice of Greenland was part of an enormous glacier that covered nearly all of Canada (Figure 3.29) and the northernmost portions of the United States and Eurasia. The giant glacier reached thicknesses of 10,000 feet (its depth in Greenland today), enveloping entire mountain systems.

The weight of glaciers breaks up underlying rock and prepares the rock for transportation by the moving mass of ice. Consequently, glaciers alter landforms by weathering and erosion. Glaciers scour the land as they move, leaving surface scratches on the rocks that remain. Much of eastern Canada has been scoured by glaciers that left little soil but many ice-gouged lakes and streams. The erosional forms created by glacial scourings have a variety of names. A *glacial trough* is a deep, U-shaped valley visible only after the glacier has receded. If the valley is today below sea level, as in Norway or British Columbia, *fiords*, or arms of the sea, are formed. Some of the landforms created by scouring are shown in Figure 3.28. Figure 3.30 shows *tarns*, small lakes in the hollowed-out depression of a *cirque*. Cirques are formed by ice erosion in a glacial valley.

Glaciers create landforms when they deposit the debris they have transported. These deposits, called *glacial till*, consist of rocks, pebbles, and silt. As the great tongues of ice move forward, debris accumulates in parts of the glacier. The ice that scours valley walls and the ice at the tip of the advancing tongue are particularly filled with debris (Figure 3.31). As a glacier melts, it leaves behind hills of glacial till, called **moraines**.

Many other landforms have been formed by glaciers. The most important is the *outwash plain*, a gently sloping area in front of a melting glacier. The melting along a broad front sends thousands of small streams running out from the glacier in braided fashion, streams that deposit neatly stratified glacial till. Outwash plains, which are essentially great alluvial fans, cover a wide area and provide new, rich parent-material for soil formation. Most of the midwestern part of the United States owes some of its soil fertility to relatively recent glacial deposition.

Before the end of the most recent ice age, there were at least three previous advances that occurred during the million years of the Pleistocene period. Firm evidence is not available on whether we have emerged from the cycle of ice advance and retreat. The factors concerning the earth's changing temperature, which are discussed in the next chapter, must be considered before it is possible to assess the likelihood of a new ice advance. For the first

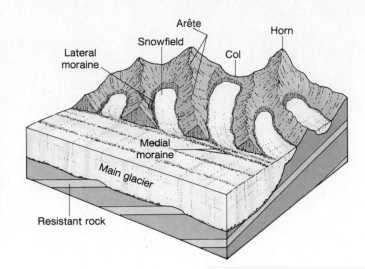

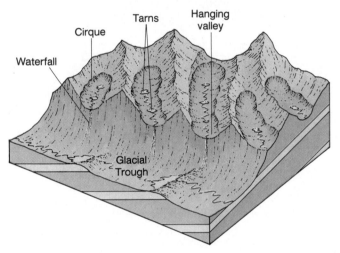

Figure 3.28

The evolution of alpine glacial landforms. Frost shattering and ice movement carve *cirques,* the irregular bottoms of which may contain lakes (*tarns*) after the glacial melt. Where cirque walls adjoin from opposite sides, knifelike ridges called *arêtes* are formed, interrupted by overeroded passes or *cols.* The intersection of three or more arêtes creates a pointed peak, or *horn.* Rock debris falling from cirque walls is carried along by the moving ice. *Lateral moraines* form between the ice and the valley walls; *medial moraines* mark the union of such debris where two valley glaciers join.

half of this century, the world's glaciers were melting faster than they were building up. Current trends are not clear, although there is the fear that the greenhouse effect (discussed in Chapter 5) is warming the earth and will cause the seas to rise.

Waves and Currents

While glacial action is intermittent in earth history, the breaking of ocean waves on continental coasts and on islands is unceasing and causes considerable change in coastal landforms. As waves reach the shallow water close to shore, they are forced to become higher until a breaker is formed, as shown in Figure 3.32. The uprush of water

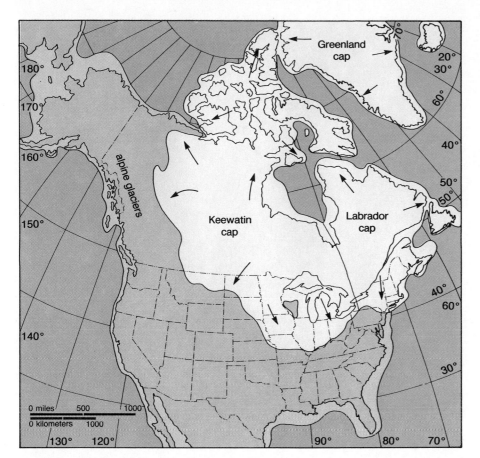

Figure 3.29
Farthest extent of glaciation in the Northern Hemisphere. Separate centers of snow accumulation and ice formation developed. Large lakes were created between the western mountains of North America and the advancing ice front. To the south, huge rivers carried away glacial meltwaters. Since large volumes of moisture were trapped as ice on the land, sea levels were lowered and continental margins were extended.

Figure 3.30
Tarns in glacial cirques.

Figure 3.31
Arêtes and lateral and medial moraines.

Permafrost

In 1577, on his second voyage to the New World in search of the Northwest Passage, Sir Martin Frobisher reported finding ground in the far north that was frozen to depths of "four or five fathoms, even in summer" and that the frozen condition "so combineth the stones together that scarcely instruments with great force can unknit them." The permanently frozen ground, now termed *permafrost*, underlies perhaps a fifth of the earth's land surface. In the lands surrounding the Arctic Ocean, its maximum thickness has been reported in thousands of feet.

For almost 300 years after Frobisher's discovery, little attention was paid to this frost phenomenon. But in the 19th and 20th centuries, during the building of the Trans-Siberian railroad, the construction of buildings associated with the discovery of gold in Alaska and the Yukon Territory, and the development of the oil pipeline that now connects Prudhoe Bay in northern

These Alaskan railroad tracks have been warped by the effect of melting permafrost.

Alaska with Valdez in southern Alaska, attention has focused on the unique nature of permafrost.

Uncontrolled construction activities lead to the thawing of permafrost. This in turn produces unstable ground susceptible to soil movement and landslides, ground subsidence, and frost heaving. Scientists have found that successful use requires that there be the least possible disturbance of the

frozen ground. The Alaska pipeline was built above the earth's surface in order to reduce the likelihood that the relatively warm oil would disturb the permafrost and thus destroy the pipeline. It was necessary to preserve the insulating value of the ground surface as much as possible, thus the vegetation mat was not removed and a coarse gravel fill was added along the surface below the pipeline.

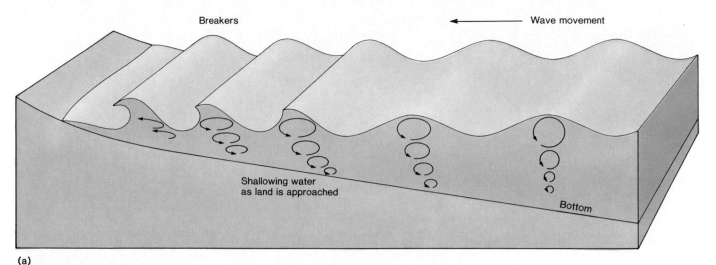

(a)

(b)

Figure 3.32
Formation of waves and breakers. (a) As the offshore swell approaches the gently sloping beach bottom, sharp-crested waves form, build up to a steep wall of water, and break forward in plunging surf. (b) Evenly spaced breakers form as successive waves touch bottom along a regularly sloping shore.

not only carries sand for deposition but also erodes the landforms at the coast, while the backwash carries the eroded material away. This type of action results in different kinds of landforms, depending on conditions.

If the land at the coast is well above sea level, the wave action causes cliffs to form. The cliffs then erode at a rate dependent on the resistance of the rock to the constant assault from the salt water. During storms, a great deal of power is released by the forward thrust of the waves, and much weathering and erosion take place. Landslides are a hazard during coastal storms, and they occur particularly in areas where weak sedimentary rock or glacial till exists.

Beaches are formed by the deposition of sand grains contained in the water. The sand originates from the vast amount of coastal erosion and from streams (Figure 3.33). *Longshore currents*, which move roughly parallel to the shore, transport the sand, forming beaches and *sandspits*.

The more sheltered the area, the better the chance for a beach to build. The backwash of waves, however, takes sand away from the beaches if no longshore current exists. As a result, *sandbars* develop a short distance away from the shoreline. If the sandbars become large enough, they eventually close off the shore, creating a new coastline that encloses *lagoons* or *inlets*. *Salt marshes* very often develop in and around these areas. Figure 3.34 shows one kind of area partially carved by waves.

Coral reefs, made not from sand but from coral organisms growing in shallow tropical water, are formed by the secretion of lime in the presence of warm water and sunlight. Reefs, consisting of millions of colorful skeletons, develop short distances offshore. Off the coast of Australia lies the most famous coral reef, the Great Barrier Reef. *Atolls*, found in the south Pacific, are reefs formed in shallow water around a volcano that has since been covered or nearly covered by water.

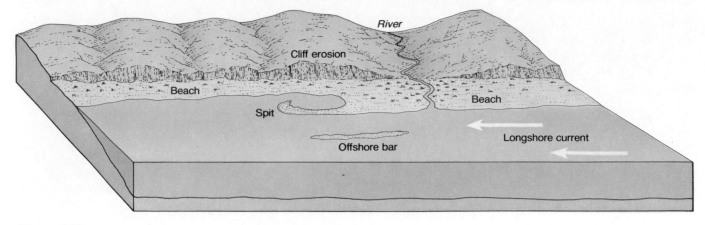

Figure 3.33
The cliffs behind the shore are eroded by waves during storms and high water. Sediment from the cliff and the river forms the beach deposit; the longshore current moves some sediment downcurrent to form a *spit*. Offshore *sandbars* are created from material removed from the beach and deposited by retreating waves.

Figure 3.34
Sea cliffs, headlands, embayments, and offshore erosional remnants are typical of cliffed shorelines, such as this one along the coast of Oregon. Beaches occur only near the mouths of rivers or in embayments.

Wind

Unlike in humid areas where the effect of wind is confined mainly to sandy beach areas, in dry climates, wind is a powerful agent of weathering, erosion, and deposition. The limited vegetation in dry areas leaves exposed particles of sand, clay, and silt subject to movement by wind. Thus, many of the sculptured features found in dry areas result from mechanical weathering, that is, from the abrasive action of sand and dust particles as they are blown against rock surfaces. Sand and dust storms occurring in a drought-stricken farm area may make it unusable for agriculture. Inhabitants of Oklahoma, Texas, and Colorado suffered greatly in the 1930s when their farmlands became the "Dust Bowl" of the United States.

Several types of landforms are produced by wind-driven sand. Figure 3.35 depicts one of these. Although sandy deserts are much less common than gravelly deserts, their characteristic landforms are better known. Most of the Sahara, the Gobi, and the western U.S. deserts are covered not with sand but with rocks, pebbles, and gravel. Each also has a small portion (and the Saudi Arabian desert a large area) covered with sand blown by wind into a series of waves or *dunes*. Unless vegetation stabilizes them, the dunes move as sand is blown from their windward faces onto and over their crests. One of the most distinctive sand desert dunes is the crescent-shaped *barchan*. Along seacoasts and inland lakeshores, in both wet and dry climates, wind can create sand ridges that may reach a height of 300 feet (90–100 m). Sometimes coastal communities and farmlands are threatened or destroyed by moving sand.

Another kind of wind-deposited material, silty in texture and pale yellow or buff in color, is called **loess**. Encountered usually in midlatitude westerly wind belts, it covers extensive areas in the United States (Figure 3.36), central Europe, central Asia, and Argentina. It has its

Figure 3.35
The prevailing wind from the left has given these transverse dunes a characteristic gentle windward slope and a steep, irregular leeward slope.

Cape Cod on the Move

The Pilgrims, who landed close to the tip of Cape Cod, first saw a thick forest region with dense undergrowth. In less than 100 years the area was cleared for farms and the trees used for firewood and materials for buildings and ships. With nothing in their path, huge sand dunes have advanced from the beach areas. Today the dunes threaten historic buildings and the main highway, and they have reduced a lake to a fraction of its former size. With no

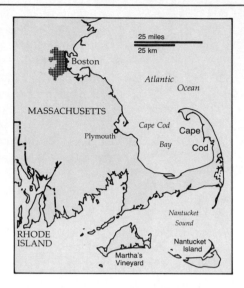

vegetation to thwart the dunes, sand blows up the windward side, rolls over the top, and collects on the leeward side, thus encouraging a steady movement. The dunes advance by up to 20 feet (6 m) a year. Today there are 58 dunes, some of which are 80 feet (25 m) high. An experimental planting program is designed to halt their advance. Beach grass is to be planted on about half of the dunes area, or about 500 acres (200 hectares).

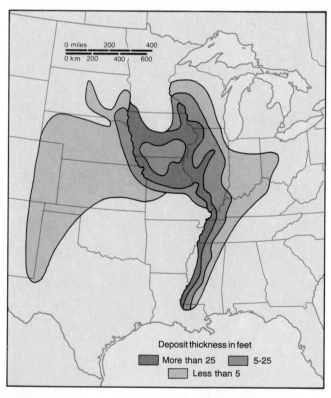

Figure 3.36
Location of windblown silt deposits, including *loess*, in the United States. The thicker layers, found in the upper Mississippi valley area, are associated with the wind movement of glacial debris. Farther west, in the Great Plains, wind-deposited materials are sandy in texture, not loessial.

greatest development in northern China, where it covers hundreds of thousands of square miles, often to depths of more than 100 feet (30 m). The wind-borne origin of loess is confirmed by its typical occurrence downwind from extensive desert areas, though major deposits are assumed to have resulted from wind erosion of nonvegetated sediment deposited by meltwater from retreating glaciers. Because rich soils usually form from loess deposits if climatic circumstances are appropriate, these areas are among the most productive agricultural lands in the world.

Landform Regions

Every piece of land that is not covered by buildings and other structures contains clues as to how it has changed over time. The geomorphologist interprets these clues, studying such things as earth materials and soils, the availability of water, drainage patterns, evidences of erosion, and glacial history. The scale of analysis might be as small as a stream or as large as a *landform region*. This is a large section of the earth's surface where there is a great deal of homogeneity among the types of landforms that characterize it. The map on the inside back cover of this book shows in a general way the kinds of landform regions found in the different parts of the world.

Note how the mountain belts generally coincide with plate boundaries not found beneath the sea (Figure 3.3) and with the earthquake-prone areas (Figure 3.6). Vast plains exist in North and South America, Europe, Asia, and Australia. Many of these regions were created under former seas and appeared as land when seas contracted. These and the smaller plains areas are the drainage basins for some of the great rivers of the world, such as the Mississippi-Missouri, Amazon, Volga, Nile, Ganges, and Tigris-Euphrates. The valleys carved by these rivers and the silt deposited by them are among the most agriculturally productive in the world. The plateau regions are many and varied. The African plateau region is the largest. Much of the African landscape is characterized by low mountains and hills whose base is several thousand feet above sea level. Generally quiet from the standpoint of tectonic activity, Africa is largely made up of geologically ancient continental blocks that have been in an advanced stage of erosion for millions of years.

The landscape affecting and affected by human action is more than landforms, moving continents, and an occasional earthquake. Except at times of natural disaster, these elements of the physical world are for most of us quiet, accepted background. More immediately affecting our lives and fortunes are the great patterns of climate that help define the limits of the economically possible at present levels of technology, the daily changes of weather that affect the success of picnic and crop yield alike, and the patterns of vegetation and soils related in nature to the realms of climate. We turn our attention to these elements of the natural environment in the next chapter.

Summary

In the most recent 200 million of earth's 4.7 billion years, continental plates broke away from Pangaea and drifted on the asthenosphere to their present positions. Rocks, the materials that constitute the earth's surface, are classified as igneous, sedimentary, and metamorphic. At or near plate intersections, tectonic activity is particularly in evidence in two forms. Diastrophism, such as faulting, results in earthquakes and, on occasion, tsunamis. Volcanism pours lava onto the earth's surface.

The building up of the earth's surface is balanced by the three gradational processes—weathering, gravity transfer, and erosion. Weathering, both mechanical and chemical, prepares materials for transport by acting to disintegrate rocks, and it is instrumental in the development of soils. Talus slopes and soil-creep are examples of the effect of gravitational transfer. The erosional agents of running water, groundwater, glaciers, waves and currents, and wind move materials to new locations. Examples of landforms created by the collection of eroded materials are alluvial fans, deltas, natural levees, moraines, and sand dunes.

Key Words

alluvium	lithosphere
asthenosphere	loess
chemical weathering	mechanical weathering
continental drift	metamorphic rocks
diastrophism	mineral
erosional agents	moraine
faults	Pangaea
floodplain	plate tectonics
folds	sedimentary rocks
glacier	subduction
gradational processes	tsunami
gravity transfer	volcanism
igneous rocks	warping
karst topography	water table

For Review

1. What evidence makes the theory of *plate tectonics* plausible?

2. What is the meaning of and the name of the process that occurs when two plates collide?

3. In what ways may rocks be classified? List three classes of rocks according to their origin. In what ways can they be distinguished from one another?

4. Explain what is meant by *gradation* and *volcanism*.

5. What is meant by *folding, joint,* and *faulting?*

6. Draw a diagram indicating the varieties of ways *faults* may occur.

7. With what earth movements are earthquakes associated? What is a *tsunami* and how does it develop?

8. What is the distinction between *mechanical* and *chemical weathering?* Is weathering responsible for landform creation? In what way do glaciers engage in mechanical weathering?

9. Explain the origin of the various landforms one usually finds in desert environments.

10. How do *glaciers* form? What landscape characteristics are associated with glacial erosion? With glacial deposition?

11. What landform features can you identify on Figure 3.14? What were the agents of their creation?

12. How are alluvial fans, deltas, natural levees, and moraines formed?

13. How is groundwater erosion differentiated from surface water erosion?

14. How are the processes that bring about change due to waves and currents related to the processes that bring about change by the force of wind?

15. What processes account for the landform features of the area in which you live?

Selected References

Butzer, K. W. *Geomorphology from the Earth*. New York: Harper and Row, 1976.

Chorley, Richard J. *Water, Earth, and Man*. London: Methuen, 1969.

Davies, J. L. *Geographical Variation in Coastal Development*. New York: Longman, 1977.

Doerr, Arthur H. *Fundamentals of Physical Geography*. Dubuque, Ia.: Wm. C. Brown, 1990.

Embleton, C., and C. A. M. King. *Glacial Geomorphology*. New York: Wiley, 1975.

Gabler, R., et al. *Essentials of Physical Geography*. 3d ed. Philadelphia: Saunders, 1987.

Goudie, A. S. *Environmental Change*. New York: Oxford University Press, 1977.

McKnight, Tom L. *Physical Geography: A Landscape Appreciation*. 3d ed. Englewood Cliffs, N.J.: Prentice-Hall, 1990.

Marsh, William M. *Earthscape: A Physical Geography*. New York: Wiley, 1987.

Muller, Robert A., and Theodore M. Oberlander. *Physical Geography Today*. 3d ed. New York: Random House, 1984.

Pitty, A. F. *The Nature of Geomorphology*. London: Methuen, 1982.

Press, Frank, and Raymond Siever. *Earth*. 4th ed. San Francisco: Freeman, 1984.

Progress in Physical Geography. Vols. 1–13, various issues.

Ritter, Dale F. *Process Geomorphology*. 2d ed. Dubuque, Ia.: Wm. C. Brown, 1986.

Scientific American Editors. *Continents Adrift and Continents Aground*. San Francisco: Freeman, 1976.

———. *Earthquakes and Volcanoes: A Scientific American Book*. San Francisco: Freeman, 1980.

Strahler, Arthur N., and Alan H. Strahler. *Elements of Physical Geography*. 4th ed. New York: Wiley, 1990.

Thompson, Russell D., et al. *Processes in Physical Geography*. New York: Longman, 1986.

Thorn, Colin E. *An Introduction to Theoretical Geomorphology*. Winchester, Mass.: Allen & Unwin, 1988.

4

Physical Geography: Weather and Climate

Typhoon winds, Tahiti.

LIKE so many before, it first appeared on an early June day in 1985 as a tiny blip on Indian radar screens. This one was recognized as a small circulation pattern developing in the Bay of Bengal south of the crowded country of Bangladesh. It was a hot humid day on the mainland. By late afternoon, it was called a tropical depression and seen to be the possible start of a major storm system. The Indian meteorologists warned the Bangladesh government that a killer storm, a hurricane (called a *typhoon* or *cyclone* in Asia), was headed for the densely settled islands at the mouth of the Ganges River (see Figure 4.1). Hourly warnings were broadcast on radio and television that all coastal residents must seek shelter immediately.

Unfortunately, most of the impoverished people were too poor to own radios, telephones, or television and did not hear the warning. By late that night, what was to become the most damaging storm in Bangladesh since 1970 came ashore. The afflicted area, an already flat landscape, was stripped of all huts, houses, and other buildings. People were swept away by the waves. At week's end it was estimated that about 20,000 people had died, thousands were injured, and a quarter million were left homeless. Those who survived had clung to floating bamboo rooftops or pieces of driftwood. Many eventually died because supplies and medicine could not be transported to people marooned on islands. The storm had destroyed roads and bridges. There was no shelter, no clothing, and no food.

The power of hurricanes is concentrated in a narrow path. Whether such meteorological events occur in Asia or North America, they do great damage. The lives of all those in the paths of storms are affected. Tropical storms are one type, certainly an extreme type, of weather phenomenon. Most people are "weather watchers": they watch the TV forecasts with great interest and plan their lives around weather events. In this chapter we review that subsection of physical geography concerned with weather and climate. It deals with normal, patterned phenomena out of which such an abnormality as the Bangladesh hurricane occasionally emerges.

A weather forecaster describes current conditions for a limited area, such as a metropolitan area, and predicts future weather conditions. If the elements that make up the **weather**, such as temperature, wind, and precipitation, are recorded at specified moments in time, such as every hour, an inventory of weather conditions can be developed. By finding trends in data that have been gathered over an extended period of time, we can speak about typical conditions. These characteristic circumstances describe the **climate** of a region. Weather is a moment's view of the lower atmosphere, while climate is a description of typical weather conditions in an area or place over a period of time. Geographers analyze the differences in weather and climate from location to location in order to understand how climatic elements affect human occupance of the earth.

In geography, we are particularly interested in the physical environment that surrounds us. That is why the **troposphere**, the lowest layer of the earth's atmosphere, attracts our attention. This layer, extending about 6 miles (10 km) above the ground, contains virtually all of the air, clouds, and precipitation of the earth.

In this chapter, we try to answer the questions that are usually raised about the characteristics of the lower atmosphere. By discussing the answers to these questions from the viewpoint of averages or average variations, we attempt to give a view of the earth's climatic differences, a view held to be most important for understanding the way people use the land. Climate is a key to understanding, in a broad way, the distribution of world population. People have great difficulty living in areas that are, on the average, very cold, very hot, very dry, or very wet. They are also negatively affected by huge storms or flooding. In this chapter we first discuss the elements that constitute weather conditions and then describe the various climates of the earth.

Air Temperature

Perhaps the most fundamental question about weather is: Why do temperatures vary from place to place? The answer to that question requires a discussion of a number of concepts that help to focus on the way heat accumulates on the earth's surface.

Energy from the sun, called *solar energy*, is transformed into heat, primarily at the earth's surface and secondarily in the atmosphere. Not every part of the earth or its overlying atmosphere receives the same amount of solar energy. At any given place, the amount of incoming solar radiation, or **insolation**, available depends on the intensity and duration of radiation from the sun. These are determined by (1) the angle at which the sun's rays strike the earth and (2) the number of daylight hours. These two fundamental factors plus the following five modifying variables determine the temperature at any given location: (1) the amount of water vapor in the air, (2) the degree of cloud cover (or cover in general), (3) the nature of the surface of the earth, (4) the elevation above sea level, and (5) the degree of air movement. Let us look at these factors briefly.

Earth Inclination

The earth, as Figure 4.2 indicates, does not spin on an axis that is perpendicular to a line connecting the center of the earth and the sun. Rather, the earth's axis is tilted about 23.5° away from the perpendicular; every 24 hours the earth rotates once on that axis, as shown in Figure 4.3.

While rotating, the earth is slowly revolving around the sun in a nearly circular annual orbit. The axis of the earth, that is, the imaginary line connecting the North Pole to the South Pole, always remains in the same position. This is to say that there is parallelism of the earth's axis during rotation and revolution (Figure 4.4). If the earth were not tilted from the perpendicular, the solar energy received *at a given latitude* would not vary during the course of the year. The rays of the sun would strike the equator most

directly, and as the distance away from the equator became greater, the rays would strike the earth at ever-increasing angles, diminishing the intensity of the energy and giving climates a latitudinal standardization (Figures 4.5 and 4.6).

Because of the inclination, however, the location of highest incidence of incoming solar energy varies during the course of the year. When the Northern Hemisphere is tilted directly toward the sun, the vertical rays are felt as far north as 23.5°N latitude. This position of the earth occurs about June 21, the summer *solstice* for the Northern Hemisphere and the winter solstice for the Southern. About December 21, when the vertical rays of

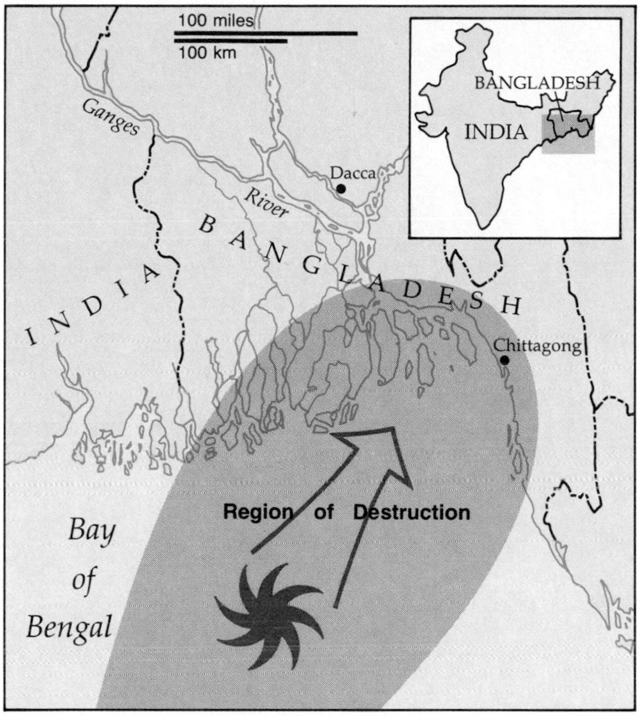

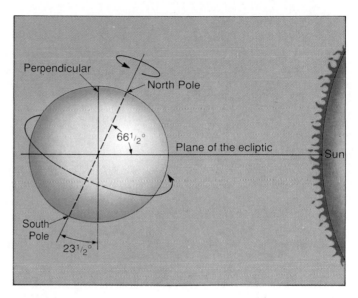

Figure 4.1

Path of the 1985 hurricane (cyclone, typhoon) in the Bay of Bengal. Dacca and Chittagong are the principal cities of Bangladesh.

Figure 4.2

The earth spins on an axis tilted about 23½° from the perpendicular or 66½° from the *plane of the ecliptic,* an imaginary plane that contains the lines connecting the center of the earth at all times of the year with the center of the sun.

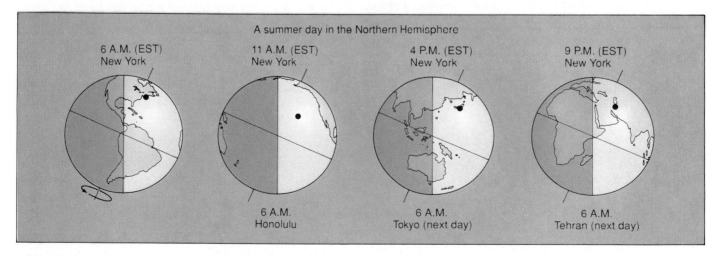

Figure 4.3

The process of the 24-hour rotation of the earth on its axis.

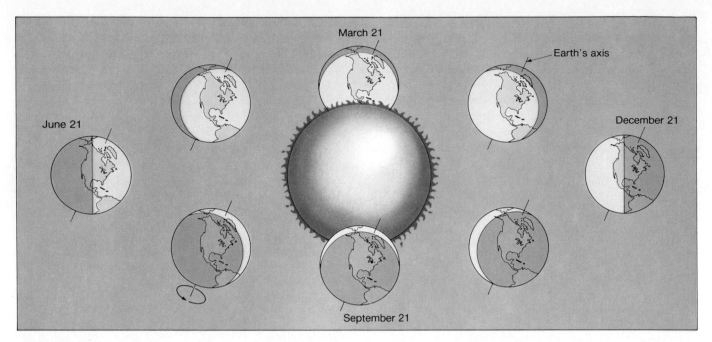

Figure 4.4
The process of the yearly revolution of the earth about the sun.

Figure 4.5
Notice in the lower diagram that as the earth revolves, the north polar area in June is bathed in sunshine for 24 hours, while the south polar areas are dark. The most intense of the sun's rays are felt north of the equator in June and south of the equator in December. None of this is true in the untilted case.

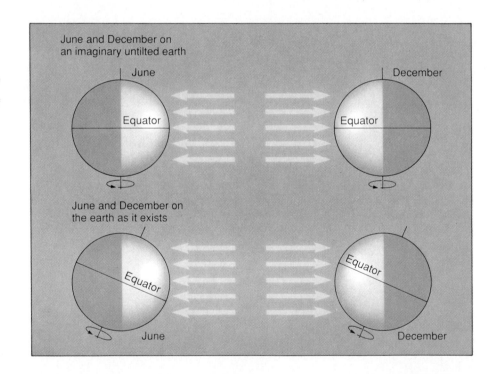

the sun strike near 23.5°S latitude, it is the beginning of summer in the Southern Hemisphere and the onset of winter in the Northern. During the rest of the year, the position of the earth relative to the sun results in direct rays migrating from about 23.5°N to 23.5°S and back again. On about March 21 and September 21 (the spring and autumn *equinoxes*), the vertical rays of the sun strike the equator.

The tilt of the earth also means that the length of the day and night varies during the year. One-half of the earth is always illuminated by the sun, but only at the equator is it light for 12 hours each day of the year. As distance away from the equator becomes greater, the hours of daylight or darkness increase, depending on whether the direct rays of the sun are north or south of the equator. In the summer, daylight increases to the maximum of 24

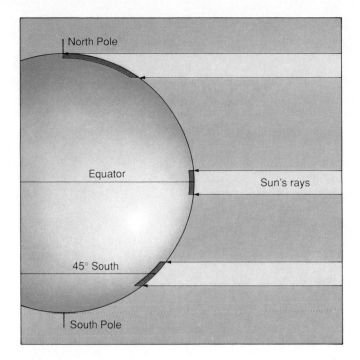

Figure 4.6
Three equal, imaginary rays from the sun are shown striking the earth at different latitudes at the time of the equinox. As distance increases away from the equator, the rays become more diffused, showing how the sun's intensity is diluted in the high latitudes.

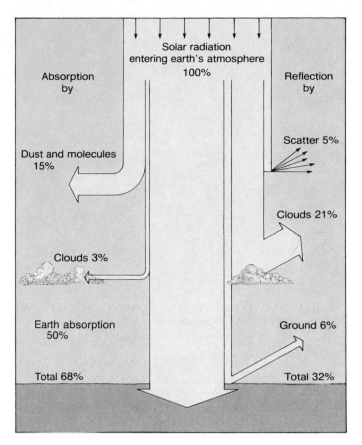

Figure 4.7
Consider the incoming solar radiation as 100%. The portion that is absorbed into the earth (50%) is eventually released to the atmosphere and then reradiated into space. Notice that the outgoing radiation is equal to 100%, showing that there is an energy balance on the earth.

hours in the summer polar region, and during the same period, nighttime finally reaches 24 hours in length in the other polar region.

Because of the 24-hour daylight, it would seem that much solar energy should be available in the summer polar region, but this is not the case. The angle of the sun is so narrow (the sun is low in the sky) that the solar energy is spread over a wide surface. By contrast, the combination of relatively long days and sun angles close to 90° makes an enormous amount of energy available to areas in the neighborhood of 15° to 30° north and south latitude during each hemisphere's summer.

Reflection and Reradiation

Much of the insolation potentially receivable is, in fact, sent back to outer space or diffused in the troposphere in a process known as **reflection**. Clouds, which are dense concentrations of suspended, tiny water or ice particles, reflect a great deal of energy. Light-colored surfaces, especially snow cover, also serve to reflect large amounts of solar energy.

Energy is lost through reradiation as well as reflection. In the **reradiation** process, the earth acts as a communicator of energy. As indicated in Figure 4.7, the shortwave energy that is absorbed into the land and water is returned to the atmosphere in the form of longwave terrestrial radiation. On a clear night, when no clouds can

block or diffuse the movement, temperatures become lower and lower as the earth reradiates as heat the energy it has received and stored during the day.

Some kinds of earth surface material, especially water, store solar energy more effectively than others. Because water is transparent, solar rays can penetrate a great distance below its surface. If water currents are present, the heat is distributed even more effectively. On the other hand, land surfaces are opaque, and so all of the energy received from the sun is concentrated at the surface. Land, having more heat available at the surface, reradiates its energy faster than does water. Air is heated by this process of reradiation from the earth and not directly by energy from the sun passing through it. Thus, because land heats and cools much more rapidly than water, the extremes of hot and cold temperatures recorded on earth occur on land and not the sea.

Temperatures are moderated by the presence of large bodies of water near land areas. Note in Figure 4.8 that coastal areas have lower summer temperatures and higher winter temperatures than places at the same distance from the equator that are not seacoasts. Land areas affected by

January

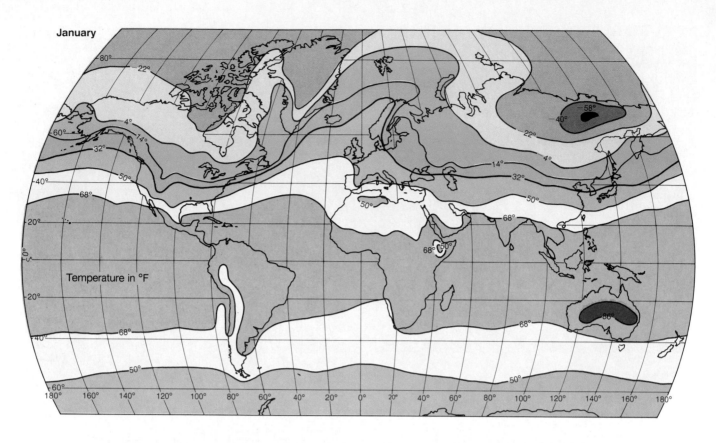

Temperature in °F

July

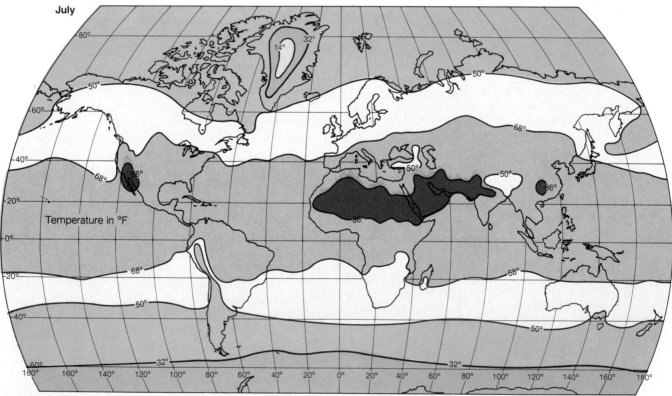

Temperature in °F

Figure 4.8
At a given latitude, water areas are warmer than land areas in winter and cooler in summer. *Isotherms* are lines of equal temperature.

the moderating influences of water are considered *marine* environments; those areas not affected by nearby water are *continental* environments.

Temperatures vary in a cyclical way from day to day. In the course of a day, as incoming solar energy exceeds the energy lost through reflection and reradiation, temperatures begin to rise. The ground stores some heat, and temperatures continue to rise until the angle of the sun becomes so narrow that the energy received no longer exceeds that lost by the reflection and reradiation processes. Not all of the heat is lost during the night, but long nights appreciably deplete the stored energy.

The Lapse Rate

We might think that as we move vertically away from the earth toward the sun, temperatures would increase; such is not true within the troposphere. The earth is a body absorbing and reradiating heat, and therefore temperatures are usually warmest at the earth's surface and lower as elevation increases. Note on Figure 4.9 that this temperature **lapse rate** (the rate of change of temperature with altitude in the troposphere) averages about 3.5°F per 1000 feet (6.4°C per 1000 m). For example, the difference in elevation between Denver and Pikes Peak is about 9000 feet (2700 m), which would normally result in a 32°F (17°C) difference in temperature. Jet planes flying at an altitude of 30,000 feet (9100 m) are moving through air about 100°F (56°C) colder than ground temperatures.

The normal lapse rate does not always hold, however. Rapid reradiation sometimes causes temperatures to be higher above the earth's surface than at the surface itself. This particular condition, in which air at lower altitudes is cooler than air aloft, is called a **temperature inversion**. An inversion is important because of its effect on air movement. Warm air at the surface, which would normally rise, may be blocked by the still warmer air of the temperature inversion. Thus the surface air is trapped, and if it is filled with automobile exhaust emissions or smoke, a serious smog condition may develop (see Figure 4.10). Because of the condition of the nearby mountains, Los Angeles, pictured in Figure 4.11, often experiences temperature inversions, causing the sunlight to be reduced to a dull haze.

The effect of air movement on temperature is made clear in the following section on air pressure and winds.

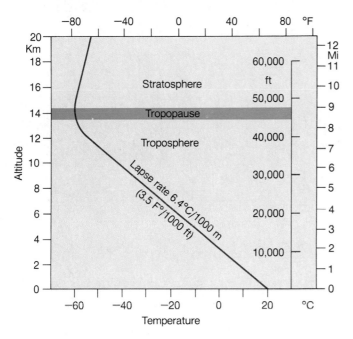

Figure 4.9
The temperature lapse rate under typical conditions.

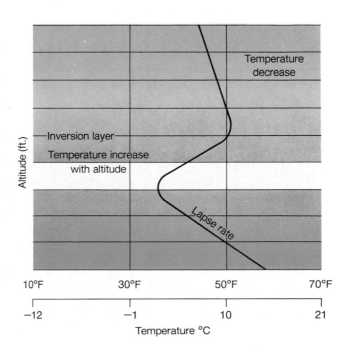

Figure 4.10
Temperature inversion. Subsiding air has forced up the temperature at a height of several hundred feet.

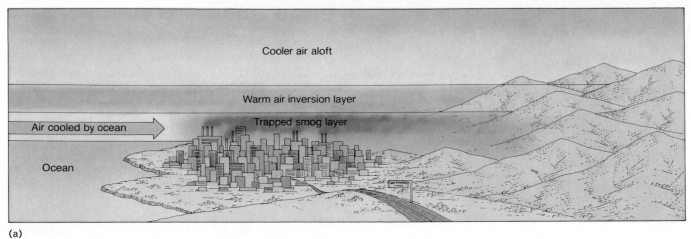

(a)

(b)

Figure 4.11

Smog in the Los Angeles area. (a) Below the inversion layer, stagnant air holds increasing amounts of pollutants, caused mainly by automobile exhausts. (b) Afternoon smog in Los Angeles.

The Donora Tragedy

A heavy fog settled over the valley town of Donora, Pennsylvania, in late October 1948. Stagnant, moisture-filled air was trapped in the valley by surrounding hills and by a temperature inversion that held the cooler air, gradually filling with smoke and fumes from the town's zinc works, against the ground under a lid of lighter, warmer upper air. For five days the smog increased in concentration; the sulfur dioxide emitted from the zinc works continually converted to deadly sulfur trioxide by contact with the air. Both old and young, with and without past histories of respiratory problems, reported to doctors and hospitals difficulty in breathing and unbearable chest pains. Before the rains washed the air clean nearly a week after the smog buildup, 20 were dead and hundreds hospitalized. A normally harmless, water-saturated inversion had been converted to deadly poison by a tragic union of natural weather processes and human activity.

Air Pressure and Winds

A second fundamental question about weather and climate concerns **air pressure**: How do differences in air pressure from place to place affect weather conditions? The answer to that question requires that we first explain why there are differences in air pressure.

Air is a gaseous substance whose weight affects air pressure. If it were possible to carve out 1 cubic inch of air at the surface of the earth and weigh it along with all the other cubic inches of air above it, under normal conditions the total weight would come to about 14.7 pounds (6.67 kg) of air as measured at sea level. Actually this is not very heavy when you consider the dimensions of the 1-inch column of air: 1 inch by 1 inch by about 6 miles, or about 220 cubic feet (6.2 m³). The weight of air 3 miles (4.8 km) above the earth's surface, however, is considerably less than 14.7 pounds because there is correspondingly less air above it. So, air is heavier and air pressure is higher close to the earth's surface. It is a physical law that for equal amounts of cold and hot air, the cold air is heavier. That is why hot-air balloons, filled with lighter air, can rise into the atmosphere. A cold morning is characterized by relatively heavy air, but as afternoon temperatures rise, the air becomes lighter.

As early as the 17th century, it was discovered that when a column of mercury is contained within a tube, normal atmospheric pressure at sea level is sufficient to balance the weight of the mercury column to a height of 29.92 inches (76 cm). *Barometers* of varying types are used to record changes in air pressure, and barometric readings in inches of mercury are a normal part, along with recorded temperatures, of every weather report. Since air pressure at a given location changes as air warms and cools, barometers record a drop in atmospheric pressure when air is heating and a rise in pressure when air is cooling.

In order to visualize the effect of air movements on weather, it is useful to think of air as a liquid made up of two fluids of different densities (representing light air and heavy air), such as water and gasoline. If the fluids are introduced into a tank at the same time, the lighter liquid will move to the top. This result represents the vertical motion of air. The heavier liquid spreads out horizontally over the bottom so that it is the same thickness everywhere. This flow represents horizontal movement of air, or the wind. Air seeks equilibrium by evening out pressure imbalances that result from heating and cooling (see Figure 4.13). Air races from heavy, cold air locations to light, warm air locations. Thus the greater the differences in air pressure between places, the greater the wind.

Pressure Gradient Force

Because of differences in the earth's surface—water, snow cover, dark green forests, cities, and so on—and the other factors mentioned above that affect energy receipt and retention, zones of high and low air pressure develop. Sometimes these high- and low-pressure zones cover entire continents, but usually they are considerably smaller—several hundred miles wide—and within these regions small differences are noted over short distances. When pressure differences exist between areas, a **pressure gradient** is formed.

In order to balance out the pressure differences, air from the heavier high-pressure areas flows to the low-pressure zones. The heavy air stays close to the earth's surface as it moves, produces winds, and helps to speed the movement of warm air upward. The velocity, or speed, of the wind is in direct proportion to pressure differences. As depicted in Figure 4.12, winds are caused by pressure differences that induce airflow from points of high to points of low pressure. If distances beween high and low pressure are short, pressure gradients are steep and wind velocities are great. If different pressures are far apart, the equalization process results in more gentle air movements.

The Convection System

As you know, the temperature of the room you are in is lower near the floor than near the ceiling because warm air rises and cool air descends. The circulatory motion of descending cool air and ascending warm air is known as **convection**. A convectional wind system results from the flow of air that replaces warm, rising air and the rapid movement of replacement air.

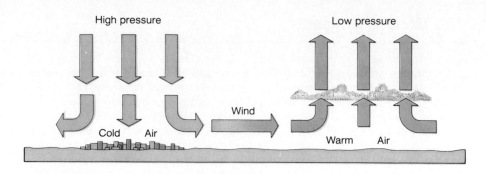

Figure 4.12
If the distance between centers of high pressure and low pressure is short, the wind is strong. If the pressure differences are great, winds are again strong. The strongest winds are produced by extreme pressure differences over very short distances.

A good example of a convectional system is land and sea breezes. Close to a large body of water, the daytime heating differential between land and water is great. As a result, the warmer air over the land rises vertically, only to be replaced by cooler air from over the sea. At night, just the opposite occurs. The water is now warmer than the land, which has reradiated much of its heat, and the result is a land breeze toward the sea. These two winds make seashore locations in warm climates particularly comfortable.

Mountain and Valley Breezes

Gravitational force causes the heavy, cool air that accumulates over snow in mountainous areas to descend into lower valley locations, as suggested in Figure 4.13. Consequently, valleys can become much colder than the slopes, and there is a temperature inversion. Slopes are the preferred sites for agriculture in mountainous regions because the cold air from **mountain breezes** can cause freezing conditions in the valleys. In densely settled narrow valleys where industry is concentrated, air pollution can become particularly dangerous. Mountain breezes usually occur during the night; **valley breezes**, caused by warm air moving up slopes in mountainous regions, are usually a daytime phenomenon. The canyons of southern California are the scene of strong mountain and valley breezes, and in the dry season they become dangerous areas for the spread of brush and forest fires.

Coriolis Effect

In the process of moving from high to low pressure, the wind appears to veer toward the right in the Northern Hemisphere and toward the left in the Southern Hemisphere, no matter what the compass direction of the path. This apparent deflection is called the **Coriolis effect**. Were it not for this effect, winds would move in exactly the direction specified by the pressure gradient.

To illustrate the impact of the Coriolis effect upon winds, a familiar example may be helpful. Think of a line of ice skaters holding hands while skating in a circle with one of the skaters near the center of the circle. This skater turns slowly, while the skater at the outside of the circle

must skate very rapidly in order to keep the line straight. In a similar way, the equatorial regions of the earth are rotating at a much faster rate than the areas around the poles. Next, suppose that the center skater threw a ball directly toward the skater at the end of the line. By the time the ball arrived, it would pass behind the outside skater. If the skaters are going in a counterclockwise direction, as the earth appears to be moving viewed from the position of the North Pole, the ball appears to the person at the North Pole to pass to the right of the outside skater. If the skaters are going in a clockwise direction, as the earth appears to be moving viewed from the South Pole position, the deflection is to the left. Because air (like the ball) is not firmly attached to the earth, it, too, will appear to be deflected. The air maintains its direction of movement, but the earth's surface moves out from under it. Since the position of the air is measured relative to the earth's surface, the air appears to have diverged from its straight path.

The Coriolis effect and the pressure gradient force produce spirals rather than simple straight-line patterns of wind, as indicated on Figure 4.14. The spiral of wind is the basic form of the many storms that are so very important to the earth's air circulation system. These storm patterns are discussed later in the chapter.

The Frictional Effect

Wind movement is slowed by the frictional drag of the earth's surface. The effect is strongest at the surface and declines until it becomes ineffective at about 5000 feet (1524 m) above the surface. Not only is there a slowing of the wind but wind direction is changed. Instead of following a course exactly dictated by the pressure gradient force or by the Coriolis effect, the **frictional effect** causes the wind to follow an intermediate path.

Wind Belts

Equatorial areas of the earth are zones of low pressure. Intense solar heating in these areas is responsible for a convectional effect. Note on Figure 4.15 how the warm air rises and tends to move away from the *equatorial low pressure* in both a northerly and a southerly direction. As the equatorial air rises, it cools and eventually becomes

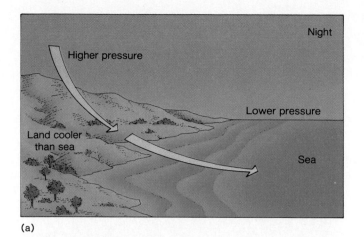

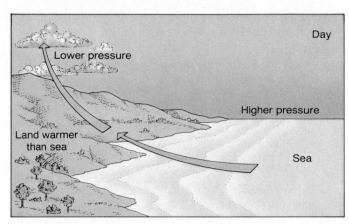

(a)

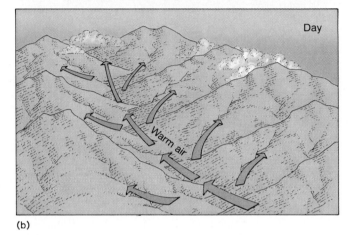

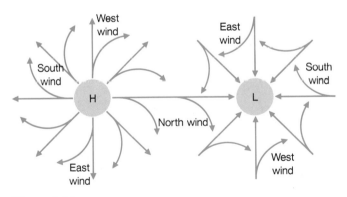

(b)

Figure 4.13
Convectional wind effects due to differential heating and cooling. (a) Land and sea breezes. (b) Mountain and valley breezes.

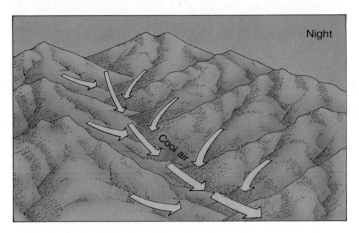

Figure 4.14
The effect of the Coriolis force on flowing air. The straight arrows indicate the paths winds would follow flowing out of an area of high (H) pressure or into one of low (L) pressure were they to follow the paths dictated by pressure differentials. The curved arrows represent the apparent deflecting effect of the Coriolis force. Wind direction—indicated on the diagram for selected curved arrows—is always given by the direction *from* which the wind is coming.

heavy. Finally, the heavy air falls, forming zones of high pressure. These areas of *subtropical high pressure* are located at about 30°N and 30°S of the equator.

When this cooled air reaches the earth's surface, it, too, moves in both a northerly and a southerly direction. The Coriolis effect, however, modifies the wind direction and creates, in the Northern Hemisphere, the belts of winds called the *northeast trades* in the tropics and the *westerlies* (really the southwesterlies) in the midlatitudes. The names refer to the direction from which the winds come. Most of the United States lies within the belt of westerlies; that is, the air usually moves across the country from southwest to northeast. There is also a series of cells of ascending air over the oceans to the north of the westerlies called the *subpolar low*. These areas tend to be cool and rainy. The *polar easterlies* connect the subpolar low areas to the *polar high*.

The general planetary air-circulation pattern is modified by local wind conditions. It should be clear that these belts move in unison as the vertical rays of the sun change position. For example, equatorial low conditions are evident in the area just north of the equator in the Northern Hemisphere summer and just south of the equator in the Southern Hemisphere summer. Recent evidence suggests that sunspot activity, which varies in a cycle averaging about 11 years, affects the surface wind direction within the wind belts. More will be said about air circulation later when the discussion turns to the movement of air masses.

The strongest flows of upper air winds (30,000 to 40,000 feet or 9 to 12 km) are the **jet streams**. These air streams, moving at 100 to 200 miles per hour from west to east in the Northern Hemisphere, circle the earth in an undulating pattern, first north then south as they move eastward. There are three to six undulations at any one time in the Northern Hemisphere, but the waves are not always continuous. These undulations or waves control the flow of air masses on the earth's surface. The more stable the undulations, the greater the repetitiveness of current conditions. These waves tend to separate cold, polar air from warm, tropical air. When a wave dips far to the south, cold air is brought equatorward and warm air moves poleward, bringing severe weather changes to the midlatitudes. The jet stream is more pronounced in the winter than in the summer.

Nowhere does the manner in which the seasonal shift takes place have such a profound effect on humanity as in densely populated south and east Asia. The wind, which in India comes from the southwest in summer, reaches the landmass after picking up a great deal of moisture over the warm Indian Ocean. As it crosses the coast mountains and then the foothills of the Himalayas in April and May, the monsoon rains begin. **Monsoon** is a term meaning seasonal winds. South and east Asia's farm economy, and particularly the rice crop, is totally dependent on summer monsoon rainwater. If the wind shift is late or the rainfall is significantly more or less than optimum, crop failure may result. The undue prolongation of the summer monsoon rains in 1978 caused disastrous flooding and crop and life loss in eastern India and Southeast Asia. The transition to dry northeast winter winds occurs gradually across the region, first becoming noticeable in the north in September. By January, most of the subcontinent is dry. Then, beginning in March in the south, the yearly cycle repeats itself.

Ocean Currents

Surface ocean currents correspond roughly to wind direction patterns because the winds of the world set the ocean currents in motion. In addition, just as differences in air pressure cause wind movements, so differences in the density of water help to force water to move. When ocean water evaporates, salt and other minerals that will not

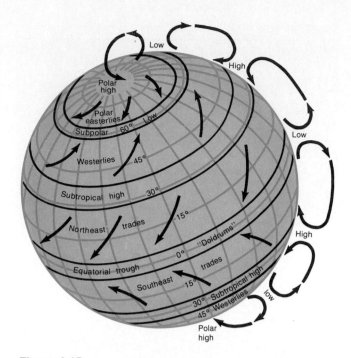

Figure 4.15
The planetary wind and pressure belts as they would develop on an earth of homogeneous surface. The named high- and low-pressure belts represent surface pressure conditions; the named wind belts are prevailing surface wind movements responding to pressure gradients and the Coriolis effect. In the upper atmosphere the conditions are reversed. Descending air, creating surface high pressure, represents the contraction of the air column through cooling and the creation of lower pressure aloft. Air rising from surface flows increases upper air pressure at those zones. The upper atmospheric airflows shown on the diagram respond to pressure contrasts at those higher altitudes. Land and water contrasts on the actual earth, particularly evident in the Northern Hemisphere, create complex distortions of this simplified pattern.

evaporate are left behind, making the water denser. High-density water exists in areas of high pressure, where descending dry air readily picks up moisture. In areas of low pressure, where rainfall is plentiful, the ocean water is of low density. Wind direction (including the Coriolis effect) and the differences in density cause water to move in wide paths from one part of the ocean to another.

There is an important difference between surface air movements and surface water movements. Landmasses are barriers to water movement, deflecting currents and sometimes forcing them to move in a direction opposite to the main current; air, on the other hand, moves freely over both land and water.

The shape of an ocean basin also has an important effect on ocean current patterns. For example, the north Pacific current, which moves from west to east, strikes the west coast of Canada and the United States and then is forced to move both north and south (although the major movement is the cold ocean current that moves south along the coast of California). In the Atlantic Ocean, however, as Figure 4.16 indicates, the current is deflected in a

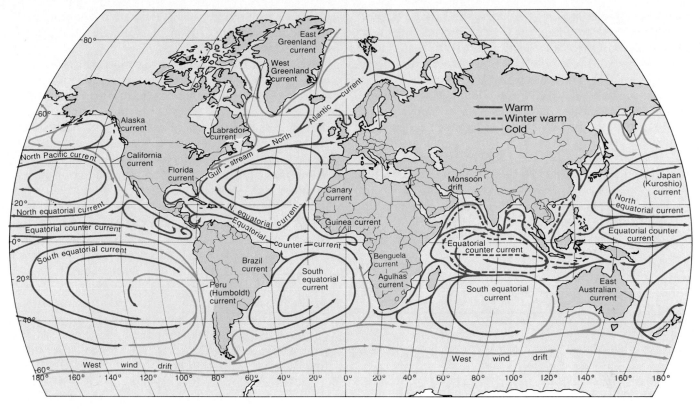

Figure 4.16

The principal surface ocean currents of the world. Notice how the warm waters of the Gulf of Mexico, the Caribbean, and the tropical Atlantic Ocean drift to northern Europe.

El Niño

El Niño, Spanish for "the child," is a term coined years ago by fishermen who noticed that the usually cool waters off the Peruvian coast were considerably warmer every three or four years around Christmas time. The fish catch was reduced significantly during those periods, and, if the fishermen had been able to make the scientific associations that present-day oceanographers and climatologists can, they would have recognized that a host of other effects follow from El Niño. In 1982–1983 an unusually severe and extended period for El Niño caused enormous damage and many deaths. Parts of South America were ravaged by floods while droughts occurred in Australia and parts of Asia. Hurricanes (typhoons) that rarely strike

Tahiti hit the island six times during that period.

The winds that usually blow from east to west over the central Pacific Ocean, from the cold ocean current to the warm waters of East Asia, slow or even reverse during periods of El Niño. The warm water builds up along the west coast of South America, and its effects are felt around the world. For example, in 1982–1983, huge downpours, coastal erosion, and flooding caused tens of millions of dollars of damage in southern California.

The phenomenon occurs every two to seven years but at different degrees of intensity. For example, El Niño occurred in 1986, but the modest warm water build-up did not cause extraordinary circumstances. The cold water

peak between El Niños is called La Niña. The last La Niña occurred in 1988, a year marked by drought in large portions of North America.

The El Niño condition is an example of the interaction of atmospheric pressure and ocean temperature. The atmospheric and oceanic states encourage each other. Under normal conditions, the contrast in temperatures across the ocean helps drive the winds, which in turn keep pushing water to the west, maintaining the contrast in water temperature. But in some years there is a *southern oscillation,* in which atmospheric pressure rises near Australia, the wind falters, and El Niño is created off South America.

northeasterly direction by the shape of the coast (Nova Scotia and Newfoundland jut far out into the Atlantic), and then it moves freely across to and past the British Isles and Norway, all the way to the extreme northwest coast of the USSR. This massive movement of warm water to northerly lands, called the **North Atlantic drift**, has enormous significance to the inhabitants of those areas. Without it, northern Europe would be much colder than it is.

Ocean currents affect not only the temperature but also the precipitation on the land areas adjacent to the ocean. A cold ocean current near a land area robs the air above the current of its potential moisture supply, thus denying moisture to nearby land. Coastal deserts of the world usually border cold ocean currents. On the other hand, warm ocean currents, such as those off India, bring moisture to the adjacent land area, especially when the prevailing winds are landward.

Our earlier question of the ways in which differences in air pressure affect weather conditions is now answered in terms of the movement of warm and cool air over various surfaces at different times of the year and different times of the day. A more complete answer about the causes of different types of weather conditions, however, requires an explanation of the susceptibility of places to receive precipitation, because rainfall and wind patterns are highly related.

Moisture in the Atmosphere

Air contains water vapor (what we feel as humidity), which is the source of all precipitation. **Precipitation** is water in any form (rain, sleet, snow, hail) deposited on the earth's surface. Ascending air can easily expand because there is less pressure on it. When the heat of the lower air spreads out through a larger volume in the troposphere, the mass of air becomes cooler. Cool air is less able than warm air to hold water vapor (Figure 4.17).

The air is said to be *saturated* when the air contains so much water vapor that it will condense (change from a gas to a liquid) and form droplets if fine particles, called *condensation nuclei*, are present. These particles, mostly dust, pollen, smoke, and salt crystals, are nearly always available. At first, the tiny water droplets are usually too light to fall. When there are so many droplets that they coalesce into drops and become too heavy to remain suspended in air, they fall as rain. If temperatures are below the freezing point, the water vapor changes to ice crystals instead of water droplets. When enough ice crystals are present, snow is formed (Figure 4.18).

Rain droplets or ice crystals in large numbers form clouds, which are supported by slight upward movements of air. The form and altitude of clouds depend on the amount of water vapor in the air, the temperature, and the wind movement. Descending air in high-pressure zones usually yields cloudless skies. Whenever warm, moist air rises, clouds form. The most dramatic cloud formation is probably the *cumulonimbus*, pictured in Figure 4.19. This

Figure 4.17

The water-carrying capacity of air and relative humidity. The actual water in the air (water vapor) divided by the water-carrying capacity (×100) equals the relative humidity. The solid line represents the maximum water-carrying capacity of air at different temperatures.

Figure 4.18

As warm air rises, it cools. As it cools, its water vapor condenses and clouds form. If the water content of the air is greater than the air's capacity to retain moisture, some form of precipitation is likely to occur.

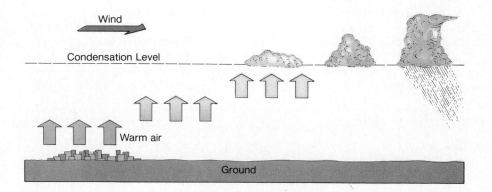

(a)

(b)

(c)

Figure 4.19
Cloud types: (a) cumulonimbus, (b) stratus, (c) cirrus.

is the anvil-head cloud that often accompanies heavy rain. Low, gray *stratus* clouds appear more often in cooler seasons than in the warmer months. The very high, wispy *cirrus* clouds that may appear in all seasons are made entirely of ice crystals.

The amount of water vapor in the air compared to the amount when condensation would begin at a given temperature is called the air's **relative humidity**. The warmer the air, the greater the amount of water vapor it can contain. If the relative humidity is 100%, the air is completely saturated with water vapor. A value of 60% on a hot day means that the air is extremely humid. A 60% reading on a cold day, however, tells us that although the air contains relatively large amounts of water vapor, it holds in absolute terms much less than on the hot, muggy day. This example demonstrates the point that relative humidity is meaningful only if we keep the air temperature in mind.

If dew or frost is on the ground in the morning, it means that during the night the temperature dropped to the level at which condensation took place (Figure 4.17). The critical temperature for condensation is the **dew point**. Foggy or cloudy conditions on the earth's surface imply that the dew point has been reached and that the relative humidity is 100%.

Types of Precipitation

When large masses of air rise, precipitation may take place. It can be one of three types: (1) convectional, (2) orographic, or (3) cyclonic or frontal precipitation.

Convectional precipitation results from rising, heated, moisture-laden air. As the air rises, it cools; when its dew point is reached, condensation and precipitation occur, as Figure 4.20 shows. This process is typical of summer storms or showers in tropical and continental climates. Usually the ground is heated during the morning and the early afternoon. The warm air that accumulates begins to rise, and cumulus clouds, then cumulonimbus clouds develop. Finally, there is lightning, thunder, and heavy rainfall, which, since the storm is moving, may affect each part of the ground for only a brief period. It is not unusual for these convectional storms to occur in late afternoon or early evening.

If quickly rising air currents violently circulate the air within the cloud, ice crystals may form near the top of the cloud. These ice crystals may begin to fall when they are large enough, but a new updraft can force them back up, with a subsequent enlargement of the pieces of ice. This process may occur again and again, until the ice can no longer be sustained by the updrafts and pieces of ice fall to the ground as hail.

Orographic precipitation, depicted in Figure 4.21, occurs as warm air is forced to rise because hills or mountains stand in the way of moisture-laden winds. This type of precipitation is typical of areas where mountains and hills are situated close to oceans or large lakes. Saturated air from over the water blows onshore, rising as the land rises. Again, the process of cooling, condensation, and precipitation takes place. The *windward* side (the side toward which the wind blows) of the hills and mountains receives a great deal of precipitation. The other side, called the *leeward*, and the adjoining regions downwind are very often dry. The air that passes over the mountains or hills descends and warms, and as we have seen, descending air does not produce precipitation and warming air absorbs moisture from surfaces over which it passes. A graphic depiction of the great differences in rainfall over very short distances is shown on the map of the state of Washington in Figure 4.22.

Cyclonic or **frontal precipitation**, the third type, is common to the midlatitudes, where cool and warm air masses meet. It also occurs in the tropics, though less frequently, as the originator of hurricanes and typhoons. The most extreme form of cyclonic storm is the tornado, which is more windstorm than rainstorm. In order to understand cyclonic or frontal precipitation, one must first visualize the nature of air masses and the way cyclones develop.

Air masses are large bodies of air with similar temperature and humidity characteristics throughout; they form over a **source region**. Source regions include large

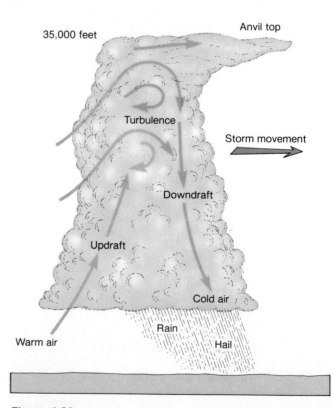

Figure 4.20
When warm air laden with moisture rises, a cumulonimbus cloud may develop and convectional precipitation occur. The turbulence within the system creates a downdraft of cold upper-altitude air.

areas of uniform surface and relatively consistent temperatures, such as the cold land areas of northern Canada, the north central part of the Soviet Union, or the warm tropical water areas in any of the oceans close to the equator. Source regions for North America are shown in Figure 4.23. During a period of a few days or a week, an air mass may form in a source region. For example, in northern Canada, in the fall, when snow has already covered the vast subarctic landscape, cold, heavy, and dry air develops over the frozen land surface. A further discussion of air masses as regional entities may be found in Chapter 13.

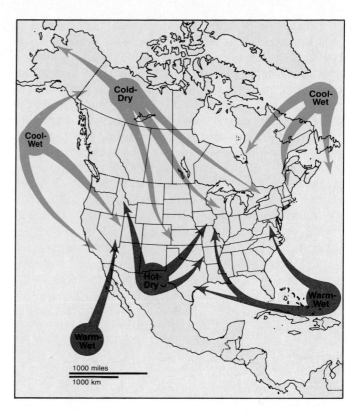

Figure 4.23

Source regions for air masses in North America. The United States and Canada, lying between major contrasting air-mass source regions, are subject to numerous storms and changes of weather.

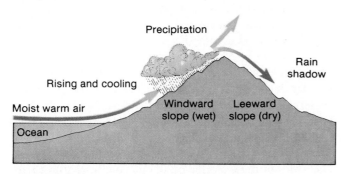

Figure 4.21

Orographic precipitation. Surface winds may be raised to higher elevations by hills or mountains lying in their paths. If such orographically lifted air is sufficiently cooled, precipitation occurs. Descending air on the leeward side of the upland barrier becomes warmer, its capacity to retain moisture is increased, and water absorption rather than release takes place.

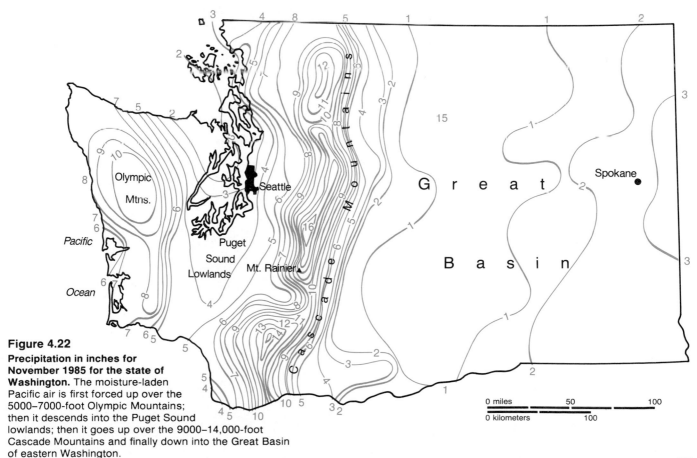

Figure 4.22

Precipitation in inches for November 1985 for the state of Washington. The moisture-laden Pacific air is first forced up over the 5000–7000-foot Olympic Mountains; then it descends into the Puget Sound lowlands; then it goes up over the 9000–14,000-foot Cascade Mountains and finally down into the Great Basin of eastern Washington.

When it is large, this continental polar air mass begins to move toward the lighter, warmer air to the south. The leading edge of the tongue of air is called a **front**. The front in this case separates the cold, dry air from whatever air is found in its path. If a warm, moist air mass is in the path of the polar air mass, the heavier cold air hugs the ground and forces the lighter air up above it. The rising moist air condenses, and frontal precipitation occurs. On the other hand, if it is warm air that is on the move, it slides over the cold air pushing the cold air back, and again, precipitation occurs. In the first case, when cold air moves toward warm air, cumulonimbus clouds form and precipitation is brief and heavy. As the front passes, temperatures drop appreciably, the sky clears, and the air becomes noticeably drier. In the second case, when warm air is moving over cold air, steel-gray nimbostratus (*nimbo* means "rain") clouds form, and the precipitation is steady and long lasting. As the front passes, warm, muggy air becomes characteristic of the area. Figure 4.24 summarizes the movement of fronts.

Storms

When two air masses come into contact, there is the possibility that storms will develop. If the contrasts in temperature and humidity are sufficiently great, or if the wind directions of the two masses that touch are opposite, a wave might develop in the front as shown in Figure 4.25. Once established, waves may enlarge and become wedgelike. One wedge is the cooler air moving along the surface, and the other wedge is the warmer air moving up over the cold air. In both wedges, the warm air rises, and a low-pressure center forms. Considerable precipitation is accompanied by counterclockwise winds around the low-pressure area. A large system of air circulation centered on a region of low atmospheric pressure, such as the one described, is called a midlatitude **cyclone**.

A cyclone may be a weak storm or one of great intensity. The tropical storm is likely to begin over warm tropical waters, far enough from the equator for the Coriolis effect to be significant. Here a wave that is not associated with a front may form. If conditions are such that a wedge develops, an intense tropical storm may grow, fed by the energy embodied in the rising moist, warm air. This storm is called a **hurricane** in the Atlantic region and a *typhoon* in the Pacific area. Figure 4.26 shows the paths of hurricanes in the Atlantic and their wind movements. The winds of these storms move in a counterclockwise direction, converging near the center, and rising in several

Figure 4.24
In this diagram, the cold front has recently passed over city A and is heading in the direction of city B. The warm front is moving away from city B.

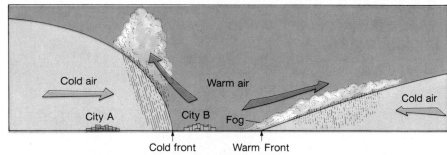

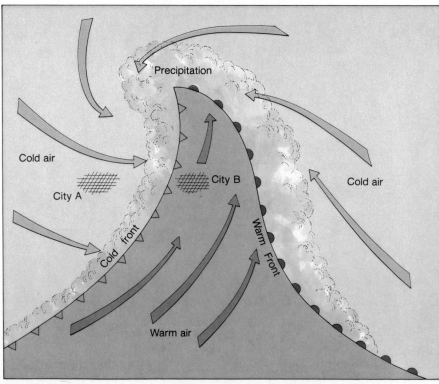

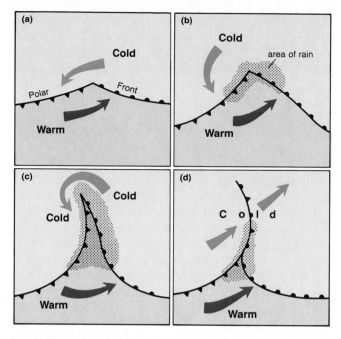

Figure 4.25

When wedges of warm and cold air develop along a low-pressure trough in the midlatitudes, there is the possibility of cyclonic storm formation. A wave begins to form along the polar front (a). In the Northern Hemisphere, cold air begins to turn in a southerly direction, while warm air moves north (b). The meeting lines of these unlike air masses are called *fronts*. Cold air, generally moving faster than warm air, begins to overtake the warm air, forcing it to rise, and, in the process, the storm deepens (c). Eventually, two sections of cold air join; the warm air forms a pocket overhead (d), removing it from its energy and moisture source. The cyclonic storm dissipates as the polar front is reestablished.

concentric belts. At the center, called the *eye*, the air descends, and, as a result, gentle breezes and relatively clear skies are the rule. Over land, these storms lose their warm-water energy source and subside quickly. If they move far into the colder northern waters, they are pushed or blocked by other air masses and also lose their energy source and abate.

The most violent storm of all is the **tornado**, but it is also the smallest (Figure 4.27). A typical tornado is less than 100 feet in diameter. Tornadoes are spawned in the huge cumulonimbus clouds that sometimes travel in advance of a cold front along a squall line. The central part of the United States in spring or fall, when adjacent air masses contrast the most, is the scene of many of these funnel-shaped killer clouds. Although winds are estimated to reach 500 miles per hour, these storms are small and usually travel on the ground for less than a mile, so that they affect only limited areas.

Climate

We have traced some of the causes of weather changes that occur as air from high-pressure zones flows toward low-pressure areas, fronts pass and waves develop, dew points are reached, and sea breezes arise. As we have noted, in some parts of the world these changes occur more rapidly and more often than in other parts.

Weather conditions from day to day can be explained by the principles we have developed. However, one

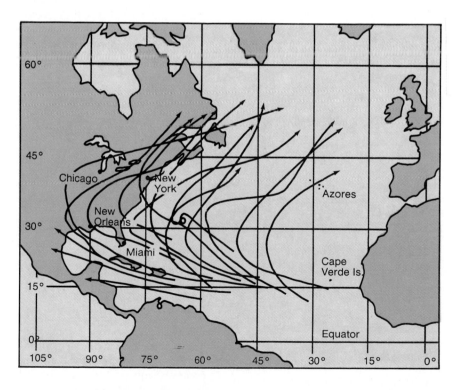

Figure 4.26

Tracks of typical hurricanes. In the United States, the most vulnerable areas are the Gulf coast of Florida; Cape Hatteras, North Carolina; Long Island, New York; and Cape Cod, Massachusetts.

cannot understand the effect of the weather elements—temperature, precipitation, air pressure, and winds—unless one is conscious of the earth's surface features. Weather forecasters in each location on earth must deal with the elements of weather in the context of the local environment.

The complexities of daily weather conditions may be summarized by statements about climate. The climate of an area is a generalization based on daily and seasonal weather conditions. Are the summers warm on the average? Is there a tendency toward heavy snow in the winter? Are the winds normally from the southeast? Are the climatic averages typical of daily weather conditions, or are the variations from day to day or week to week so great that one should speak of average variations rather than just averages? These are the questions we must ask in order to form an intelligent description of the differences in conditions from place to place.

The two most important elements that differentiate weather conditions are temperature and precipitation.

While air pressure is also an important element, differences in air pressure are hardly noticeable without the use of a barometer. Thus we may regard warm, moderate, cold, and very cold temperatures as characteristic of a place or a region. In addition, high, moderate, and low precipitation are good indicators of the degree of humidity or aridity in a place or region. We shall take these two scales, define the terms more precisely, and map the areas of the world having various combinations of temperatures and precipitation.

Because extreme seasonal changes do occur, two global climatic maps are shown in Figure 4.28, one for winters (a) and one for summers (b). Maps could have been developed for each of the four seasons, or for the 12 months, but these two give a rather good, though brief, description of climatic differences. Bear in mind that a summer map of the world is a combination of the climates of the Northern Hemisphere in July, August, and September and of the Southern Hemisphere in January, February, and March because the seasons in the two hemispheres are reversed.

Soil Types

In the previous chapter it was mentioned that the gradational processes of mechanical and chemical weathering create soils. Recall, too, that soil type is much more a function of the climate of the region in which it occurs than it is of the kind of bedrock below it. The higher the temperatures and more abundant the moisture, the more accelerated are the weathering processes. The deepest soils are found in the warmest, wettest environments, while the soil is only thinly developed in arctic regions. Since soil is made up not only of slowly disintegrating rock but also of more rapidly decaying overlying growth, vegetation influences the type of soil that evolves more than does bedrock. The water that drains through soil carries with it chemicals in solution and tiny particles. A great deal of rainfall in a warm climate has a devastating effect on the supply of plant nutrients. In such circumstances, even though soil is deep, it loses its nutrient richness to the ever-present groundwater.

Although broad classes of soil types extend over large areas of the earth's surface, the individual characteristics of soil in a given place may be markedly different over very short distances. Variation is due to the following soil-forming factors:

- the chemical composition of the rock that provides the basic material (the *geologic* factor)

Soil Classification

Soil Type	Brief Description
Inceptisols	Poorly developed; form in cold climates and some river valleys.
Spodosols	Acidic; form beneath coniferous forests; lack humus.
Alfisols	Gray-brown; rich in plant nutrients derived from fallen leaves of deciduous trees.
Ultisols	Reddish; develop in warm, wet/dry regions beneath forest vegetation.
Oxisols	Red, yellow, and yellowish-brown just below a darkened surface layer indicative of extreme chemical weathering.
Mollisols	Dark brown to black; form beneath grasses; extremely rich in nutrients.
Vertisols	High clay content; form under grasses in climates with pronounced wet and dry periods.
Entisols	Poorly developed, thin, sandy soils, often found in mountain environments.
Aridisols	Light color; sandy, desert soils.
Histosols	Soils form in the arctic as peat; consist of plant remains that accumulate in water.

- the temperature and moisture conditions (the *climatic* factor)
- the slope of the land and thus the drainage conditions (the *topographic* factor)
- the organisms in the soil, such as ants, worms, algae, fungi, and bacteria (the *living organisms* factor)
- the overlying plants and plant roots that provide the link between soil nutrients and plants (the *vegetation* factor)

- the time necessary to bring these factors together to form soils (the *chronologic* factor)
- in addition, humans and animals have done much to alter soils.

Soils vary in *humus content* (decomposed organic matter); *color* (dark colors usually indicate high humus content; light colors indicate highly leached soils in wet areas, and alkaline soils in dry areas); *texture* (sand is the

Figure 4.27

Tornado. In the United States, these violent storms occur most frequently in the central and south central part of the country (especially in Oklahoma, Kansas, and the Texas Panhandle), where cold polar air very often meets warm, moist Gulf air.

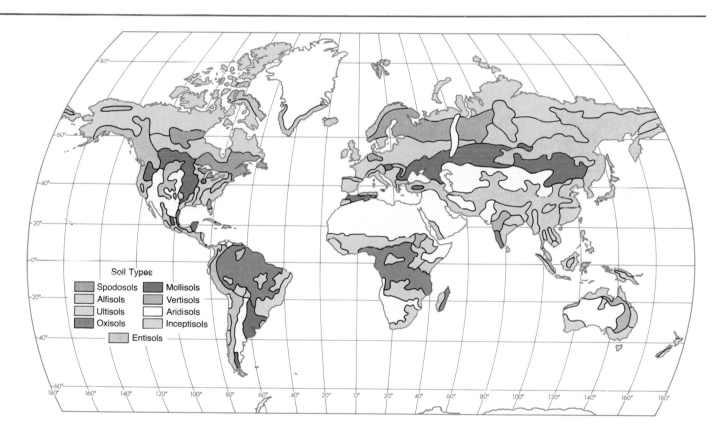

Soil Types

- Spodosols
- Alfisols
- Ultisols
- Oxisols
- Entisols
- Mollisols
- Vertisols
- Aridisols
- Inceptisols

largest soil particle type, followed by silt, and then clay—the most agriculturally productive soil is an equal combination of the three, called *loam*); *structure* (the capacity to hold water and air, and the ability to allow water to pass through); and *acidity/alkalinity* (soils balanced between being very acidic and very alkaline are most productive agriculturally).

In Chapter 3, we described the three horizons of soil. Clearly, it would take a very detailed system to identify each of the various combinations of soil horizon thicknesses, soil factors, and soil characteristics. The table and map present the most general classification system. One must keep in mind that each of the names given represents a wide range of soil characteristics. These names are used later in the chapter to describe the typical, general types of soils found in the various climates of the world.

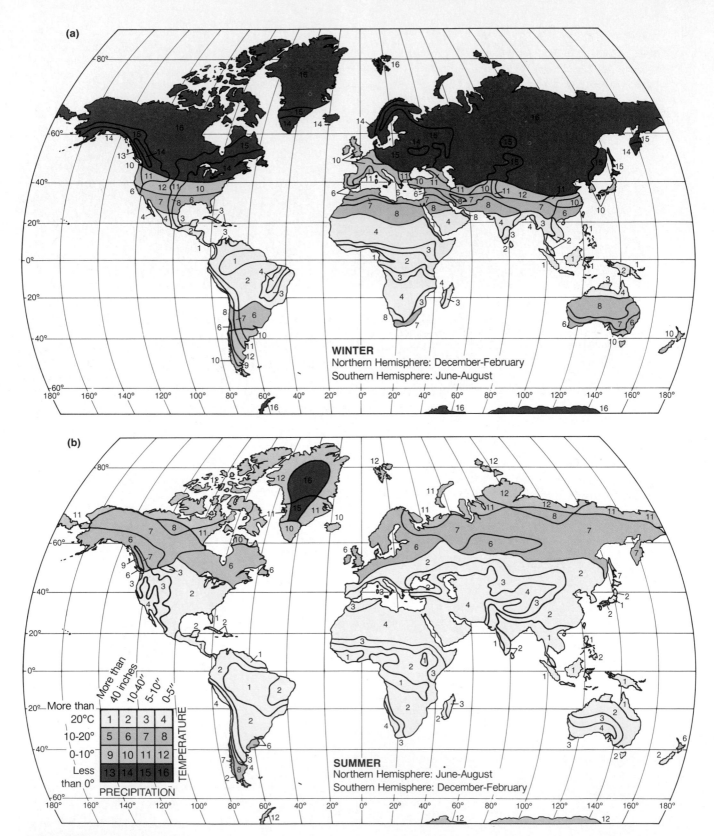

Figure 4.28

These maps combine temperature and precipitation data to display seasonal variations in the basic components of climate. In reading the maps, remember that the winter climate shown in (a) represents the winter season for both the Northern and Southern Hemispheres. This means that December, January, and February data were used for the north latitudes and June,

July, and August data were used for the south latitudinal part of the world. The result is a winter view of the world, one of greatly varying temperatures and small amounts of precipitation. The summer view (b) shows a world of nearly universally warm temperatures and large amounts of precipitation.

Natural Vegetation

Each climate is typified by a particular mixture of **natural vegetation**, that is, the plant life that would exist in each area if humans did not directly interfere with the growth process. Natural vegetation, little of which remains today in areas of human settlement, has close interrelationships with soils, landforms, groundwater, elevation, and other features of habitat including animals. The tie between *biomes*—the major structural subdivisions of assemblages of plants and animals—and climate is particularly close. Indeed, the earliest maps of the world climatic zones were based not upon statistical variation in recorded temperatures, pressures, precipitation, and other measures of the atmosphere but upon observed variation in vegetative regions. The discussion of climates, therefore, benefits from a visualization of regional differences by reference to the type of mature plant communities that, in nature, developed within those climates.

The accompanying map shows the general pattern of the earth's natural vegetation regions. In the hotter parts of the world, where rainfall is heavy and scattered throughout the year, the vegetation type is *tropical rain forest*. *Forests*, in general, are made up of trees growing close enough together that there is a continuous and overlapping leaf canopy. In the tropics, the forest consists of hundreds of tree species in any small area. Because the canopy blocks the sun's rays, there is only a sparse undergrowth. When the tropical rainfall is seasonal, *savanna* vegetation occurs. This is a low grassland with occasional patches of forests or individual trees.

Mediterranean or *chaparral* vegetation is found in the hot summer/mild, damp winter midlatitudes characterizing California, Australia, and the Mediterranean Sea regions. This type of vegetation consists mainly of shrubs and trees of limited size, such as the live oak. Together they form a low, dense vegetation that is green in the wet season and brown in the dry season. Most dryland areas support some vegetation. The *semidesert* and *desert* vegetation is made up of dwarf trees and shrubs, and various types of cactus, though in gravelly and sandy areas, there are virtually no plants.

In areas of modest year-round rainfall in temperate parts of the world, such as central North America, southern South America, and south-central Asia, the type of vegetation most often found is *prairie* or *steppe*. These are extensive grasslands usu-

ally growing on high humus soils. When the rainfall is higher in temperate areas, the natural vegetation turns to *deciduous woodlands*. These are types of trees, such as oak, elm, and sycamore, that lose their leaves during the cold season.

Beyond the temperate zones, in the northerly regions that have mild summers and very cold winters, one finds the *coniferous forests*. Usually only a few species predominate, such as pine and spruce trees. Still farther north the forests yield to *tundra* vegetation, a complex mix of very low-growing shrubs, mosses, and grasses.

At various elevations in mountain areas it is possible to see many of these vegetation types. In the tropics, for example in Peru, moving from east to west, at first one is in the Amazon River's tropical rain forests, which are then followed by deciduous forests of the low Andes Mountains. As one enters the valleys in the high Andes, coniferous forests appear, followed by the mosses and rocks above the treeline. Descending onto the west slopes, grasslands give way to chaparral areas on the low slopes and, finally, semidesert and desert vegetation appears at the base of the mountains close to the Pacific Ocean.

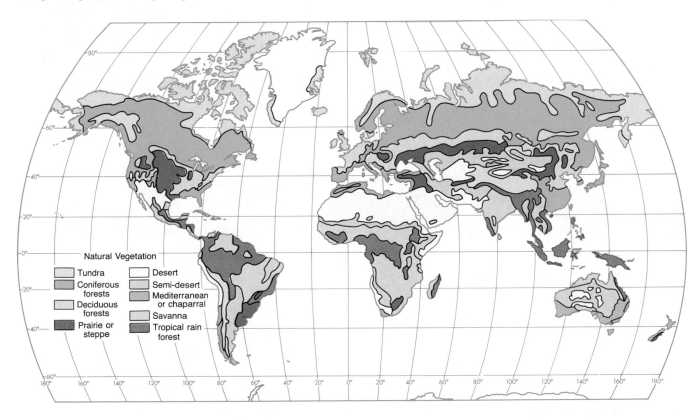

Natural Vegetation
- Tundra
- Coniferous forests
- Deciduous forests
- Prairie or steppe
- Desert
- Semi-desert
- Mediterranean or chaparral
- Savanna
- Tropical rain forest

Figure 4.29 depicts the various climates of the world. It is based on the type of information presented in Figure 4.28. Note that in rugged mountain areas we use the symbol *M* to indicate that climate at any one location may differ appreciably from nearby areas depending on such factors as elevation, latitude, southward- or northward-facing slopes, exposure to moisture-laden winds, and so on.

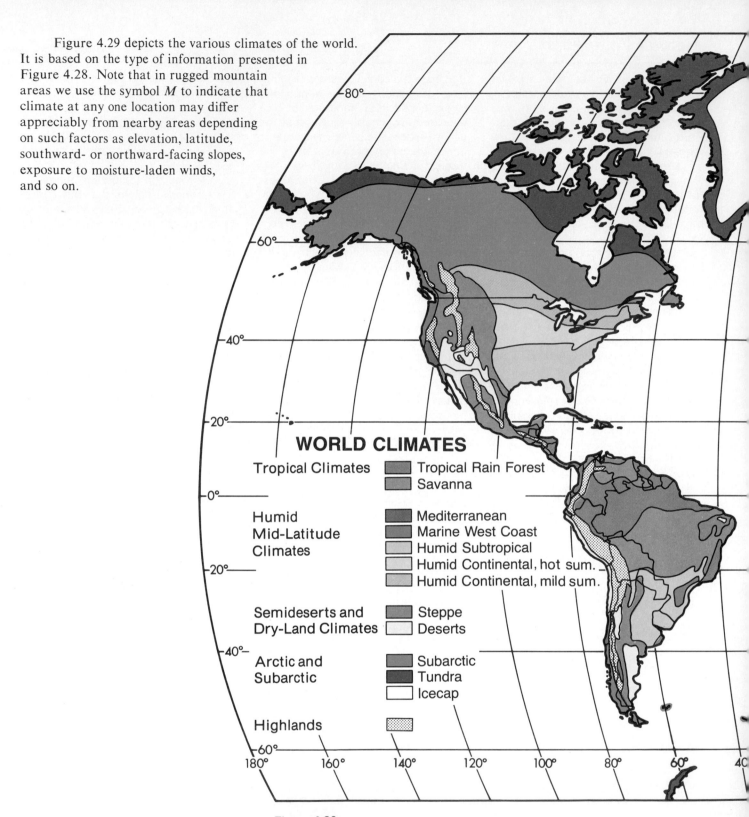

WORLD CLIMATES

Tropical Climates
- Tropical Rain Forest
- Savanna

Humid Mid-Latitude Climates
- Mediterranean
- Marine West Coast
- Humid Subtropical
- Humid Continental, hot sum.
- Humid Continental, mild sum.

Semideserts and Dry-Land Climates
- Steppe
- Deserts

Arctic and Subarctic
- Subarctic
- Tundra
- Icecap

Highlands

Figure 4.29
Climates of the world.

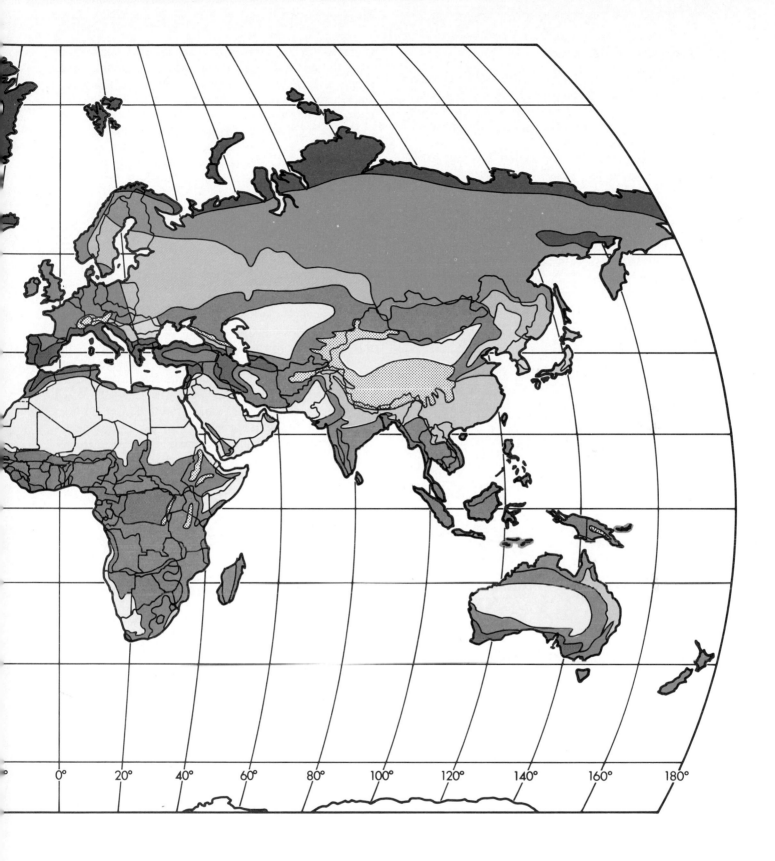

The Tropical Climates

Tropical climates are generally associated with the earth areas lying between the farthest north and farthest south lines of the vertical rays of the sun—the *Tropic of Cancer* and the *Tropic of Capricorn*. The location of tropical climates is given in Figure 4.30.

The numbers following the section headings below refer to the keys on Figure 4.28. The first figure represents typical winter conditions, and the second is for summer. Each represents idealized conditions.

Tropical Rain Forest (1;1)

The areas that straddle the equator are located generally within the equatorial low-pressure zone. These regions are called **tropical rain forest**. They are warm, wet climates in both winter and summer (Figure 4.31). Their rainfall usually comes in the form of daily convectional thunderstorms. Although most days are sunny and hot, by afternoon cumulonimbus clouds form and convectional rain falls.

The natural vegetation is tropical rain forest, which is still present but declining rapidly because of burning in large areas such as the Amazon Basin of South America and the Zaire River Basin of Africa. Tall, dense forests of broadleaf trees and heavy vines predominate. Among the hundreds of species of trees that are found here, there are dark woods, light woods, as well as spongy softwoods, such as balsa, and hardwoods, such as teak and mahogany

(Figure 4.32). Rain forests also extend some distance away from the equator along coasts where prevailing winds supply a constant source of moisture to coastal uplands and where the orographic effect provides enough precipitation for heavy vegetation to develop.

The typical soils of these regions are *oxisols*. As a result of the rapid weathering, most of the soil nutrients necessary for cultivated plants are absent. Only with large inputs of fertilizers can the soils be made to sustain continued agricultural use.

Savanna (3;1)

As the sun's vertical rays become more distant from the equator in summer, the equatorial low-pressure zone follows the sun's path. Thus the areas to the north and south of the rain forest are wet in the summer months, although still hot, but are dry the remainder of the year because the moist equatorial low has been replaced by the dry air of the subtropical highs. These areas are known as **savanna** lands because of the kind of natural vegetation found there. Although the natural vegetation of savanna areas appears to have been a form of scrub forest, the savanna is now recognized as a grassland with widely dispersed trees. The tendency toward a more forested cover has been discouraged through the periodic clearing by burning engaged in by local agriculturalists and hunters. Sometimes, savannas seem to have been purposely designed, for they have a parklike look, as in Figure 4.33a. The east African

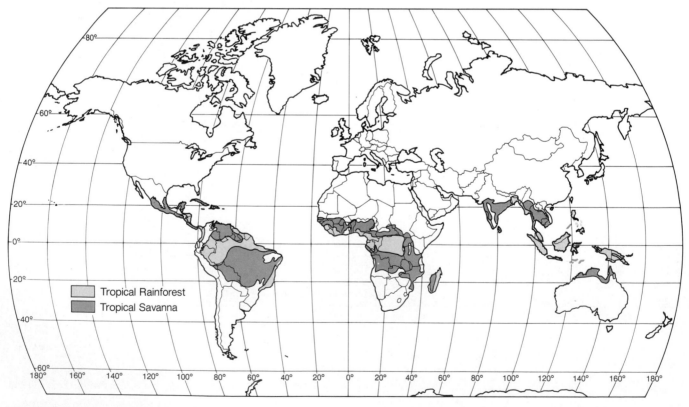

Figure 4.30
The location of tropical climates.

region of Kenya and Tanzania contains well-known grasslands and fire-resisting species of trees where large animals, such as giraffes, lions, and elephants, roam. The *campos* and *llanos* of South America are other huge savanna areas.

The wetter portions of the savanna lands are often underlain by *ultisols*, soils that develop in warm, wet-dry regions beneath forest vegetation (Figure 4.33b). These soils are weak in nutrients for cultivated plants, but respond well when lime and fertilizers are applied. In the drier parts of the savanna, the characteristic soils are *vertisols*. These form under grasses in warm climates. When it rains, the surface becomes plasticlike, and some of the soil slides into the cracks that formed in the dry season. Consequently, vertisols are difficult to till and are used productively as grazing land.

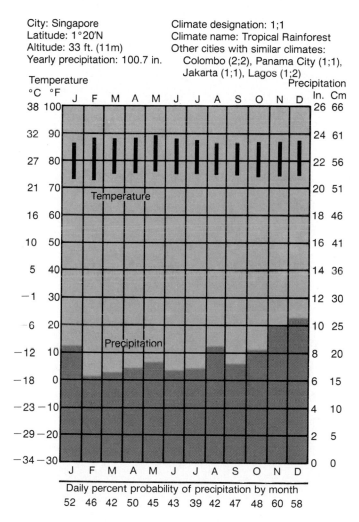

City: Singapore
Latitude: 1°20'N
Altitude: 33 ft. (11m)
Yearly precipitation: 100.7 in.

Climate designation: 1;1
Climate name: Tropical Rainforest
Other cities with similar climates:
    Colombo (2;2), Panama City (1;1),
    Jakarta (1;1), Lagos (1;2)

Daily percent probability of precipitation by month
52 46 42 50 45 43 39 42 47 48 60 58

Figure 4.31
This and succeeding climate charts (*climagraphs*) show average daily high and low temperatures for each month, the average precipitation for each month, and the probability of precipitation on any particular day in a designated month. For Singapore, the average daily high temperature in August is 87° F (30.5° C), the low is 75° F (24° C). The rainfall for the month, on average, is 8.4 inches (21 cm), and on a given day in August, there is a 42% chance of rainfall.

Figure 4.32
Tropical rain forest. The vegetation is characterized by tall, broadleaf, hardwood trees and vines.

(a)

(b)

Figure 4.33
The parklike landscapes of grasses and trees characteristic of
the (a) drier and (b) wetter tropical savanna.

There are usually no marked breaks between one kind of climate and another; instead there are zones of transition. These transitional zones are typical of plains and plateaus (they are not so gradual in mountainous regions). Between the tropical rain forest and the savanna, there are forests that are less dense than the rain forest itself.

A special case needs mentioning. In Asia, when summer monsoon winds carry water-laden air to the mainland, far more rain falls on the hills, mountains, and adjacent plains than in the savanna (notice the pattern of precipitation on Figure 4.34). As a result, the vegetation is much denser, even though the winters are dry. Jungle growth and large forests are the natural vegetation. Much of this vegetation, however, no longer exists because people have been using the land for rice and tea production for many generations.

Hot Deserts (7;4)

Eventually the grasses shorten, and desert shrubs become evident on the poleward side of the savannas. Here we approach the belt of subtropical high pressure that brings considerable sunshine and hot summer weather but very little precipitation. Note the almost total lack of rainfall on Figure 4.35. The precipitation that does fall is of the convectional variety, but it is sporadic. As conditions become drier, there are fewer and fewer drought resistant shrubs, and in some areas only gravelly and sandy deserts exist, as suggested by Figure 4.36. The great hot deserts of the world, such as the Sahara, Arabian, Australian, and Kalahari deserts, are all the products of high-pressure zones. Often the driest parts of these deserts are on the west coast, where cold ocean currents are found, as Figure 4.28 indicates. The soils, *aridisols*, respond well to cultivation when irrigation is made available. Earlier mention was made of the relationship between cold ocean currents and deserts.

The Humid Midlatitude Climates

Figure 4.37 gives the location of a number of climate types that are all humid, that is, not having desert conditions in winter or summer or both, and having winter temperatures well below those of the tropical climates. These climate types would be neatly defined, paralleling the lines of latitude, were it not for mountain ranges, warm or cold ocean currents, and particularly land-water configurations. These factors cause the greatest variations in the middle latitudes.

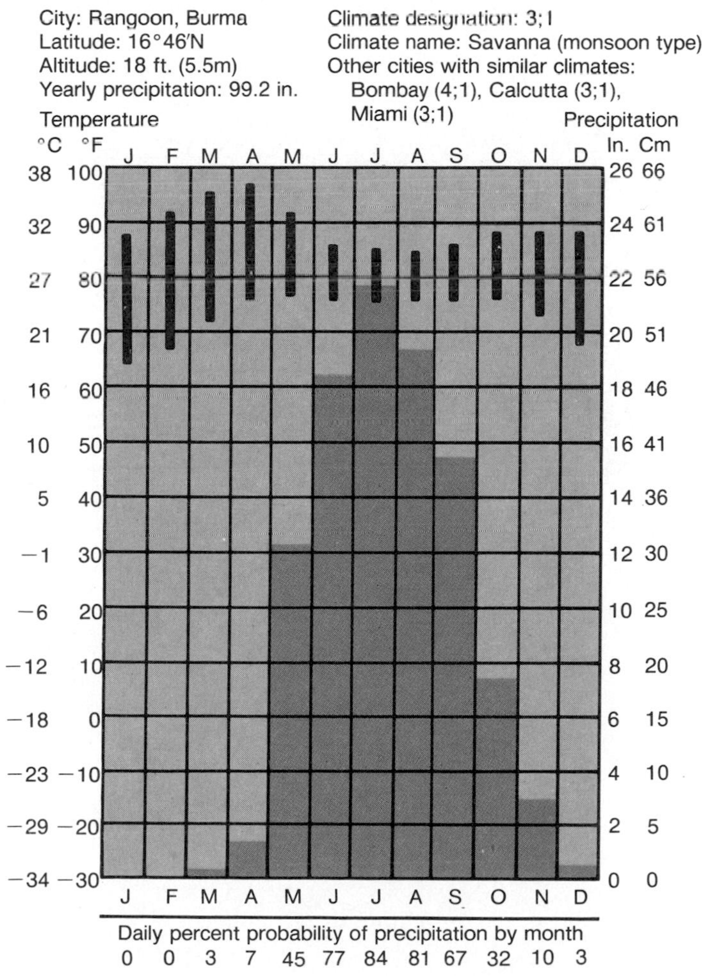

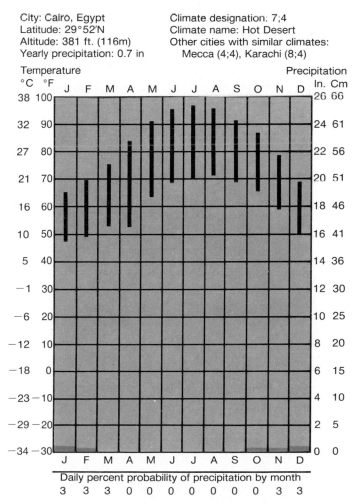

City: Rangoon, Burma
Latitude: 16°46'N
Altitude: 18 ft. (5.5m)
Yearly precipitation: 99.2 in.
Climate designation: 3;I
Climate name: Savanna (monsoon type)
Other cities with similar climates: Bombay (4;1), Calcutta (3;1), Miami (3;1)

Daily percent probability of precipitation by month
0 0 3 7 45 77 84 81 67 32 10 3

Figure 4.34
Climagraph for Rangoon, Burma. (See Figure 4.31 for explanation.)

City: Cairo, Egypt
Latitude: 29°52'N
Altitude: 381 ft. (116m)
Yearly precipitation: 0.7 in
Climate designation: 7;4
Climate name: Hot Desert
Other cities with similar climates: Mecca (4;4), Karachi (8;4)

Daily percent probability of precipitation by month
3 3 3 0 0 0 0 0 0 0 3 3

Figure 4.35
Climagraph for Cairo, Egypt. (See Figure 4.31 for explanation.)

Figure 4.36
Death Valley, California. Devoid of stabilizing vegetation, desert sands are constantly rearranged in complex dune formations.

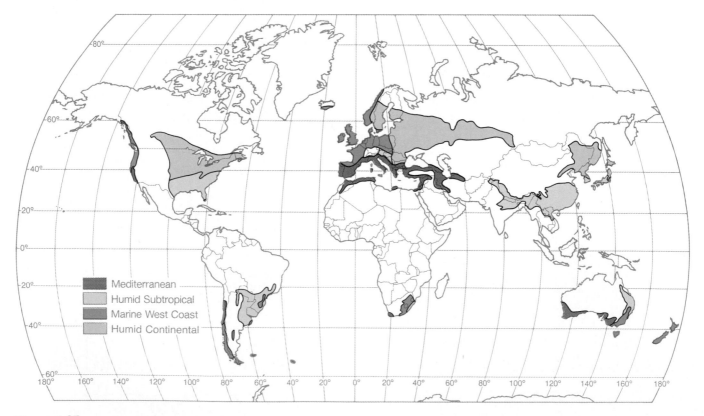

Mediterranean
Humid Subtropical
Marine West Coast
Humid Continental

Figure 4.37
The location of humid midlatitude climates.

Mediterranean Climate (6;3)

Midlatitude winds are generally from the west in both the Northern and the Southern Hemispheres, and a significant amount of the precipitation is generated by frontal systems. Thus it is important to know whether the water is cold or warm near land areas. Several climatic zones are noticeable in the middle latitudes. They are all marked by warm summer temperatures, except in those areas cooled by westerly winds from the ocean. To the poleward side of the hot deserts, there is a transition zone between the subtropical high and the moist westerlies zones. Here, cyclonic storms bring rainfall only in the winter, when the westerlies shift toward the equator. The summers are dry and hot as the subtropical highs shift slightly poleward (Figure 4.38), and the winters are not cold. These conditions describe the **Mediterranean climate**, which is often found on west coasts of continents in middle latitudes. Southern California, the Mediterranean area, western Australia, the tip of South Africa, and central Chile in South America are characterized by such a climate. In these areas, where there is enough precipitation, shrubs and small deciduous trees (trees that lose their leaves once a year), such as scrub oak, grow, as shown in Figure 4.39.

The Mediterranean climate area—long and densely settled in southern Europe, the Near East, and North Africa—has more moisture and a greater variety of vegetation and soil types than is found in the desert. Clear, dry air predominates, and winters are relatively short and mild. Plants and flowers grow all year around. Even though the summers are hot, the nights are usually cool and clear. Much of the vegetation of this area is in the form of crops.

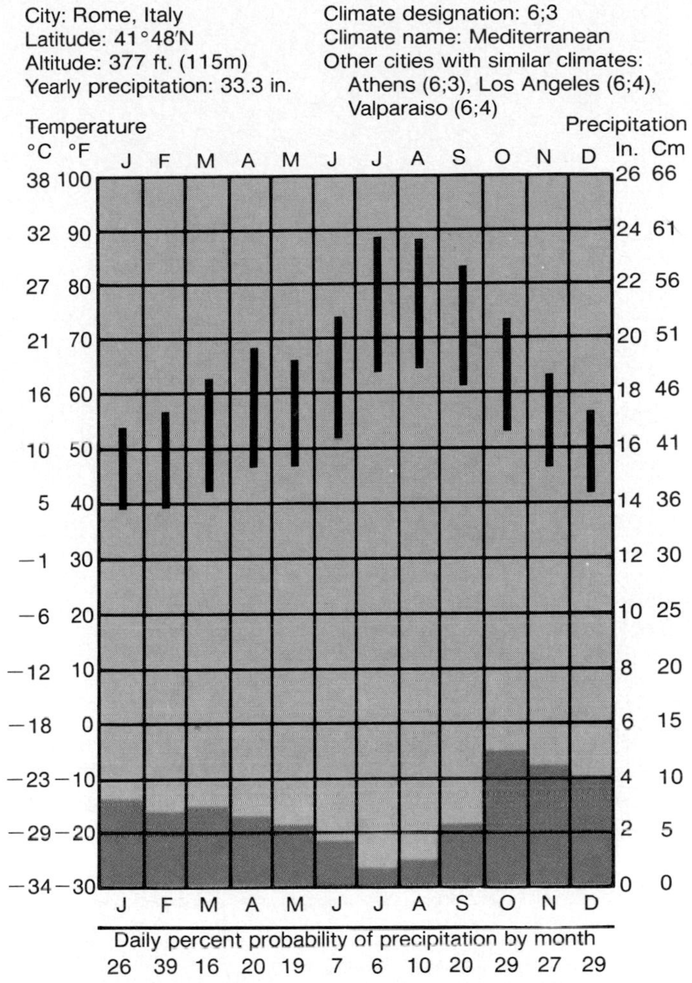

Marine West Coast Climate (10;6)

Closer to the poles, but still within the westerly wind belt, are areas of **marine west coast climate**. Here, cyclonic storms play a relatively large role. In the winter, more rain falls and cooler temperatures prevail than in the Mediterranean zones. Compare the patterns in Figures 4.38 and 4.40. Little rain falls in the summer in the transitional zone just poleward of the Mediterranean climate, but closer to the poles, rainfall increases appreciably in summer and even more in winter. Marine winds from the west moderate both summer and winter temperatures. Thus summers are pleasantly cool, and winters, though cold, do not normally produce freezing temperatures.

This climate affects relatively small land areas in all but one region. Because northern Europe has no great mountain belt to thwart the west-to-east flow of moist air, the marine west coast climate stretches well across the continent to Poland. At that point, cyclonic storms originating in the Arctic regions begin to be felt. Northern Europe's moderate climate, then, owes its existence to a lack of mountains and to a relatively warm ocean current whose influence is felt for about 1000 miles, from Ireland to central Europe.

The orographic effect from the mountains in areas such as the northwestern United States, western Canada,

City: Rome, Italy
Latitude: 41°48'N
Altitude: 377 ft. (115m)
Yearly precipitation: 33.3 in.

Climate designation: 6;3
Climate name: Mediterranean
Other cities with similar climates:
Athens (6;3), Los Angeles (6;4), Valparaiso (6;4)

Daily percent probability of precipitation by month
26 39 16 20 19 7 6 10 20 29 27 29

Figure 4.38
Climagraph for Rome, Italy. (See Figure 4.31 for explanation.)

and southern Chile produces enormous amounts of precipitation, very often in the form of snow on the windward side. Vast coniferous forests (needle-leaf trees, such as pine, spruce, and fir) cover the mountains' lower elevations. Because the mountains prevent moist air from continuing to the leeward side, midlatitude deserts are found to the east of these marine west coast areas.

The main soils of marine west coast areas are *spodosols*. They are strongly acidic and low in plant nutrients. This is a result of the acidic humus that develops from the fallen needles of the coniferous trees. Fertilizers must be used for farming in order to neutralize the acidity.

Humid Subtropical Climate (6;2)

On the east coasts of continents, the transition is from the equatorial climate to the **humid subtropical climate**. Here, convectional summer showers and winter cyclonic storms are the sources of precipitation. As illustrated by Figure 4.41, this climate is one of hot, moist summers and moderate, moist winters. In the fall, on occasion, hurricanes born in tropical waters strike the coastal areas.

Figure 4.39
Vegetation typical of an area with a Mediterranean climate.
Trees such as scrub oak are short and scattered.

City: Vancouver, Canada
Latitude: 49°17'N
Altitude: 45 ft. (14m)
Yearly precipitation: 57.7 in.

Climate designation: 10;6
Climate name: Marine West Coast
Other cities with similar climates:
    Seattle (10;6), London (10;6),
    Paris (10;6),

City: Charleston, S.C.
Latitude: 32°46'N
Altitude: 16 ft. (5m)
Yearly precipitation: 47.5 in.

Climate designation: 6;2
Climate name: Humid Subtropical
Other cities with similar climates:
    Canton (6;2), Sydney (6;2),
    New Orleans (6;2)

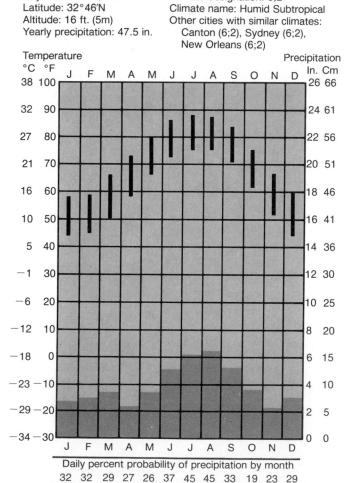

Daily percent probability of precipitation by month
65 61 55 47 39 37 23 26 30 52 63 71

Daily percent probability of precipitation by month
32 32 29 27 26 37 45 45 33 19 23 29

Figure 4.40
Climagraph for Vancouver, Canada. (See Figure 4.31 for explanation.)

Figure 4.41
Climagraph for Charleston, South Carolina. (See Figure 4.31 for explanation.)

The generally even distribution of rainfall allows for forests of deciduous, hardwood trees, such as oak and maple, whose leaves turn orange and red before falling in autumn. In addition, conifers become mixed with deciduous trees as a second-growth forest.

The transition poleward to the continental climates is accompanied by increasingly colder winters and shorter summers. In this direction as well, cyclonic storms become more responsible for rainfall than convectional showers, but at this point, the region can no longer be described as humid subtropical; rather it is described as *humid continental* (see next section). Southern Brazil, the southeastern United States, and southern China all have the humid subtropical climate.

Beneath the deciduous forests of both the humid subtropical and the humid continental climates are *alfisols*. The A-horizon is generally gray-brown in color. The soils, usually rich in plant nutrients, derived from the strongly basic humus created by the fallen leaves of deciduous trees, retain moisture during the hot days of summer, allowing for productive agriculture.

Humid Continental Climate (10;2)

The air masses that drift toward the equator from their origin close to the poles and the air masses that drift toward the poles from the tropics produce frontal precipitation. Whenever warmer air or marine air is in the way of the cold continental air masses, or vice versa, frontal storms develop. We speak of the climates that these air masses influence as **humid continental**. Figures 4.42 and 4.43 show the range and the dominance of winter temperatures within this climatic type. The continental climate may be contrasted to the marine west coast climate, the former having prevailing winds from the land, the latter from the sea. Coniferous forests become more plentiful in the direction of the poles, until temperatures become so low that trees are denied an adequate growing season and are therefore stunted (Figure 4.44). Along with the coniferous forests are the infertile *spodosols*. The transition from the very cold air masses of the winter to the occasional convectional storms of the summer means that there are four distinct seasons.

City: Chicago, Illinois
Latitude: 41°52′N
Altitude: 595 ft. (181m)
Yearly precipitation: 33.3 in.

Climate designation: 14;2
Climate name: Humid Continental (warm
Other cities with similar climates:
New York (10;2), Berlin (11;2),
Warsaw (15;2)

City: Moscow, USSR
Latitude: 55°46′N
Altitude: 505 ft. (154m)
Yearly precipitation: 21.8 in.

Climate designation: 15;6
Climate name: Humid Continental (cool summer)
Other cities with similar climates:
Montreal (14;6), Winnipeg (15;6),
Leningrad (15;6)

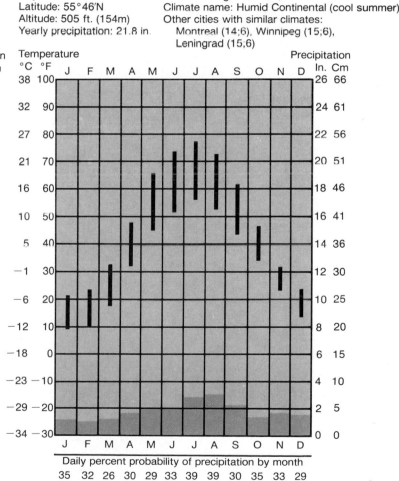

Figure 4.42
Climagraph for Chicago, Illinois. (See Figure 4.31 for explanation.)

Figure 4.43
Climagraph for Moscow, USSR. (See Figure 4.31 for explanation.)

Three huge areas of the world are characterized by a humid continental climate: (1) the north and central United States and southern Canada, (2) most of the European portion of the Soviet Union, and (3) northern China. Because there are no land areas at a comparable latitude in the Southern Hemisphere, this climate is not represented there. In fact, the only nonmountain cold climate in the Southern Hemisphere is the polar climate of Antarctica.

Midlatitude Semideserts and Dry-Land Climates (10;4)

The locations of these climates are shown in Figure 4.45. In the interior of the continents, behind mountains that block the west winds, or in lands far from the reaches of moist tropical air, there are extensive regions of *semi-desert* conditions. Figure 4.46 illustrates typical temperature and precipitation patterns in these areas.

Occasionally a summer convectional storm or a frontal system with some moisture still available occurs. These moderately dry lands are called **steppes**. The natural vegetation is grass, although desert shrubs, pictured in Figure 4.47, are found in the drier portions of the steppes. Although rain is not plentiful, the soils are rich because the grasses return nutrients to the soil. The soils, *mollisols*, have a dark brown to black A-horizon and are among the most naturally fertile soils in the world. As a

Figure 4.44
In the extensive region of east central Canada and the area around Moscow, USSR, the summers are long and warm enough to support a dense coniferous forest. Farther north, growth is less luxuriant.

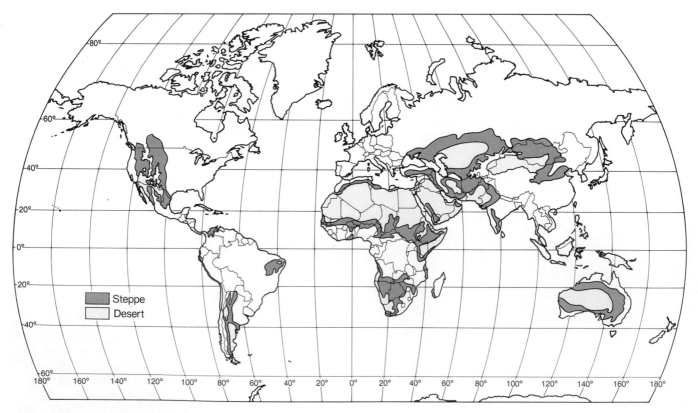

Figure 4.45
The location of steppe and desert climates.

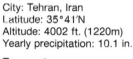

City: Tehran, Iran
Latitude: 35°41′N
Altitude: 4002 ft. (1220m)
Yearly precipitation: 10.1 in.

Climate designation: 10;4
Climate name: Midlatitude Dryland
Other cities with similar climates:
    Salt Lake City (12;4), Ankara (10;4)

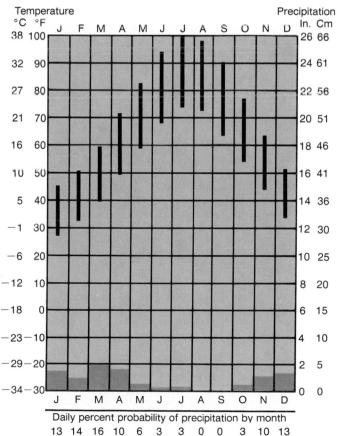

Daily percent probability of precipitation by month
13 14 16 10 6 3 3 0 0 3 10 13

Figure 4.46
Climagraph for Tehran, Iran. (See Figure 4.31 for explanation.)

Figure 4.47
Desert shrubs in the midlatitude drylands of northern Mexico.

result, humans have made the steppes of the United States, Canada, Soviet Union, and China among the most productive agricultural regions of the world. The steppes are also known for their hot, dry summers and for biting winter winds that sometimes bring blizzards.

Subarctic and Arctic Climates (16;7)

Toward the north and into the interior of the North American and Eurasian landmasses, colder and colder temperatures prevail (Figures 4.48 and 4.49). Trees become stunted, and eventually only mosses and other cool-weather plants of the type shown in Figure 4.50 will grow. Because long cold winters predominate, the ground is frozen most of the year. Very often the word **tundra** is used to describe

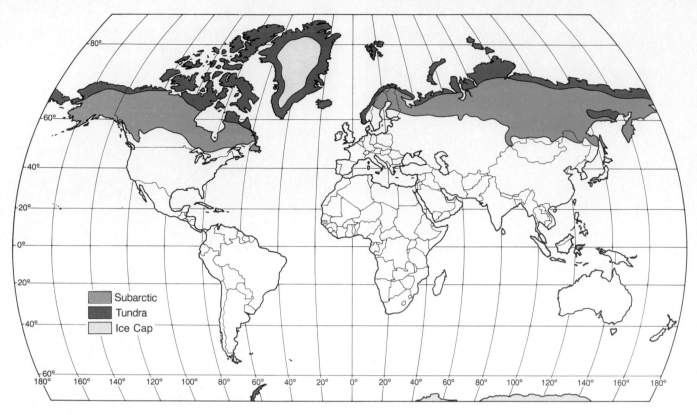

Figure 4.48
The location of arctic and subarctic climates.

the northern boundary zone beyond the treed subarctic regions. A few cool summer months, with an abundant supply of mosquitoes, break up the monotony of extremely cold but not very snowy conditions. Strong easterly winds blow snow, which, combined with ice fogs and little winter sunlight, contributes to a very bleak climate indeed. Alaska, northern Canada, and the northern USSR are covered with the stunted trees of the subarctic climate or by the bleak, treeless expanse of the tundra. Antarctica and Greenland, however, are icy deserts.

Soils in these vast arctic regions are varied, but perhaps most characteristic are *histosols*. These soils tend to be peat or muck, consisting of plant remains that accumulate in water. In forested areas they will tend more toward spodosols. They form in areas of poor drainage both in the arctic and the midlatitudes.

These thumbnail sketches of climatic conditions throughout the world give us the basic patterns of the larger regions. On any given day, conditions may be quite different from those discussed or mapped, but it is the physical climatological processes in general that concern us. We can deepen our knowledge of climates by applying our understanding of the elements of weather.

City: Fairbanks, Alaska
Latitude: 64°51′N
Altitude: 440 ft. (134m)
Yearly precipitation: 12.4 in.

Climate designation: 16;6
Climate name: Subarctic
Other cities with similar climates:
 Yellowknife (15;6), Yakutsk (16;7)

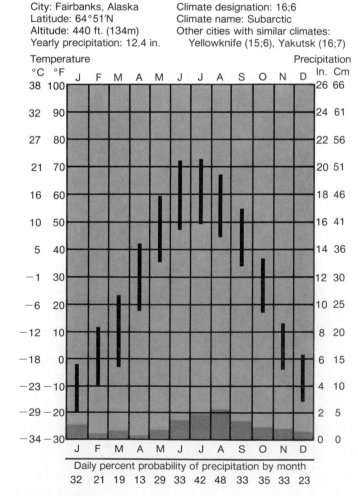

Daily percent probability of precipitation by month
32 21 19 13 29 33 42 48 33 35 33 23

Figure 4.49
Climagraph for Fairbanks, Alaska. (See Figure 4.31 for explanation.)

Figure 4.50
Tundra vegetation in the Ruby Range of the Northwest Territories, Canada.

Climatic Change

It has been stressed that climates are only averages of perhaps greatly varying day-to-day conditions. Figure 4.51 gives an idea of the global variation in precipitation from year to year. Temperatures are less changeable than precipitation on a year-to-year basis, but they too vary. How can we account for these variations? Scientists in research stations all over the world are investigating this question. The data they use range from daily temperature and precipitation records to calculations of the position of the earth relative to the sun. Because day-to-day records for most places go back only 50–100 years, scientists look for additional information in rock formations, the chemical composition of earth materials, and astronomical changes.

The fact that a glacier covered most of Canada and one-third of the United States about 20,000 years ago, as well as at three other times in the last 1 million years, has raised the question of periodic changes in climate. By analyzing layers of fossil microorganisms on the ocean floor, scientists have determined that for the last 400,000 years, there have been climatic cycles of 100,000 years' duration. Some scientists believe that this 100,000-year cycle corresponds to the cyclical changes in the earth's orbit

around the sun, and that such changes are responsible for the succession of the ice ages. When the orbit is nearly circular, the earth experiences relatively cold temperatures, and when it is elliptical, as it is now, the earth is exposed to more total solar radiation and thus experiences warmer temperatures.

Another cycle of 42,000 years corresponds to the tilt of the earth relative to the orbital plane. The tilt varies from 22.1° to 24.5°. A low tilt position—that is, a more perpendicular position of the earth—is accompanied by periods of colder climate. Cooler summers are thought to be critical in the formation of ice sheets.

Even though the last ice age ended about 11,000 years ago, and the earth is now in one of its warmest periods, scientific evidence suggests that there will be substantially more ice on earth 3000 years from now. Climate can change more abruptly than this time scale suggests, however. Relatively small changes in upper-air wind movements can change a climate significantly in decades. Today, polar air is more dominant than tropical air. Changes in the amounts of precipitation or in the reliability of precipitation, as well as temperature changes, can have an enormous impact on agriculture and patterns of

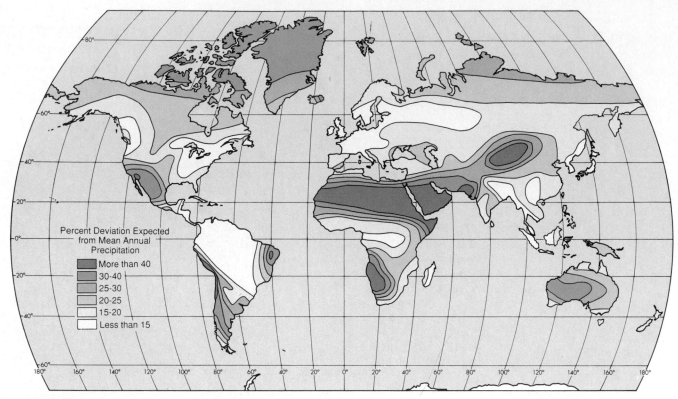

Figure 4.51
The world pattern of precipitation variability. Note that regions of low total precipitation tend to have high variability. In general, the lower the amount of long-term average annual precipitation, the lower is the probability that the average will be recorded in any single year.

Our Inconstant Climates

In recent years archaeologists and historians have been presenting evidence that ancient seats of power—Sumeria in the Middle East, Mycenae in southern Greece, or Mali in Africa—may have fallen not to barbarians but to unfavorable alterations in the climates under which they came to power. Between A.D. 1550 and 1850, a "little ice age" descended on the Northern Hemisphere. Arctic ice expanded, glaciers advanced, and drier areas of the earth were desiccated. Crop failures were common; Iceland lost one-quarter of its population between 1753 and 1759, and 1816 was the "year without a summer" in New England, when snow fell in June and frost came in July.

A pronounced warming trend began about 1890 and peaked by 1945. During that period, the margin of agriculture was extended northward, the pattern of commercial fishing shifted poleward, and the reliability of crop yields increased. The equable climate, plus medical advances, presumably did much to encourage accelerating growth in the world's population.

Now, ample evidence tells us climate is again changing. From the 1940s to the 1970s, the mean temperature of the globe declined; the growing season in England became two weeks shorter; disastrous droughts occurred in Africa and Asia; unexpected freezes altered crop patterns in Latin America.

The recent global cooling trend altered atmospheric circulation and rainfall. The *circumpolar vortex*, high-altitude winds circling the poles from west to east, moved nearer the equator than in previous decades. Its stability occasionally prevented the expected northward movement of monsoonal seasonal winds in the Northern Hemisphere and widened and relocated desert conditions associated with the subtropical high. In recent years, however, scientists have uncovered substantial evidence that indicates that the cooling trend may be more than offset by the effects of increased carbon dioxide in the atmosphere (see Chapter 5).

human settlement. In India, monsoon winds have fluctuated greatly in intensity recently, and in north Africa, summer rains have periodically failed for a number of years.

Great volcanic eruptions that spread dust particles around the world may also produce climatic change because the dust blocks the sun's rays to some extent. For three years after the volcano Krakatoa near Java erupted in 1883, there was a noticeable decline in temperatures throughout the world.

Finally, some scientists are concerned about the effect of human civilization on climatic change. There is evidence that industrial and automobile emissions are changing the gaseous mixture in the atmosphere. These changes are among the topics discussed in Chapter 5, dealing with environmental concerns.

Summary

In this chapter we introduced various concepts and terms that are used to understand weather and climate. We identified solar energy as the great generator of the main weather elements of temperature, moisture, and atmospheric pressure. Spatial variation in these elements is caused both by the earth's broad physical characteristics, such as greater solar radiation at the equator than at the poles, and by local physical characteristics, such as the effect of water bodies or mountains on local weather conditions.

Climate regions help us to simplify the complexities that arise from such special conditions as monsoon winds in Asia or cold ocean currents off the west coast of South America. In just a few descriptive sentences, it is possible to identify the essence of the wide variations in weather conditions that may exist at a place over the course of a year. When one says that Seattle is in a region of marine west coast climate, for example, images of the average daily conditions communicate to us not only the facts of weather at a place but also the reasons for conditions being as they are. In addition, knowledge of climate tells us about the conditions within which one carries out life's daily tasks.

Weather and climate are among the building blocks that help us to better understand the role of humans on the surface of the earth. Although in the following chapters attention will be devoted mainly to the characteristics of human cultural landscapes, one should keep in mind that the physical landscape significantly affects human behavior.

Key Words

air mass	monsoon
air pressure	mountain breeze
climate	natural vegetation
convection	North Atlantic drift
convectional precipitation	orographic precipitation
Coriolis effect	precipitation
cyclone	pressure gradient
cyclonic or frontal	reflection
precipitation	relative humidity
dew point	reradiation
El Niño	savanna
frictional effect	source region
front	steppe
humid continental climate	temperature inversion
hurricane	tornado
insolation	tropical rain forest
jet stream	troposphere
lapse rate	tundra
marine west coast climate	valley breeze
Mediterranean climate	weather

For Review

1. What is the difference between *weather* and *climate*?

2. What determines the amount of *insolation* received at a given point? Does all solar energy potentially receivable actually reach the earth? If not, why?

3. How is the atmosphere heated? What is the *lapse rate*, and what does it indicate about the atmospheric heat source? Describe a *temperature inversion*.

4. What is the relationship between atmospheric pressure and surface temperatures? What is a *pressure gradient*, and of what concern is it in weather forecasting?

5. In what ways do land and water areas respond differently to equal insolation? How are these responses related to atmospheric temperatures and pressures?

6. Draw and label a diagram of the planetary wind and pressure system. Account for the occupance and character of each wind and pressure belt. Why are the belts latitudinally ordered?

7. What is *relative humidity*? How is it affected by changes in air temperatures? What is the *dew point*?

8. What are the three types of large-scale *precipitation*? How does each occur?

9. What are *air masses*? What is a *front*? Describe the development of a cyclonic storm, showing how it is related to air masses and fronts.

10. What factors were chiefly responsible for today's weather?

11. Summarize the distinguishing temperature, moisture, vegetation, and soil characteristics of each type of climate.

12. What is the climate at Tokyo, London, São Paulo, Leningrad, and Bangkok?

Selected References

Barry, R. G. *Atmosphere, Weather and Climate.* 5th ed. New York: Methuen, 1987.

———. *Mountain Weather and Climate.* New York: Methuen, 1982.

Briggs, D. *Soils.* London: Butterworth, 1974.

Bryson, Reid A., and Thomas J. Murray. *Climates of Hunger: Mankind and the World's Changing Weather.* Madison: University of Wisconsin Press, 1977.

Cole, Franklyn W. *Introduction to Meteorology.* New York: Wiley, 1980.

Griffiths, John F., and Dennis M. Driscoll. *Survey of Climatology.* Columbus, Ohio: Charles E. Merrill, 1982.

Hays, J. D., J. Imbrie, and N. J. Shackleton. "Variations in the Earth's Orbit: Pacemaker of the Ice Ages." *Science 194* (Dec. 10, 1976): 1121–32.

Henderson-Sellers, Ann, and Peter J. Robinson. *Contemporary Climatology.* New York: Wiley, 1986.

Lydolph, Paul E. *Weather and Climate.* Totowa, N.J.: Rowman and Allanheld, 1985.

McKnight, Tom L. *Physical Geography: A Landscape Appreciation.* 3d ed. Englewood Cliffs, N.J.: Prentice-Hall, 1990.

Muller, R. A., and Theodore M. Oberlander. *Physical Geography Today.* 3d ed. New York: Random House, 1984.

Oliver, John E., and John J. Hidore. *Climatology: An Introduction.* Columbus, Ohio: Charles E. Merrill, 1984.

Ruffner, James A., and Frank E. Bair. *The Weather Almanac.* 4th ed. Detroit: Gale Research Co., 1984.

Simmons, I. G. *Biogeographical Processes.* London: Allen & Unwin, 1982.

Simpson, R. H., and H. Riehl. *The Hurricane and Its Impact.* Baton Rouge: Louisiana State University Press, 1981.

Stewart, George R. *Storm.* New York: Random House, 1941.

Strahler, Arthur N., and Alan H. Strahler. *Elements of Physical Geography.* 4th ed. New York: Wiley, 1990.

Trewartha, Glenn T., and Lyle H. Horn. *An Introduction to Climate.* 5th ed. Madison: University of Wisconsin Press, 1981.

Weatherwise. Issued six times a year by Weatherwise, Inc., 230 Nassau St., Princeton, N.J. 08540.

C H A P T E R

5

Human Impact on the Environment

Pulp mill in LaTuque, Quebec.

TIMES BEACH, Missouri, used to be a tight-knit, working-class community along the Meramec River southwest of St. Louis. Now the people are gone. Empty houses and stores, abandoned cars, and rusting refrigerators stand silent. "Thanks for coming," says the sign on the Easy Living Laundromat to nonexistent customers. Wildflowers bloom on overgrown lawns. Flies buzz, squirrels chase each other up trees, and an occasional coyote prowls the streets. Were they allowed, visitors might think they were seeing an uncannily realistic Hollywood stage set. Instead, the ghost town is a symbol of what happens when hazardous wastes contaminate a community.

The trouble began in the summer of 1971, when Times Beach hired a man to spread oil on its 10 miles (16 km) of dirt roads to keep down the dust. Unfortunately, his chief occupation was removing used oil and other waste from a downstate chemical factory, and it was a mixture of these products that he sprayed all over town that summer and the next. The tens of thousands of gallons of purple sludge were contaminated with dioxin, the most toxic substance ever manufactured.

Effects of the dioxin exposure appeared almost immediately in two riding arenas that had been sprayed. Within days, hundreds of birds and small animals died, kittens were stillborn, and many horses became ill and died. Then residents began reporting a variety of medical disorders, including miscarriages, seizure disorders, liver impairment, and kidney cancer. Not until 1982 did the Environmental Protection Agency (EPA), alerted to high levels of dioxin in wastes stored at the chemical plant, make a thorough investigation of Times Beach. They found levels of the dioxin compound TCDD as high as 300 parts per billion in some of the soil samples; 1 part per billion is the maximum concentration deemed safe. The EPA purchased every piece of property in town and ordered its evacuation (Figure 5.1).

The story of Times Beach is one of many that could have been selected to illustrate how people can affect the quality of the water, air, and soil on which their existence depends.

Terrestrial features and ocean basins, elements of weather and characteristics of climate, flora, and fauna comprise the building blocks of that complex mosaic called the *environment*, or the totality of things that in any way may affect an organism. Humans exist within a natural environment—the sum of the physical world—that they have modified by their individual and collective actions. Forests have been cleared, grasslands plowed, dams built, and cities constructed. On the natural environment, then, has been erected a cultural environment, modifying, altering, or destroying the balance of nature that existed before human impact was expressed. This chapter is concerned with the interrelation between humans and the natural environment that they have so greatly altered.

Since the beginnings of agriculture, humans have changed the face of the earth, have distorted delicate balances and interplays of nature, and, in the process, have both enhanced and endangered the societies and the economies that they have erected. The essentials of the natural balance and the ways in which humans have altered it are not only our topics here but are also matters of social concern that rank among the principal domestic and international issues of our times. As we shall see, the fuels we consume, the raw materials we use, the products we create, and the wastes we discard all contribute to the harmful alteration of the **biosphere**, the thin film of air, water, and earth within which we live.

Ecosystems

The biosphere is composed of three interrelated parts: (1) the *troposphere*, some 6–7 miles (9.5–11.25 km) thick; (2) the *hydrosphere*, surface and subsurface waters in oceans, streams, lakes, glaciers, or groundwater —much of it locked in ice or earth and not immediately available for use; and (3) the upper reaches of the earth's *crust*, a few thousand feet at most, containing the soils that support plant life, the minerals that plants and animals require to exist, and the fossil fuels and ores that humans exploit. The biosphere is an intricately interlocked system, containing all that is needed for life, all that is available for living things to use, and, presumably, all that ever will be available. The ingredients of the biosphere must be, and are, constantly recycled and renewed in nature. Plants purify the air, the air helps to purify the water, the water and the minerals are used by plants and animals and are returned for reuse.

The biosphere, therefore, consists of two intertwined components: (1) a nonliving outside (solar) energy source and requisite chemicals and (2) a living world of

Figure 5.1
Times Beach, Missouri.

plants and animals. In turn, the biosphere may be subdivided into specific **ecosystems**, functional units that consist of all the organisms (plants and animals) and physical features (air, water, soil, and chemicals) existing together in a particular area. Figure 5.2 depicts one kind of ecosystem and in crude form suggests the most important principle concerning all ecosystems: everything is interconnected. Any intrusion or interruption in the balance that has been naturally achieved inevitably results in undesirable effects elsewhere in the system. Each organism occupies a specific *niche*, or place, within an ecosystem. In the energy exchange system, each organism plays a definite role; individual organisms survive because of other organisms that also live in that environment. The problem lies not in recognizing the niches but in anticipating the chain of causation and the readjustments of the system consequent on disturbing the occupants of a particular niche.

Food Chains and Renewal Cycles

Life depends on the energy and nutrients flowing through an ecosystem. The transfer of energy and materials from one organism to another is one link in a **food chain**, defined as a sequence of organisms, such as green plants, herbivores, and carnivores, through which energy and materials move within an ecosystem. Some food chains have only two links, as when human beings eat rice; most have only three or four. Because the ecosystem in nature is in a continuous cycle of integrated operation, there is no start or end to a food chain. There are, simply, nutritional transfer stages in which each lower level in the food chain transfers part of its contained energy to the next-higher-level consumer.

The *decomposers* pictured in Figure 5.2 are essential in maintaining food chains and the cycle of life. They cause the disintegration of organic matter—animal carcasses and droppings, dead vegetation, wastepaper, and so on. In the process of decomposition, the chemical nature of the material is changed, and the nutrients contained within it become available for reuse by plants or animals. *Nutrients*, the minerals and other elements that organisms need for growth, are never destroyed; they keep moving from living to nonliving things and back again. Our bodies contain nutrients that were once part of other organisms, perhaps a hare, a hawk, or an oak tree.

The idea of a cycle is important in furthering our understanding of the natural renewal of the elements essential to life, which include carbon, oxygen, hydrogen, and phosphorus. Because the supply of these materials is fixed, they must be continuously recycled through food chains. Figures 5.3 and 5.4 show two important cycles, the phosphorus cycle and the hydrologic cycle.

Phosphorus is an element essential to the survival of many species of plants and animals. It is an important component of DNA and of bones and teeth. Plants absorb phosphate from the soil, and other organisms acquire it by eating plants. It is returned to the soil through animal excretions or through the decay of plants and animals. When phosphate that has been used as an agricultural fertilizer is carried to rivers and eventually to the sea through runoff, only a small percentage is returned to the land through fish catches and the excretions (guano) of seabirds, which eat fish. Most of it remains in deep ocean waters until geologic processes cause uplifting of the ocean floor.

Ecosystems change constantly whether people are present or not, but humans have affected them more than has any other species. The impact of humans on ecosystems was small at first, with low population size, energy consumption, and technological levels. It has increased so rapidly and pervasively as to present us with widely recognized and varied ecological crises. Some of the effects of humans on the natural environment are the topic of the remainder of this chapter.

Impact on Water

The supply of water is constant. The system by which it continuously circulates through the biosphere is called the **hydrologic cycle** (Figure 5.4). In that cycle, water may change form and composition, but under natural environmental circumstances, it is purified in the recycling process and is again made available with appropriate properties to the ecosystems of the earth. *Evaporation* and

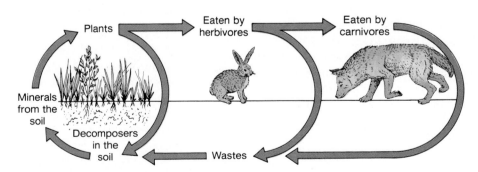

Figure 5.2
A simplified example of an ecosystem illustrating the interdependence of organisms and the physical environment.

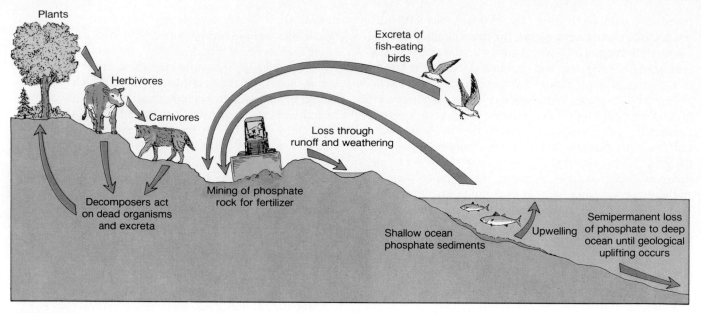

Figure 5.3

The phosphorus cycle. The natural phosphorus cycle is
exceedingly slow and has been upset by human activity.
Currently, phosphorus is being washed into the sea faster than it
is being returned to the land, primarily because it is being mined
for use as a fertilizer.

transpiration (the emission of water vapor from plants)
are the mechanisms by which water is redistributed. Water
vapor collects in clouds, condenses, and then falls again
to earth. There it is reevaporated and retranspired, only
to fall once more as precipitation.

Humans have a discernible impact on the hydrologic cycle by affecting the speed of water recirculation
and by hastening the return of water to the sea. Massive
withdrawals of water from surface supplies by thermal
power plants or by industry for cooling purposes can ab-
breviate the normal cycle by prematurely converting water
from a liquid to its vaporous state. Comparable accelerated vaporization occurs through evaporation from reservoirs, particularly in arid regions. The amount of usable
water available at some point in the cycle is thereby reduced.

Because the need for water will continue to grow as
the world population increases, many scientists fear a
coming water crisis. Agriculture currently accounts for
almost 80% of all the water used by people, most of it for
irrigation. The Food and Agriculture Organization (FAO)
of the United Nations has estimated that the amount of
water used for irrigation will have to double to feed all
those who will be alive in the year 2000. Regional water
shortages will affect developed as well as less developed
countries.

For example, agriculture in large parts of Kansas,
Nebraska, Colorado, and Texas draws on the water stored
in a vast underground formation called the Ogallala
Aquifer (Figure 5.5). (An *aquifer* is water-bearing, porous

rock lying between impermeable layers.) The water bed
is less than 200 feet (60 m) thick in part of the area; yet,
because large amounts are withdrawn for irrigated agriculture, the water table falls from 2 to 5 feet each year.
Preliminary studies indicate that if present rates of
pumping continue, as much as 40% of the irrigated acreage
in portions of the Great or High Plains states will be lost
by the year 2020.

Urbanization and industrialization also place
growing demands on water supplies. Some 150 tons
(40,000 gallons) of water are used to produce a ton of steel;
250 tons are needed to make a ton of paper. As more and
more countries industrialize, pressure on the water supply
will increase.

People's dependence on water has long led to efforts
to control its supply. Such manipulation has altered the
quantity and the quality of water in rivers and streams.

Modification of Streams

To prevent flooding, to regulate the water supply for agriculture and urban settlements, or to generate power,
people have for thousands of years manipulated rivers by
constructing dams, canals, and reservoirs. Although they
generally have achieved their purposes, these structures
can have unintended environmental consequences. These
include reduction in the sediment load downstream, followed by a reduction in the amount of the nutrients available for crops and fish; an increase in the salinity of the
soil; and subsidence (discussed later in this chapter).

Channelization, another method of modifying river
flow, is the construction of embankments and dikes and

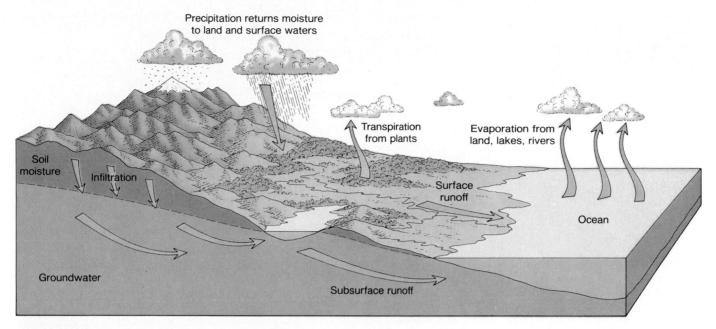

Figure 5.4

The hydrologic cycle. The sun provides energy for the evaporation of fresh and ocean water. The water is held as vapor until the air becomes supersaturated. Atmospheric moisture is returned to the earth's surface as solid or liquid precipitation to complete the cycle. Because precipitation is not uniformly distributed, moisture is not necessarily returned to areas in the same quantity as it has evaporated from them. The continents receive more water than they lose. The excess returns to the seas as surface water or groundwater. A global water balance, however, is always maintained.

the straightening, widening, and/or deepening of channels to control floodwaters or to improve navigation. Many of the great rivers of the world, including the Nile and the Huang He, are lined by embankment systems. Like dams, these systems can have unforeseen consequences. They reduce the natural storage of floodwaters, can aggravate flood peaks downstream, and can cause excessive erosion. Some of the possible ecological effects of channelization are shown in Figure 5.6.

Channelization and dam construction are deliberate attempts to modify river regimes, but other types of human action also affect river flow. Urbanization, for example, has significant hydrologic impacts, including a lowering of the water table, pollution, and increased flood runoff. Likewise, the removal of forest cover increases runoff, promotes flash floods, lowers the water table, and hastens erosion. Nevertheless, the primary adverse human impact on water is felt in the area of water quality. People withdraw water from lakes, rivers, or underground deposits to use for drinking, bathing, agriculture, industry, and many other purposes. Although the water that is withdrawn returns to the water cycle, it is not always returned in the same condition as at the time of withdrawal. Water, like other segments of the ecosystem, is subject to serious problems of pollution.

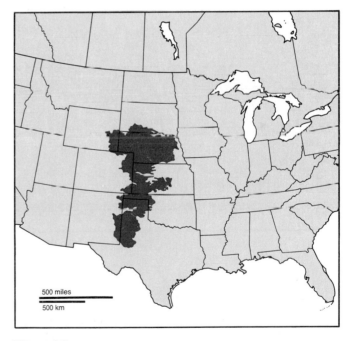

Figure 5.5

The giant Ogallala Aquifer, the country's largest underground water supply. Some 20 million acres (8 million hectares) of land are now watered by pumping from the aquifer, which supports about one-quarter of the country's cotton crop and a great deal of its corn and wheat. About 40% of the country's cattle are fed grain grown in the High Plains region. Approximately 300 gallons of water in the field are needed to grow enough wheat for one loaf of bread, 4200 gallons to produce one pound of beef— figures that help explain why the aquifer is being depleted faster than it can be replenished by nature.

Figure 5.6

Comparison of a natural channel with a channelized stream. Some of the possible effects of channelization are depicted. Dredging and filling (the deposition of dredged materials to create new land) are other ways in which people modify the configuration of water bodies.

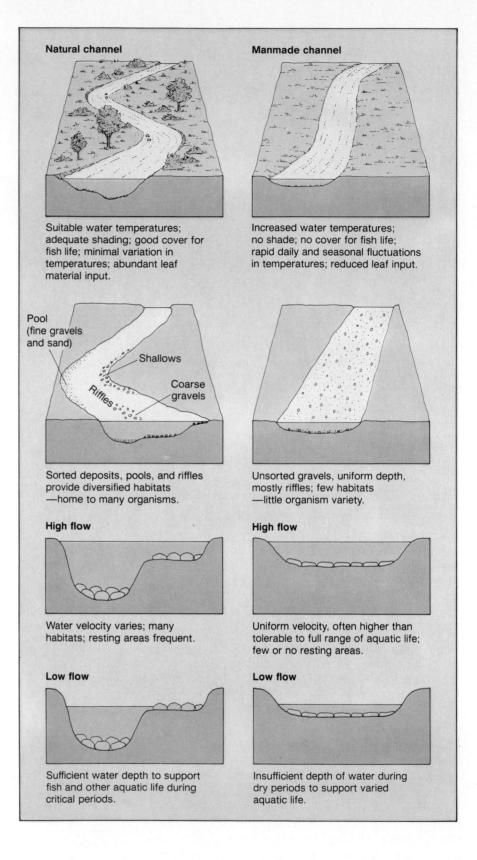

Natural channel

Suitable water temperatures; adequate shading; good cover for fish life; minimal variation in temperatures; abundant leaf material input.

Pool (fine gravels and sand)

Shallows

Riffles

Coarse gravels

Sorted deposits, pools, and riffles provide diversified habitats —home to many organisms.

High flow

Water velocity varies; many habitats; resting areas frequent.

Low flow

Sufficient water depth to support fish and other aquatic life during critical periods.

Manmade channel

Increased water temperatures; no shade; no cover for fish life; rapid daily and seasonal fluctuations in temperatures; reduced leaf input.

Unsorted gravels, uniform depth, mostly riffles; few habitats —little organism variety.

High flow

Uniform velocity, often higher than tolerable to full range of aquatic life; few or no resting areas.

Low flow

Insufficient depth of water during dry periods to support varied aquatic life.

Blueprint for Disaster: Stream Diversion and the Aral Sea

It used to be the fourth largest lake in the world, covering an area of 27,000 square miles, (69,900 km²) larger than the state of West Virginia. Now the Soviet Union's Aral Sea is only the sixth largest lake, and it may cease to exist by the year 2010, having been converted instead into a number of lakes. The level of the sea has dropped 40 feet (12 m) since 1960 and the volume of the water is only one-third of what it used to be.

The shrinkage of the lake is just one consequence of diverting nearly all the water from its primary sources, the Amu Darya and the Syr Darya, to irrigate the agricultural fields of Turkmenia and Uzbekistan. These are some of the others:

• As the shoreline has receded, it has left behind 10,000 square miles (26,000 km²) of salty desert waste. The high concentrations of salt, fertilizers, and pesticides are slowly turning the Aral into a dead sea. Because the water is too salty for most fish species to endure, the commercial fishing industry has already collapsed. Once a fishing port, Muinak is now 30 miles from the sea.

• In addition, wind storms whip up the salty grit from the dried-up seabed and deposit it on cropland hundreds of miles away. Thus the agricultural crops for which the Aral Sea was sacrificed are themselves at risk. This is not a

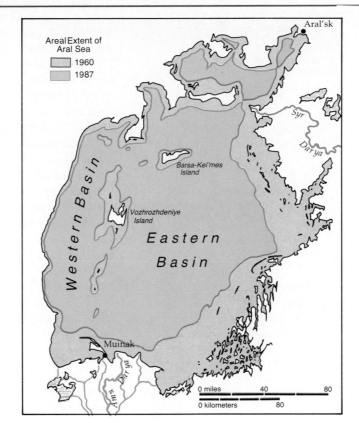

Areal Extent of Aral Sea
☐ 1960
☐ 1987

minor concern: the irrigated land provides virtually all the cotton and at least one-fourth of the fruit, vegetables, and rice produced in the Soviet Union.

• Finally, the rivers that feed the Aral Sea have become sluggish sewers, contaminated by industrial and agricultural wastes and sewage. Al-

though the rivers teem with viruses causing dysentery, typhoid, and other diseases, millions of people depend on them for drinking water. Over 65,000 cases of hepatitis have been reported in the region in the last 15 years, and child mortality rates are among the highest in the world.

The Meaning of Water Pollution

As a general definition, **environmental pollution** by humans means the introduction into the biosphere of wastes that, because of their volume or their composition or both, cannot be readily disposed of by natural recycling processes. In the case of water, the central idea is that pollution exists when water composition has been so modified by the presence of one or more substances that either it cannot be used for a specific purpose or it is less suitable for that use than it was in its natural state. Pollution is brought about by the discharge into water of substances that cause unfavorable changes in its chemical or physical nature or in the quantity and quality of the organisms living in the water. Pollution is a relative term. Water that is not suitable for drinking may be completely satisfactory for cleaning streets. Water that is too polluted for fish may provide an acceptable environment for certain water plants.

Human activity is not the only cause of water pollution. Leaves that fall from trees and decay, animal wastes, oil seepages, and other natural phenomena may affect water quality. There are natural processes, however, to take care of such pollution. Organisms in water are able to degrade, assimilate, and disperse such substances in the amounts in which they naturally occur. Only in rare instances do natural pollutants overwhelm the cleansing abilities of the recipient waters. What is happening now is that the quantities of wastes discharged by humans often exceed the ability of a given body of water to purify itself. In addition, humans are introducing pollutants, such as metals or inorganic substances, that cannot be broken down at all by natural mechanisms or that take a very long time to break down. Table 5.1 shows the kinds of water pollutants associated with different sources.

Table 5.1

Sources and Types of Water Pollutants

Source of waste	Kind of pollutant
Municipalities and residences	Human wastes, detergents, garbage, trash
Urban drainage	Suspended sediment, fertilizers, pesticides, road salt
Agriculture	
Crop production	Fertilizers, herbicides, pesticides, erosion sediment
Irrigation return flow	Mineral salts and erosion sediment
Industry	Wide range of pollutants, including biodegradable wastes in the paper and food-processing industries, heat discharge in a variety of activities, nondegradable wastes in the chemical and iron and steel industries, and radioactivity
Mining	Acids, sediment, metal wastes, culm (coal dust)
Electric power production	Heat and nuclear wastes
Recreation and navigation	Human wastes, garbage, fuel wastes

As long as there are people on earth, there will be pollution. Thus the problem is one not of eliminating pollution but of controlling it. This is particularly true in the technologically advanced countries. The more developed and affluent the country, the more resources it uses and the more it pollutes.

Agricultural Sources of Water Pollution

On a worldwide basis, agriculture probably contributes more to water pollution than does any other single activity. In the United States, agriculture is estimated to be responsible for about two-thirds of stream pollution. Agricultural runoff carries three main types of pollutants: fertilizers, biocides, and animal wastes.

Fertilizers

Agriculture is a chief contributor of *excess nutrients* to water bodies. Pollution occurs when nitrates and phosphates that have been used in fertilizers and are present in animal manure drain into streams and rivers, eventually accumulating in ponds, lakes, and estuaries. The nutrients hasten the process of **eutrophication**, or the enrichment of waters by nutrients. Eutrophication occurs

naturally when nutrients in the surrounding area are washed into the water, but when the sources of enrichment are artificial, as is true of commercial fertilizers, the body of water may become overloaded with nutrients. The end result may be an oxygen deficiency in the water.

Scientists have estimated that as many as one-third of the medium- and large-size lakes in the United States have been affected by accelerated eutrophication. Symptoms of a eutrophic lake are prolific weed growth, large masses of algae, fish kills, rapid accumulation of sediments on the lake bottom, and water that has a foul taste and odor.

Figure 5.7 illustrates one form that overfertilization of a water body can take. Algae and other plants are stimulated to grow abundantly. When they die, the level of dissolved oxygen in the water decreases, primarily because of the bacteria acting on the dead and decomposing vegetation. Fish and plants that cannot tolerate the poorly oxygenated water are eliminated. In addition to being potentially lethal for fish, eutrophication affects the suitability of water for drinking and bathing, because excess nutrients have been shown to pose a health hazard to humans.

Biocides

The herbicides and pesticides used in agriculture are another source of chemical pollution of water bodies. Runoff from farms where such *biocides* have been applied contaminates both ground and surface waters. One of the problems connected with the use of biocides is that the long-term effects of such usage are not always immediately known. DDT, for example, was used for many years before people discovered its effect on birds, fish, and water plant life. One of the dilemmas that we are facing is that we have not yet discovered how to balance our need for increased agricultural yields in the short run—and the manipulation of the environment that it entails—with what is in our best interest in the long run.

Animal Wastes

A final agricultural source of chemical pollution is animal wastes, especially in countries where animals are raised intensively. This is a problem particularly in feedlots, where animals are crowded together at maximum densities to be fattened before slaughter. Large feedlots, such as the one pictured in Figure 5.8, may produce as much waste as would a large city. It is estimated that animal wastes in the United States total about 1.5 billion tons per year, with feedlots generating about half of the total. If not treated properly, the manure pollutes both soil and water with infectious agents and excess nutrients.

Other Sources of Water Pollution

As is evident from Table 5.1, agriculture is only one of the human activities that contributes to water pollution. Other sources are industry, mining, urban drainage, and sewage.

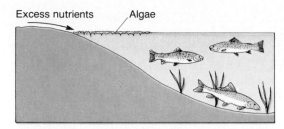

Excess nutrients   Algae

Nitrates and phosphates enrich the lake, stimulating the growth of blue-green algae.

Figure 5.7
Eutrophication is hastened by artificial sources of nutrients. Although eutrophication is primarily a result of agricultural activities, additional sources of nitrates in both surface and underground water supplies are urban drainage, industrial wastewater, and septic tanks.

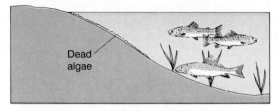

Dead algae

When the algae die and are decomposed by oxygen-consuming bacteria, the level of dissolved oxygen in the lake decreases.

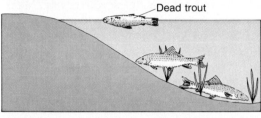

Dead trout

Fish that require less oxygen to survive (such as carp) increase in numbers while fish requiring more oxygen die.

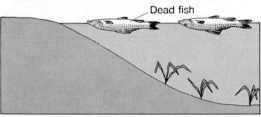

Dead fish

Fish kills lower the oxygen content even more, until no fish can survive.

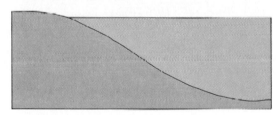

Diversity of plant and animal life continues to decrease until the lake is suitable only for bacteria that do not require oxygen.

Figure 5.8
A feedlot in Kansas. The sanitary disposal of organic wastes generated by such concentrations of animals is a problem only recently addressed by environmental protection agencies.

Human Impact on the Environment   **129**

Industry

In developed countries, industry probably contributes as much to contamination of the water supply as does agriculture. In the United States, about half the water used daily is used by industry. Many industries dump organic and inorganic wastes into bodies of water. These may be acids, highly toxic minerals, such as mercury or arsenic, or, in the case of petroleum refineries, toxic organic chemicals. The nuclear power industry has caused some water pollution when radioactive material has seeped from tanks in which wastes have been buried, either at sea or underground.

Such pollution can have a variety of effects. Organisms not adapted to living in contaminated water may die; the water may become unsuitable for domestic use or irrigation; or the wastes may reenter the food chain, with deleterious effects on humans. A particularly unfortunate example occurred in Japan in 1953. A chemical plant in Minamata Bay that used mercury chloride in its manufacturing process discharged the waste mercury into the bay, where it settled with the mud. Fish that fed on organisms in the mud absorbed the mercury and concentrated it; the fish were in turn eaten by humans. Over 700 people died, and at least 9000 others suffered deformity or other permanent disability.

A similar contamination of the water supply with mercury is occurring today in the Amazon River and its tributaries. Because mercury attaches itself to gold, an estimated half-million Brazilian prospectors use the toxic liquid to separate gold from the accompanying mud. Each year, hundreds of tons of mercury are poured in or right next to the rivers, poisoning the water and the fish that swim in it.

Among the pollutants that have been discharged into the water supply in the United States are **polychlorinated biphenyls** (PCBs), a family of related chemicals used as lubricants in pipelines and in a wide variety of electrical devices, paints, and plastics. During the manufacturing process, companies have dumped PCBs into rivers, from which they have entered the food chain. Several states have banned commercial fishing in lakes and rivers where fish have higher levels of PCBs than are considered safe. Although not all of the effects of PCBs on human health are known, they have been linked to birth defects, damage to the immune system, liver disease, and cancer. In 1977, the Environmental Protection Agency banned the direct discharge of PCBs into U.S. waters, but immense quantities remain in water bodies.

The petroleum industry is a significant contributor to the chemical pollution of water. Oceans are becoming increasingly contaminated by oil. Although massive oil spills like that from the tanker *Exxon Valdez* in March 1989 command public attention, the greatest single source of petroleum pollution in open ocean regions is the discharge of tank flushings and ballast from tanker holds.

Other sources include tanker collisions and explosions, and seepage from offshore oil installations. Recent studies indicate that the Gulf of Mexico, the site of extensive offshore drilling, is among the most seriously polluted major bodies of water in the world.

Acid precipitation (sometimes called acid rain), a by-product of emissions from factories, power plants, and automobiles, has affected the water quality and ecology of thousands of lakes and streams in the world. Because the precipitation is caused by pollutants in the air, it is discussed later in this chapter.

Many industrial processes, as well as electric power production, require the use of water as a coolant (Figure 5.9). **Thermal pollution** occurs when water that has been heated is returned to the environment and has adverse effects on the plants and animals in the water body. Many plants and fish cannot survive changes of even a few degrees in water temperature. They either die or migrate. The species that depend on them for food must also either die or migrate. Thus the food chain has been disrupted. In addition, the higher the temperature of the water, the less oxygen it contains, which means that only lower-order plants and animals can survive.

Mining

Surface mining for coal, iron, copper, gold, and other substances contributes to contamination of the water supply through the wastes it generates. Rainwater reacts with the wastes, and dissolved minerals seep into nearby water bodies. The exact chemical changes produced depend on the composition of the coal or ore slag heaps and the reaction of the minerals with sediments or river water. In addition to altering the quality of the water, the contaminants have secondary effects on plant and animal life. Each year, for example, thousands of animals and migratory birds die in such western states as Arizona, Nevada, and California after drinking cyanide-laced waters at gold mines. The toxic water is poured over mounds of crushed rock to leach the gold and then settles into ponds and lakes that attract wildlife.

Urban Drainage

A host of pollutants derives from the activities associated with urbanization. The use of detergents has increased the phosphorus content of rivers, and salt (used for deicing roads) increases the chloride content of runoff. Water runoff from urban areas contains contaminants from garbage, animal droppings, litter, vehicle drippings, and the like. Because the sources of pollution are so varied, the water supply in any single area is often affected by diverse contaminants. This diversity complicates the problem of controlling water quality.

Contaminated drinking-water wells have been found in more than half of the states. Hundreds of wells in the New York metropolitan area have been closed in recent years because of chemical contamination, and thousands

Figure 5.9
Cooling towers at Three Mile Island nuclear plant near Harrisburg, Pennsylvania. Heated wastewaters are often significantly warmer than the waters into which they are discharged, disrupting the growth, reproduction, and migration of fish populations.

more may be closed in coming years. Chemicals have reached groundwater by seeping into aquifers from landfills, from ruptured gasoline and fuel-oil storage tanks, from septic tanks, and from fields sprayed with pesticides and herbicides. The pollution of aquifers is particularly troublesome because, unlike surface waters, groundwater lacks natural cleansing properties; it can remain contaminated for centuries.

Sewage

Sewage can also be a major water pollutant, depending on how well it is treated before being discharged. Raw, untreated human waste contains viruses responsible for dysentery, polio, hepatitis, spinal meningitis, and other diseases. In general, sewage causes less water pollution in countries such as the United States than in countries where sewage treatment is either not practiced or is not thorough.

North America is not without such sources of water pollution, however. Only half of the U.S. population lives in communities with sewage-treatment plants that meet the minimum goals set by the federal Clean Water Act. The aged sewer system that serves Boston and 42 nearby communities (soon to be replaced) discharges millions of gallons of raw sewage from some 2 million people into Boston Harbor annually. A number of cities in New Jersey and New York, including New York City, dump millions of gallons of treated sewage, called *sludge*, into the Atlantic Ocean each year, a practice that is supposed to end by 1992. And because pollution does not observe international borders, the waters and beaches of San Diego are regularly contaminated by the 15 million gallons of raw sewage that daily flow down the Tijuana River to the Pacific Ocean, where currents carry it northward.

Controlling Water Pollution

In recent years, concern over increased levels of pollution has brought about major improvements in the quality of some surface waters, both in the United States and abroad. The federal government in 1972 took the lead in regulating water pollution with the enactment of the Clean Water Act. Its objective was "to restore and maintain the chemical, physical, and biological integrity of the nation's waters." Congress established uniform nationwide controls for each category of major polluting industry and directed the government to pay most of the cost of new sewage-treatment plants. Since 1972 such plants have been built to serve over 80 million Americans, and industries have spent billions of dollars to comply with the Clean Water Act by reducing organic waste discharges.

The gains have been impressive. Many rivers and lakes that were ecologically dead or dying are now thriving. Once dumping grounds for all kinds of human and industrial waste, the Hudson, Potomac, Cuyahoga, and

Trinity rivers are cleaner, more inviting, and more productive than before, and they now support fishing, swimming, and recreational boating. Similarly, Seattle's Lake Washington and the Great Lakes are healthier than they were two decades ago. Recently, authorities have announced ambitious plans to clean up the waters of Chesapeake Bay, the country's largest estuary, and to undo much of the damage that has been inflicted on Florida's Everglades by improving the water quality of the Kissimmee River and Lake Okeechobee.

Environmental awareness in other countries has also prompted legislation and action. For example, the river Thames in southern England, which had become seriously contaminated by sewage and industrial wastes, is now cleaner than it has been in centuries. The enforcement of stringent pollution-control standards has halted the downward trend in quality. Algae and seaweed, fish and wildfowl have returned to the river in abundance.

Even the Mediterranean Sea is on its way to gradual recovery. When the 18 countries that border the sea signed the Convention for the Protection of the Mediterranean Sea against Pollution in 1976, all coastal cities dumped their untreated sewage into the sea, tankers spewed oily wastes into it, and tons of phosphorus, detergents, lead, and other substances contaminated the waters. Now many cities have built or are building sewage-treatment plants, ships are prohibited from indiscriminate dumping, and national governments are beginning to enforce control of pollution from land-based sources.

Such gains should not mislead us. While some of the most severe problems have been attacked, serious pollution still plagues about one-fourth of our country's rivers and streams, lakes and reservoirs. The solution to water pollution lies in the effective treatment of municipal and industrial wastes; the regulation of chemical runoff from agriculture, mining, and forestry; and the development of less polluting technologies. Although pollution-control projects are expensive, the long-term costs of pollution are even higher.

Impact on Air and Climate

The **troposphere**, the thin layer of air just above the earth's surface, contains all the air that we breathe. Every day thousands of tons of pollutants are discharged into the air by cars and incinerators, factories and airplanes. Air is polluted when it contains substances in sufficient concentrations to have a harmful effect on living things.

Air Pollutants

Truly clean air has probably never existed. Just as there are natural sources of water pollution, so there are substances that pollute the air without the aid of humans. Ash from volcanic eruptions, marsh gases, smoke from forest fires, and wind-blown dust are natural sources of air pollution.

Normally these pollutants are of low volume and they are widely dispersed throughout the atmosphere. On occasion, a major volcanic eruption may produce so much dust that the atmosphere is temporarily altered. The eruption in 1883 of a volcano on the island of Krakatoa, west of Java, produced dust that lingered in the area for over a year. In general, however, the natural sources of air pollution do not have a significant, long-term effect on air, which, like water, is able to cleanse itself.

Far more important than naturally occurring pollutants are the substances that people discharge into the air. These pollutants result primarily from burning fossil fuels (coal, gas, and oil) and other materials. Fossil fuels are burned in power plants that generate electricity, in many industrial plants, in home furnaces, and in cars, trucks, buses, and airplanes. Scientists estimate that about three-quarters of all air pollutants come from burning fossil fuels. The remaining pollutants largely result from industrial processes other than fuel burning, incinerating solid wastes, forest and agricultural fires, and the evaporation of solvents. Figure 5.10 depicts the major sources and types of air pollutants.

Air pollution is a global problem. A 1988 study by the World Health Organization concluded that most of the world's 1.8 billion city dwellers breathe bad air. Sulfur dioxide levels are considered unacceptable for some 625 million people in developing countries alone, and levels of smoke, dust, and other particulates are unacceptable for 1.25 billion. Also in 1988, the EPA released a report

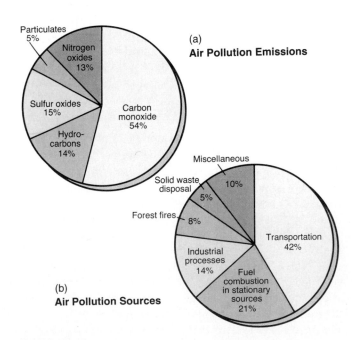

Figure 5.10

(a) Emissions of the principal air pollutants in the United States. Carbon monoxide, one of the most lethal gases known, comes primarily from vehicle exhausts. **(b) Sources of air pollution.** If pollution from vehicles were eliminated, emissions would be reduced by 42%.

showing that in this country over 2 billion pounds of toxic chemicals are emitted into the air annually by industry alone (Figure 5.11).

Factors Affecting Air Pollution

Many factors affect the type and the degree of air pollution found at a given place. Those over which people have relatively little control are climate, weather, wind patterns, and topography. These determine whether pollutants will be blown away or are likely to accumulate. Thus a city on a plain is less likely to experience a buildup than is a city in a valley.

Unusual weather can alter the normal patterns of pollutant dispersal. A *temperature inversion* magnifies the effects of air pollution. Under normal circumstances, air temperature decreases away from the earth's surface. A stationary layer of warm, dry air over a region, however, will prevent the normal rising and cooling of air from below. As described in Chapter 4, the air becomes stagnant during an inversion. Pollutants accumulate in the lowest layer instead of being blown away, so that the air becomes more and more contaminated. Normally inversions last for only a few hours, although certain areas experience them much of the time. Temperature inversions occur often in Los Angeles in the fall and Denver in the winter (Figure 5.12). If an inversion lingers long enough, say, several days, it can contribute to the accumulation of air pollutants to levels that seriously affect human health.

The air pollutants generated in one place may have their most serious effect in areas hundreds of miles away. Thus the worst effects of the air pollution that originates in New York City are felt in Connecticut and parts of Massachusetts. The chemical reaction that produces smog takes a few hours, and by that time air currents have carried the pollutants away from New York. In a similar fashion, New York is the recipient of pollutants produced in other places. Much of the acid rain that affects New England and eastern Canada originates in the coal-fired power plants along the lower Great Lakes and in the Ohio Valley that use extremely high smokestacks to disperse sulfurous emissions. And the coal-based industries in the Soviet Union and Europe produce sulfate, carbon, and other pollutants that are transported by air currents to the land north of the Arctic Circle, where they result in a contamination known as Arctic haze.

Other factors that affect the type and degree of air pollution at a given place are the levels of urbanization and industrialization. Population densities, traffic densities, the type and density of industries, and home heating practices all help to determine the kinds of substances discharged into the air at a single point. In general, the more urbanized and industrialized a place is, the more responsible it is for pollution. The United States may contribute as much as one-third of the world's air pollution, a figure roughly equivalent to the proportion of the world's fossil fuel and mineral resources consumed in this country.

Figure 5.12
The brown cloud that hovers over Denver when temperature inversions keep air pollutants from dispersing. Denver frequently has the worst carbon monoxide level in the country from mid-November to mid-January, when cold winter air over the city is trapped by warm, still air at higher altitudes. The city has recently embarked on a campaign to reduce air pollution.

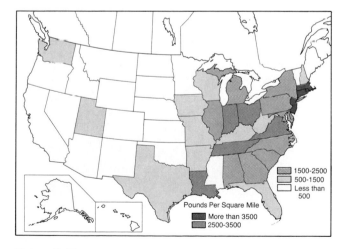

Pounds Per Square Mile

More than 3500
2500-3500
1500-2500
500-1500
Less than 500

Figure 5.11
Millions of pounds of toxic industrial pollutants released into the air in each state in 1988. The EPA survey on which this map is based did not include pollution from automobiles, releases from toxic waste dumps, or pollution from companies that emit less than 75,000 pounds of toxic substances per year.

Human Impact on the Environment   **133**

The sources of pollution are so many and varied that we cannot begin to discuss them all in this chapter. Instead, we will discuss three types of air pollution and their associated effects.

Acid Rain

Although acid *precipitation* is a more precise description, **acid rain** is the term generally used for pollutants, chiefly oxides of sulfur and nitrogen, that are created by burning fossil fuels and that change chemically as they are transported through the atmosphere and fall back to earth as rain, snow, fog, or dust. When sulfur dioxide is absorbed into water vapor in the atmosphere, it becomes sulfuric acid. Sulfur dioxide contributes about two-thirds of the acid in the rain. About one-third comes from nitrogen oxides, transformed into nitric acid in the atmosphere.

When washed out of the air by rain, snow, or fog, the acids change the *pH factor* (the measure of acidity/alkalinity on a scale of 0 to 14) of soil and water, setting off a chain of chemical and biological reactions (Figure 5.13). The average pH of normal rainfall is 5.6, slightly acidic, but acid rainfalls with a pH of 2.4—approximately the acidity of vinegar and lemon juice—have been recorded. It is important to note that the pH scale is logarithmic, so that 4.0 is ten times more acidic than 5.0, and *100* times more acidic than 6.0. Acid rain also coats the ground with particles of aluminum and toxic heavy metals, such as cadmium and lead.

Once the pollutants are airborne, winds can carry them hundreds of miles, depositing them far from their source. Approximately half of the acid rain that falls on eastern Canada originates in the United States, chiefly from coal-burning power plants in the Midwest (Figure 5.14). This issue has become a sore point in U.S.-Canadian relations because the effects of acid precipitation on water bodies, forests, and wildlife are deadly. The problem is that the acidity of a lake or stream need not increase much before it begins to disrupt the food chain.

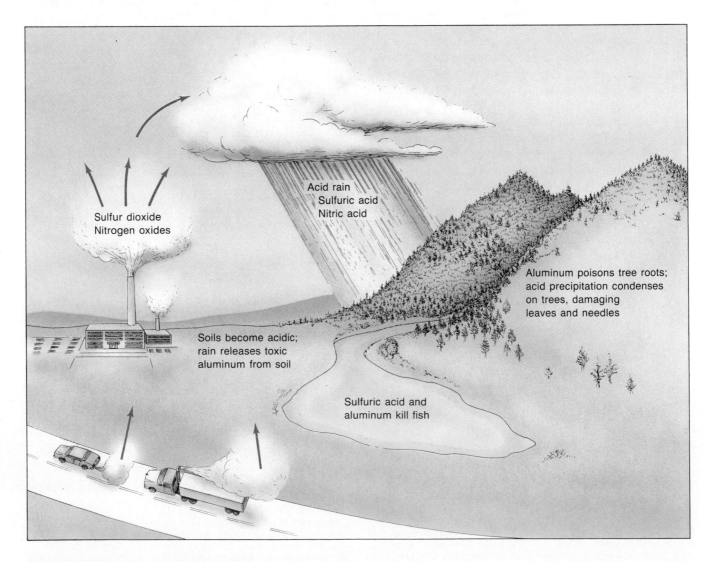

Figure 5.13
The formation and effects of acid rain. The acids in the precipitation damage soil, vegetation, and water.

Canada's Department of the Environment reports that acid rain has already damaged some 14,000 lakes in that country, rendering them almost fishless, and another 150,000 are in peril. Acid rains have also been linked to the disappearance or decline of fish populations in Scandinavia. Some 20% of Sweden's lakes are said to have been damaged by the pollution, and much of Norway's fish population has been exterminated.

While acid rain has for some years been recognized as a threat to freshwater organisms, recent evidence suggests that it may also damage marine life in coastal salt water. In this case the damage stems not from acidity but from the nitrogen in the rain, which causes eutrophication (discussed earlier in this chapter). Nitrogen is a nutrient that stimulates the excessive growth of algae. The algae in turn use up the supply of dissolved oxygen and reduce the amount of sunlight that penetrates the surface of the water.

As Figure 5.15 indicates, acid rain also affects soils and vegetation. It leaches toxic constituents like aluminum salts from the soil and kills microorganisms in the soil that break down organic matter and recycle nutrients through the ecosystem. Extensive forest damage has occurred in parts of North America, northern and western Europe, the USSR, and China. One can also see the corrosive effects of atmospheric acid on marble and limestone sculptures and buildings and on metals, such as iron and bronze.

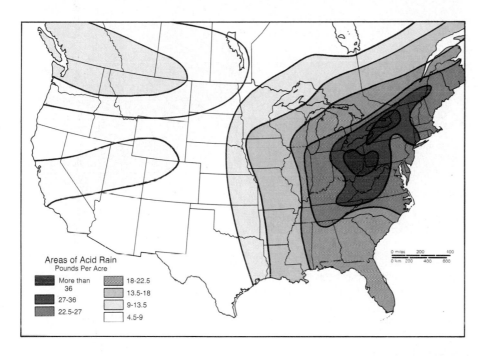

Figure 5.14
Where acid rain falls. In general, the areas that receive the most acid rain in the eastern United States and Canada are those that are least able to tolerate it. Their surface waters tend to be acidic rather than alkaline and are unable to neutralize the acids deposited by rain or snow.

Areas of Acid Rain
Pounds Per Acre

- More than 36
- 27-36
- 22.5-27
- 18-22.5
- 13.5-18
- 9-13.5
- 4.5-9

Figure 5.15
Dead trees on Mount Mitchell, North Carolina. A combination of acid rain and ozone pollution has caused extensive damage to forests along the crest of the Appalachians from Maine to Georgia. The pollution weakens trees to the point where they cannot survive such natural stresses as temperature extremes, high winds, drought, or insects. In addition the high levels of lead and other heavy metals in the soil make it difficult for the forests to regenerate.

The Tall Smokestack Paradox

The dramatic increase in acid precipitation in recent decades is partly the result of an effort to curb air pollution. Written before concern about acid rain became widespread, the U.S. Clean Air Act of 1970 restricts the deposit of specific pollutants over the surrounding countryside and sets standards only for ground-level air quality.

Most of the millions of tons of sulfur and nitrogen oxides that are released into the atmosphere each year come from "stationary source" fuel combustion, primarily coal- and oil-burning power plants. In order to keep air in local communities clean enough to meet the air-quality standards, industries have since the early 1970s been building ever-taller smokestacks that discharge sulfur dioxide and other pollutants into the upper atmosphere. Stacks 1000 feet (305 m) high are now a common sight at utility plants and factories; previously, stacks 200–300 feet (60–90 m) high were the norm. The situation is not unlike disposing of garbage by throwing it over your backyard fence. It still comes down, but not in your yard. Ironically, the farther and higher the noxious emissions go, the longer they have to combine with other atmospheric components and moisture and form acids; thus, the taller stacks have directly aggravated the acid rain problem.

Recognizing the problem that the stacks created, the Environmental Protection Agency in June 1985 issued rules discouraging the use of tall smokestacks to disperse emissions.

Plants now must seek ways to reduce rather than simply disperse emissions, for example, installing scrubbers on their stacks or using low-sulfur coal.

Ozone Pollution

While sulfur oxides are the chief cause of acid rain, oxides of nitrogen are responsible for the formation of *photochemical smog*. It is created when nitrogen oxides react with the oxygen present in water vapor in the air to form nitrogen dioxide. In the presence of sunlight, nitrogen dioxide reacts with hydrocarbons from automobile exhausts and industry to form new compounds, such as **ozone**. Warm, dry weather and poor air circulation promote ozone formation. The hotter and sunnier the weather, the more ozone and smog are created. In general, therefore, more ozone is produced during the summer months than during the rest of the year.

As many as 76 million Americans, or about one-third of the population, live in areas where ozone pollution exceeds the limit established by the federal Clean Air Act, 0.12 parts of ozone per million parts of air. The climate and topography of California are particularly conducive to ozone pollution. Its valleys are encircled by mountains that help hold air pollutants in the basins. When temperature inversions occur, the pollutants are effectively trapped, unable to escape to the stratosphere. Ozone levels in Los Angeles exceed the acceptable level nearly half the days in the year, and sometimes reach three times the acceptable level. Other metropolitan areas subject to ozone pollution are shown in Figure 5.16.

Ozone pollution is associated with respiratory problems. It has long been known to aggravate coughing and breathing problems for asthmatics, but recent studies indicate that exposure to ozone over a period of 6–8 hours also causes respiratory problems in healthy people. Chronic exposure to smog causes permanent damage to lungs, aging them prematurely, and is believed to increase the incidence of such respiratory ailments as pneumonia and emphysema. Because children have smaller breathing passages and less developed immune systems, they are especially susceptible to damage from the polluted air. In

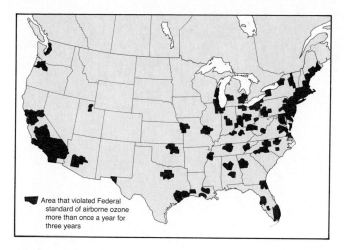

Figure 5.16

Urban areas where ozone levels violated the federal standard more than once a year for three years. By far the worst ozone pollution in the country occurs in the large Los Angeles basin. In 1988, it exceeded the federal ozone standard nearly half the year, 172 days.

addition to its effects on humans, ozone harms forests and agricultural crops (refer to Figure 5.15). Plants grown under smoggy conditions yield less and mature more slowly than normal.

Depletion of the Ozone Layer

Ozone, the same chemical that is a noxious pollutant near the ground, is essential in the stratosphere. There, some 6–15 miles above the ground, ozone forms a protective blanket, the **ozone layer**, which shields all forms of life on earth from overexposure to lethal ultraviolet (UV) radiation from the sun. Mounting evidence indicates that emissions from a variety of chemicals are destroying the ozone layer. Most important are a family of synthetic chemicals developed in 1931 known as **chlorofluorocarbons (CFCs)**. Commercially valuable, CFCs are used as coolants for refrigerators and air conditioners, as aerosol spray propellants, and as a component in foam packaging materials. Also implicated are *halons*, used in fire extinguishers.

After the gases are released, they rise through the lower atmosphere and, after a period of 7–15 years, reach the stratosphere (Figure 5.17). There UV light breaks their chemical bonds and releases chlorine and bromine atoms. Over time, a single one of these atoms can destroy tens of thousands (if not a potentially infinite number) of ozone molecules.

Every Southern Hemisphere spring, beginning around September, the atmosphere over the Antarctic loses more and more ozone. In 1987, researchers discovered what is popularly termed a "hole" as big as the continental United States in the ozone layer over Antarctica, extending northward as far as populated areas of South America (Figure 5.18). It is not truly a hole, but depletion in the ozone layer is as much as 60%. The hole disappears

in the summer, when the winds change and the ozone-deficient air mixes with the surrounding atmosphere. A less dramatic but still serious depletion of the ozone shield occurs over the North Pole.

A depleted ozone layer allows more UV radiation to reach the earth's surface. Although the exact consequences of that increase won't be known for years, it is almost certain to cause a dramatic rise in the incidence of skin cancers. Some fear it may also damage the immune system of humans and other animals by impairing the cells that fight viral infections and parasitic disease. Because UV radiation also causes cell and tissue damage in plants, it is likely to have an adverse effect on crops. The most serious damage may occur in oceans. Increased amounts of UV affect the photosynthesis and metabolism of the microscopic plants called phytoplankton that flourish just below the surface of the Antarctic Ocean and are the base of the marine food chain, as well as playing a central role in the earth's CO_2 cycle.

Although 80 countries have indicated intent to ban production of CFCs by the end of the century, the destructive effects will continue for decades. The two most widely used forms of CFCs stay in the stratosphere breaking down ozone molecules for up to 100 years. Thus, even if use of the chemicals were to stop tomorrow, it would take the planet a century to replenish the ozone already lost.

In addition to their effect on the ozone layer, CFCs may have an impact on climate by contributing to the "greenhouse effect," the subject we explore next.

The Greenhouse Effect

In recent years, researchers have begun to realize that air pollution has an effect on the major factors that control the temperature of the earth's surface. Scientists have issued numerous warnings of a greenhouse effect that could cause earth temperatures to rise with significant impacts on earth's ecosystems. The theory of the **greenhouse effect** is that gases released by combustion concentrate in the atmosphere, where they function as an insulating barrier, absorbing infrared radiation that would otherwise be reflected back into the upper atmosphere. In other words, like glass in a greenhouse, the gases admit incoming solar radiation but retard its reradiation back into space (Figure 5.19). This causes a gradual warming of both the earth's surface and the lower atmosphere.

Carbon dioxide (CO_2), created mostly by burning fossil fuels, is the most plentiful of the gases and is thought to be responsible for about half of the warming. Equally important, taken together, is the accumulation of three other types of gases: (1) methane, from agriculture, swamps, and waste dumps; (2) nitrous oxides, from motor vehicles, industry, and chemical fertilizers; and (3) chlorofluorocarbons and halons, widely used industrial chemicals. Although these gases may be present in small amounts, some of them trap heat thousands of times more

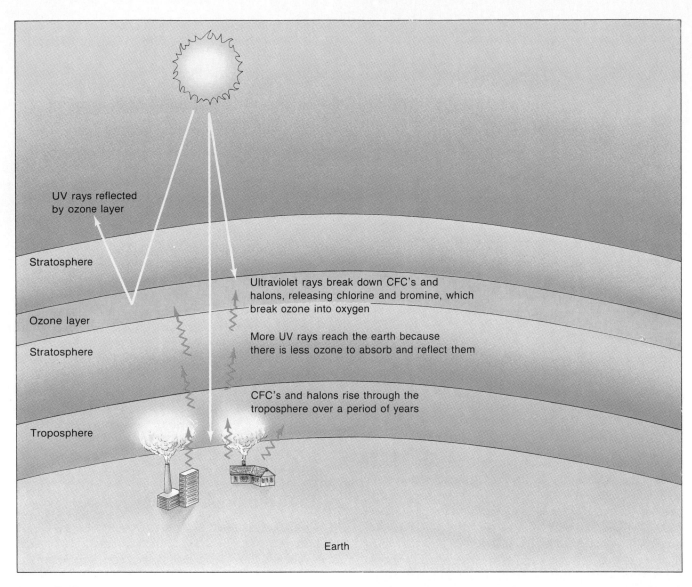

Within the figure:

UV rays reflected by ozone layer

Stratosphere

Ozone layer

Stratosphere

Troposphere

Ultraviolet rays break down CFC's and halons, releasing chlorine and bromine, which break ozone into oxygen

More UV rays reach the earth because there is less ozone to absorb and reflect them

CFC's and halons rise through the troposphere over a period of years

Earth

Figure 5.17

How ozone is lost. CFCs and halons released into the air rise through the troposphere without breaking down (as most pollutants do) and eventually enter the stratosphere. Once they reach the ozone layer, ultraviolet rays break them down, releasing chlorine (from CFCs) and bromine (from halons). These elements in turn disrupt ozone molecules, breaking them up into molecular oxygen and thus depleting the ozone layer.

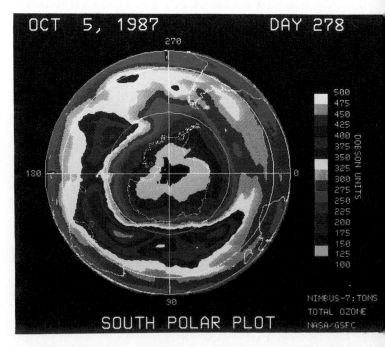

OCT 5, 1987   DAY 278

270

180   0

500
475
450
425
400
375
350
325
300
275
250
225
200
175
150
125
100

DOBSON UNITS

90

NIMBUS-7:TOMS
TOTAL OZONE
NASA/GSFC

SOUTH POLAR PLOT

Figure 5.18

Ozone levels over the Southern Hemisphere in October 1987. The black and lavender colors over Antarctica show the areas of greatest depletion.

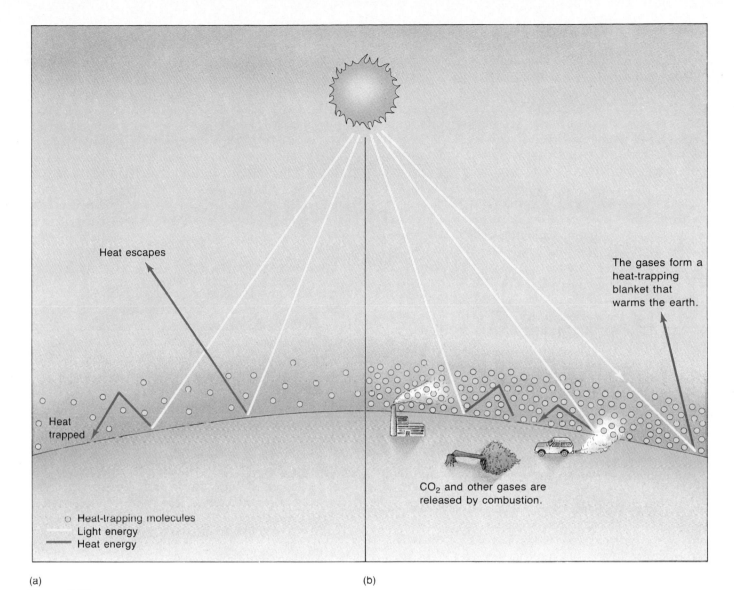

(a)   (b)

Figure 5.19

How the greenhouse effect works. When there is a low level of CO_2 and other gases in the air, as in (a), incoming solar radiation strikes the earth's surface, and the earth radiates the energy back into space as infrared light (heat). The greenhouse effect, depicted in (b), is the result of the billions of tons of CO_2 and other gases that are released into the air each year. They form a blanket that deflects the energy downward, preventing it from escaping into the atmosphere.

Legend (in figure):
○ Heat-trapping molecules
— Light energy
— Heat energy

Heat escapes

The gases form a heat-trapping blanket that warms the earth.

Heat trapped

CO_2 and other gases are released by combustion.

effectively than does CO_2. Fluorocarbon 12, for example, has 20,000 times the capacity of CO_2 to trap heat, and fluorocarbon 11 has 17,500 times the capacity of CO_2.

As the Industrial Revolution gained momentum in Europe and North America during the 19th century, the concentration of CO_2 in the atmosphere rose from its preindustrial level of about 274 parts per million (ppm) to over 340 ppm. (Just since 1958, concentrations of CO_2 have increased from 315 ppm.) Each year, we send about 5.5 billion tons of CO_2 into the atmosphere, only half of which is absorbed by the oceans and forests. Burning of the tropical rain forests not only adds to the emissions but also means the loss of trees that naturally absorb CO_2. The methane concentration in the lower atmosphere has already more than doubled from its preindustrial level and

is currently increasing by just over 1% per year. The carbon monoxide concentration also seems to be increasing at a rate of slightly over 1% per year.

When CO_2 concentrations reach about 550 ppm (double pre-Industrial Revolution levels), average annual global temperatures are expected to rise by 4–9°F (2–5°C). Predictions vary as to when this doubling will occur. The year 2050 is commonly cited, but because gases other than CO_2 are contributing to the greenhouse effect, the warming may occur as early as 2030.

Proponents of the greenhouse effect theory believe that human activity has already put enough of the various gases into the atmosphere that there is no way to avoid a significant rise in temperature in the next century. That is, some warming is inevitable even if all emissions were to stop today, because the greenhouse gases are already in the atmosphere. Consequently, serious environmental and economic damage are a foregone conclusion, although the worst problems are not expected to appear until the next century.

Researchers have developed various mathematical models to simulate the effect of greenhouse gases on the earth. The warming, they predict, will not be uniform. It will be greater at higher latitudes than in equatorial regions, and it will produce significant changes in sea level, precipitation, and vegetation. The sea level is expected to rise some 1–4 feet, both as a result of some ice-cap and glacial melting and from *thermal expansion* of the water (water expands as its temperature increases). Most coastal marshes and swamps would be inundated by salt water; coastal erosion would increase. Water quality would decline as aquifers became polluted by salt. Such low-lying regions as the North American Gulf Coast, the Netherlands, the Nile Delta, Bangladesh, and much of Southeast Asia could lose substantial amounts of land. Many major ports might be flooded.

Warming of lakes and oceans would speed evaporation, causing more active convection currents in the atmosphere and thus fiercer storms. Important regional changes in precipitation would occur, with some areas receiving more precipitation, others less. Polar and equatorial regions might get heavier rainfall, while the mid-latitudes become drier.

Changes in temperature and precipitation would affect soils and vegetation. The composition of forests would change, as some areas became less favorable for certain species of plants, more hospitable to others. Hotter, drier weather would reduce crop yields in some areas, such as the corn and wheat belts of the Midwest. Conversely, more northerly agricultural regions, such as parts of Canada and the USSR, might become more productive.

It should be noted that not all scientists agree with these predictions. Some argue that global temperatures might stabilize or even decrease as the concentration of greenhouse gases increases. A hotter atmosphere, they say, would increase evaporation, sending up more water vapor that could condense into clouds. The increased cloud cover might reflect so much sunlight that it would slow the rate at which the earth would be heated. Others contend that the increased evaporation would produce more rainfall. As it falls, the rain would cool the land and subsequently cool the air over the land.

Finally, some researchers believe the geological record shows that large fluctuations in global temperature have always occurred independently of human activity, never as a result of it. These fluctuations are caused by such unpredictable events as variations in solar radiation, shifts in the earth's orbit and in ocean currents, meteoric activity, and volcanic eruptions.

Impact on Land and Soils

People have affected the earth wherever they have lived. Whatever we do or have done in the past to satisfy our basic needs has had an impact on the landscape. To provide food, clothing, shelter, transportation, and defense, we have cleared the land and replanted it, rechanneled waterways, and built roads, fortresses, and cities. We have mined the earth's resources, logged entire forests, terraced mountainsides, even reclaimed land from the sea. The nature of the changes made in any single area depends on what was there to begin with and how people have used the land. Some of these changes are examined in the following sections.

Landforms Produced by Excavation

Although we tend to think of landforms as "givens," created by natural processes over millions of years, people have played and continue to play a significant role in shaping local physical landscapes. Some features are created deliberately, others unknowingly or indirectly. Pits, ponds, ridges, trenches, subsidence depressions, canals, and reservoirs are the chief landform features resulting from excavations. Some date back to neolithic times, when people dug into chalk pits to obtain flint for toolmaking. Excavation has had its greatest impact within the last two centuries, however, as earth-moving operations have been undertaken for mining; for building construction and for agriculture; and for the construction of transport facilities such as railways, ship canals, and highways.

Surface mining, which involves the removal of vegetation, topsoil, and rocks from the earth's surface in order to get at the resources underneath, has perhaps had the greatest environmental impact. Open-pit mining and strip mining are the methods most commonly used.

Open-pit mining is used primarily to obtain iron and copper, sand, gravel, and stone. As Figure 5.20 indicates, an enormous pit remains after the mining has been completed because most of the material has been removed for processing. *Strip* mining is being increasingly employed in this country as a source of coal; more coal per year now comes from strip mines than from underground mines. Phosphate is also mined in this way. A trench is dug, the material is excavated, and another trench is dug, the soil and waste rock being deposited in the first trench, and so on. Unless reclamation is practiced, the result is a ridged landscape.

Landscapes marred by vast open pits or unevenly filled trenches are one of the most visible results of surface mining. Thousands of square miles of land have been affected, with the prospect of thousands more to come as the amount of surface mining increases. Damage to the aesthetic value of an area is not the only liability of surface mining. If the area is large, wildlife habitats are disrupted, and surface and subsurface drainage patterns are

(a)

(b)

Figure 5.20

(a) Aerial view of the Bingham Canyon open-pit copper mine in Utah. The pit is over 1500 feet (450 m) deep. The operations cover more than 1000 acres (405 hectares). (b) About 150 square miles (390 km^2) of land surface in the United States are lost each year to the strip mining of coal and other resources; far more is disrupted worldwide. Besides altering the topography, strip mining interrupts surface and subsurface drainage patterns, destroys vegetation, and places sterile and frequently highly acidic subsoil and rock on top of the new ground surface.

disturbed. In the United States in recent years, concern over the effect of strip mining has prompted federal and state legislation to increase regulation and to stop the worst abuses. Strip-mining companies are now expected to restore mined land to its original contours and to replant vegetation.

Landforms Produced by Dumping

Excavation in one area often leads to the creation, via dumping, of landforms nearby. Both surface and subsurface mining produce tons of waste and enormous spoil piles. In fact, in terms of tonnage, mining is the single greatest contributor to solid wastes, with about 2 billion tons per year left to be disposed of in this country alone. The normal custom is to dump waste rocks and mill tailings in huge heaps near the mine sites. Unfortunately this practice has secondary effects on the environment.

Carried by wind and water, dust from wastes pollutes the air, and dissolved minerals pollute nearby water sources. Occasionally the wastes cause greater damage, as happened in Wales in 1966, when slag heaps from the coal mines slid onto the village of Aberfan, burying over 140 schoolchildren. Such tragedies call attention to the need for less potentially destructive ways of disposing of mine wastes.

Another example of the combined effect of excavating and filling on the landscape is the agricultural terrace characteristic of parts of Asia. In order to retain water and increase the amount of arable land, terraces are cut into the slopes of hills and mountains. Low walls protect the patches of level land. The *tells* of the Middle East, pictured in Figure 5.21, are another type of landform produced by the accumulation of waste material.

Human impact on land has been particularly strong in areas where land and water meet. Dredging and filling operations undertaken for purposes of water control create landscape features like embankments and dikes. In many places, the actual shape of the shoreline has been altered, as builders in need of additional land have dumped solid wastes into landfills. In the Netherlands, millions of acres of land have been reclaimed from the sea by the building of dikes to enclose polders and canals to drain them. Farming practices in river valleys have had significant effects on deltas. For example, increased sedimentation has often extended the area of land into the sea.

Formation of Surface Depressions

The extraction of material from beneath the ground can lead to **subsidence**, the settling or sinking of a portion of the land surface. Many of the world's great cities are sinking because of the removal of *fluids* (groundwater, oil, and gas) from beneath them. Cities threatened by such subsidence are located on unconsolidated sediments (New Orleans, Bangkok), coastal marshes (Venice, Tokyo), or lake beds (Mexico City). When the fluids are removed, the sediments compact and the land surface sinks. Because many of the cities are on coasts or estuaries and are often only a few feet above sea level, subsidence makes them more vulnerable to flooding from the sea.

Groundwater abstraction has created serious subsidence in many places. Recent evidence indicates that the withdrawal of trillions of gallons of water from a 4500-square-mile (11,650-km²) area of Arizona has resulted in widespread ground subsidence and the formation of more than a hundred earth fissures, jagged ground cracks that can be as much as 9 miles (14.5 km) long and 400 feet (120 m) deep.

The removal of *solids* (such as coal, salt, and gold) by underground mining may result in the collapse of land over the mine. *Sinkholes* or *pits* (circular, steep-walled depressions) and *sags* (larger and shallower depressions) are two types of landscape features produced by such collapse. If surface drainage patterns are disrupted, subsidence *lakes* may form in the depressions. Subsidence has become a more serious problem as towns and cities have expanded over mined-out areas.

Figure 5.21

Tell Hesi, northeast of Gaza, Israel. Here and elsewhere in the Middle East, the debris of millennia of human settlement gradually raised the level of the land surface, producing *tells*, or occupation mounds. The city literally was constantly rebuilt at higher elevations upon the accumulation of refuse of earlier occupants. In some cases, the striking landforms may rise hundreds of feet above the surrounding plains.

As one might expect, subsidence damages structures built on the land, including buildings, roads, and sewage lines. A dramatic example occurred in Los Angeles in 1963, when subsidence caused the dam at the Baldwin Hills Reservoir to crack. In less than 2 hours, the water emptied into the city, resulting in millions of dollars worth of property damage (Figure 5.22). The withdrawal of groundwater from beneath Mexico City has led to severe though differential subsidence. One of the reasons the 1985 earthquake in that city was so damaging was that subsidence had weakened building structures.

Soils

By design or by accident, people have brought about many changes in the physical, chemical, and biochemical nature of the soil and altered its structure, fertility, and drainage characteristics. The exact nature of the changes in any area depends on past practices as well as on the original nature of the land.

Over much of the earth's surface, the thin layer of topsoil upon which life depends is only a few inches deep, usually less than a foot (30 cm). Below it, the lithosphere

Figure 5.22
Subsidence caused cracking and emptying of the Baldwin Hills Reservoir of Los Angeles in December 1963.

is a complex mixture of rock particles, inorganic mineral matter, organic material, living organisms, air, and water. Under natural conditions, soil is constantly being formed by the physical and chemical decomposition of rock material and by the decay of organic matter. It is simultaneously being eroded, for **soil erosion**—the removal of soil particles, usually by wind or running water—is as natural a process as soil formation and occurs even when land is totally covered by forests or grass. Under most natural conditions, however, the rate of soil formation equals or exceeds the rate of soil erosion, so that soil depth and fertility tend to increase with time.

When land is cleared and planted to crops, or when the vegetative cover is broken by overgrazing or other disturbances, the process of erosion accelerates. When its rate exceeds that of soil formation, the topsoil becomes thinner and eventually disappears, leaving behind only sterile subsoil or barren rock. At that point, the renewable soil resource has been converted through human impact into a nonrenewable and dissipated asset. Carried to the extreme of bare rock hillsides or wind-denuded plains, erosion spells the total end of agricultural use of the land.

Such massive destruction of the soil resource could endanger the survival of the civilization it has supported. For the most part, however, farmers devise ingenious ways to preserve and even improve the soil resource upon which their lives and livelihoods depend. Farming skills have not declined in recent years, but pressures upon farmlands have increased with population growth. Farming has been forced higher up onto steeper slopes, more forest land has been converted to cultivation, grazing and crops have been pushed farther and more intensively into semiarid areas, and existing fields have had to be worked more intensively and less carefully. Many traditional agricultural systems and areas that were ecologically stable and secure as recently as 1950, when world population stood at 2.5 billion, are disintegrating under the pressures of more than 5 billion.

The pressure of growing population numbers is having an especially destructive effect on tropical rain forests. Expanded demand for fuel and commercial wood, and a midlatitude market for beef that can be satisfied profitably by replacing tropical forest with cleared grazing land are responsible for some of the loss, but the major cause of *deforestation* is clearing the land for crops. Extending across parts of Asia, Africa, and Latin America, the tropical rain forests are the most biologically diverse places on earth, but some 40,000 square miles (over 100,000 km^2) are being destroyed every year. About 45% of their original expanse has already been cleared or degraded. Deforestation is discussed in more detail in Chapter 11, but it is important to note here that accelerated soil erosion quickly removes tropical forest soils from deforested areas. Lands cleared for agriculture almost immediately become unsuitable for that use partially because of soil loss (Figure 5.23).

Figure 5.23
Wholesale destruction of tropical forests guarantees
environmental degradation so severe that the forest can never
naturally regenerate itself. Exposed soils quickly deteriorate in
structure and fertility and are easily eroded.

Figure 5.24
A drought-stricken area in Burkina Faso, one of the countries in
the Sahel region of Africa where *desertification* has been
accelerated by both climate and human activity. The cultivation of
marginal land, overgrazing by livestock, and climatic change
during the 1970s and 1980s have led to the destruction of native
vegetation, to erosion, and to the expansion of the deserts of the
world.

The tropical rain forests can succumb to deliberate
massive human assaults and be irretrievably lost. With
much less effort, and with no intent to destroy or alter the
environment, humans are similarly affecting the arid and
semiarid regions of the world. The process is called **desertification**, the spread of desertlike landscapes into arid
and semiarid environments. Although climatic change can
in some instances be a contributing cause, desertification
is usually charged to increasing human pressures exerted
through overgrazing, deforestation for fuel wood, clearing
of original vegetation for cultivation, and burning (Figure
5.24).

Whatever the reason, the process begins in the same
way: the disruption or removal of the native cover of
grasses and shrubs through farming or overgrazing. If the
disruption is severe enough, the original vegetation cannot
reestablish itself, and the exposed soil is made susceptible
to erosion during the brief, heavy rains that dominate precipitation patterns in semiarid regions. Water runs off the
land surface instead of seeping in, carrying soil particles
with it. When the water is lost through surface flow rather
than seepage downward, the water table is lowered. Eventually, even deep-rooted bushes are unable to reach
groundwater, and all natural vegetation is lost. The process is accentuated when too many grazing animals pack
the earth down with their hooves, blocking the passage of
air and water through the soil. When both plant cover and
soil moisture are lost, desertification has occurred.

It happens with increasing frequency in many areas
of the earth as pressures upon the land continue (Figure
5.25). Africa is most at risk; the United Nations has estimated that 40% of that continent's nondesert land is in
danger of desertification. But nearly a third of Asia and
a fifth of Latin America's land are similarly endangered.
In countries where desertification is particularly extensive
and severe (Algeria, Ethiopia, Iraq, Jordan, Lebanon,
Mali, and Niger), per capita food production declined by
some 40% between 1950 and the mid-1980s. The resulting

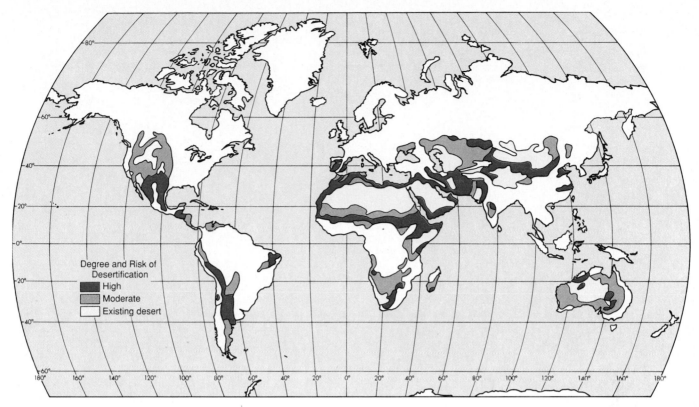

Figure 5.25

Desertification threatens about one-third of the world's land surface, some 18.5 million square miles (48 million km²). Once it occurs in its extreme form, large area desertification is probably irreversible.

threat of starvation spurs populations of the affected areas to increase their farming and livestock pressures on the denuded land, further contributing to their desertification.

Accelerated soil erosion is not limited to Africa and Asia. Indeed, in recent years, soil erosion in the United States has been at an all-time high (Figure 5.26). Wind and water are blowing and washing soil off pasturelands in the Great Plains, ranches in Texas, and farms in the Southeast. The country's croplands lose almost 2 billion tons of soil per year to erosion, an average annual loss of over 4 tons per acre. In some areas, the average is 15–20 tons per acre. Of the roughly 413 million acres (167 million hectares) of land that are intensively cropped in this country, over one-third are losing topsoil faster than it can be replaced naturally (it can take up to a century to replace an inch of topsoil). In parts of Illinois and Iowa where the topsoil was once a foot deep, less than half of it remains. Every hour about 40,000 tons of topsoil wash into the Mississippi River.

Like most processes, soil erosion has secondary effects. As the soil quality and quantity decline, croplands become less productive and yields drop. Streams and reservoirs experience accelerated siltation. In countries where the topsoil is heavily laden with agricultural chemicals, erosion-borne silt pollutes water supplies. The danger of floods increases as bottomlands fill with silt, and the costs of maintaining navigation channels grow.

Accelerated erosion is a primary cause of agricultural soil deterioration, but in arid and semiarid areas, salt accumulation is a contributing factor. In many parts of the world, irrigation has led to excessive salinity of the soil. These regions are naturally salty because evapotranspiration exceeds precipitation. Since irrigation water tends to move slowly, and thus to evaporate more rapidly, the salts carried in the water are absorbed into the ground. Over time, increased salinity makes the soil less productive. Thousands of once-fertile acres have been abandoned in Iran and Iraq; over 25% of the irrigated areas of Pakistan, Syria, and Egypt are affected by such salting up, or **salinization.** In the United States, the 250-mile-long (400 km) San Joaquin Valley of California is experiencing a similar buildup of harmful salty residues. Because the valley is a major supplier of fruits and vegetables to the rest of the country, the conversion of over 1 million acres (405,000 hectares) to a barren salt flat would have an impact well beyond California's borders.

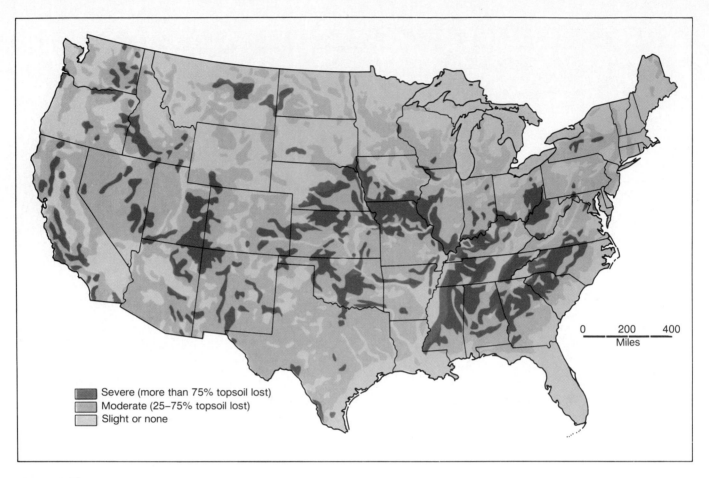

Figure 5.26
Soil erosion in the United States. Although many activities (mining, construction, and urbanization) contribute to erosion, agriculture and deforestation are particularly significant. In recent years, erosion has been most severe not in the southwestern Dust Bowl but in the moist, rolling-hill regions of western

Mississippi, western Tennessee, and Missouri. Each of these areas loses about 10 tons of topsoil per acre of cropland annually. However, soil resource stress and depletion affect all parts of the country.

Severe (more than 75% topsoil lost)
Moderate (25–75% topsoil lost)
Slight or none

0   200   400
Miles

Impact on Plants and Animals

People have affected plant and animal life on the earth in several different ways. When human impact is severe enough, a species becomes *extinct*, that is, it no longer exists. Although fossil records show that extinction is a natural feature of life on earth, scientists estimate that in recent history the rate of extinction has increased a million times—an increase due to human activity. Species that are in danger of becoming extinct are called *endangered*, while a *threatened* species is one that is likely to become endangered within the foreseeable future. In the United States, about 500 species of plants and animals are on the list of endangered and threatened species that is maintained by the federal government and updated yearly; another 4000 species are candidates for the list.

In this section we examine some of the ways people have modified plant and animal life.

Introduction of New Species

The deliberate or inadvertent introduction of a plant or animal into an area where it did not previously exist can have damaging and unforeseen consequences. Introduced species have often left behind their natural enemies—predators and diseases—giving them an advantage over native species that are held in check by biological controls. The rabbit, for example, was purposely introduced into Australia in 1859. The original dozen pairs multiplied to a population in the thousands in only a few years and, despite programs of control, to an estimated 1 billion a hundred years later (Figure 5.27). Inasmuch as five rabbits eat about as much as one sheep, a national problem had been created. Rabbits had denuded much of the grassland on which sheep could graze.

Plants and animals found on islands are particularly prone to extinction. Because island species often occur on just one or a few islands, the loss of only a few individuals can be devastating to small populations. Goats, used as

Maintaining Soil Productivity

In much of the world, increasing population numbers are largely responsible for accelerated soil erosion. In the United States in the 1970s and 1980s, economic conditions contributed to a rate of soil erosion equal to that registered during the Dust Bowl era of the 1930s. Federal tax laws and the high farmland values of the 1970s encouraged farmers to plow virgin grasslands and to tear down windbreaks to increase their cultivable land and yields. The secretary of agriculture exhorted farmers to plant all of their land, "from fencerow to fencerow," to produce more grain for export. Land was converted from cattle grazing to corn and soybean production as livestock prices declined.

When prices of both land and agricultural products declined in the 1980s, farmers felt impelled to produce as much as they could in order to meet their debts and make any profit at all. To maintain or increase their productivity, many neglected conservation practices, plowing under marginal lands and using fields for the same crops every year.

Conservation techniques were not forgotten, of course. They were both practiced by many and persistently advocated by farm organizations and soil conservation groups. Techniques to reduce erosion by holding the soils in place are well-known. They include contour plowing, terracing, strip-cropping and crop rotation, erecting windbreaks and constructing water diversion channels, and practicing no-till farming (allowing crop residue such as cut corn stalks to remain on the soil surface throughout the winter). In addition, farmers can be paid to idle marginal, highly erodible land. Only by employing such practices can the country maintain the long-term productivity of soil, the resource base upon which all depend.

No-till farming

Strip-cropping

food for whaling crews, were imported to Guadalupe Island off the coast of Baja, California, in the 19th century. Without predators to keep their numbers low, the goats reproduced unchecked, with a devastating effect on the indigenous vegetation.

Introduced plants as well as animals can alter vegetative patterns. The Asiatic chestnut blight, for instance, has destroyed most of the native American chestnut trees in the United States, trees with significant commercial as well as aesthetic value. The cause was the importation of

Figure 5.27
Rabbits converging on a water hole in Australia during a drought. Deliberately imported to Australia, the rabbit became an economic burden and environmental menace, competing with sheep for grazing land and stimulating soil erosion.

chestnut trees from China to the United States. They carried a fungus fatal to the American chestnut tree but not to the Asiatic variety, which is largely immune to it.

The water hyacinth entered North America in 1884, when bulbs were given as souvenirs at an exposition in New Orleans. Before the end of the century, the spread of the plant had become a matter of concern. In one growing season, a single plant can produce over 60,000 offshoots. To the consternation of those who sail and fish, swamps, lakes, and canals in the southern United States have become clogged with hyacinths, which also affect fish and plankton.

These are just a few of the many examples that illustrate an often-ignored ecological truth: plant and animal life are so interrelated that when people introduce a new species to a region, whether by choice or by chance, there may be unforeseen and far-reaching consequences.

Habitat Disruption

One of the main causes of extinction has been the loss or alteration of habitats for wildlife. By clearing forest land, draining wetlands, extending farmland, and building cities, people modify or destroy the habitats in which plants and animals have lived. Tidal marshes have been subjected to dredging and filling for residential and industrial development. The loss of such areas reduces the essential habitat of fish, crustaceans, and mollusks. The whooping crane has been virtually eliminated in the United States because the marshes where it nested were drained, and roads and canals brought intruders into its habitat. Its comeback, sought by breeding programs in the United States and Canada, is still uncertain.

Many people fear that as countries in Africa and South America become more industrialized and more urbanized and expand their areas under cultivation, there will be an increasingly negative impact on wildlife. It is already known that in Africa, wild animals are vanishing fast, in part the victims of habitat destruction. In Botswana, for example, 250,000 antelope and zebra died in a decade, disoriented by fences erected to protect cattle. As selected animal species decline in numbers, balances among species are upset and entire ecosystems are disrupted.

The destruction of the world's rain forests, the most biologically diverse places on earth, is by some estimates causing the extinction of about 1000 plant and animal species every year. For every plant that becomes extinct, about 20 animals that depend on the plant also vanish.

Hunting and Commercial Exploitation

Another way in which people have affected plants and animals is through deliberate destruction. We have overhunted and overfished, for food, fur, hides, jewelry, and trophies. In the past, unregulated hunting harmed wildlife all over the world and was responsible for the destruction of many populations and species. Beaver, sea otter, alligator, and buffalo are among the species brought to the edge of extinction in the United States by thoughtless exploitation. Under protective legislation, their populations are now increasing, but hunting in developing countries still poses a threat to a number of species.

Three African animals whose existence is threatened by hunting, most of it illegal, are the elephant, rhinoceros, and mountain gorilla. Prized for its ivory tusks, the African elephant has been ruthlessly slaughtered. Ten million elephants are estimated to have been alive in the 1930s. By 1979 the population had dropped to 1.5 million, and a decade later, to only one-half million (Figure 5.28). The rhinoceros, killed for its horn, is now an endangered species. Worldwide, the population has declined from 70,000 in 1970 to about 10,000 today. Only 400 mountain gorillas are believed to exist, most of them in the Virunga Mountains of Uganda and Rwanda.

Figure 5.28
Elephant killed by poachers in Kenya. Although habitat alteration has contributed to the decline of the elephant population, the greatest threat comes from illegal poaching. By their voracious consumption of vegetation, elephants help shape ecosystems in both savanna woodlands and rain forests. As elephants disappear, so will many species that share its habitat, including zebras, gazelles, and giraffes.

Poisoning and Contamination

Humans have also affected plant and animal life by poisoning or contamination. In the last several years, we have become acutely conscious of the effect of insecticides, rodenticides, and herbicides, known collectively as *biocides*. The best known and most widely used has been DDT, although there are now thousands of compounds in use.

DDT was first used during World War II to kill insects that carried diseases, such as malaria and yellow fever. In the years following the war, tons of DDT and other biocides were used, sometimes to combat disease, sometimes to increase agricultural yields. Insects, after all, can destroy a significant percentage of a given crop, either when it is in the field or after it has been harvested.

In the last ten years, some of the side effects of these biocides have been well enough documented for us to question their indiscriminate use. Once used, a biocide settles into the soil, where it may remain or may be washed into a body of water. In either case, it is absorbed by organisms living in the soil or the mud. Through a process known as **biological magnification**, the biocide accumulates and is concentrated at progressively higher levels in the food chain (Figure 5.29). By remaining as residue in fatty tissues, a very small amount of a biocide produces unexpected effects, with predators accumulating larger amounts than their prey. The higher the level of an organism in the food chain, the greater the concentration of DDT will be, and that concentration may be lethal. Robins and other small birds die when they eat earthworms that have ingested DDT that has settled into the earth after being sprayed on trees.

DDT also causes a decrease in the thickness of the eggshells of some of the larger birds, causing a greater number of eggs to break than normally would. Peregrine falcons, bald eagles, and brown pelicans were among the birds nearly made extinct by this disruption of the reproductive process.

Although the use of DDT has declined as its effects have become apparent, other chlorinated hydrocarbon compounds have been developed and are in wide use. Indeed, the use of pesticides in the United States has more than doubled in the last 25 years. More than 2 billion pounds (900 million kg) containing more than 600 active ingredients are applied each year. The pesticides pollute water supplies, often contaminate the crops they are meant to protect, and sometimes sicken the farm workers who apply them. In addition, they too often are only temporarily effective.

Biocides may in fact exacerbate the problem their use is designed to eradicate. By altering the natural processes that determine which insects in a population will

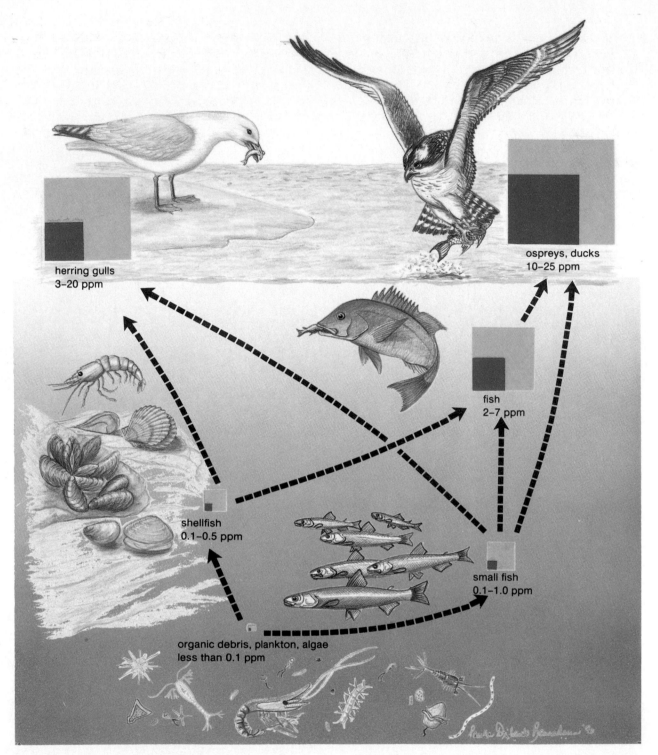

Figure 5.29

A simplified example of biological magnification. Although the level of DDT in the water and mud may be low, the impact on organisms at the top of the food chain can be significant. In this example, birds at the top of the chain have concentrations of residues as much as 250 times greater than the concentration in the water. PCBs and radioisotopes such as strontium-90 and cesium-137 undergo magnification in the food chain just as insecticides do.

herring gulls
3–20 ppm

ospreys, ducks
10–25 ppm

fish
2–7 ppm

shellfish
0.1–0.5 ppm

small fish
0.1–1.0 ppm

organic debris, plankton, algae
less than 0.1 ppm

survive, biocides spur the development of resistant species. If all but 5% of the mosquito population in an area are killed by an insecticide, the ones that survive are the most resistant individuals, and they are the ones that will produce the succeeding generations.

There are now insects whose total resistance to certain pesticides has led some scientists to conclude that the entire process of insecticide development may be self-defeating. Despite the enormous growth in the use of pesticides, crop loss to insect and weed pests has actually grown. According to Department of Agriculture figures, 32% of crops were lost to pests in 1945; 40 years later, such losses had increased to 37%.

Pesticides have also been known to increase problems by destroying the natural enemies of the intended target, leaving it to breed unchallenged (Figure 5.30). Examples include the tobacco budworm and the brown planthopper, which were relatively minor pests before intensive crop spraying destroyed rival pests. The budworm severely cut cotton production in Mexico and Texas in the 1960s. The brown planthopper continues to affect rice growing in Asia.

Solid-Waste Disposal

Modern technologies and the societies that have developed them produce enormous amounts of solid wastes that must be disposed of. The rubbish heaps of past cultures suggest that humankind has always been faced with the problem of ridding itself of materials it no longer needs. The problem for advanced societies, with their ever-greater variety, amount, and durability of refuse, is even more serious: how to dispose of the solid wastes produced by residential, commercial, and industrial processes. Although these account for much less tonnage than the wastes produced by mining or agriculture, they are everywhere, a problem with which each individual and each municipality must deal.

Municipal Waste

The wastes that communities must somehow dispose of include newspapers and beer cans, toothpaste tubes and old television sets, broken refrigerators and rusted cars. American communities are now facing twin crises in disposing of these wastes: the sheer volume of trash and the toxic nature of much of it. Solid-waste disposal is a greater problem in the United States than in any other country, for we throw away more trash per person than any country in the world. As Figure 5.31 indicates, the amount of household waste has doubled in the last 30 years, and currently stands at some 160 million tons per year, or about 1300 pounds (600 kg) per person. Solid-waste disposal costs are now the second largest expenditure of most local governments, ranking just behind education. Americans generate more than twice as much waste per person as do Japanese and Europeans, four times as much as Pakistanis or Indonesians.

Our volume of trash is the result of three factors—affluence, packaging, and open space. Craving convenience, Americans rely on disposable goods that they throw away after very limited use. Thus, although readily available substitutes are more economical, we annually throw out 16 billion baby diapers, 2 billion razors, 1.6 billion pens, and a million tons of paper towels and napkins. People in

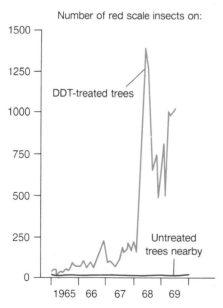

Number of red scale insects on:

Figure 5.30
Using DDT to destroy sap-sucking red scale insects on lemon trees had the opposite effect, actually increasing the numbers of the pest. Spraying an infested crop kills perhaps 90% of the pests but also the insects that eat the pests. With their food abundant and their predators rare, the remaining pests recover faster than their enemies, whose prey is now scarce and harder to locate.

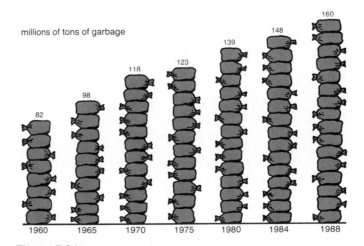

millions of tons of garbage

Figure 5.31
Municipal solid waste generated each year in the United States, in millions of tons. The figures do not include waste from industry or agriculture.

less affluent countries repair and recycle a far greater proportion of domestic products. In addition, nearly all consumer goods are encased in some sort of wrapping, whether it be paper, cardboard, plastic, or foam. An astounding one-third of the yearly volume of trash consists of these packaging materials. Finally, the United States has traditionally had ample space in which to dump unwanted materials. Countries that ran short of such space decades ago have made greater progress in reducing the volume of waste.

Although ordinary household trash does not meet the governmental designation of **hazardous waste**—defined as discarded material that may pose a substantial threat to human health or to the environment when improperly stored, transported, or disposed of—much of it is hazardous nonetheless. Products containing toxic chemicals include paint thinners and removers, furniture polishes, bleaches, oven and drain cleaners, used motor oil, and garden weed killers and pesticides.

Countries use various methods of disposing of solid wastes, and each has its own impact on the environment. Loading wastes onto barges and dumping them in the sea, long a practice for coastal communities, inevitably pollutes the ocean. Open dumps on land are a menace to public health, for they harbor disease-carrying rats and insects. Burning combustibles discharges chemicals and particulates into the air. In the United States, three methods of solid-waste disposal are employed: landfills, incineration, and recycling (Figure 5.32).

Landfills

An estimated 80% of U.S. municipal solid waste is deposited in *sanitary landfills*, where each day's waste is compacted and covered by a layer of soil (Figure 5.33). "Sanitary" is a deceptive word. There are currently no federal standards to which local landfills must adhere, and while some communities and states regulate the environmental impact of dumps, many do not. Even if no commercial or industrial waste has been dumped at the site, most landfills eventually produce *leachate* liquids that contaminate the groundwater. The EPA estimates that only 15% of the country's landfills have liners to prevent contaminants from seeping into groundwater supplies. Indeed, over two-thirds of the dumps lack any type of system to monitor groundwater quality. Leachate occurs when precipitation entering the landfill interacts with the decomposing materials. Thus, heavy metals are leached from batteries and old electrical parts, while vinyl chlorides come from the plastic in household products. Typical leachate contains more than 40 organic chemicals, many of them poisonous.

Municipal landfill capacity is shrinking dramatically. The number of landfills in the United States fell from 18,000 in the late 1970s to 6000 in 1989, increasing pressure on those that remained open. A third of these will be closed within the next five years, either because they are full or because they pose a threat to the environment.

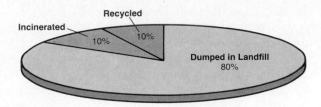

Figure 5.32
Methods of solid waste disposal in the United States.

Public opposition has made it increasingly difficult to site new landfills, forcing communities to look for alternative methods of waste disposal. A few in the northeast have resorted to long-distance, even interstate hauling, paying other communities to take in their garbage, but the chief effect has been to increase interest in incineration and recycling.

Incineration

The quickest way to reduce the volume of trash is to burn it, a practice that was common at this country's open dumps until it was halted by the Clean Air Act of 1970. Concern over air pollution also forced the closure of old, inefficient *incinerators* (facilities designed to burn waste), providing an impetus to designing a new generation of incinerators. By 1988, more than 100 municipal incinerators were operating in the United States, mostly in the northeast, burning about 10% of the national total of trash, and another 200 are expected to be operating soon. Connecticut burns about half of its household trash; and Massachusetts, New Jersey, and New York may soon do likewise. A Dade County facility burns 25% of the waste from 26 municipalities, including Miami. Most municipal incinerators are of the waste-to-energy type, which use extra-high (1800°F) temperatures to reduce trash to ash and simultaneously generate electricity or steam that is then sold to help pay operating costs.

A decade ago incinerators were hailed as the ideal solution to overflowing landfills, but it has become apparent that they pose environmental problems of their own by generating toxic pollutants in both air emissions and ash. Air emissions from incinerator stacks have been found to contain an alphabet soup of highly toxic elements, ranging from a (arsenic) to z (zinc), and including, among others, cadmium, dioxins, lead, and mercury, as well as significant amounts of such gases as carbon monoxide, sulfur dioxide, and nitrogen oxides. Emissions can be kept to acceptably low limits by installing electrostatic precipitators, filters, and scrubbers to capture pollutants before they are released into the outside air, although the devices add significantly to the cost of the plant. A greater problem is created by the concentration of toxins in the ash residue of burning. Incinerators typically reduce trash by only 75%. One-fourth remains as ash, which must then be buried in a landfill. Ideally the ash would be treated as hazardous waste, to lessen the danger of contaminating groundwater supplies.

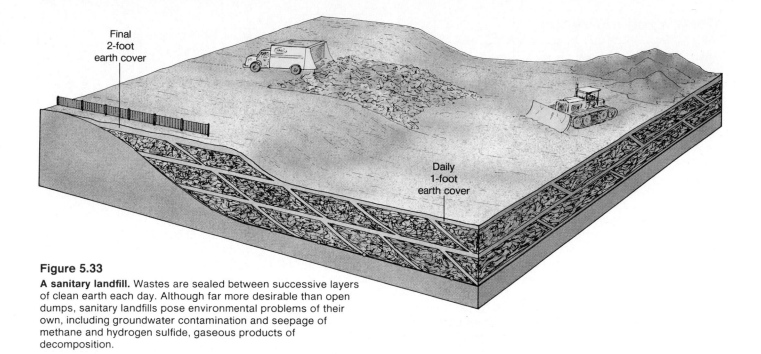

Figure 5.33

A sanitary landfill. Wastes are sealed between successive layers of clean earth each day. Although far more desirable than open dumps, sanitary landfills pose environmental problems of their own, including groundwater contamination and seepage of methane and hydrogen sulfide, gaseous products of decomposition.

Recycling

Awareness of the problems posed by landfills and incinerators has spurred interest in *recycling*, which reduces the amount of waste needing disposal by making a portion of it available for reuse. Thousands of businesses and at least 600 communities in the United States have initiated recycling programs for a variety of materials. Most programs collect paper, aluminum beverage cans, and glass. Some also compact leaves and other yard waste. Although recycling of plastics is more difficult, plastic milk and soda bottles are easily recycled.

At present about 10% of the country's waste is recycled, a figure that is likely to increase in the near future. By 1989, ten states had passed mandatory recycling laws, and the EPA has recommended a national goal of 25% waste recycling by 1992. The figure is not unattainable; Japan recycles more than half of its waste, and Western European countries about 30%.

Recycling has a beneficent impact on the environment. It saves natural resources by making it possible to cut fewer trees, burn less oil, mine less ore. Because it takes less energy to make things out of recycled material than out of virgin materials, recycling saves energy. It reduces the pollution of air, water, and land that stems from the manufacture of new materials and from other methods of waste disposal, and saves increasingly scarce landfill space for materials that cannot be recycled. These include hazardous wastes.

Hazardous Waste

The EPA has classified more than 400 substances as hazardous, posing a threat to human health or the environment. Industrial wastes have grown steadily more toxic.

Currently about 10% of industrial waste materials are considered hazardous.

Every facility that either uses or produces radioactive materials generates *low-level waste*. Nuclear power plants produce about half the total low-level waste in the form of used resins, filter sludges, lubricating oils, and detergent wastes. Industries that manufacture radiopharmaceuticals, smoke alarms, radium watch dials, and other consumer goods produce low-level waste consisting of machinery parts, plastics, and organic solvents. Research establishments, including universities and hospitals, also produce radioactive waste materials. While some low-level waste will lose its contamination within months, some will remain radioactive for centuries.

High-level waste is nuclear waste with a relatively high level of radioactivity. It consists primarily of spent power reactor fuel assemblies—termed "civilian waste"—and waste generated as a by-product of the manufacture of nuclear weapons, or "military waste." The volume of high-level waste is not only great but increasing rapidly. Approximately one-third of a reactor's rods need to be disposed of every year.

By the end of the 1980s, nearly 70,000 spent-fuel assemblies were being stored in the containment pools of the country's commercial nuclear power reactors, awaiting more permanent disposition. Some 6000 more are added annually. "Spent fuel" is a misleading term: the assemblies are removed from commercial reactors not because their radiation is spent but because they have become too radioactive for further use. The assemblies will remain radioactively "hot" for thousands of years.

Unfortunately, no satisfactory method for disposing of any hazardous waste has yet been devised. Some wastes

Commerce in Poison

Early in March 1988, A.S. Bulk Handling Inc., a Norwegian shipping company, dumped 15,000 tons of a substance described as "raw material for bricks" in an abandoned quarry on the resort island of Kassa, just off Conakry, the capital of Guinea. Barefoot children played on the debris, but weekend vacationers from the mainland soon noticed that the island's vegetation had begun to shrivel. A government investigation revealed that the material was toxic incinerator ash from the city of Philadelphia, the first shipment under a contract to dispose of 85,000 tons of waste in the country. Guinea ordered the waste removed and its Norwegian broker jailed, and in July the ash was returned to Philadelphia.

That return shipment was the one bright spot in an otherwise dark picture. Between 1986 and 1988, over 3 million tons of hazardous waste were transported from Western Europe and the United States to countries in Africa and Eastern Europe, an irresponsible and reckless way to dispose of dangerous refuse. As the cost of toxic waste disposal in Western Europe and the United States soared as high as $2500 per ton, waste brokers sought cheaper sites in poorer countries. Nearly every West African country received offers from companies seeking to dispose of refuse. Waste brokers made enormous profits, for their offers to governments and private businesses ranged from $3 to $50 per ton.

It is evident why industrialized countries want to export waste, but why would a country like Guinea accept the shipments? Partly for economic reasons: poverty and a huge foreign debt. In addition, receiving countries have not been accurately informed about the nature of the wastes or the dangers they present. Incinerator ash, chlorinated solvents, PCBs, lead paint waste, and other hazardous substances have been described in disposal contracts by such phrases as "ordinary industrial waste," "complex organic matter," "non-explosive, non-radioactive, non-self-combusting materials," and, in the Guinea case, "raw material for bricks."

Exporting waste simply shifts toxic materials from one country to another, where they may do more damage to health or the environment. Waste that is deadly in the United States and Europe will be even more so in countries that lack the technology to dispose of it safely, and where the disposal sites have not been studied for their geological suitability. It is especially dangerous to dump wastes in areas with high rainfall, such as many of the West African countries, because it is likely to contaminate groundwater supplies.

Irate at this latest type of exploitation of poor countries, the Organization of African Unity in 1988 adopted a resolution condemning the dumping of all foreign wastes on that continent. Under the sponsorship of the United Nations, 117 countries in March of 1989 adopted a treaty aimed at regulating the trade in waste. The Basel Convention on the Control of Transboundary Movements of Hazardous Wastes and Their Disposal hopes to prevent the export of such wastes to unsafe or inadequate sites by requiring exporters to receive consent from receiving countries before shipping the waste. The treaty, which will become effective 90 days after 20 countries have formally adopted it, also requires both exporting and receiving countries to insure that the waste is disposed of in an environmentally sound manner. It remains to be seen how effectively the treaty is applied.

have been sealed in protective tanks and dumped at sea. Investigations have shown, however, that the tanks may be moved from the original dumping site by strong currents and may be crushed by water pressure, causing leakage of the wastes. Even without such physical damage, the life expectancy of the containers must be presumed to be far shorter than the half-life of their radioactive contents. Other radioactive wastes have been placed in tanks and buried in the earth. Millions of cubic feet of both high- and low-level waste are stored at various locations in the United States, one of which is shown in Figure 5.34. Several of these storage areas have experienced leakages, with seepage of waste into the surrounding soil and groundwater.

Another method of waste disposal, used for chemicals as well as for radioactive wastes, has been to inject them into deep steel- or concrete-lined wells. Because underground injection poses a threat of groundwater contamination and may contribute to earth tremors, the injection of wastes into or above strata that contain aquifers is being phased out.

The most common method of disposal of low-level waste is in landfills, often the local municipal dump, where the waste chemicals may leach through the soil and into the groundwater. By EPA estimates, the United States contains at least 25,000 legal and illegal dumps with hazardous waste. As many as 2000 are deemed potential ecological disasters. Some scientists contend that a safe dump exists only on paper, that even following the federal government's suggested guidelines for the best possible hazardous waste landfill will yield an insecure one (Figure 5.35).

Figure 5.34

Storage tanks under construction in Hanford, Washington. Built to contain high-level radioactive wastes, the tanks are shown before they were encased in concrete and buried underground.

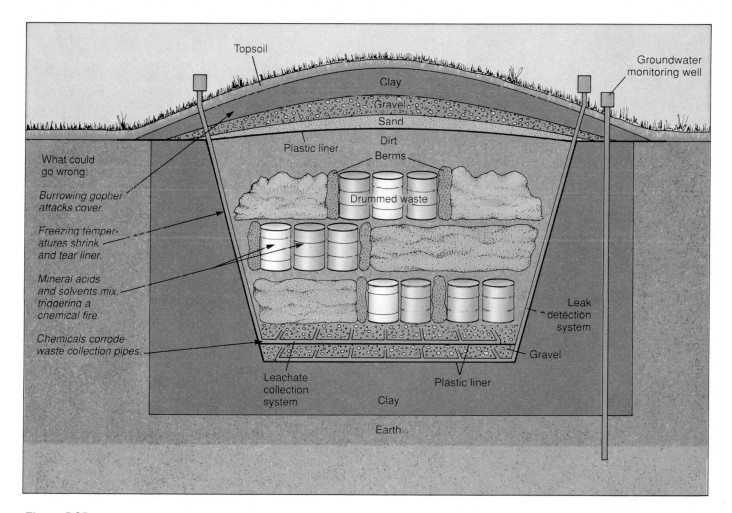

Topsoil
Clay
Gravel
Sand
Plastic liner
Dirt
Berms
Drummed waste

Groundwater monitoring well

What could go wrong:

Burrowing gopher attacks cover.

Freezing temperatures shrink and tear liner.

Mineral acids and solvents mix, triggering a chemical fire.

Chemicals corrode waste collection pipes.

Leak detection system

Gravel

Leachate collection system

Plastic liner

Clay

Earth

Figure 5.35

The closest thing to what experts consider a secure landfill for hazardous waste is a pit 60 feet (18 m) deep, sealed in a plastic liner to keep contaminated liquids from seeping into the soil. Toxic waste is placed in drums and covered with dirt. Leachate, the liquid mixture of wastes that seep to the bottom of the pit, passes into perforated collection pipes and is pumped to the surface for treatment. But landfills fail easily. Bulldozers can tear the plastic liner and leachate disintegrate it. The waste may crush collection pipes, and debris clog the perforations. If erosion breaches the landfill's protective cover, rain may mix with waste, overloading the collection system or causing the landfill to overflow.

Solid waste will never cease to be a problem, but its impact on the environment can be lessened by reducing the volume of waste that is generated, eliminating or reducing the production of toxic residues, halting irresponsible dumping, and finding ways to reuse the resources that waste contains. Until then, current methods of waste disposal will continue to pollute soil, air, and water.

Summary

Humans are part of the natural environment and depend, literally, for their lives on the water, air, and other resources contained in the biosphere. But people have subjected the intricately interconnected systems of that biosphere—the troposphere, the hydrosphere, and the earth's crust—to profound and frequently unwittingly destructive alteration. All human activities have effects on the environment, effects that are complex and never isolated. An external action that impinges on any part of the web of nature inevitably triggers chain reactions, the ultimate consequences of which appear never to be fully anticipated. The simple desire to suppress summer dust in the town of Times Beach, Missouri, led in the end to the demise of the community.

Efforts to control the supply of water alter both the quantity and quality of water in streams, and structures such as dams and reservoirs often create unintended side effects. Pollutants associated with agriculture, industry, and other activities have degraded the quality of freshwater supplies, although regulatory efforts have brought about major improvements in some areas in recent years.

Combustion of fossil fuels has contributed to serious problems of air pollution. Some manifestations of that pollution, such as acid rain and depletion of the ozone layer, are matters of global concern. If the theory of the greenhouse effect is valid, air pollution will alter normal patterns of temperature and precipitation sometime during the next century.

Activities such as agriculture and mining have long helped to shape local landscapes, producing a variety of landforms, and have also contributed to degradation of air, water, and soil. In this century, pressure exerted by world population growth and economic expansion have accelerated deforestation, soil erosion, and desertification over much of the world.

People affect other living things—plants and animals—by importing them to areas where they did not previously exist, disrupting their habitats, hunting, and using biocides to eradicate them. The greatest human impact is occurring in the tropical rain forests, where agriculture, industrialization, and urbanization are contributing to the extinction of at least 1000 species a year.

Finally, all the common methods of disposing of the wastes that humans produce release contaminants into the surrounding environment, providing further evidence of the fact that people cannot manipulate, distort, pollute, or destroy any part of the ecosystem without diminishing its quality or disrupting its structure.

Key Words

acid rain	hazardous waste
biological magnification	hydrologic cycle
biosphere	ozone
channelization	ozone layer
chlorofluorocarbons (CFCs)	polychlorinated biphenyls
desertification	(PCBs)
ecosystem	salinization
environmental pollution	soil erosion
eutrophication	subsidence
food chain	thermal pollution
greenhouse effect	troposphere

For Review

1. Sketch and label a diagram of the *biosphere*. Briefly indicate the content of its component parts. Is that content permanent and unchanging? Explain.

2. How are the concepts of *ecosystem*, *niche*, and *food chain* related? How does each add to our understanding of the "web of nature"?

3. Draw a diagram of or briefly describe the *hydrologic cycle*. How do people have an impact on that cycle? What effect does urbanization have on it?

4. Is all environmental pollution the result of human action? When can we say that pollution of a part of the biosphere has occurred?

5. Describe the chief sources of water pollution. What steps have the United States and other countries taken to control water pollution?

6. What factors affect the type and degree of air pollution found at a place? What is *acid rain*, and where is it a problem? Describe the relationship of *ozone* to *photochemical smog*. Why has the ozone layer been depleted?

7. What causes the greenhouse effect? What impact might it have on the environment?

8. What kinds of landforms has excavation produced? Dumping? What are the chief causes and effects of subsidence?

9. How have people diminished the amount and productivity of soil? What types of areas are particularly subject to desertification? Why has soil erosion accelerated in recent decades?

10. Briefly describe the chief ways that humans affect plant and animal life. What is meant by *biological magnification*? Why may the use of biocides be self-defeating?

11. What methods do communities use to dispose of solid waste? What ecological problems does solid-waste disposal present? How does the government define *hazardous waste*, and how is it disposed of?

Selected References

Abrahamson, Dean E., ed. *The Challenge of Global Warming.* Covelo, Calif.: Island Press, 1989.

Bailes, Kendall E., ed. *Environmental History: Critical Issues in Comparative Perspective.* Lanham, Md.: University Press of America, 1985.

Bolin, Bert, et al., eds. *The Greenhouse Effect, Climatic Change, and Ecosystems.* SCOPE 29. New York: Wiley, 1986.

Brown, Lester R., et al. *State of the World.* Washington, D.C.: Worldwatch Institute. Annual.

Brown, Lester R. and Edward C. Wolf. *Soil Erosion: Quiet Crisis in the World Economy.* Worldwatch Paper 60. Washington, D.C.: Worldwatch Institute, 1984.

Clark, William C., and R. E. Munn, eds. *Sustainable Development of the Biosphere.* International Institute for Applied Systems Analysis. Cambridge: Cambridge University Press, 1986.

Dix, H. M. *Environmental Pollution: Atmosphere, Land, Water, and Noise.* New York: Wiley, 1981.

Dregne, Harold E. *Desertification of Arid Lands.* New York: Harwood Academic Publishers, 1983.

Drew, David. *Man-Environment Processes.* London: Allen & Unwin, 1983.

Ehrlich, Anne H., and Paul R. Ehrlich. *Earth.* London: Methuen, 1987.

Elsom, Derek. *Atmospheric Pollution: Causes, Effects and Control Policies.* New York: Basil Blackwell Inc., 1987.

French, Hilary F. *Clearing the Air: A Global Agenda.* Worldwatch Paper 94. Washington, D.C.: Worldwatch Institute, 1990.

Goldman, Benjamin A., et al. *Hazardous Waste Management: Reducing the Risk.* Council on Economic Priorities. Washington, D.C.: Island Press, 1986.

Goudie, Andrew. *The Human Impact on the Natural Environment.* 2d ed. Oxford, England: Basil Blackwell Ltd., 1986.

Longman, K. A., and J. Jenik. *Tropical Forest and Its Environment.* 2d ed. New York: Longman Scientific & Technical/Wiley, 1987.

Louma, Samuel N. *Introduction to Environmental Issues.* New York: Macmillan, 1984.

McCormick, John. *Acid Earth: The Global Threat of Acid Pollution.* London: Earthscan/International Institute for Environment and Development, 1985.

Miller, G. Tyler, Jr. *Living in the Environment.* 6th ed. Belmont, Calif.: Wadsworth, 1990.

Neal, Homer, and J. R. Schubel. *Solid Waste Management and the Environment.* Englewood Cliffs, N.J.: Prentice-Hall, 1987.

Postel, Sandra. *Air Pollution, Acid Rain, and the Future of Forests.* Worldwatch Paper 58. Washington, D.C.: Worldwatch Institute, 1984.

Pye, Veronica, Ruth Patrick, and John Quarles. *Groundwater Contamination in the United States.* Philadelphia: University of Pennsylvania Press, 1983.

ReVelle, Charles, and Penelope ReVelle. *The Environment.* 3d ed. Boston: Jones and Bartlett, 1988.

Scientific American, 261, no. 3 (September 1989). Special issue.

Shea, Cynthia P. *Protecting Life on Earth: Steps to Save the Ozone Layer.* Worldwatch Paper 87. Washington, D.C.: Worldwatch Institute, 1988.

Sheridan, David. *Desertification of the United States.* Council on Environmental Quality. Washington, D.C.: Government Printing Office, 1981.

Southwick, Charles H., ed. *Global Ecology.* Sunderland, Mass.: Sinauer Associates, Inc., 1985.

Thomas, William, ed. *Man's Role in Changing the Face of the Earth.* Chicago: University of Chicago Press, 1956.

Wellburn, Alan. *Air Pollution and Acid Rain.* New York: Wiley, 1988.

Worster, Donald, ed. *The Ends of the Earth: Perspectives on Modern Environmental History.* Cambridge: Cambridge University Press, 1988.

P A R T

2

The Culture–Environment Tradition

The Crow country. The Great Spirit put it exactly in the right place; while you are in it, you fare well; whenever you get out of it, whichever way you travel, you fare worse. . . . The Crow country is in exactly right place. It has snowy mountains and sunny plains; all kinds of climates and good things for every season. When the summer heats scorch the prairies, you can draw up under the mountains, where the air is sweet and cool. . . . In the autumn when your horses are fat and strong from the mountain pastures, you can go down on the plains and hunt the buffalo or trap beaver on the streams. And when winter comes on, you can take shelter in the woody bottoms along the rivers.

The Crow country is exactly in the right place. Everything good is found there. There is no country like the Crow country.

Such was the opinion of Arapoosh, chief of the Crows, speaking of the Big Horn basin country of Wyoming in the early 19th century. In the 1860s, Captain Raynolds reported to the secretary of war that the basin was "repelling in all its characteristics, surrounded on all sides by mountain ridges [and presenting] but few agricultural advantages."

In the four chapters of Part I, our primary concern was with the physical landscape. But while the physical environment may be described by process and data, it takes on human meaning only through the filter of culture. Arapoosh and Raynolds viewed the same landscape, but from the standpoints of their separate cultures and conditionings. Culture is like a piece of tinted glass, affecting and distorting our view of the earth. Culture conditions the way people think about the land, the way they use and alter the land, and the way they interact with one another upon the land.

Such conditioning is the focus of the culture–environment tradition of geography, a tradition still concerned with the landscape but not in the physical science sense of Part I of this book. In Part II, humans are the focus. Of course, the physical environment is always in our minds as we develop the notion of cultural difference and reality.

Our landscapes, however, take on an added dimension and become human rather than purely physical.

The four chapters in this part of our study, therefore, concentrate upon the "people" portion of geography's environment–culture–people relationship. In Chapter 6, "Population Geography," we start with the basics of human populations—their numbers, compositions, distributions, growth trends, and the pressures they exert upon the resources of the lands they occupy. These quantitative and distributional aspects of peoples are important current concerns, as frequent popular reference to a "population explosion," public debate about legal and undocumented immigration, and speculation about population growth and food availability attest. More fundamentally, of course, numbers and locations of people are the essential background to all other understandings of human geography.

People are, however, more than numbers in a global counting parlor. They are, separately, individuals who think, react, and behave in response to the physical and social environments they occupy. In turn, those thoughts, actions, and responses are strongly conditioned by the standards and structures of the cultures to which the individuals belong. The world is a mosaic of culture groups and human landscapes that invite geographic study. Chapter 7, "Cultural Geography," introduces that study by examining the components and subsystems of culture, the ways culture changes, and the key variables in defining a culture and in producing variations in culture from place to place.

One of those variables, politics, is explored in Chapter 8, "Political Geography." Political systems and processes strongly influence the form and distribution of many elements of culture. Economic and transportation systems coincide with national boundaries. Political regulations, whether detailed zoning codes or broad environmental protection laws, have a marked effect on the cultural landscape. In some countries, even such apparently nonpolitical matters as religion, literature, music, and fine arts may be affected by governmental support of certain forms of expression and rejection and prohibition of others. In an increasingly interrelated world community, the international aspects of political association create supranational patterns of culture distinct from traditional linguistic, religious, or ethnic affiliations.

Patterns of human spatial behavior and the factors that account for the manner in which people use space are the subject of Chapter 9, "Behavioral Geography." The way people view the environment is important, for human actions are guided as much by how things are perceived to be arranged and structured as by the objective reality of their location and content. The individual and group behaviors of humans create the cultural landscapes—those of production, resource utilization, and urban settlement—that are the topics of Part III, "The Locational Tradition." Chapter 9, therefore, may be seen as an entity in itself and as a bridge to other subfields of the discipline.

Our theme in Part II, however, is people and the collective and personal cultural landscapes they create or envision. Let us begin with people themselves before we move on to consider their cultures and behaviors.

Fishermen on stilts along the south coast of Sri Lanka.

C H A P T E R

Population Geography

Sugamo district, Tokyo, Japan.

"ZERO, possibly even negative [population] growth" was the 1972 slogan proposed by the prime minister of Singapore, an island country in Southeast Asia. His country's population, which stood at 1 million at the end of the Second World War, had doubled by the mid-1960s. In support, the government decreed "Boy or girl, two is enough" and refused maternity leaves and access to health insurance for third or subsequent births. Abortion and sterilization were legalized, and children born fourth or later in a family were to be discriminated against in school admissions policy. In response, birth rates by the mid-1980s fell to below the level necessary to replace the population, and abortions were terminating more than one-third of all pregnancies,

"At least two. Better three. Four if you can afford it" was the national slogan proposed by that same prime minister in 1986, reflecting fears that the stringencies of the earlier campaign had gone too far. From concern that overpopulation would doom the country to perpetual Third World poverty, Prime Minister Lee Kuan Yew was moved to worry that population limitation would deprive it of the growth potential and national strength implicit in a youthful, educated work force adequate to replace and support the present aging population.

The policy reversal in Singapore reflects an inflexible population reality: the structure of the present determines the content of the future. The size, characteristics, and growth trends of today's populations help shape the well-being of peoples yet unborn, but whose numbers and distributions are now being determined. The numbers, age, and sex distribution of people; patterns and trends in their fertility and mortality; their density of settlement and rate of growth all affect and are affected by the social, political, and economic organization of a society. Through them, we begin to understand how the people in a given area live, how they may interact with one another, how they use the land, what pressure there is on resources, and what the future may bring.

Population geography provides the background tools and understandings of those interests. It focuses on the number, composition, and distribution of human beings in relation to variations in the conditions of earth space. It differs from *demography,* the statistical study of human population, in its concern with *spatial* analysis—the relationship of numbers to area. In seeking to describe and account for the uneven distribution of people over the earth's surface, population geographers study the temporal and spatial variation of the components of population change. Birth and death rates, fertility rates, and growth rates are among the data they examine as they study the spatial associations of population with both physical landscape elements and cultural geographic characteristics. Population geographers are also concerned with the significance of spatial differences and of current population trends in relation to the kind and quality of life that people lead—that is, the relationship of societies to such matters as type of economic development, level of living, supply of food and other resources, and conditions of health and well-being.

These are topics thrust into prominence in recent years by a "population explosion," dramatic increases in human numbers that have occurred in many parts of the globe. Those increases and the association between them and their supporting earth resources are fundamental expressions of the human–environmental relationships that are the substance of geographic inquiry.

Population Growth

Sometime during the spring of 1990 a human birth raised the earth's population to 5.3 billion people. In 1961 there were about 3 billion. That is, over the years between those two dates, the world's population grew on average by more than 80 million people annually, or some 220,000 per day. The rate has risen in recent years. By 1990, 93 million people were being added annually, or more than 250,000 per day. It is generally agreed that we will see a year 2000 population of 6.3 billion (the present century began with fewer than 2 billion) inhabitants. Eventually, most analysts assume, world population will stabilize at between 8 and 11 billion persons, with the majority of the growth occurring in countries now considered "less developed" (Figure 6.1). We will return to these projections and the difficulties inherent in making them later in this chapter.

Just what is implied by numbers in the millions and billions? With what can we equate the 1990 population of Gabon in Africa (a little over 1 million) or of China (a little over 1 billion)? Unless we have some grasp of their scale and meaning, our understanding of the data and data manipulations of the population geographer can at best be superficial. It is difficult to appreciate how vast a number is 1 million or 1 billion, and how great the distinction between them is. Some examples offered by the Population Reference Bureau may help you to visualize their immensity and implications.

A 1-inch stack of U.S. paper currency contains 233 bills. If you had a *million* dollars in thousand-dollar bills, the stack would be 4.3 inches high. If you had a *billion* dollars in thousand-dollar bills, your pile of money would reach 357 feet—about the length of a football field.

You had lived a *million* seconds when you were 11.6 days old. You won't be a *billion* seconds old until you are 31.7 years of age.

The supersonic airplane, the Concorde, could theoretically circle the globe in only 18.5 hours at its cruising speed of 1340 miles (2156 km) per hour. It would take 31 days for a passenger to journey a *million* miles on the Concorde, while a trip of a *billion* miles would last 85 years.

The implications of the present numbers and the potential increases in population are of vital current social, political, and, importantly, ecological concern. Population numbers were much smaller some 11,000 years ago, when continental glaciers began their retreat, people spread to formerly unoccupied portions of the globe, and human experimentation with food sources initiated the Agricultural Revolution. The 5 or 10 million people who then constituted all of humanity obviously had considerable potential to expand their numbers. In retrospect, we see that the natural resource base of the earth had a population-supporting capacity far in excess of the pressures exerted upon it by early hunting and gathering groups.

Some would maintain that despite present numbers or even those we can reasonably anticipate for the future, the adaptive and exploitive ingenuity of humans is in no danger of being taxed. Others, however, compare the earth to a self-contained spaceship and declare with chilling conviction that a finite vessel cannot bear an ever-increasing number of passengers. They point to recurring problems of malnutrition and starvation (though these are realistically more a matter of failures of distribution than of inability to produce enough foodstuffs worldwide). They cite dangerous conditions of air and water pollution, the nearing exhaustion of many mineral resources and fossil fuels, and other evidences of strains on world resources as foretelling the discernible outer limits of population growth.

Why are we suddenly confronted with what seems to many an insoluble problem—the apparently unending tendency of humankind to increase in numbers? On a worldwide basis, populations grow only one way: the number of births in a given period exceeds the number of deaths. Ignoring for the moment regional population changes resulting from migration, we can conclude that the observed and projected dramatic increases in population must result from the failure of natural controls to limit the number of births or to increase the number of deaths, or from the success of human ingenuity in circumventing such controls when they exist. The implications of such considerations will become clearer after we define some terms important in the study of world population and explore their significance.

Figure 6.1

World population numbers and projections. (a) Following two centuries of slow growth, world population began explosive expansion after World War II. Both United Nations and U.S. Bureau of the Census projections are for a global population of about 6.2 billion in A.D. 2000. (b) The greatest numerical growth will occur in Asia, and sub-Saharan Africa will show the highest percentage increase, but the relative population rankings of major world regions will remain the same.

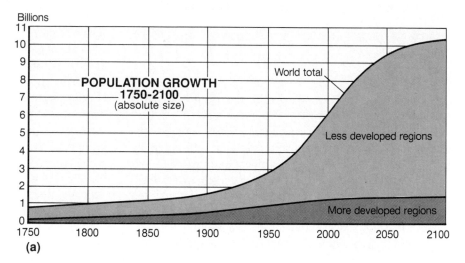

(a)

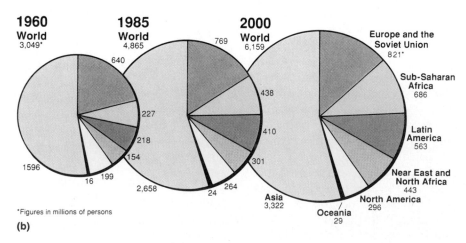

(b)

Some Population Definitions

Demographers employ a wide range of measures of population composition and trends, though all of their calculations start with a count of events: of individuals in the population, of births, deaths, marriages, and so on. To those basic counts demographers bring refinements that make the figures more meaningful and useful in population analysis. Among them are *rates* and *cohort measures.*

Rates simply record the frequency of occurrence of an event during a given time frame for a designated population—for example, the marriage rate as the number of marriages performed per 1000 population in the United States last year. **Cohort measures** refer data to a population group unified by a specified common characteristic—the age cohort of 1–5 years, perhaps, or the college class of 1994 (Figure 6.2). Basic 1990 counts and rates useful in the analysis of world population and population trends have been reprinted with the permission of the Population Reference Bureau as an appendix to this book. Examination of them will document the discussion that follows.

Birth Rates

The **crude birth rate** (CBR), often referred to simply as the *birth rate,* is the annual number of live births per 1000 population. It is "crude" because it relates births to total population without regard to the age or sex composition of that population. A country with a population of 2 million and with 40,000 births a year would have a crude birth rate of 20 per 1000.

$$\frac{40,000}{2,000,000} = 20 \text{ per } 1000$$

The birth rate of a country is, of course, strongly influenced by the age and sex structure of its population, by the customs and family size expectations of its inhabitants, and by its adopted population policies. Since these conditions vary widely, recorded national birth rates vary—in 1989, from a high of 52 per 1000 in African countries to the low of 10 per 1000 in Japan and Italy. Although birth rates greater than 30 per 1000 are considered *high,* more than half of the world's people live in countries with rates that are equally high or higher (Figure 6.3). In these countries the population is predominantly agricultural and rural, and a high proportion of the female population is young. They are found chiefly in Africa, southern Asia, and Latin America.

Birth rates of less than 20 per 1000 are reckoned *low* and are characteristic of industrialized, urbanized countries. Most European countries, the Soviet Union, Anglo-America, Japan, Australia, and New Zealand have such rates as, importantly, do a few developing states that have adopted stringent family planning programs. As recently as 1986, China was in this category, but later relaxation of family size controls has resulted in a jump in its birth rates (see "China's Way"). *Transitional* birth rates (between 20 and 30 per 1000) characterize some, mainly smaller, "developing" countries.

As the recent population histories of Singapore, recounted at the opening of this chapter, and China indicate, birth rates are subject to change. The low birth rates of European countries and of some of the areas that they colonized are usually ascribed to industrialization, urbanization, and in recent years, maturing populations. While restrictive family planning policies in China rapidly reduced the birth rate from over 33 per 1000 in 1970 to 18

Figure 6.2
Whatever their differences may be by race, sex, or ethnicity, these babies will forever be clustered demographically into a single *birth cohort.*

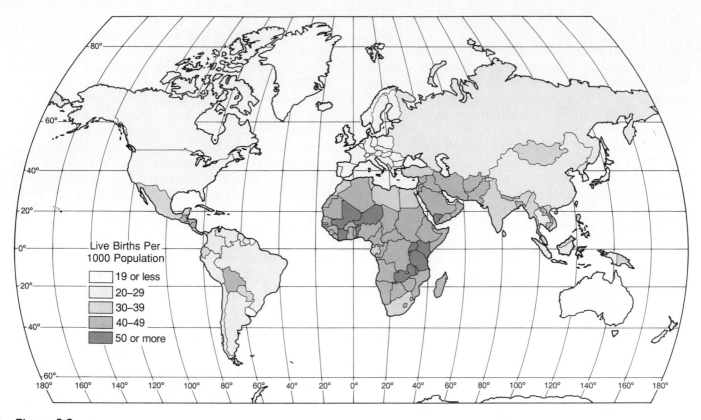

Figure 6.3

Crude birth rates. The map suggests a degree of precision that is misleading in the absence of reliable, universal registration of births. The pattern shown serves, however, as a generally useful summary of comparative reproduction patterns if class divisions are not taken too literally. Reported or estimated population data vary annually, so this and other population maps may not agree in all details with the figures recorded in the Appendix.

Live Births Per 1000 Population

	19 or less
	20–29
	30–39
	40–49
	50 or more

per 1000 in 1986, Japan experienced a 15-point decline in the decade 1948–1958 with little governmental intervention.

The stage of economic development is closely related to variations in birth rates among countries, although rigorous testing of this relationship proves it to be imperfect (Figure 6.3). As a group, the more developed states of the world showed a crude birth rate of 15 per 1000 in 1990; less developed countries (excluding China) registered 35 per 1000. Religious and political beliefs can also affect the rates. The convictions of many Roman Catholics and Muslims that their religion forbids the use of artificial birth control techniques often lead to high birth rates among believers (but dominantly Catholic Italy shows Europe's lowest birth rate, and Islam itself does not prohibit contraception). Similarly, some governments of both Western and Eastern Europe—concerned about birth rates too low to sustain present population levels—subsidize births in an attempt to raise those rates. Regional variations in percentage contributions to world population growth are summarized in Figure 6.4.

Fertility Rates

Crude birth rates may display such regional variability because of variations in age and sex composition or disparities in births among the reproductive-age, rather than total, population. A more accurate statement than the birth rate of the amount of reproduction in the population is the **total fertility rate** (TFR) (Figure 6.5). This rate tells us the average number of children that would be born to each woman if during her childbearing years she bore children at the current year's rate for women that age. The fertility rate minimizes the effects of fluctuation in the population structure and is thus a more reliable figure for regional comparative and predictive purposes than the crude birth rate.

A total fertility rate of 2.1 is necessary just to replace present population. On a worldwide basis, the TFR for 1990 was 3.5. The more developed countries recorded a 2.0 rate, while less developed states (excluding China) had a collective TFR of 4.0, down from 5.0 in the mid-1980s. Despite the rate disparity between the two classes of states, the fertility rates for many less developed countries have dropped a third or more since the early 1960s

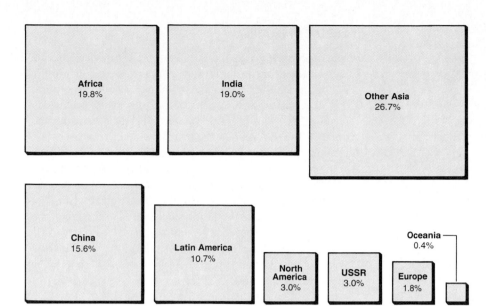

Figure 6.4
Percentage contributions to world population growth, by region, 1980–1990. Birth rate changes affecting different-sized regional populations are altering the world pattern of population increase. Between 1965 and 1975, China's contribution to world growth was two and a half times that of Africa. Between 1980 and 1990, Africa's numerical growth was 1.3 times that of China. China added 65 million more people to world population than did India between 1970 and 1980. In the next decade, India's growth exceeded that of China by nearly 30 million.

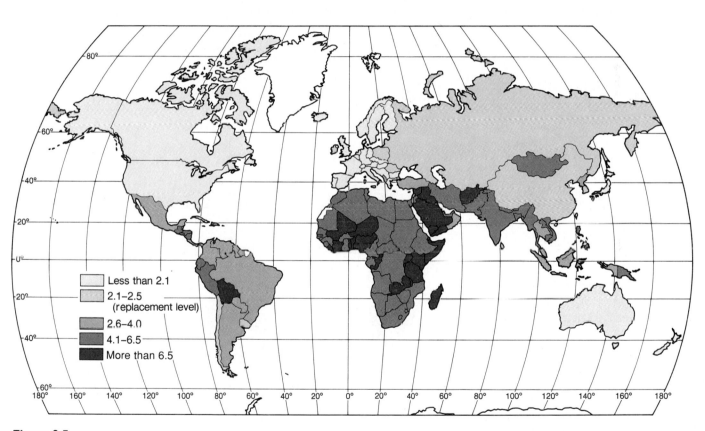

Figure 6.5

Total fertility rate indicates the average number of children that would be born to each woman if, during her childbearing years, she bore children at the same rate that women of those ages attained in a given year. Since the TFR is age-adjusted, two countries with identical birth rates may have quite different fertility rates and therefore different prospects for growth. Depending upon mortality conditions, a TFR of 2.1 to 2.5 children per family is considered the "replacement level," at which a population will eventually stop growing.

(Table 6.1). Substantial reductions occurred in several Asian and many Latin American countries. China's drop from a TFR of 5.8 births per woman in 1970 to 2.4 in 1989 was both most impressive and demographically most significant because of China's status as the world's foremost state in population. However, fertility rates are still well over 6 per woman in much of Africa and in many smaller Asian countries (see Appendix).

Table 6.1

Fertility Change in Selected Less Developed Countries and Regions

Region or country	Total fertility rate		
	Early 1960s	Late 1980s	% Change
East Asia	5.3	2.3	−56.6
China	5.9	2.4	−59.3
South Asia	6.1	4.8	−21.3
Afghanistan	7.0	6.9	−1.4
Bangladesh	6.7	5.8	−13.4
India	5.8	4.3	−25.9
Nepal	5.9	6.0	+1.7
Thailand	6.4	2.7	−57.8
Africa	6.7	6.3	−6.0
Egypt	7.1	5.3	−25.4
Kenya	8.2	8.1	−1.2
Nigeria	6.9	6.6	−4.3
Latin America	5.9	3.6	−39.0
Brazil	6.2	3.4	−43.2
Guatemala	6.8	5.6	−17.6
Mexico	6.7	3.8	−43.3

Death Rates

The **crude death rate** (CDR), also called the **mortality rate**, is calculated in the same way as the crude birth rate: the annual number of events per 1000 population. In general, the death rate, like the birth rate, varies with national levels of economic development. Highest rates (over 20 per 1000) are found in the less developed countries, particularly in Africa, and the lowest (less than 10) in developed countries, though the correlation is far from exact (Figure 6.6). Dramatic reductions in death rates occurred in the years following World War II as antibiotics, vaccination, and pesticides to treat diseases and control disease carriers were disseminated to almost all parts of the world. Distinctions between more developed and less developed countries in mortality were reduced but by no means erased, as death rates in many tropical, newly independent countries fell to a quarter or less of what they had been as late as the 1940s.

Like crude birth rates, death rates are meaningful for comparative purposes only when we study identically structured populations. Countries with a high proportion of elderly people, such as Austria and West Germany, would be expected to have higher death rates than those with a high proportion of young people, such as Spain,

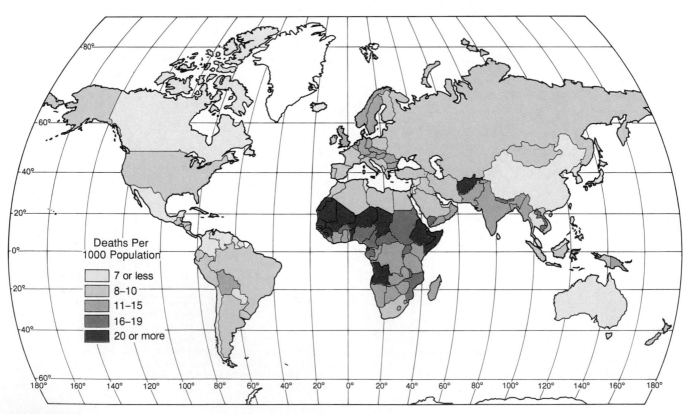

Figure 6.6

Crude death rates show less worldwide variability than do birth rates displayed in Figure 6.3, which is the result of widespread availability of at least minimal health protection measures and a generally youthful population in developing countries, where death rates are frequently lower than in "old age" Europe.

assuming equality in other national conditions affecting health and longevity. The pronounced youthfulness of populations in developing countries is an important factor in the recently reduced mortality rates of those areas.

To overcome that lack of comparability, death rates can be calculated for specific age groups. The *infant mortality rate,* for example, is the ratio of deaths of infants aged 1 year or under per 1000 live births:

$$\frac{\text{deaths age 1 year or less}}{1000 \text{ live births}}$$

Infant mortality rates are significant because it is at these ages that the greatest declines in mortality have occurred, largely as a result of the increased availability of health services. The drop in infant mortality accounts for a large part of the decline in the general death rate in the last few decades, for mortality during the first year of life is usually greater than in any other year.

Two centuries ago, it was not uncommon for 200–300 infants per 1000 to die in their first year. Even today, despite significant declines in those rates over the last half century in many individual countries (Figure 6.7), striking world regional and national contrasts remain. For all of Africa, infant mortality rates reached 109 per 1000, and individual African states (for example, Ethiopia and Sierra Leone) showed rates above 150 in 1990. In contrast, infant mortality rates in North America and Western Europe were in the 6–10 range.

The time and space variations in infant mortality and in crude death rates are obviously not due to differences in the innate ability of different national groups to survive. They are essentially the result of differential access to the health maintenance technologies known and widely adopted since the 1940s but not yet universally available. Penicillin and other antibiotics, as well as DDT and other pesticides, protect, preserve, and prolong lives that before their introduction would have been lost. Modern medicine and sanitation have increased life expectancy and altered age-old relationships between birth and death rates. The availability and employment of these modern methods, however, have varied regionally, and the developed countries have been most able to benefit from them. In such underdeveloped and impoverished areas as much of sub-Saharan Africa, the chief causes of death are those no longer of concern in more developed lands: infectious diseases, such as tuberculosis and malaria; intestinal infections; and among infants and children especially malnutrition and dehydration from diarrhea.

Population Pyramids

Another means of comparing populations is through the **population pyramid**, a graphic device that represents a population's age and sex composition. The term *pyramid* describes the diagram's shape for many countries in the 1800s, when the display was created: a broad base of younger age groups and a progressive narrowing toward the apex as older populations were thinned by death. Now many different shapes are encountered, each reflecting a different population history (Figure 6.8). By grouping several generations of people, the pyramids highlight the impact of "baby booms," population-reducing wars, birth rate reductions, and external migrations.

A rapidly growing country such as Mexico has most people in the lowest age cohorts; the percentage in older age groups declines successively, yielding a pyramid with markedly sloping sides. Typically, female life expectancy is reduced in older cohorts of less developed countries, so that for Mexico the proportion of females in older age groups is lower than in, for example, the United States or Sweden. In the latter, a wealthy country with a very slow

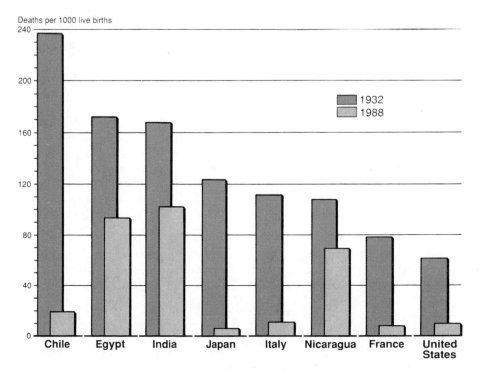

Deaths per 1000 live births

1932
1988

Chile   Egypt   India   Japan   Italy   Nicaragua   France   United States

Figure 6.7

Infant mortality rates for selected countries. Dramatic declines in the rate have occurred in all countries, a result of international programs of health care delivery aimed at infants and children in developing states. Nevertheless, the decreases have been proportionately greatest in the urbanized, industrialized countries, where sanitation, safe water, and quality health care are more widely available.

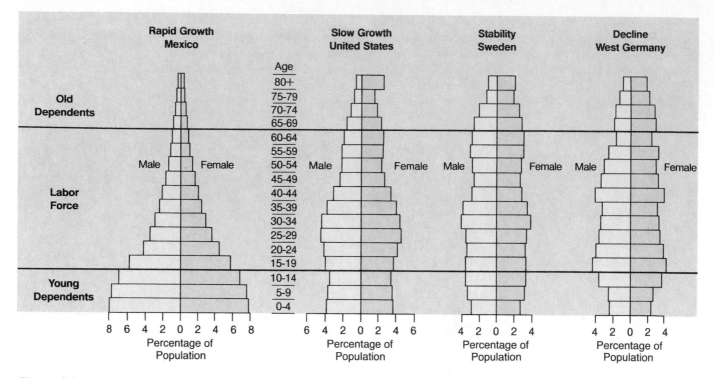

Figure 6.8

Four patterns of population structure. These diagrams show that population "pyramids" assume many shapes. The age distribution of national populations both records the present and foretells the future. In countries like Mexico, social costs related to the young are important and economic expansion is vital to provide employment for new entrants in the labor force. West Germany's negative growth means a future with fewer workers to support a growing demand for social services for the elderly. The U.S. pyramid is for 1987.

China's Way

"Every stomach comes with two hands attached" was Chairman Mao's explanation for his permissive attitude toward population growth. An ever-larger population is "a good thing" he announced in 1965 when the birth rate was 37 per 1000. At the start of Mao's rule in 1949, China had an estimated 540 million stomachs. At his death in 1976, population had risen to 852 million, the consequence of a drop in infant mortality rates, longer life spans, and—during the 1950s—the government's failure to recognize the impact of unchecked birth rates on the economy and on the livelihood and well-being of the population.

During the 1970s, when it became evident that population growth was consuming more than half of the annual increase in gross national product, China introduced a well-publicized campaign advocating the "two-child family" and providing services, including abortions, supporting that program. In response, China's growth rate dropped by 1975 to 15.7 per 1000,

down from 28.5 per 1000 just 10 years earlier. Mao's death cleared the way for an even more forceful program of birth restriction. "One couple, one child" became the slogan of a vigorous population control drive launched in 1979. "Husband and wife," stated the new constitution of 1982, "have a duty to practice family planning."

The "one child" program was consistently pushed, its message carried by billboards, newspapers, radio, and television and its achievement assured in a tightly controlled society by both incentives and penalties. Late marriages were encouraged. Free contraceptives, abortions, and sterilizations were provided. Cash awards and free medical care were promised to parents who agreed to limit their families to just a single child. Penalties for second births, including fines equal to 15% of family income for 7 years, were levied. At the campaign's height in 1983, the government ordered the sterilization of either husband or wife

for couples with more than one child. Social pressures in the form of public criticism and even, reportedly, forced abortions helped guarantee compliance with the birth limitation goals of the campaign. Tragically, infanticide—particularly the exposure or murder of female babies—was a reported means both of conforming to a one-child limit and of increasing the chances that the one child would be male.

By 1986, China's growth rate had fallen to 1%, far below the 2.4% registered among the rest of the world's less developed countries. Its birth rate stood at 18 per 1000, and its year 2000 population was projected by Chinese observers to be no more than 1.2 billion, the goal set at the outset of family planning in 1971. However, the 1987 jump in the birth rate to 21 (and a 1990 growth rate of 1.4%) reflected a relaxation of stringent controls, an increase in the number of young people of childbearing age, and changes in age at marriage.

rate of growth, the population is nearly equally divided among the age groups, giving a "pyramid" with almost vertical sides. Among older cohorts, as West Germany shows, there may be an imbalance between men and women because of the greater life expectancy of the latter. As a group, developing countries have a generalized population pyramid distinctly different from the collective pattern shown by more developed countries (Figure 6.9).

The population pyramid provides a quickly visualized demographic picture of immediate practical and predictive value. For example, the percentage of a country's population in each age group strongly influences demand for goods and services within that national economy. A country with a high proportion of young has a high demand for educational facilities and certain types of health delivery services. Additionally, of course, a large portion of the population is too young to be employed (Figures 6.9 and 6.10). On the other hand, a population with a high percentage of elderly people also requires medical goods and services specific to that age group, and these people must be supported by a smaller proportion of workers. The **dependency ratio** is a simple measure of the number of dependents, old or young, that each 100 people in the productive years (usually 20–64) must support. Population pyramids give quick visual evidence of that ratio.

Over the next several decades, its population pyramid tells us, the United States will likely become more and more like Sweden in demographic structure. The proportion of elderly will increase, the proportion of young

people will decrease, and the median age of the population will rise. Social services and costs related to an aging population will become greater national concerns (Figure 6.11). Mexico, on the other hand, faces continuing problems of expansion of educational facilities, training of

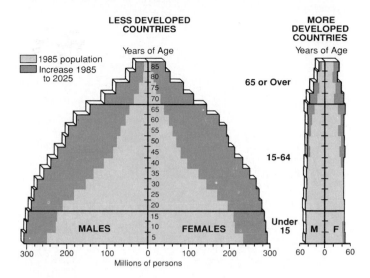

Figure 6.9

Generalized population pyramids. Less developed countries show a much younger age profile than do more developed countries. About 37% (1989) of their population is below age 15. In contrast, the proportion of population above age 65 in the more developed countries is almost three times that of the developing world. The projected results of those differences in 2025 are indicated.

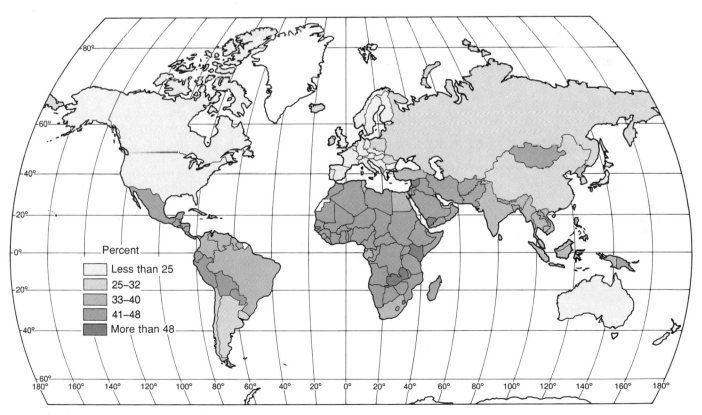

Figure 6.10

Percentage of population under 15 years of age. A high proportion of a national population under 15 increases the dependency ratio of that country and promises future population growth as the youthful cohorts enter childbearing years.

Figure 6.11

As these Dutch senior citizens exemplify, Europe is an aging continent with an ever-growing proportion of the elderly dependent upon the financial support of a reduced working-age population. Rapidly growing developing countries, in contrast, face increasing costs for the needs of the very young.

teachers, creation of jobs, and provision of health services required by a young and growing population.

Natural Increase

Knowledge of their sex and age distributions also enables countries to forecast their future population levels, though the reliability of projections decreases with increasing length of forecast (Figure 6.12). Thus, a country with a high proportion of young people will experience a high rate of natural increase unless there is a very high mortality rate among infants and juveniles or fertility and birth rates change materially. The **rate of natural increase** of a population is derived by subtracting the crude death rate from the crude birth rate. *Natural* means that increases or decreases due to migration are not included. If a country had a birth rate of 22 per 1000 and a death rate of 12 per 1000 for a given year, the rate of natural increase would be 10 per 1000. This rate is usually expressed as a percentage, that is, as a rate per 100 rather than per 1000. In the example given, the annual increase would be 1%.

Doubling Times

The rate of increase can be related to the time it takes for a population to double, that is, the **doubling time.** Table 6.2 shows that it would take 70 years for a population with a rate of increase of 1% (approximately the rate of growth of the Soviet Union in the late 1980s) to double. A 2% rate of increase means that the population will double in only 35 years. How could adding only 20 people per 1000 cause a population to grow so quickly? The principle is the same as that used to compound interest in a bank. Table 6.3 shows the number yielded by a 2% rate of increase at the end of successive 5-year periods.

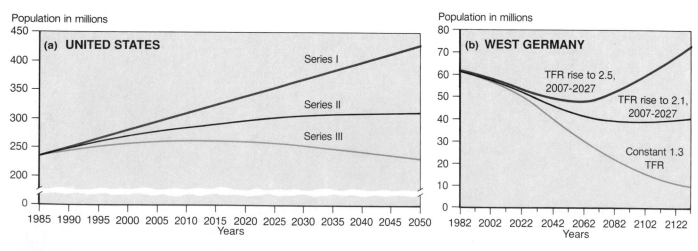

Figure 6.12

Possible population futures: United States and West Germany.
(a) U.S. population projections to year 2050. Population projections often prove inaccurate because birth and death rates and the number of immigrants are constantly changing. The middle series projection assumes middle levels of fertility, mortality, and immigration. Depending on the assumptions, U.S. population in 2050 might range from 231 million to 429 million.
(b) The West German projections are based solely upon varying

assumptions about that country's total fertility rate in 1983, the world's lowest at 1.3 births per woman in that year. If that rate remained constant, population would drop to just under 10 million by 2132. Even if the total population rate rose to the replacement level of 2.1 children during the years 2007–27 (the middle projection), the population decrease would still amount to a one-third reduction from the 61.5 million of 1989.

For the world as a whole, the rates of increase have risen over the span of human history. Therefore, the doubling time has decreased. Note in Table 6.4 how the population of the world has doubled in successively shorter periods of time. It may reach 9.5 billion during the first half of the 21st century if the present rate of growth continues (Figure 6.1). In countries with high rates of increase (Figure 6.13), the doubling time is less than the 39 years projected for the world as a whole (at growth rates displayed in 1990). Should world fertility rates decline, population doubling time will correspondingly increase (Figure 6.14).

Table 6.2
Doubling Time in Years at Different Rates of Increase

Annual percentage increase	Doubling time (years)
0.5	140
1.0	70
2.0	35
3.0	24
4.0	17
5.0	14
10.0	7

Table 6.3
Population Growth Yielded by a 2% Rate of Increase

Year	Population
0	1000
5	1104
10	1219
15	1345
20	1485
25	1640
30	1810
35	2000

Table 6.4
Population Growth and Approximate Doubling Times Since A.D. 1

Year	Estimated population	Doubling time (years)
1	250 million	
1650	500 million	1650
1850	1 billion	200
1930	2 billion	80
1975	4 billion	45
?	8 billion	?

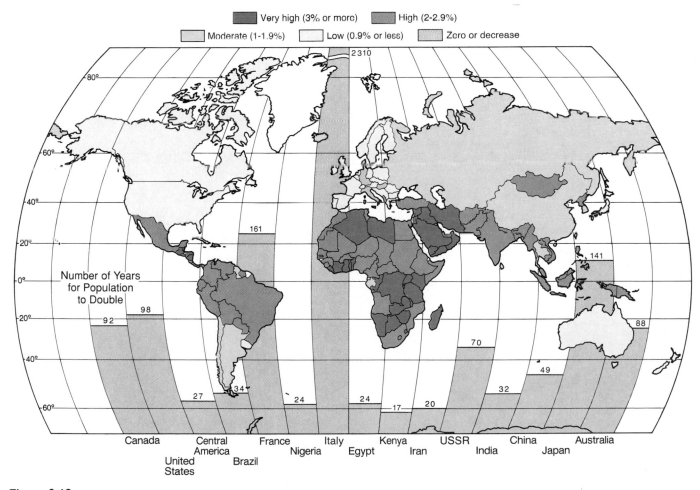

Figure 6.13

Annual rates of natural increase. The world's 1989 rate of natural increase (1.8%) would mean a doubling of population in 39 years. Many individual continents and countries, of course, deviate widely from the global average rate of growth. Africa as a whole has the highest rates of increase, followed by Central and South America. Europe and North America are prominent among the low-growth areas, with such countries as Italy actually experiencing single year negative growth and showing doubling times measured in millennia.

Here, then, lies the answer to the question posed earlier. Even small annual additions accumulate to large total increments because we are dealing with geometric or exponential (1, 2, 4, 8) rather than arithmetic (1, 2, 3, 4) growth. The ever-increasing base population has reached such a size that each additional doubling results in an astronomical increase in the total. One observer, G. Tyler Miller, has offered a graphic illustration of the inevitable consequences of such doubling, or **J-curve**, growth. He asks that we mentally fold in half a page of this book. We have, of course, doubled its thickness. We can—mentally, though certainly not physically—continue the doubling process several times without great effect, but 12 such doublings would yield a thickness of 1 foot (30.5 cm) and 20 doublings would result in a thickness of nearly 260 feet (79 m). From then on, the results of further doubling are astounding. Doubling the page only about 52 times, Miller reports, would give a thickness reaching from the earth to the sun. Rounding the bend on the J-curve, which world population has done (Figure 6.15), poses problems and has implications for human occupance of the earth of a vastly greater order of magnitude than ever faced before.

The Demographic Transition

The theoretical consequence of exponential population growth cannot be realized. Some form of braking mechanism must necessarily operate to control totally unregulated population growth. If voluntary population limitation is not undertaken, involuntary controls of an unpleasant nature may be set in motion.

One attempt to summarize an observed relationship between population growth and economic development is the **demographic transition model,** which traces the changing levels of human fertility and mortality presumably associated with industrialization and urbanization. Over time, the model assumes, high birth and death rates will gradually be replaced by low rates (Figure 6.16). The *first stage* of that replacement process—and of the demographic transition model—is characterized by high birth and high but fluctuating death rates.

As long as births only slightly exceed deaths, even when the rates of both are high, the population will grow only slowly. This was the case for most of human history

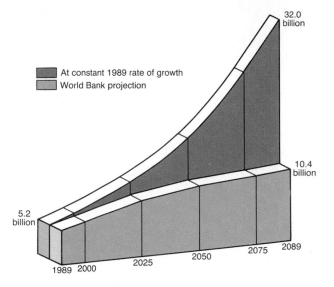

Figure 6.14

The purpose of the "doubling time" calculation is to illustrate the long-range effect of growth rates upon populations. It should never be used to suggest a prediction of future population size, for population growth reflects not just birth rates, but death rates, age structure, and migration. Demographers generally assume that high present growth rates will gradually be reduced. Therefore, if population does double, it will take longer than is suggested by "doubling time" based on the current rate.

Figure 6.15

World population growth 8000 B.C. to A.D. 2000. Notice that the bend in the J-curve begins in about the mid-1700s when industrialization started to provide new means to support the population growth made possible by revolutionary changes in agriculture and food supply. Improvements in medical science and nutrition served to reduce death rates near the opening of the 20th century in the industrializing countries.

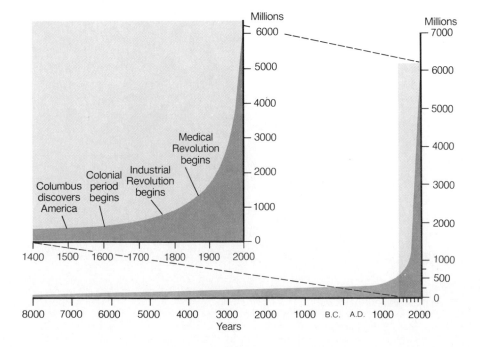

until about A.D. 1750. Demographers think that it took from approximately A.D. 1 to A.D. 1650 for the population to increase from 250 million to 500 million, a doubling time of more than a millennium and a half. Growth was not steady, of course. There were periods of regional expansion that were usually offset by sometimes catastrophic decline. Wars, famine, and other disasters took heavy tolls. For example, the bubonic plague (the Black Death), which swept across Europe in the 14th century, is estimated to have killed over one-third of the population of that continent. The first stage of the demographic transition model is no longer found in any country. In 1990, the highest death rates—found in several African and Asian countries—were in the low 20s per 1000; but the birth rates in many of those same countries were even higher, near or above 50 per 1000.

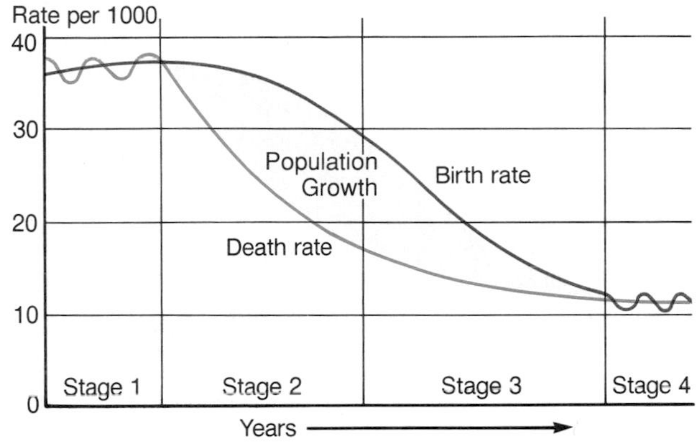

The Western Experience

The demographic transition model was developed to explain the population history of Western Europe. That area entered a *second stage* with the industrialization that began about 1750. Its effects—declining death rates accompanied by continuing high birth rates—have been dispersed worldwide even without universal conversion to an industrial economy. Rapidly rising populations during the second demographic stage result from dramatic increases in life expectancy. That, in turn, reflects falling death rates due to advances in medical and sanitation practices, improved foodstuff storage and distribution, a rising per capita income, and the urbanization that provides the environment in which sanitary, medical, and food distributional improvements are concentrated (Figure 6.17). Birth rates do not fall as soon as death rates; ingrained cultural patterns change more slowly than technologies. In many societies, large families arc considered advantageous. Children contribute to the family by starting to work at an early age and by supporting their parents in old age.

Many countries in Latin America and parts of southern and southeastern Asia display the characteristics of this second stage in the population model. Pakistan,

Figure 6.16

Stages in the demographic transition. During the first stage, birth and death rates are both high, and population grows slowly. When the death rate drops and the birth rate remains high, there is a rapid increase in numbers. During the third stage, birth rates decline and population growth is less rapid. The fourth stage is marked by low birth and death rates and, consequently, by a low rate of natural increase.

Figure 6.17

Vienna, Austria, in the 1870s. A modernizing Europe experienced improved living conditions and declining death rates during the 19th century.

with a birth rate of 43 and a death rate of 13, and Bolivia, with respective rates of 38 and 12 (1990 estimates), are typical. The annual rates of increase of such countries are near 30 per 1000, and their populations will double in about 25 years. Such rates, of course, do not mean that the full impact of the Industrial Revolution has been worldwide; they do mean that the underdeveloped societies have been beneficiaries of the life preservation techniques associated with it.

The *third stage* follows when birth rates decline as people begin to control family size. The advantages that having many children bring in an agrarian society are not so evident in urbanized, industrialized cultures. In fact, such cultures may view children as economic liabilities rather than assets. When the birth rate falls and death rate remains low, the population size begins to level off. Chile, Sri Lanka, and Thailand are among the many countries now displaying the low death rates and transitional birth rates of the third stage.

The demographic transition ends with a *fourth* and final stage. Essentially all European countries, Canada, Australia, and Japan are among the 30 or so states that have entered this phase. Because it is characterized by very low birth and death rates, it yields very slight percentage increases in population. Population doubling times may be as long as a thousand or more years if those present low birth rates continue. In a few countries, such as Denmark, Germany, and Hungary, death rates have begun to equal or exceed birth rates, and populations are declining. (See "Europe's Population Dilemma.")

The demographic transition model describes the experience of northwest European countries as they went from rural-agrarian societies to urban-industrial ones. It may not fully reflect the prospects of contemporary developing countries. In Europe, church and municipal records, some dating from the 16th century, show that people tended to marry late or not at all. In England before the Industrial Revolution as many as half of all women in the 15–50 age cohort were unmarried. Infant mortality was high, life expectancy low. With the coming of industrialization in the 18th and 19th centuries, immediate factory wages instead of long apprenticeship programs permitted earlier marriage and more children. Since improvements in sanitation and health came only slowly, death rates remained high. Around 1800, 25% of Swedish infants died before their first birthday. Population growth rates remained below 1% per year in France throughout the 19th century.

Beginning about 1860, first death rates and then birth rates began their dramatic decline. The mortality revolution came first as an *epidemiologic transition* echoed the demographic transition with which it is associated. Many formerly fatal epidemic diseases become endemic, that is, essentially continual within a population. As people developed partial immunities, mortalities associated with the diseases decreased. Improvements in animal husbandry, crop rotation, and other agricultural practices, and new foodstuffs (the potato was an early example) from overseas colonies raised the level of health and disease resistance of the European population in general. At the same time, sewage systems and sanitary water supplies became common in larger cities, and general levels of hygiene improved everywhere (Figure 6.18). Deaths due to infectious, parasitic, and respiratory diseases and to malnutrition declined, and those related to a maturing and aging population increased. Western Europe passed from a first stage "Age of Pestilence and Famine" to an ultimate "Age of Degenerative and Man-made Diseases." However, recent increases in drug- and antibiotic-resistant diseases, pesticide resistance of disease-carrying insects, and such new scourges of both the less developed and more developed countries as AIDS (acquired immune deficiency syndrome) cast doubt on the finality of that "ultimate" stage.

The striking reduction in death rates was echoed by similar declines in birth rates as European societies began to alter their traditional concepts of ideal family size. In cities, child labor laws and mandatory schooling meant that children became a burden, not a contribution, to family economies. As "poor-relief" legislation and other forms of public welfare substituted for family support structures, the insurance value of children declined. Family consumption patterns altered as the Industrial Revolution made more widely available goods that served consumption desires, not just basic living needs. Children hindered rather than aided the achievement of the age's promise of social mobility and life-style improvement. Perhaps most important, and by some measures preceding and independent of the implications of the Industrial Revolution, were changes in the status of women and in their spreading conviction that control over childbearing was within their power and to their benefit.

A World Divided

The demographic transition model described the presumed inevitable course of population events from the high birth and death rates of premodern (underdeveloped) societies to the low and stable rates of advanced (developed) countries. It failed to anticipate, however, that by the 1990s many developing societies would apparently be locked in the second stage of the model, unable to realize the economic gains and social changes necessary to progress to the third stage of falling birth rates. The population history of Europe was apparently not inevitably or fully applicable to Third World countries of the middle and late 20th century.

Death rates, of course, could be, and were, lowered quickly and dramatically through the introduction of Western technologies of medicine and public health. These

Figure 6.18
Pure piped water replacing individual or neighborhood wells and sewers and waste treatment plants instead of privies became increasingly common in urban Europe and North America during the 19th century. Their modern successors, such as the San José, California, treatment plant shown here, helped complete the *epidemiologic transition* in developed countries.

Europe's Population Dilemma

Although international dismay may be expressed over rising world populations, some European states are facing an opposite domestic concern. Europe's population is older than that of any other continent, and for many of its countries, population is stagnating or declining. The fertility rates of 23 of the continent's 30 countries in 1989 were below the **replacement level**—the level of fertility at which populations replace themselves—of 2.1 per 1000. None of the large countries of Western Europe is at present replacing its population through natural increase. With 1989 fertility rates of 1.3 to 1.4, Austria, Italy, and West Germany stand at the bottom of the international reproduction scale. Not surprisingly, West Germany also has the oldest population in the world, with a small proportion of young and a large share of middle-aged and retired persons in its population pyramid. Most of Eastern Europe has the same problem of declining population growth, reduced work-age cohorts, and an aging citizenry. Europe's population will begin to decline in the 1990s.

Spatially, the continent's fertility is peripheral. Catholic Ireland and Poland have fertility rates a little above the replacement level. Only Albania and Turkey, where high Muslim reproduction rates are the rule, are far above the replacement fertility point. The Soviet Union, with a fertility rate of 2.5, owes much of its growth to its Central Asian Muslim, not its European, population. "In demographic terms," France's prime minister remarked, "Europe is vanishing."

The national social and economic consequences of population stability or reduction are not always perceived by those who advocate **zero population growth,** a condition achieved when births plus immigration equal deaths plus emigration. An exact equation of births and deaths means an increasing proportion of older citizens, fewer young people, and a rise in the median age of the population. Actual population decline, now the common European condition, exaggerates those consequences. Already, schools are closing and universities cut back in the face of permanently reduced demand. By the end of the century, the unemployment of the 1980s may be replaced by a shortage of workers, particularly skilled workers. Governments will have to provide pensions and social services for the one-quarter of their citizens older than 60 and pay for them by taxes on a diminishing work force. Already in West Germany there are four pensioners for every ten workers. By 2030, the numbers will be equal.

France, which grants substantial payments and services to parents to help in child rearing, is the only West European state with an official probirth policy. East Europe is much more active in encouraging growth. Three children per family has become the adopted goal of several Eastern bloc countries that fear for national survival if present trends continue. Parental subsidies, housing priorities, and generous maternity leaves at full salaries are among the inducements for multiple-child families. In Romania, before its recent revolution, contraceptives were banned and abortions permitted only on doctor's orders in the hope that birth rates might be increased. In general, despite some short-term successes, Europeans of east and west have ignored both blandishments and restrictions. The consequences will only gradually emerge over the next several decades.

included antibiotics, insecticides, sanitation, immunization, infant and child health care, and eradication of smallpox. Such imported technologies and treatments accomplished in a few years what it took Europe 50 or 100 years to experience. Sri Lanka, for example, sprayed extensively with DDT to combat malaria, and life expectancy jumped from 44 years in 1946 to 60 only 8 years later. With similar public health programs, India also experienced a steady reduction in its death rate after 1947. Simultaneously, with international sponsorship, food aid cut the death toll of developing states during drought and other disasters. The dramatic decline in mortality, which emerged only gradually throughout the European world but occurred so rapidly in contemporary developing countries, has been the most fundamental demographic change in human history.

Corresponding reductions in birth rates have been harder to achieve and depend less on supplied technology and assistance than they do on social acceptance of the idea of fewer children and smaller families (Figure 6.19). Simply put, fertility falls as incomes increase, health improves, and (particularly, female) educational levels rise, conditions that make parents desire fewer children. But not all women or cultures want smaller families. Surveys of several African countries indicate a continuing preference among women for large families of between six and nine children. And not all societies have adopted the birth restriction programs so effectively followed in Singapore, China, Thailand, and many other countries.

The consequence is a world polarized demographically into two distinct groups. Roughly half of the world's countries—voluntarily or through national plan—have limited their rates of natural increase to about 0.8% annually. The other half are growing, on average, at triple that rate. In both instances, the established pattern tends to become self-reinforcing. Low growth permits the expansion of personal income and accumulation of capital that enhance the quality and security of life and make large families less attractive or essential. When the population doubles each generation, as it must at the fertility rates of the high-growth half of the divided world, a different reinforcing mechanism operates. Population growth consumes in social services and assistance the investment capital that might promote economic expansion. Increasing populations place ever-greater demands on limited soil, forest, water, grassland, and cropland resources. Those pressures may, through human-induced deforestation and desertification, for example, consume the environmental base itself. Productivity declines and population-supporting capacities are so diminished as to make difficult or impossible the economic progress upon which the demographic transition depends.

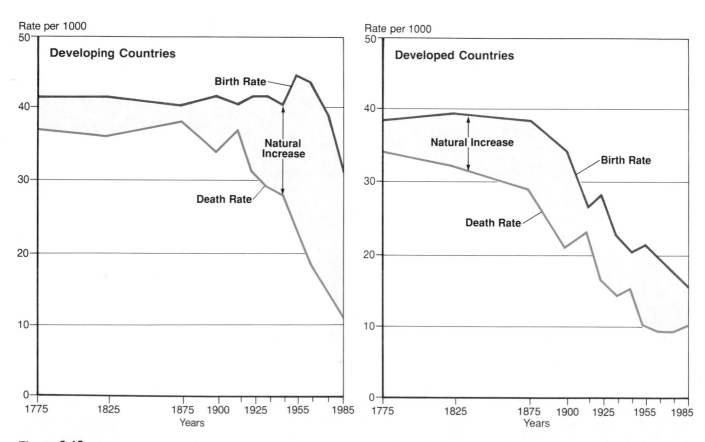

Figure 6.19

World birth and death rates. The "population explosion" after World War II (1939–45) reflected the effects of drastically reduced death rates in developing countries without compensating reductions in births. Mortality declines in European and western societies had been more gradual, with birth rate reductions keeping pace.

The Demographic Equation

Births and deaths among a region's population—natural increases or decreases—tell only part of the story of population change. Migration involves the movement of people from one residential location to another. When that relocation occurs across political boundaries, it affects the population structure of both the origin and destination jurisdictions. The **demographic equation** summarizes the contribution made to regional population change over time by the combination of *natural change* (difference between births and deaths) and *net migration* (difference between in-migration and out-migration). On a global scale, of course, all population change is accounted for by natural change. The impact of migration on the demographic equation increases as the population size of the areal unit studied decreases.

Population Relocation

In the past, emigration proved an important device for relieving the pressures of rapid population growth in at least some European countries (Figure 6.20). For example, in one 90-year span, 45% of the natural increase in the population of the British Isles emigrated, and between 1846 and 1935 some 60 million Europeans of all nationalities left that continent. Despite recent massive movements of economic and political refugees across Asian, African, and Latin American boundaries, emigration today provides no comparable relief valve for Third World countries (Table 6.5). Total population numbers are too great to be much affected by migrations of even millions of people. A more detailed treatment of the processes and patterns of international and intranational migrations as expressions of spatial interaction is presented in Chapter 9.

Immigration Impacts

Where, however, cross-border movements are massive enough, migration may have a pronounced impact upon the demographic equation and result in significant changes in the population structures of both the origin and destination regions. Past European and African migrations, for example, not only altered, but substantially created, the population structures of new, sparsely inhabited lands of colonization in the Western Hemisphere and Australasia. In some decades 30% to more than 40% of population increase in the United States was accounted for by immigration (Table 6.6). Similarly, eastward-moving Slavs colonized underpopulated Siberia and overwhelmed native peoples.

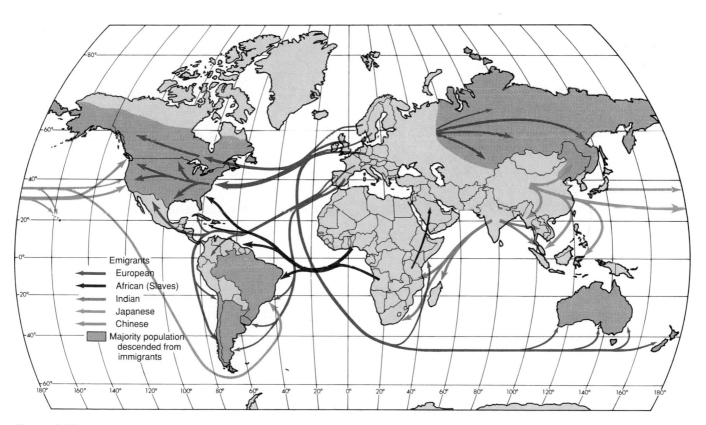

Figure 6.20

Principal migrations of recent centuries. The arrows suggest the major free and forced international population movements since about 1700. The shaded areas on the map are regions whose present population is more than 50% descended from these immigrants of recent centuries.

Table 6.5

Percentage of Natural Population Increase That Has Permanently Emigrated

Period	Europe	Asiaª	Africa	Latin Americaª
1851–1880	11.7	0.4	*b	0.3
1881–1910	19.5	0.3	*b	0.9
1911–1940	14.4	0.1	*b	1.8
1940–1960	2.7ᶜ	0.1	0.1	1.0
1960–1970	5.2	0.2	0.3	1.0
1970–1980	4.0	0.5		2.5

ªThe periods from 1850 to 1960 report emigration only to the United States.
*b = less than 0.1 percent.
ᶜEmigration only to the United States.

Table 6.6

Immigration in U.S. Population Growth

Decade	Percentage increase in foreign-born population	Immigration as % of total population growth
1870–1880	20	29
1880–1890	39	43
1890–1900	12	32
1900–1910	31	42
1910–1920	3	17
1920–1930	3	22

Migrants are rarely a representative cross section of the population group they leave, and they add an unbalanced age and sex component to the group they join. A recurrent research observation is that emigrant groups are heavily skewed in favor of young singles. Whether males or females dominate the outflow seems to vary with circumstances, though in international flows males nearly always seem to exceed females. In West Africa, males held a significant dominance over females among international migrants (Table 6.7).

At the least, then, the receiving country will have its population structure altered by an outside increase in its younger age and, probably, male cohorts. The results are both immediate, in altered population structure, and potential, in future impact on reproduction rates and excess of births over deaths. The origin area will have lost a portion of its young, active members of childbearing years. It perhaps will have suffered distortion in its young adult sex ratios, and it certainly will have recorded a statistical aging of its population. The receiving society will likely experience increases in births associated with the youthful newcomers and, in general, have its average age reduced.

World Population Distribution

The millions and billions of people of our discussion are not uniformly distributed over the earth. The most striking feature of the world population distribution map (Figure 6.21) is the very unevenness of the pattern. Some land areas are nearly uninhabited, others are sparsely settled, and still others contain dense agglomerations of people. More than half the world's people are found—unevenly concentrated, to be sure—in rural areas. More than 40%

are urbanites, however, and a constantly growing proportion are residents of very large cities of 1 million or more.

Earth regions of apparently very similar physical makeup show quite different population numbers and densities, perhaps the result of differently timed settlement or of settlement by different cultural groups. Had North America been settled by Chinese instead of Europeans, for example, it is likely that its western sections would be far more densely settled than they now are. Europe, inhabited thousands of years before North America, contains about twice the population of the United States on less land.

We can draw certain generalizing conclusions from the uneven but far from irrational distribution of population shown in Figure 6.21. First, about 90% of all people live north of the equator and 75% of the total dwell in the midlatitudes between 20° and 60° North. Second, a large majority of the world's inhabitants occupy only a small part of its land surface. Over half the people live on about 5% of the land, two-thirds on 10%, and almost nine-tenths on less than 20%. Third, people congregate in lowland areas; their numbers decrease sharply with increases in elevation. Temperature, length of growing season, slope and erosion problems, even oxygen reductions at very high altitudes, all appear to limit the habitability of higher elevations. One estimate is that between 50% and 60% of all people live below 650 feet (200 meters), a zone containing less than 30% of total land area. Nearly 80% reside below 1650 feet (500 meters).

Fourth, although low-lying areas are preferred settlement locations, not all such areas are equally favored. Continental margins have attracted densest settlement. About two-thirds of world population is concentrated within 300 miles (500 km) of the ocean, much of it on alluvial lowlands and river valleys. Latitude, aridity, and

Table 6.7

Number of Males per 100 Females among Working-age West African Immigrants to Selected Countries in the mid-1970s

Country	Males
Gambia	181
Ghana	166
Ivory Coast	177
Liberia	164
Senegal	119
Sierra Leone	187
Togo	88
Burkina Faso	79
All countries	156

elevation, however, limit the attractiveness of many seafront locations. Low temperatures and infertile soils of extensive coastal lowlands of higher latitudes of the Northern Hemisphere have restricted settlement there despite the low elevations of the Arctic coastal plain and the advantages of ocean proximity. Mountainous or desert coasts are sparsely occupied at any latitude, and some tropical lowlands and river valleys that are marshy, forested, and disease infested are unevenly settled.

Within the sections of the world generally conducive to settlement, four areas contain great clusters of population: East Asia, South Asia, Europe, and northeastern United States-southeastern Canada. The *East Asia* zone, which includes Japan, China, Taiwan, and South Korea, is areally the largest cluster. The four countries forming it contain 25% of all people on earth, and China alone accounts for one in five of the world's inhabitants. The *South Asia* cluster is composed primarily of countries associated with the Indian subcontinent—Bangladesh, India, Pakistan, and the island state of Sri Lanka—though some might add to it the Southeast Asian countries of Cambodia (Kampuchea), Myanmar (Burma), and Thailand. The four core countries alone account for another one-fifth, 20%, of the world's 5.3 billion inhabitants. The South and the East Asian concentrations are thus home to nearly one-half of the world's people.

Europe—southern, western, and eastern through much of European USSR—is the third extensive world population concentration, with another 13% of its inhabitants. Much smaller in extent and total numbers is the cluster in *northeastern United States and adjacent Canada*. Other such smaller but pronounced concentrations are found around the globe: on the island of Jawa (Java) in Indonesia, along the Nile River in Egypt, and in discontinuous pockets in Africa and Latin America.

The term **ecumene** is applied to permanently inhabited areas of the earth's surface. The ancient Greeks used the word, derived from their verb "to inhabit," to describe their known world between what they believed to be the unpopulated searing southern equatorial and permanently frozen northern polar reaches of the earth. Clearly, natural conditions are less restrictive than Greek geographers believed. Both ancient and modern technologies have rendered habitable areas that natural conditions make forbidding. Irrigation, terracing, diking, and draining are among the methods devised to extend the ecumene locally (Figure 6.22).

At the world scale, the ancient observation of habitability appears remarkably astute. The **nonecumene**, or *anecumene,* the uninhabited or very sparsely occupied zone, does include the permanent ice caps of the Far North and Antarctica and large segments of the tundra and coniferous forest of northern Asia and North America. But the nonecumene is not continuous, as the ancients supposed. It is discontinuously encountered in all portions of the globe and includes parts of the tropical rain forests of equatorial zones, midlatitude deserts of both Northern and Southern Hemispheres, and high mountain areas. Even parts of these unoccupied or sparsely occupied districts have localized dense settlement nodes or zones based on irrigation agriculture, mining and industrial activities, and the like. Perhaps the most anomalous case of settlement in the nonecumene world is that of the dense population in the Andes Mountains of South America and the plateau of Mexico. Here, Native Americans found temperate conditions away from the dry coast regions and the hot, wet Amazon basin. The fertile high basins have served a large population for more than a thousand years.

Even with these locally important exceptions, the nonecumene portion of the earth is extensive. Some 35–40% of all the world's land surface is inhospitable and without significant settlement. This is, admittedly, a smaller proportion of the earth than would have qualified as uninhabitable in ancient times or even during the last century. Since the end of the Ice Age some 11,000 years ago, humans have steadily expanded their areas of settlement.

Population Density

Margins of habitation could only be extended, of course, as humans learned to support themselves from the resources of new settlement areas. The numbers that could be sustained in old or new habitation zones were, and are, related to the resource potential of those areas and the cultural levels and technologies possessed by the occupying populations. We express the relationship between number of inhabitants and the area they occupy as **population density.**

Figure 6.21
World population density.

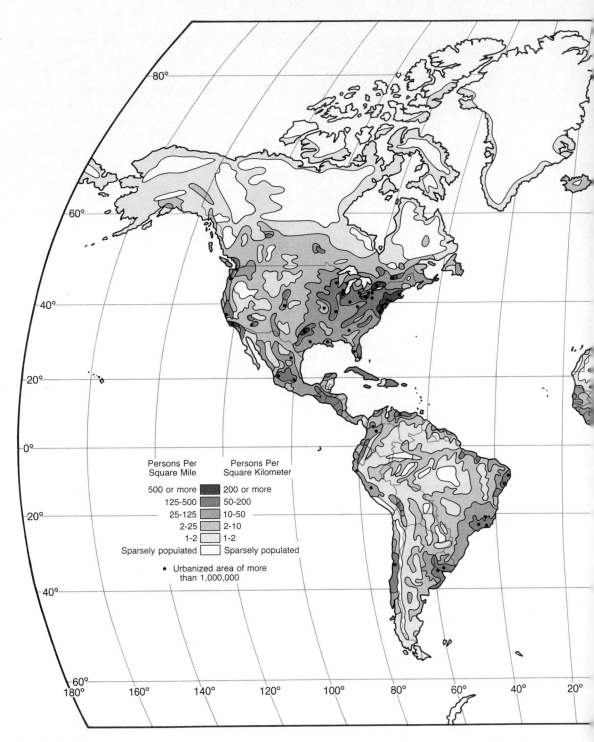

Persons Per Square Mile

500 or more	■
125-500	■
25-125	■
2-25	□
1-2	□
Sparsely populated	□

Persons Per Square Kilometer

200 or more	■
50-200	■
10-50	■
2-10	□
1-2	□
Sparsely populated	□

• Urbanized area of more than 1,000,000

Density figures are useful, if sometimes misleading, representations of regional variations of human distribution. The **crude density,** or **arithmetic density,** of population is the commonest and least satisfying expression of that variation. It is the calculation of the number of people per unit area of land, usually within the boundaries of a political entity. It is an easily reckoned figure. All that is required is information on total population and total area, both commonly available for national or other political units. It can, however, be misleading and may obscure more of reality than it reveals. It is an average, and a country may contain extensive regions that are only sparsely populated or largely undevelopable along with intensively settled and developed districts. A national average density figure reveals nothing about either class of territory. In general, the larger the political unit for which crude or arithmetic population density is calculated, the less useful is the figure.

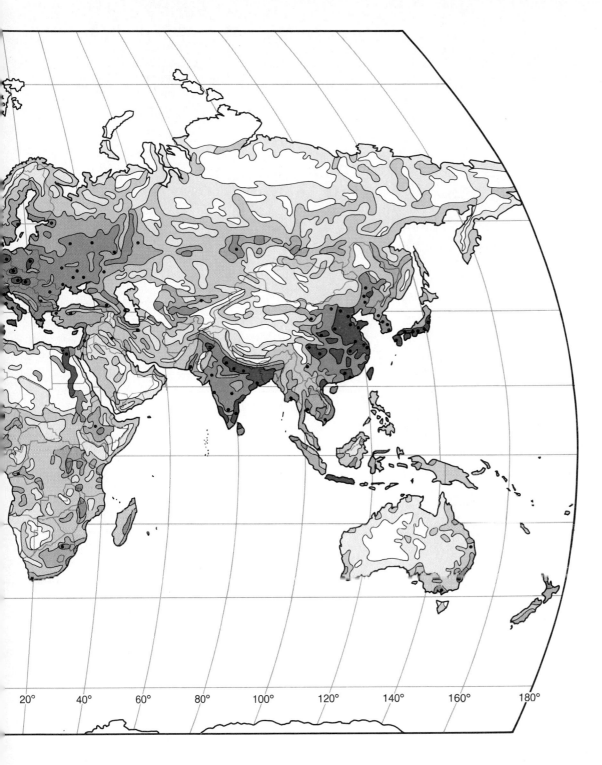

Various modifications may be made to refine density as a meaningful abstraction of distribution. Its descriptive precision is improved if the area in question can be subdivided into comparable regions or units. Thus, it is more revealing to know that in 1988 New Jersey had a density of 1034 and Wyoming of 5 persons per square mile (399 and 2 per km²) of land area than to know only that the figure for the conterminous United States (48 states) was 81 per square mile (31 per km²). If large, sparsely populated Alaska is added, the U.S. density figure drops below 68 per square mile.) The calculation may also be modified to provide density distinctions between classes of population—rural versus urban, for example. Rural densities in the United States rarely exceed 300 per square mile (115 km²), while portions of major cities can have tens of thousands of people per square mile.

Figure 6.22
Terracing of hillsides is one device to extend a naturally limited productive area. The technique is effectively used on the densely settled Indonesian island of Bali.

Table 6.8
Comparative Densities for Selected Countries

Country	Area (thousand sq mi)	1989 Population (millions)	Crude density	Physiological density
Australia	2967.9	16.8	6	94
Bangladesh	55.6	114.7	2063	3286
Canada	3851.8	26.3	7	137
China	3705.4	1103.9	298	2708
Egypt	386.7	54.8	142	7117
India	1269.3	835.0	658	1290
Iran	636.3	53.9	85	941
Japan	143.7	123.2	857	6588
Nigeria	356.7	115.3	323	951
United Kingdom	94.5	57.3	606	2091
United States	3615.1	248.8	69	344
USSR	8649.5	289.0	33	334

Another revealing refinement of crude density relates population not simply to total national territory but to that area of a country that is or may be cultivated, that is, to *arable* land. The resulting figure is the **physiological density** and is, in a sense, an expression of population pressure differentially exerted on agricultural land. It is readily apparent from Table 6.8 that there are differences between countries in physiological density and that the contrasts between crude and physiological densities of and between countries point up actual settlement pressures that are not revealed by crude or arithmetic densities alone. But their calculation depends on uncertain definitions of arable and cultivated land, assumes that all arable land is equally productive and comparably used, and includes only one part of a country's resource base.

Overpopulation

It is an easy and common step from concepts of population density to assumptions about overpopulation or overcrowding. It is wise to remember that **overpopulation** is a value judgment reflecting an observation or conviction that an environment or territory is unable to support its present population. (A related but opposite concept of *underpopulation* refers to the circumstance of too few people to develop the resources of a country or region sufficiently to improve the level of living of its inhabitants.)

Overpopulation is not the necessary and inevitable consequence of high density of population. Tiny Monaco, a principality in southern Europe about half the size of New York's Central Park, has a crude density of nearly 40,000 people per square mile (15,000 per km²). The sizable (604,000 square miles) Mongolian People's Republic, between China and the Soviet Union, has 4 persons per square mile (1.5 per km²); and Iran, only slightly larger, has 87 (34 per km²). Macao, an island possession of Portugal off the coast of China, has more than 58,000 persons per square mile (22,000 per km²); the Falkland Islands off the Atlantic Coast of Argentina count at most one person for every 2.5 square miles (6.5 km²) of territory. No conclusions about conditions of life, levels of income, adequacy of food, or prospects for prosperity can be drawn from these density comparisons.

Overcrowding is a reflection not of numbers per unit area, but of the **carrying capacity** of land—the number of people an area can support given the prevailing technology. A region devoted to efficient, energy-intensive commercial agriculture that makes heavy use of irrigation, fertilizers, and biocides can support more people at a higher level of living than one engaged in the slash-and-burn agriculture described in Chapter 10. An industrial society that takes advantage of resources, such as coal and iron ore, and has access to imported food will not feel population pressure at the same density levels as a country with rudimentary technology.

Since carrying capacity is related to level of economic development, maps such as Figure 6.21, displaying present patterns of population distribution and density, do not suggest a correlation with conditions of life. Many industrialized, urbanized countries have lower densities and higher levels of living than do less-developed ones. Densities in the United States, where there is a great deal of unused and unsettled land, are considerably lower than those in Bangladesh, where essentially all land is arable and which (with over 2000 persons per square mile) is the most densely populated nonisland state in the world. At the same time, many African countries have low population densities and low levels of living, whereas Japan combines both high densities and wealth.

Overpopulation can be equated with levels of living or conditions of life that reflect a continuing imbalance between numbers of people and carrying capacity of the land. One measure of that imbalance might be the unavailability of food supplies sufficient in caloric content to meet individual daily energy requirements and so balanced as to satisfy normal nutritional needs. Unfortunately, as Figure 6.23 suggests, dietary insufficiencies—with long-term adverse implications for life expectancy, physical vigor, and mental development—are most likely to be encountered in the developing countries, where much of the population is in the younger age cohorts (Figure 6.10).

If those developing countries simultaneously have rapidly increasing population numbers dependent upon domestically produced foodstuffs, the prospects must be for continuing undernourishment and overpopulation. Much of sub-Saharan Africa finds itself in this circumstance. Africa's per capita food production decreased 20% between 1960 and 1985, and a further 30% drop is predicted over the following quarter century as the population–food gap widens (Figure 6.24). Africa is not alone. The international Food and Agriculture Organization (FAO) projects that by A.D. 2000, no less than 65 separate countries with nearly 30% of the population of

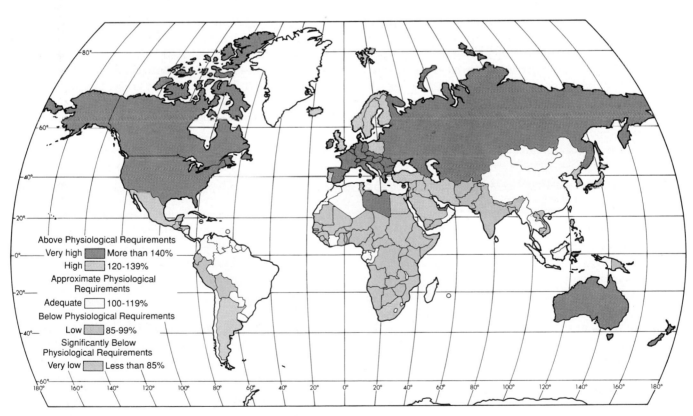

Figure 6.23

Percentage of required dietary energy supply received daily. If the world's food supply were evenly divided, all would have an adequate diet. Each person's share would be between 2600 and 2700 calories. The UN Food and Agricultural Organization specifies 2360 calories as the daily minimum necessity. For comparison, U.S. daily calorie availability is about 3700.

the developing world will be unable to feed their inhabitants from their own national territories at the low level of agricultural technology and inputs apt to be employed.

In the contemporary world, of course, insufficiency of domestic agricultural production to meet national caloric requirements cannot be considered a measure of overcrowding or poverty. Only a few countries are agriculturally self-sufficient. A number of those at or close to that achievement, such as China and India, are not yet among the developed countries of the world. Japan, a leader among the advanced states, is the world's biggest food importer and supplies from its own production only 40% of the calories its population consumes. Its physiological density is high, as Table 6.8 indicates, but it obviously does not rely on an arable land resource for its present development. Largely lacking in either agricultural or industrial resources, it nonetheless ranks well on all indicators of national well-being and prosperity. For countries such as Japan, a sudden cessation of the international trade that permits the exchange of industrial products for imported food and raw materials would be disastrous. Domestic food production could not maintain the dietary levels now enjoyed by their populations and they, more starkly than many underdeveloped countries, would be "overpopulated."

Urbanization

Pressures upon the land resource of countries are increased not just by their growing populations but by the reduction of arable land caused by such growth. More and more of world population growth must be accommodated not in traditional rural areas but in cities. During the years 1980 through 1984, while world population grew by over 300 million, the economically active population in agriculture remained constant. This may suggest that the food needs of the world can be met by an ever-smaller proportion of the total work force. Less optimistically, it may mean that the world has run out of arable land able to support growing rural populations. In either case, the fact remains that in a world where some 44% of employment is agricultural, most new entrants into the labor pool must seek their livelihood elsewhere than on the land.

Increasingly, of course, that means residence and the hope for employment in the city. Largely because of population increases, the number and size of cities everywhere is growing. In 1950, less than 30% of the world's population lived in urban areas; by 1990 over 41% of a much larger total population were urban dwellers (Figures 6.25 and 12.2). As a result of the attractions of cities in both the promise of jobs and access to health, welfare, and other public services, the *urbanization* (transformation from rural to urban status) of population in developing countries is increasing dramatically. In the late

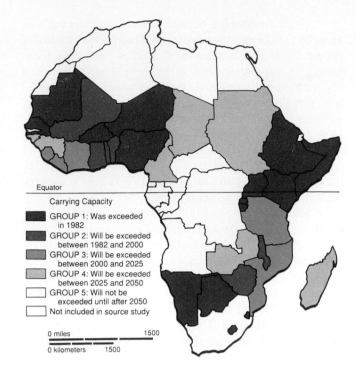

Figure 6.24
Carrying capacity and potentials in sub-Saharan Africa. The map assumes that all cultivated land is used for growing food; food imports are insignificant; and agriculture is conducted by low technology methods.

1980s, some one-third of their inhabitants were urban, and collectively the less developed areas contained over 60% of the world's urban population.

The sheer growth of those cities in people and territory has increased pressures on arable land and adjusted upward both arithmetic and physiological densities. Urbanization consumes millions of acres of cropland each year. In Egypt, for example, urban expansion and development between 1965 and 1985 took out of production as much fertile soil as the Aswan Dam made newly available. Some of these cities, which are surrounded by concentrations of people living in uncontrolled settlements, slums, and shantytowns (Figure 6.26), are among the most densely populated areas in the world. They face massive problems in providing housing, jobs, education, and adequate health and social services for their residents. These and other matters of urban geography are the topics of Chapter 12.

Population Data and Projections

Population geographers, demographers, planners, governmental officials, and a host of others rely on detailed population data to make their assessments of present national and world population patterns and to estimate future conditions. Birth rates and death rates, rates of fertility and natural increase, age and sex composition of the population, and other data are all necessary ingredients for their work.

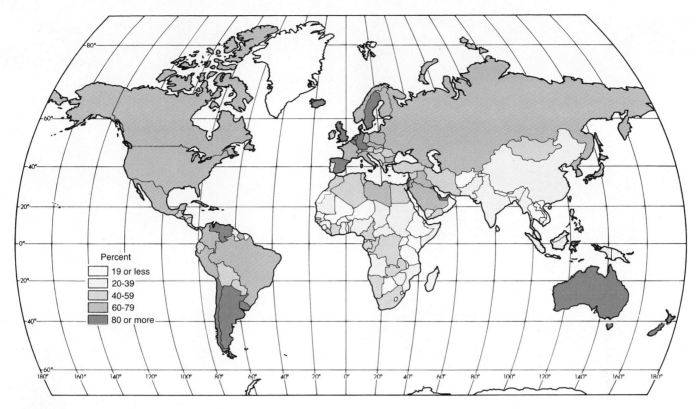

Figure 6.25

Percentage of national population that is classified as urban.
Urbanization has been particularly rapid in the developing
continents. In 1950, only 17% of Asians and 15% of Africans
were urban; at the end of the 1980s, about one-third of both
Asians and Africans were city dwellers.

Population Data

The data that students of population employ come primarily from the United Nations Statistical Office, the World Bank, the Population Reference Bureau, and ultimately, from national censuses and sample surveys. Unfortunately, the data as reported may on occasion be more misleading than informative. For much of the developing world, a national census is a massive undertaking. Isolation and poor transportation, insufficiency of funds and trained census personnel, high rates of illiteracy limiting the type of questions that can be asked, and populations suspicious of all things governmental serve to restrict the frequency, coverage, and accuracy of population reports.

However derived, detailed data are published by the major reporting agencies for all national units even when those figures are poorly based on fact or are essentially fictitious. For years, data on the total population, birth and death rates, and other vital statistics for Somalia were regularly reported and annually revised. The fact was, however, that Somalia had never had a census and had no system whatsoever for recording births. Seemingly precise data were regularly reported, as well, for Ethiopia. When that country had its first-ever census in 1985, at least one data source had to drop its estimate of the country's birth rate by 15% and increase its figure for Ethiopia's total population by more than 20%.

Fortunately, census coverage on a world basis is improving. Almost every county has now had at least one census of its population, and most have been subjected to periodic sample surveys (Figure 6.27). However, only about 10% of the developing world's population live in countries with anything approaching complete systems for registering births and deaths. It has been estimated that 40% or less of live births in Indonesia, Pakistan, India, or the Philippines are officially recorded. It appears that deaths are even less completely reported than births throughout Asia. And whatever the deficiencies of Asian states, African statistics are still less complete and reliable. It is, of course, on just these basic birth and death data that projections about population growth and composition are founded.

Even the age structure reported for national populations, so essential to many areas of population analysis, must be viewed with suspicion. In many societies, birthdays are not noted, nor are years recorded by the Western calendar. Non-Western ways of counting age also confuse

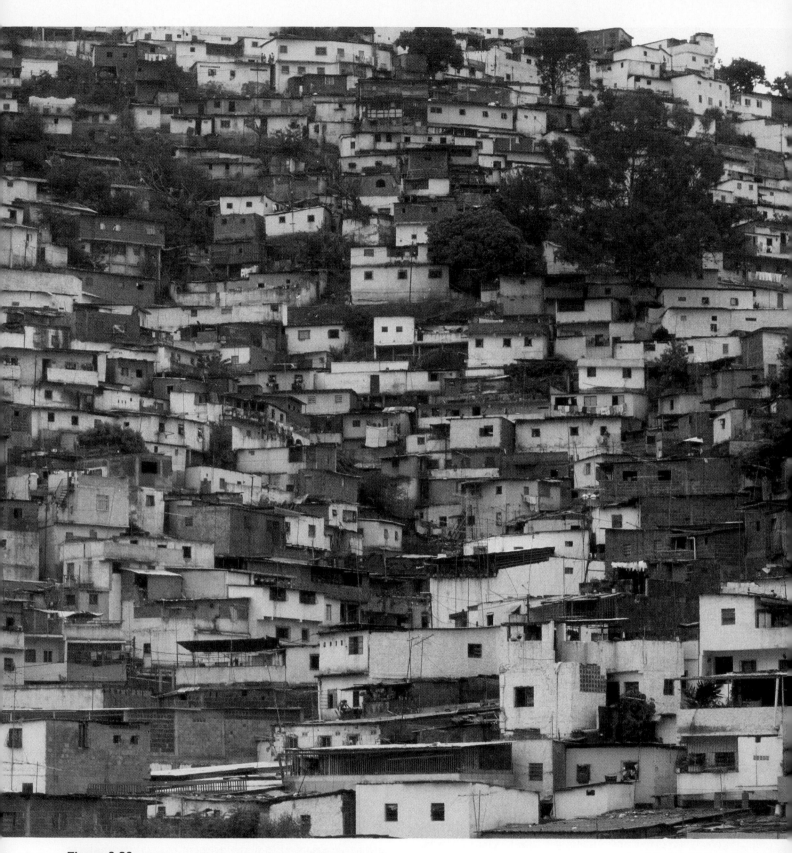

Figure 6.26
Millions of people of the developing world live in shantytown
settlements on the fringes of large cities, without benefit of
running water, electricity, sewage systems, or other public
services. The residents of this *barrio* rising above Caracas,
Venezuela, are part of their number.

Figure 6.27
Taking the census in rural China. The sign reads, "Third National Census. Mobile Registration Station."

the record. The Chinese, for example, consider a person to be 1 year old at birth and increase that age by 1 year each [Chinese] New Year's Day. Bias and error arise from the common tendency of people after middle age to report their ages in round numbers ending in 0. Also evident is a bias toward claiming an age ending in the number 5 or as an even number of years. Inaccuracy and noncomparability of reckoning added to incompleteness of survey and response conspire to cloud national comparisons in which age or the implications of age are important ingredients.

Population Projections

For all their inadequacies and imprecisions, current data reported for country units form the basis of **population projections,** estimates of future population size, age, and sex composition based on current data. Projections are not forecasts, and demographers are not the social science equivalent of meteorologists. Weather forecasters work with a myriad of accurate observations applied against a known, tested model of the atmosphere. The demographer, in contrast, works with sparse, imprecise, and missing data applied to human actions that will be unpredictably responsive to stimuli not yet evident.

Population projections, therefore, are based on assumptions for the future applied to current data that are, themselves, frequently suspect. Since projections are not

predictions, they can never be wrong. They are simply the inevitable result of calculations about fertility, mortality, and migration applied to each age cohort of a population now living, and the making of birth rate, survival, and migration assumptions about cohorts yet unborn. Of course, the perfectly valid *projections* of future population size and structure resulting from those calculations may be dead wrong as *predictions* .

Since those projections are invariably treated as scientific expectations by a public that ignores their underlying qualifying assumptions, agencies such as the UN that estimate the population of, say, Africa in the year 2025, do so by not one but three projections: high, medium, and low. (See "World Population Projections.") Each is completely valid but based on different assumptions of growth rate. For areas as large as Africa, the medium projection is assumed to benefit by compensating errors and statistically predictable behaviors of very large populations. For individual African countries and smaller populations, the medium projection may be much less satisfying. The usual tendency in projections is to assume that something like current conditions will be applicable in the future. Obviously, the more distant the future, the less likely is that assumption to remain true. The resulting observation should be that the farther into the future one wishes to project the population structure of small areas, the greater is the implicit and inevitable error (see Figure 6.12).

While the need for population projections is obvious, demographers face difficult decisions regarding the assumptions they use in preparing them. Assumptions must be made about the future course of birth and death rates and, in some cases, about migration.

Demographers must consider many factors when projecting a country's population. What is the present level of the birth rate, of literacy, and of education? Does the government have a policy to influence population growth? What is the status of women?

Along with these must be weighed the likelihood of socioeconomic change, for it is generally assumed that as a country "develops," a preference for smaller families will cause fertility to fall to the replacement level of about two children per woman. But when can one expect this to happen in less developed countries? And for the majority of more developed countries with fertility currently below replacement level, can one assume that fertility will rise to avert eventual disappearance of the population and, if so, when?

Predicting the pace of fertility decline is most important, as illustrated by the United Nations long-range projections for Africa. As with many projections, these are issued in a "series" to show the effects of different assumptions. The UN "low" projection for Africa assumes that replacement-level fertility will be reached in 2030, which would put the continent's population at 1.4 billion in 2100. If attainment of replacement-level fertility is delayed until 2065, as in the "high" variant, the population is projected to be 4.4. billion in 2100. That difference of 3 billion should serve as a warning that using population projections requires caution and consideration of *all* the possibilities.

Population Controls

All population projections include an assumption that at some point in time population growth will cease and approach the replacement level. Without that assumption, future numbers become unthinkably large. For the world at unchecked present growth rates, there would be 1 trillion people three centuries from now, 4 trillion four centuries in the future, and so on. Although there is reasonable debate about whether the world is now overpopulated and about what either its optimum or maximum sustainable population should be, totals in the trillions are beyond any reasonable expectation.

Population pressures do not come from the amount of space humans occupy. It has been calculated, for example, that the entire human race could easily stand within the boundaries of the state of Delaware. The problems stem from the food, energy, and other resources necessary to support the population and from the impact on the environment of the increasing demands and the technologies required to meet them. Rates of growth currently prevailing in many countries make it nearly impossible for them to achieve the kind of social and economic development they would like. The demographic equilibrium between high birth rates and high death rates has been upset, and at the present time human population is growing at a rate that is impossible to sustain.

Clearly, at some point population will have to stop increasing as fast as it has been. That is, either the self-induced limitations on expansion implicit in the demographic transition will be adopted or an equilibrium between population and resources will be established in more dramatic fashion. Recognition of this eventuality is not new. "[The evils of] pestilence, and famine, and wars, and earthquakes have to be regarded as a remedy for nations, as the means of pruning the luxuriance of the human race," was the opinion of the theologian Tertullian during the 2d century A.D.

Thomas Robert **Malthus** (1766–1834), an English economist and demographer, put the problem succinctly in a treatise published in 1798: all biological populations have a potential for increase that exceeds the actual rate of increase, and the resources for the support of increase are limited. In later publications, Malthus amplified his thesis by noting the following:

1. Population is inevitably limited by the means of subsistence.
2. Populations invariably increase with increase in the means of subsistence unless prevented by powerful checks.
3. The checks that inhibit the reproductive capacity of populations and keep it in balance with means of subsistence are either "private" (moral restraint, celibacy, and chastity) or "destructive" (war, poverty, pestilence, and famine).

The chilling consequences of Malthus's dictum that unchecked population increases geometrically while food production can increase only arithmetically have been reported throughout human history, as they are today. Starvation, the ultimate expression of resource depletion, is no stranger to the past or present. By conservative estimate, some 50 people worldwide will starve to death during the 2 minutes it takes you to read this page; half will be children under 5. They will, of course, be more than replaced numerically by new births during the same 2 minutes. Losses are always recouped. All battlefield casualties, perhaps 50 million, in all of humankind's wars over the last 300 years equal only a 7-month replacement period at present rates of natural increase.

Yet, inevitably—following the logic of Malthus, the apparent evidence of history, and our observations of animal populations—equilibrium must be achieved between numbers and support resources. When overpopulation of any species occurs, a population dieback is inevitable. The madly ascending leg of the J-curve is bent to the horizontal, and the J-curve is converted to an S-curve. It has happened before in human history, as Figure 6.28 summarizes. The top of the **S-curve** represents a population size consistent with and supportable by the exploitable resource base. When the population is equivalent to the carrying capacity of the occupied area, defined earlier, it is said to have reached a **homeostatic plateau.**

In animals, overcrowding and environmental stress apparently release an automatic physiological suppressant of fertility. Although famine and chronic malnutrition may reduce fertility in humans, population limitation usually must be either forced or self-imposed. The demographic transition to low birth rates matching reduced death rates is cited as evidence that Malthus's first assumption was wrong: human populations do not inevitably grow geometrically. Fertility behavior, it was observed, is conditioned by social determinants and not solely by biological or resource imperatives.

Although Malthus's ideas were discarded as deficient by the end of the 19th century in light of the European population experience, the concerns he expressed were revived during the 1950s. Observations of population growth in underdeveloped countries and the strain that growth placed upon their resources inspired the viewpoint that improvements in living standards could be achieved only by raising investment per worker. Rapid population growth was seen as a serious diversion of scarce resources away from capital investment and into unending social welfare programs. To lift living standards required that existing national efforts to lower mortality rates be balanced by governmental programs to reduce birth rates. **Neo-Malthusianism,** as this viewpoint became known, has been the underpinning of national and international programs of population limitation primarily through birth control and family planning (Figure 6.29).

Neo-Malthusianism has had a mixed reception. Asian countries, led by China and India, have in general—though with differing successes—adopted family planning programs and policies. In some instances, success has been declared complete. Singapore established its Population and Family Planning Board in 1965, when its fertility rate was 4.9 lifetime births per woman. By 1986, that rate had declined to 1.7, well below the 2.1 replacement level for developed countries, and the board was abolished as no longer necessary. Caribbean and South American countries, except the poorest and most agrarian, have also experienced declining fertility rates, though often these reductions have been achieved despite pronatalist views of governments influenced by the Roman Catholic church. Africa and the Middle East have generally been unresponsive to the neo-Malthusian arguments because of ingrained cultural convictions among people, if not in all governmental circles, that large families—six or seven children—are desirable. Islamic fundamentalism opposed to birth restrictions also is a cultural factor.

Other barriers to fertility control exist. When first proposed by Western states, neo-Malthusian arguments that family planning was necessary for development were rejected by many less developed countries. Reflecting both nationalistic and Marxist concepts, they maintained that institutional and structural obstacles rather than population increase hindered development. Some government leaders think that there is a correlation between population size and power and pursue pronatalist policies, as did

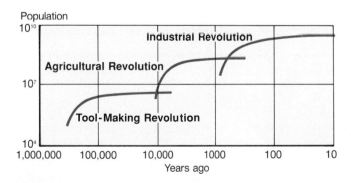

Figure 6.28

The steadily higher *homeostatic plateaus* (states of equilibrium) achieved by humans are evidence of their ability to increase the carrying capacity of the land through technological advance. Each new plateau represents the conversion of the J-curve into an S-curve. The diagram is suggestive of the process; it does not represent actual world population figures, which are shown here on a logarithmic scale. From "Human Population" by Edward S. Deevey, Jr. Copyright © 1960 by SCIENTIFIC AMERICAN, Inc. All rights reserved.

Figure 6.29

This poster in a Bangladesh family planning clinic reads: "Don't blame Allah if you have too many children."

Mao's China during the 1950s and early 1960s. And a number of American economists during the 1980s expressed the view that population growth is a stimulus, not a deterrent, to development. Since the time of Malthus, they observe, world population has grown from 900 million to 5 billion without the predicted dire consequences—proof that Malthus failed to recognize the importance of technology in raising the carrying capacity of the earth. Still higher population numbers, they suggest, are sustainable, perhaps even with improved standards of living for all.

Population Prospects

Regardless of population philosophies, theories, or cultural norms, the fact remains that in many parts of the world developing countries are showing significantly declining population growth rates. But reducing fertility levels even to the replacement level of 2.1 births per woman does not mean an immediate end to population growth. Because of the age composition of many societies, numbers of births will continue to grow even as fertility rates per woman decline. The reason is to be found in **demographic** (or **population**) **momentum**.

When a high proportion of the population is young, the product of past high fertility rates, larger and larger numbers enter the childbearing age each year. One-third or more of the people now on earth are under 15 (Figure 6.10). For Africa as a whole the figure for 1990 was 45%,

and in some individual African states half the population is in that age bracket (as it is in some Asian countries). The consequences of the fertility of these young people are yet to be realized. They will continue to be felt until the now youthful groups mature and work their way through the population pyramid. Inevitably, while this is happening, even the most stringent national policies limiting growth cannot stop it entirely. A country with a large present population base will experience large numerical increases despite declining birth rates. Indeed, the higher fertility was initially and the sharper its drop to low levels, the greater will be the role of momentum. (See "Why Momentum Matters.")

Eventually, of course, young populations grow older, and even the youthful developing countries are beginning to face the consequences of that reality. The problems of a rapidly aging population that already confront the industrialized economies are now being realized in the Third World as well. The growth rate of people aged 55 and over is three times as high in developing countries as in developed ones; in most, the rate is highest for those 75 and over. More than 1.2 million people worldwide reach the age of 55 each month; of that number, 80% live in developing countries that generally lack health, income, and social service support systems adequate to the needs of their older citizens. To the social and economic implications of their present population momentum, therefore, Third World countries must add the aging consequences of past patterns and rates of growth (Figure 6.30).

Figure 6.30

This still-active Malaysian woman is part of a rapidly aging Third-World population. By 2020, a third of Singapore citizens will be 55 or older; in 2025 China will have as large a share of its population over 60 as will Europe. Some developing countries will soon be aging faster than the developed West, but without the old-age assistance and welfare programs advanced countries have put in place.

Why Momentum Matters

When the average fertility rate of a population reaches the replacement threshold—about 2.1 births per woman—one would expect births and deaths to be in balance. Eventually they will be, but not immediately. The reason for this is *demographic momentum.*

A population's age structure reflects its past pattern of demographic events: mortality, migration, and especially, fertility. A population that has had very high fertility in the years before reaching replacement level will have a much younger age structure than a population with lower fertility before crossing the replacement threshold. The proportion of young people in a population approaching replacement-level fertility is important

because the size of the largest recently born generation as well as the size of the parent generation determine the ultimate size of the total population when births and deaths are finally in balance. Even after replacement-level fertility is reached, births will continue to outstrip deaths so long as the generation that is producing births is disproportionately larger than the older generation where most deaths occur.

China, with a 1985 population of 1.04 billion and 34% under age 15, is among the handful of less developed countries with fertility even close to replacement level. China's total fertility rate was 5 to 6 births per woman as late as the early 1970s but dropped

precipitously to 2.1 in 1984. Even with its aggressive fertility control program, China's population is expected to grow to 1.567 billion (a 51% increase over 1985) before stabilizing. Yugoslavia also recently reached replacement-level fertility, but down only from a total fertility rate of 2 to 3 births per woman. Some 24% of its 1985 population of about 23 million was under age 15. Its population is projected to stabilize at about 30 million (30% over 1985) at the end of the next century.

From: Thomas W. Merrick, with PRB staff, "World Population in Transition," *Population Bulletin*, 41, no. 2 (Washington, DC: Population Reference Bureau, 1986).

Summary

Population geography examines the numbers, composition, distribution, and trends of humans in their spatial context. Birth, death, fertility, and growth rates are important in its analyses, as are densities, migrations, and cohort measures.

Recent "explosive" increases in human numbers and the prospects of continuing population expansion may be traced to sharp reductions in death rates, increases in longevity, and the impact of demographic momentum on a youthful population largely concentrated in the developing world. Control of population numbers historically was accomplished through a demographic transition first experienced in European societies that adjusted their fertility rates downward as death rates fell and life expectancies increased. The introduction of advanced technologies of preventive and curative medicine, pesticides, and famine relief have reduced mortality rates in developing countries without always a compensating reduction in birth rates.

Although population control programs have been differentially introduced and promoted, the present 5.3 billion human beings will likely double in number by the year 2100. That growth will largely reflect increases unavoidable because of the size and youth of populations in developing countries. Eventually, a new balance between

population numbers and carrying capacity of the world will be reached, as it has always been following past periods of rapid population increase.

People are unevenly distributed over the earth. The *ecumene* (permanently inhabited portion of the globe) is discontinuous and marked by pronounced differences in population concentrations and numbers. East Asia, South Asia, Europe, and northeastern United States/southeastern Canada represent the world's greatest population clusters, though smaller areas of great density are found in other regions and continents. Since growth rates are highest and population doubling times generally shorter in world regions outside these four present main concentrations, new patterns of population concentration and dominance are taking form.

A respected geographer once commented that "population is the point of reference from which all other elements [of geography] are observed." Certainly, population geography is the essential starting point of the human component of the human–environment concerns of geography. But human populations are not merely collections of numerical units nor are they to be understood solely through statistical analysis. Societies are distinguished not just by the abstract data of their numbers, rates, and trends, but by experiences, beliefs, understandings, and aspirations, which collectively constitute that human spatial and behavioral variable called *culture.* It is to that fundamental human diversity that we next turn our attention.

Key Words

arithmetic density	mortality rate
carrying capacity	neo-Malthusianism
cohort	nonecumene
crude birth rate	overpopulation
crude death rate	physiological density
crude density	population density
demographic equation	population momentum
demographic momentum	population projection
demographic transition	population pyramid
dependency ratio	rate
doubling time	rate of natural increase
ecumene	replacement level
homeostatic plateau	S-curve
J-curve	total fertility rate
Malthus	zero population growth

For Review

1. How do the *crude birth rate* and the *fertility rate* differ? Which measure is the more accurate statement of the amount of reproduction occurring in a population?

2. How is the *crude death rate* calculated? What factors account for the worldwide decline in death rates since 1945?

3. How is a *population pyramid* constructed? What shape of "pyramid" reflects the structure of a rapidly growing country? Of a population with a slow rate of growth? What can we tell about future population numbers from those shapes?

4. What variations do we discern in the spatial pattern of the *rate of natural increase* and, consequently, of population growth? What rate of natural increase would double population in 35 years?

5. How are population numbers projected from present conditions? Are projections the same as predictions? If not, in what ways do they differ?

6. Describe the stages in the *demographic transition*. Where has the final stage of the transition been achieved? Why do some analysts doubt the applicability of the demographic transition to all parts of the world?

7. Contrast *crude population density* and *physiological density*. For what differing purposes might each be useful? How is *carrying capacity* related to the concept of density?

8. What was Malthus's underlying assumption concerning the relationship between population growth and food supply? In what ways do the arguments of *neo-Malthusians* differ from the original doctrine? What governmental policies are implicit in *neo-Malthusianism?*

9. Why is *demographic momentum* a matter of interest in population projections? In which world areas are the implications of demographic momentum most serious in calculating population growth, stability, or decline?

Selected References

Bennett, D. Gordon. *World Population Problems*. Champaign, Ill.: Park Press, 1984.

Boserup, Ester. *Population and Technological Change*. Chicago: University of Chicago Press, 1981.

Brown, Lester R., and Jodi L. Jacobson. *Our Demographically Divided World*. Worldwatch Paper 74. Washington, D.C.: Worldwatch Institute, 1986.

Clarke, John I., ed. *Geography and Population: Approaches and Applications*. Elmsford, N.Y.: Pergamon, 1984.

Clarke, John I., Mustafa M. Khogali, and Leszek A. Kosinski, eds. *Population and Development Projects in Africa*. International Geographic Union, Commission on Population Geography. New York: Cambridge University Press, 1985.

DuPaquier, J., A. Fauve-Chamoux, and E. Grebenik, eds. *Malthus Past and Present*. New York: Academic Press, 1983.

Findlay, Allan, and Anne Findlay. *Population and Development in the Third World*. London and New York: Methuen, 1987.

Grigg, David B. "Migration and Overpopulation." In *The Geographical Impact of Migration,* edited by Paul White and Robert Woods, Chapter 5. London and New York: Longman, 1980.

———. "Modern Population Growth in Historical Perspective." *Geography* 67 (1982): 97–108.

Hendry, Peter. "Food and Population: Beyond Five Billions." *Population Bulletin* 43, no. 2 (1988).

Jacobson, Jodi L. *Planning the Global Family*. Worldwatch Paper 80. Washington, D.C.: Worldwatch Institute, 1987.

Jones, Huw R. *Population Geography*. 2nd ed. London: Paul Chapman, 1990.

Keyfitz, Nathan. "The Growing Human Population." *Scientific American* 261, no. 3 (September 1989). Special issue: *Managing the Planet Earth:*118–126.

King, Timothy, and Allen Kelley. "The New Population Debate: Two Views on Population Growth and Economic Development." *Population Trends and Public Policy*. Washington, D.C.: Population Reference Bureau, 1985.

Landry, Adolphe. "Adolphe Landry on the Demographic Revolution." *Population and Development Review* 13 (1987): 731–40.

Merrick, Thomas W., with PRB staff. "World Population in Transition." *Population Bulletin* 41, no. 2 (1986).

National Academy of Sciences. *Population Growth and Economic Development Policy Questions.* Washington, D.C.: National Academy Press, 1986.

Newman, James L., and Gordon E. Matzke. *Population: Patterns, Dynamics, and Prospects.* Englewood Cliffs, N.J.: Prentice-Hall, 1984.

Omran, Abdel R. "The Epidemiologic Transition: A Theory of the Epidemiology of Population Change." *Milbank Memorial Fund Quarterly* 49 (1971): 509–38.

Peters, Gary L., and Robert P. Larkin. *Population Geography.* Dubuque, Ia.: Kendall/Hunt, 1983.

Population Reference Bureau. *Population in Perspective: Regional Views.* A Population Learning Series. Washington, D.C.: Population Reference Bureau, 1986.

———. *World Population: Toward the Next Century.* Washington, D.C.: Population Reference Bureau, 1985.

Repetto, Robert. "Population, Resources, Environment: An Uncertain Future." *Population Bulletin* 42, no. 2 (1987).

Rogers, Andrei. *Regional Population Projection Models.* Scientific Geography Series, vol. 4. Newbury Park, Calif.: Sage, 1985.

Roseman, Curtis C. *Changing Migration Patterns within the United States.* Association of American Geographers, Commission on College Geography, *Resource Papers for College Geography* No. 77–2. Washington, D.C.: Association of American Geographers, 1977.

Rotberg, Robert, and Theodore Rabb, eds. *Hunger and History: The Impact of Changing Food Production and Consumption Patterns on Society.* New York: Cambridge University Press, 1984.

Schnell, George A., and Mark Monmonier. *The Study of Population: Elements, Patterns and Processes.* Columbus, Ohio: Charles E. Merrill, 1983.

Simon, Julian. *The Ultimate Resource.* Princeton, N.J.: Princeton University Press, 1981.

Trewartha, Glenn. *A Geography of Population: World Patterns.* New York: Wiley, 1969.

———. *The Less Developed Realm: A Geography of Its Population.* New York: Wiley, 1972.

———. *The More Developed Realm: A Geography of Its Population.* New York: Pergamon, 1978.

United Nations, Population Division. *World Population Prospects: 1988.* New York: United Nations, 1989.

Van de Kaa, Dirk J. "Europe's Second Demographic Transition." *Population Bulletin* 42, no. 1 (1987).

Weeks, John R. "The Demography of Islamic Nations." *Population Bulletin* 43, no. 4 (1988).

Woods, Robert. *Population Analysis in Geography.* New York and London: Longman, 1979.

———. *Theoretical Population Geography.* London: Longman, 1982.

World Bank. *Sub-Saharan Africa: From Crisis to Sustainable Growth.* Washington, D.C.: The World Bank, 1989.

World Population: Approaching the Year 2000. Samuel H. Preston, special editor. *The Annals of the American Academy of Political and Social Science,* vol. 510 (July 1990).

C H A P T E R

7

Cultural Geography

Ceremonial dress of hill tribe, Northern Thailand.

THE Gauda's[1] son is eighteen months old. Every morning, a boy employed by the Gauda carries the Gauda's son through the streets of Gopalpur. The Gauda's son is clean; his clothing is elegant. When he is carried along the street, the old women stop their ceaseless grinding and pounding of grain and gather around. If the child wants something to play with, he is given it. If he cries, there is consternation. If he plays with another boy, watchful adults make sure that the other boy does nothing to annoy the Gauda's son.

Shielded by servants, protected and comforted by virtually everyone in the village, the Gauda's son soon learns that tears and rage will produce anything he wants. At the same time, he begins to learn that the same superiority which gives him license to direct others and to demand their services places him in a state of danger. The green mangoes eaten by all of the other children in the village will give him a fever; coarse and chewy substances are likely to give him a stomachache. While other children clothe themselves in mud and dirt, he finds himself constantly being washed. As a Brahmin [*a religious leader*], he is taught to avoid all forms of pollution and to carry out complicated daily rituals of bathing, eating, sleeping, and all other normal processes of life.

In time, the Gauda's son will enter school. He will sit motionless for hours, memorizing long passages from Sanskrit holy books and long poems in English and Urdu. He will learn to perform the rituals that are the duty of every Brahmin. He will bathe daily in the cold water of the private family well, reciting prayers and following a strict procedure. The gods in his house are major deities who must be worshipped every day, at length and with great care[2].

The Gauda and his family are not Americans, as references and allusions in the preceding paragraphs make plain. The careful reader would infer, correctly, that they are Indians—and were one to read more of the book from which this excerpt is taken, it would become evident that the Gauda lives in a village in southern India. The class structure, the religion, the language, the food, and other strands of the fabric of life mentioned in the passage place the Gauda, his family, and his village in a specific time and place. They bind the people of the region together as sharers of a common culture and set them off from those of other areas with different cultural heritages. The 5.25 billion people who were the subject of Chapter 6 are of a single human family, but it is a family differentiated into many branches each characterized by a distinctive culture.

[1]Gauda = village headman
[2]From Alan J. Beals, *Gopalpur: A South Indian Village* (New York: Holt, Rinehart and Winston, 1962).

To some writers in newspapers and the popular press, "culture" means the arts (literature, painting, music, etc.). To a social scientist, **culture** is specialized behavioral patterns, understandings and adaptations that summarize the way of life of a group of people. In this broader sense, culture is as much a part of the regional differentiation of the earth as are topography, climate, and other aspects of the physical environment. The visible and invisible evidences of culture—buildings and farming patterns, language and political organization—are elements in the spatial diversity that invites and is subject to geographic inquiry. Cultural differences in area result in human landscapes with variations as subtle as the differing "feel" of urban Paris, Moscow, and New York or as obvious as the sharp contrasts of rural Sri Lanka and the Prairie Provinces of Canada (Figure 7.1).

Since such differences exist, cultural geography exists, and one branch of cultural geography addresses a whole range of "why?" and "what?" and "how?" questions. Why, since humankind constitutes a single species, are cultures so varied? What are the most pronounced ways in which cultures and culture regions are distinguished? What were the origins of the different culture regions we now observe? How, from whatever limited areas in which single culture traits and amalgams developed, were they diffused over a wider portion of the globe? Why do cultural contrasts between recognizably distinct groups persist even in such presumed "melting pot" societies as that of the United States or in the outwardly homogeneous, long-established countries of Europe? These and similar questions are the concerns of the present chapter and, in part, of Chapter 8.

Components of Culture

Culture is transmitted within a society to succeeding generations by imitation, instruction, and example. It is learned, not biological, and has nothing to do with instinct or genes. As members of a social group, individuals acquire integrated sets of behavioral patterns, environmental and social perceptions, and knowledge of existing technologies. Although one of necessity learns the culture in which one is born and reared, one need not—indeed, cannot—learn its totality. Age, sex, status, or occupation may dictate the aspects of the cultural whole in which one becomes fully indoctrinated. A culture, that is, despite overall generalized and identifying characteristics and even an outward appearance of uniformity and conformity, displays a social structure—a framework of roles and interrelationships of individuals and established groups. Each individual learns and adheres to the rules and conventions not only of the culture as a whole but, importantly, also of those specific to the subgroup to which he or she belongs. In its turn, such a subgroup itself may have a recognized social structure.

Figure 7.1
Cultural contrasts are clearly evident between the rice paddies of central Sri Lanka and the extensively farmed fields of the Canadian prairies.

Culture is a complexly interlocked web of behaviors and attitudes. Realistically, its full and diverse content cannot be appreciated, and in fact may be wholly misunderstood, if we concentrate our attention only on limited, obvious traits. Distinctive eating utensils, the use of gestures, or the ritual of religious ceremony may summarize and characterize a culture for the casual observer. These are, however, individually insignificant parts of a much more complex structure that can be appreciated only when the whole is experienced.

Out of the richness and intricacy of human life we seek to isolate for special study those more fundamental cultural variables that give structure and spatial order to societies. We begin with **culture traits**, the smallest distinctive items of culture. Culture traits are units of learned behavior ranging from the language spoken to the tools used or the games played. A trait may be an object (a fishhook, for example), a technique (weaving and knotting of a fishnet), a belief (in the spirits who live in water bodies), or an attitude (a conviction that fish is superior to other animal protein). Such traits are the most elementary expressions of culture, the building blocks of the complex behavioral patterns of distinctive groups of peoples.

Individual culture traits that are functionally interrelated comprise a **culture complex.** The existence of such complexes is universal. Keeping cattle was a *culture trait* of the Masai of Kenya and Tanzania. Related traits included the measurement of personal wealth by the number of cattle owned; a diet containing the milk and blood of cattle; and disdain for labor unrelated to herding. The assemblage of these and other related traits yielded a *culture complex* descriptive of one aspect of Masai society (Figure 7.2). In exactly the same way, religious complexes, business behavior complexes, sports complexes, and others can easily be recognized in American or any other society.

Culture traits and complexes have spatial extent. When they are plotted on maps, the regional character of the components of culture is revealed. Geographers are interested in the spatial distribution of these individual elements, but their usual concern is with the **culture region,** a portion of the earth's surface occupied by people sharing recognizable and distinctive cultural characteristics that summarize their collective attributes or activities. Examples include the political organizations societies devise, their religions, their form of economy, and even their clothing, eating utensils, or housing. There are as many such culture regions as there are separate culture traits and complexes of population groups.

Finally, as we shall see at the conclusion of this chapter, a set of culture regions showing related culture complexes and landscapes may be grouped to form a **culture realm**. The term recognizes a large segment of the earth's surface having fundamental uniformity in its cultural characteristics and showing a significant difference in them from adjacent realms. Culture realms are, in a sense, culture regions at the broadest scale of generalization.

Interaction of People and Environment

Culture develops in a physical environment that, in its way, contributes to differences among people. In primitive societies, the acquisition of food, shelter, and clothing—all

Figure 7.2
The formerly migratory Masai of eastern Africa are now largely sedentary, partially urbanized, and frequently owners of fenced farms. Cattle formed the traditional basis of Masai culture, were the source of milk and blood in their diet, and the evidence of their wealth and social status.

parts of culture—depends on the utilization of the natural resources at hand. The interrelations of people with the environment of a given area, their perceptions and utilization of it, and their impact on it are consistent and interwoven themes of geography. They are the special concerns of geographers exploring *cultural ecology,* the study of the relationship between a culture group and that natural environment it occupies.

Environments as Controls

Geographers have long dismissed as invalid and intellectually limiting the ideas of **environmental determinism**—the belief that the physical environment by itself shapes humans, their actions, and their thoughts. Environmental conditions alone cannot account for the cultural variations that occur around the world. Levels of technology, systems of social organization, and ideas about what is true and right have no obvious relationship to environmental circumstances.

The environment does place certain limitations on the human use of territory. However, such limitations must be seen not as absolute, enduring restrictions but as relative to technologies, cost considerations, national aspirations, and linkages with the larger world. Human choices in the use of landscapes are affected by group perception of the possibility and desirability of their settlement and exploitation. These are not circumstances inherent in the land.

Possibilism is the viewpoint that people, not environments, are the dynamic forces of cultural development. The needs, traditions, and level of technology of a culture affect how that culture assesses the possibilities of an area and shape the choices that it makes. Each society uses natural resources in accordance with its culture. Changes in a group's technical abilities or objectives bring about changes in its perceptions of the usefulness of the land. Of course, there are some environmental limitations on use of area. For example, if resources for feeding, clothing, or housing ourselves within an area are lacking, or if we do not recognize them there, there is no inducement for people to occupy the territory. Environments that do contain such resources provide the framework within which a culture operates.

Human Impacts

People are also able to modify their environment, and this is the other half of the human–environment relationship of geographic concern. Geography, including cultural geography, examines both the reactions of people to the physical environment and their impact on that environment. As Chapter 5 suggested, by using it we modify our environment—in part, through the material objects we place on the landscape: cities, farms, roads, and so on (Figure 7.3). The form these take is the product of the kind of culture group in which we live. The **cultural landscape**, the earth's surface as modified by human action, is the tangible, physical record of a given culture. House types, transportation networks, parks and cemeteries, and the size and distribution of settlements are among the indicators of the use that humans have made of the land.

As a rule, the more technologically advanced and complex the culture, the greater its impact on the environment, although preindustrial societies can and frequently do exert destructive pressures on the lands they occupy. (See "Chaco Canyon Desolation.") In sprawling urban-industrial societies, the cultural landscape has come to outweigh the natural physical environment in its impact on people's daily lives. It interposes itself between "nature" and humans. Residents of the cities of such societies—living and working in climate-controlled buildings, driving to enclosed shopping malls—can go through life with very little contact with or concern about the physical environment.

The Subsystems of Culture

Understanding a culture fully is, perhaps, impossible for one who is not part of it. For analytical purposes, however, the traits and complexes of culture—its building blocks

Chaco Canyon Desolation

It is not certain when they first came, but by A.D. 1000 the Anasazi people were building a flourishing civilization in what are now Arizona and New Mexico. In the Chaco Canyon alone they erected as many as 75 towns, all centered around pueblos, huge stone-and-adobe apartment buildings as tall as five stories and with as many as 800 rooms. These were the largest and tallest buildings of North America prior to the construction of "cloud scrapers" in major cities at the end of the 19th century. An elaborate network of roads and irrigation canals connected and supported the pueblos. About A.D. 1200, the settlements were abruptly abandoned. The Anasazi, advanced in their skills of agriculture and communal dwelling, were forced to move on by the ecological disaster their pressures had brought to a fragile environment.

They needed forests for fuel and for the hundreds of thousands of logs used as beams and bulwarks in their dwellings. The pinyon-juniper woodland of the canyon was quickly depleted. For larger timbers needed for construction they first harvested stands of ponderosa pine found some 25 miles away. As early as A.D. 1030 these, too, were exhausted, and the

community switched to spruce and Douglas fir from mountaintops surrounding the canyon. When they were gone by 1200, the Anasazi fate was sealed—not only by the loss of forest, but also by the irreversible ecological changes deforestation and agriculture had caused. With forest loss came erosion that destroyed the topsoil. The surface water channels that had been built for irrigation were deepened by accelerated erosion, converting them into enlarging arroyos, useless for agriculture.

The material roots of their culture destroyed, the Anasazi turned upon themselves, and warfare convulsed the region. Smaller groups sought refuge elsewhere, recreating on reduced scale their pueblo way of life but now in nearly inaccessible, highly defensible mesa and cliff locations. The destruction they had wrought destroyed the Anasazi in turn.

Figure 7.3

The physical and cultural landscapes in juxtaposition. Advanced societies are capable of so altering the circumstances of nature that the cultural landscapes they create become controlling environments. The twin cities of Juarez, Mexico, and El Paso, Texas, are "built environments" largely unrelated to their physical surroundings.

and expressions—may be grouped and examined as subsets of the whole. The anthropologist Leslie White suggested that a culture could be viewed as a three part structure composed of subsystems that he termed *technological, sociological,* and *ideological.* In a similar classification, the biologist Julian Huxley identified three components of culture: *artifacts, sociofacts,* and *mentifacts.* Together, according to these interpretations, the subsystems—identified by their separate components—comprise the structure of culture as a whole. But they are integrated; each reacts on the others and is affected by them in turn.

The **technological subsystem** is composed of the material objects and the techniques of their use by means of which people are able to live. Such objects are the tools and other instruments that enable us to feed, clothe, house, defend, transport, and amuse ourselves. The **sociological subsystem** of a culture is the sum of those expected and accepted patterns of interpersonal relations that find their outlet in economic, political, military, religious, kinship, and other associations. The **ideological subsystem** consists of the ideas, beliefs, and knowledge of a culture and the ways in which they are expressed in speech or other forms of communication.

The Technological Subsystem

Examination of variations in culture and in the manner of human existence from place to place centers on a series of commonplace questions: How do the people in an area make a living? What resources and what tools do they use to feed, clothe, and house themselves? Is a larger percentage of the population engaged in agriculture than in manufacturing? Do people travel to work in cars, on bicycles, or on foot? Do they shop for food or grow their own?

These questions concern the adaptive strategies used by different cultures in "making a living." In a broad sense, they address the technological subsystems at the disposal of those cultures—the instruments and tools people use to feed, clothe, house, defend, and amuse themselves. Huxley called the material objects we use to fill these needs *artifacts.* For most of human history, people lived by hunting and gathering, taking the bounty of nature with only minimal dependence on weaponry, implements, and the controlled use of fire. Their adaptive skills were great but their technological level was low. They had few specialized tools, could exploit only a limited range of potential resources, and had little or no control of nonhuman sources of energy.

Their impact upon the environment was small, but at the same time the "carrying capacity" of the land discussed in Chapter 6 was everywhere low, for technologies and artifacts were essentially the same among all groups.

The retreat of the last glaciers about 11,000 years ago marked the start of a period of unprecedented cultural development. It lead from primitive hunting and gathering economies at the outset through the evolution of agriculture and animal husbandry to, ultimately, urbanization, industrialization, and the intricate complexity of the modern technological subsystem. Since not all cultures passed through all stages at the same time, or even at all, *cultural divergence* between human groups became evident.

Cultural diversity among ancient societies reflected the proliferation of technologies that followed a more assured food supply and made possible a more intensive and extensive utilization of resources. Different groups in separate environmental circumstances developed specialized tools and behaviors to exploit resources they recognized. Beginning with the Industrial Revolution of the 18th century, however, a reverse trend—toward commonality of technology—began.

Today, advanced societies are nearly indistinguishable in the tools and techniques at their command. They have experienced *cultural convergence*—the sharing of technologies, organizational structures, and even cultural traits and artifacts that is so evident among widely separated societies in a modern world united by instantaneous communication and efficient transportation. Those differences in technological traditions that still exist between developed and underdeveloped societies reflect, in part, national and personal wealth, stage of economic advancement and complexity, and, importantly, the level and type of energy used (Figure 7.4).

Many aspects of a society are related to its relative level of technological development. *Development* in that comparative sense means simply the extent to which the resources of an area or country have been brought into full productive use. The terms *standard of living* or *level of living* bring to mind some of those aspects and attributes, such as personal income, level of education, food consumption, life expectancy, and availability of health care. The complexity of the occupational structure, the degree of specialization in jobs, the ways natural resources are used, and the degree of industrialization reflect the technology available to a given society.

Not all members of a culture group, of course, are equal beneficiaries of the existing tools and rewards of production. The plaintive comment that "poor people have poor ways" or the ready distinction that we make between the "Gold Coast" and the "slum" indicate that different groups have differential access to the wealth, tools, and resources of their own society. On an international scale, we distinguish between "advanced" or "rich" countries, such as Switzerland, and "underdeveloped" or "poor" ones, such as Bangladesh, though neither class of states may wish those adjectives to be applied to its circumstances.

In technologically advanced countries, many people are employed in manufacturing or allied service trades.

(a)

(b)

Figure 7.4
(a) This Balinese farmer working with draft animals employs tools typical of the low technological levels of subsistence economies.
(b) Cultures with advanced technological subsystems use complex machinery to harness inanimate energy for productive use.

Per capita incomes tend to be high, as do levels of education. These countries wield great economic and political power. In contrast, technologically less-advanced countries have a high percentage of people engaged in agricultural production, with much of the agriculture at a subsistence level (Figure 7.5). The gross national product, or GNP (the GNP equals the total value of all goods and services produced by a country during a year), of these countries is much lower than that of industrialized states. Per capita incomes tend to be low (Figure 7.6), and illiteracy rates high.

Labels such as *advanced-less advanced, developed-underdeveloped,* or *industrial-nonindustrial* can mislead us into thinking in terms of "either-or." They may also be misinterpreted to mean, in general cultural terms, "superior-inferior." This belief is totally improper, since the terms are related solely to economic and technological circumstances and bear no qualitative relationship to such vital aspects of culture as music, art, or religion.

Properly understood, however, terms and measures of economic development can reveal important national and world regional contrasts in the implied technological subsystems of different cultures and societies. Figure 7.7 suggests a spatial grouping of countries based on a composite of developmental indexes. The map indicates that technological status is high in nearly all European countries, including the USSR, and in Japan, the United States, and Canada—the "North." Most of the less developed countries are in Latin America, Africa, and southern Asia—the "South."

It is important, however, to recognize that these implied national averages conceal internal contrasts. All countries include areas that are at different levels of development. We must also remember that technological development is a dynamic concept. It is most useful and accurate to think of the countries of the world as arrayed along an ever-changing continuum of technological subsystems.

The Sociological Subsystem

Continuum and change also characterize the religious, political, formal and informal educational, and other institutions that comprise the sociological subsystem of culture. Together these *sociofacts* define the social organizations of a culture. They regulate how the individual functions relative to the group, whether it be family, church, or state.

There are no "givens" as far as patterns of interaction in any of these associations are concerned, except that most cultures possess a variety of formal and informal ways of structuring behavior. The importance to the society of the differing behavior sets varies among, and

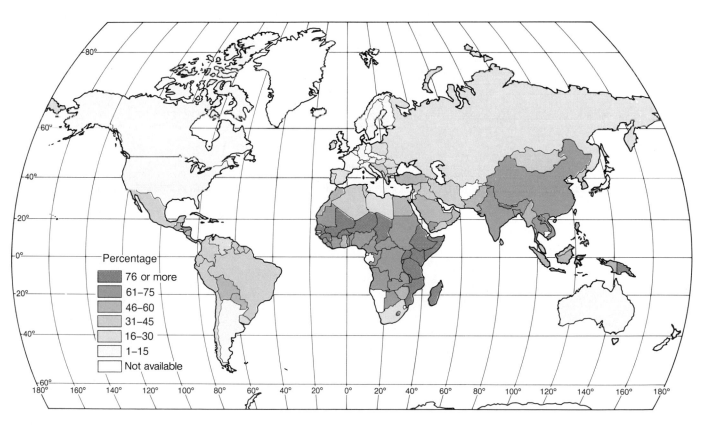

Figure 7.5
Percentage of work force engaged in agriculture. Highly developed economies are characterized by relatively low proportions of their labor forces in the agricultural sector.

Percentage
- 76 or more
- 61–75
- 46–60
- 31–45
- 16–30
- 1–15
- Not available

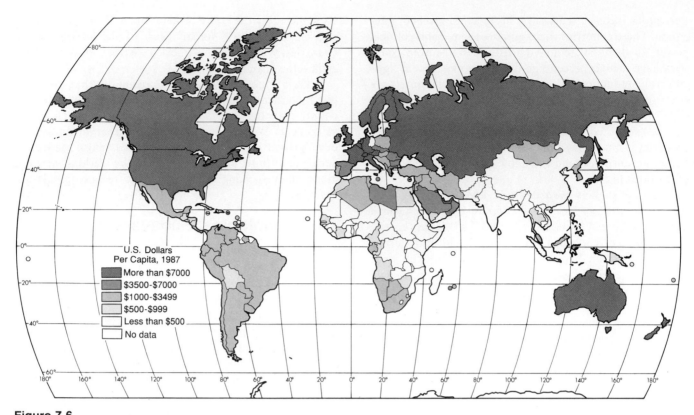

Figure 7.6

Gross national product per capita. GNP per capita is a frequently employed summary of degree of technological development, though high incomes in sparsely populated, oil-rich countries may not have the same meaning in subsystem terms as do comparable per capita values in industrially advanced states. A comparison of this map and Figure 7.5 presents an interesting study in regional contrasts. These World Bank figures are for 1987 and are in U.S. dollar equivalents.

U.S. Dollars
Per Capita, 1987

- More than $7000
- $3500-$7000
- $1000-$3499
- $500-$999
- Less than $500
- No data

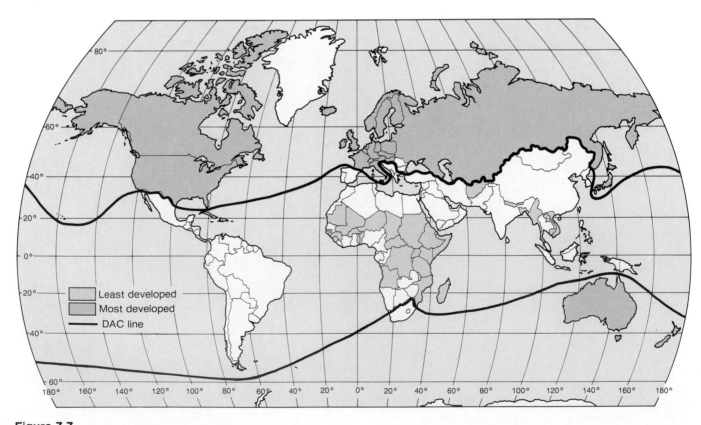

- Least developed
- Most developed
- DAC line

Figure 7.7

A multifactor recognition of ''most developed'' and ''least developed'' countries based on United Nations and World Bank data of the late 1980s. The DAC line is the division between the ''developing'' and ''developed'' countries (the ''North'' and the ''South'') recognized by the Development Assistance Committee (DAC) of the Organization for Economic Cooperation and Development (OECD).

constitutes obvious differences between, cultures. Differing patterns of behavior are learned expressions of culture and are transmitted from one generation to the next by formal instruction or by example and expectation (Figure 7.8). The story of the Gauda's son that opened this chapter illustrates the point.

Social institutions are closely related to the technological system of a culture group. The description of the cultural environment of the Gauda's son suggested through a single example the certainty that innumerable specific mixes of social structures and ways of behavior are encountered worldwide. Thus, hunter-gatherers have one set of institutions and industrial societies quite different ones. Preagricultural societies tended to be composed of small bands based on kinship ties with little social differentiation or specialization of function in the band; the San (Bushmen) of arid southern Africa and isolated rain forest groups in Amazonia might serve as modern examples (Figure 7.9).

The revolution in food production occasioned by plant and animal domestication beginning around 10,000 years ago touched off a social transformation that included increases in population, urbanization, job specialization, and structural differentiation within the society. Politically, the rules and institutions by which people were governed changed with the formation of sedentary, agricultural societies. Loyalty was transferred from the kinship group to the state; resources became possessions rather than the common property of all. Equally far-reaching changes occurred after the 18th century Industrial Revolution, leading to the complex of human social organizations that we experience and are controlled by today in "developed" states and that increasingly affect all cultures everywhere.

Culture is a complexly intertwined whole. Each organizational form or institution affects, and is affected by, related culture traits and complexes in intricate and variable ways. Systems of land and property ownership and

Figure 7.8
All societies prepare their children for membership in the culture group. In each of these settings, certain values, beliefs, skills, and proper ways of acting are being transmitted to the young people.

Figure 7.9

Hunter-gatherers practiced the most enduring life-style in human history, trading it for the more arduous life of farmers under the necessity to provide larger quantities of less diversified foodstuffs for a growing population. Among hunter-gatherers, unlike their settled farmer rivals and successors, age and sex differences, not caste or economic status, were and are the primary basis for division of labor and interpersonal relations. Here a San (Bushman) hunter of Botswana, Africa, stalks his prey. Men also help contribute to the gathered food that constitutes 80% of the San diet.

control, for example, are institutional expressions of the sociological subsystem. They are, simultaneously, explicitly central to the classification of economies and to the understanding of spatial and structural patterns of economic development, as Chapter 10 will examine. Again, for each country the adopted system of laws and justice is a cultural variable identified with the sociological subsystem but extending its influence to all aspects of economic and social organization, including the political geographic systems discussed in the following chapter.

The Ideological Subsystem

The third class of elements defining and identifying a culture comprises the ideological subsystem. This subsystem consists of ideas, beliefs, knowledge, and ways we express these things in our speech or other forms of communication. Mythologies, theologies, legend, literature, philosophy, folk wisdom, and commonsense knowledge make up this category. Passed on from generation to generation, these abstract belief systems, or *mentifacts,* tell us what we ought to believe, what we should value, and how we ought to act. Beliefs form the basis of the socialization process.

Often we know—or think we know—what the beliefs of a group are from written sources. Sometimes, however, we must depend on the actions or objectives of a group to tell us what its true ideas and values are. "Actions speak louder than words" and "Do as I say, not as I do" are commonplace recognitions of the fact that actions and words do not always coincide. The values of a group cannot be deduced from the written record alone.

Nothing in a culture stands totally alone. Changes in the ideas that a society holds may affect the sociological and technological systems just as, for example, changes in technology force changes in the social system. The abrupt alteration of the ideological structure of Russia from a monarchical, agrarian, capitalistic system to an industrialized, communistic society involved sudden, interrelated alteration in all facets of the culture system formerly observed within that country. The interlocking nature of all aspects of a culture is termed **cultural integration.**

The recognition of three distinctive subsystems of culture, while helping us to appreciate its structure and complexity, can at the same time obscure the many-sided nature of individual elements of culture. Cultural integration means that any cultural object or act may have a number of meanings. A dwelling, for example, is an artifact providing shelter for its occupants. It is, simultaneously, a sociofact reflecting the nature of the family or kinship group it is designed to house, and a mentifact summarizing a culture group's convictions about appropriate design, orientation, and building materials of dwelling units. In the same vein, clothing serves as an artifact of bodily protection appropriate to climate, available materials and techniques, or activities of the wearer.

But garments also may be sociofacts, identifying an individual's role in the social structure of the community or culture, and mentifacts, evoking larger community value systems (Figure 7.10).

Culture Change

The recurring theme of cultural geography is change. No culture is, or has been, characterized by a permanently fixed set of material objects, systems of organization, or even ideologies, although all of these may be long-enduring within a society at equilibrium with its resource base and totally isolated from an impetus for change. Such isolation and stagnation have always been rare. On the whole, while cultures are essentially conservative, they are simultaneously constantly changing.

Many individual changes, of course, are so slight that they initially will be almost unnoticed, though collectively they may substantially alter the affected culture. Think of how the culture of the United States differs today from what it was in 1940—not in essentials, perhaps, but in the innumerable electrical, electronic, and transportational devices that have been introduced and in the social, behavioral, and recreational changes they have wrought. Such cumulative changes occur because the cultural traits of any group are not independent; they are clustered in a coherent and integrated pattern. Change on a small scale will have wide repercussions as associated traits also

(b)

(a)

Figure 7.10

(a) When clothing serves primarily to cover, protect, or assist in activities, it is an *artifact*. (b) Some garments are *sociofacts*, identifying a role or position within the social structure: the distinctive "uniforms" of the soldier, the cleric, or the beribboned ambassador immediately proclaim their respective roles in a culture's social organizations. (c) The mandatory chadors of Iranian women are *mentifacts*, indicative not specifically of the role of the wearer but of the values of the culture the wearer represents.

(c)

change to accommodate the adopted adjustment. Change, both major and minor, within cultures is induced by *innovation* and *spatial diffusion*.

Innovation and Diffusion

Innovation implies changes to a culture that result from the adoption of new ideas, either those created within the social group itself or those originating elsewhere. The novelty may be an invented improvement in material technology, such as the bow and arrow or the jet engine. It may involve the development of nonmaterial forms of social structure and interaction: feudalism, for example, or Christianity.

Primitive and traditional societies characteristically are not innovative. In societies at equilibrium with their environment and with no unmet needs, change has no adaptive value and has no reason to occur. Indeed, all societies have an innate resistance to change. Complaints about youthful fads or the glorification of times past are familiar cases in point. However, when a social group is inappropriately unresponsive—mentally, psychologically, or economically—to changing circumstances and innovation, it is said to exhibit *cultural lag*.

Innovation, frequently under stress, has characterized the history of humankind. Growing populations at the end of the Ice Age necessitated an expanded food base. In response, domestication of plants and animals appears to have occurred independently in more than one world area. Indeed, a most striking fact about early agriculture is the universality of its development or adoption within a very short span of human history. Some 10,000 years ago, virtually all of humankind was supported by hunting and gathering. By no later than 2000 years ago, the majority lived by farming. Nonetheless, recognizable areas of "invention" of agriculture have been identified, as shown in Figure 7.11. From them, presumably, there was a rapid diffusion of food types, production techniques, and new modes of economic and social organization. All innovation has a radiating impact upon the web of culture; the more basic the innovation, the more pervasive its consequences.

Few innovations in human history have been more basic than the Agricultural Revolution. It affected every aspect of society. Culture altered at an accelerating pace, and change itself became a way of life. Humans learned the arts of spinning and weaving plant and animal fibers. They learned to use the potter's wheel, to fire clay, and make utensils. They developed techniques of brick making,

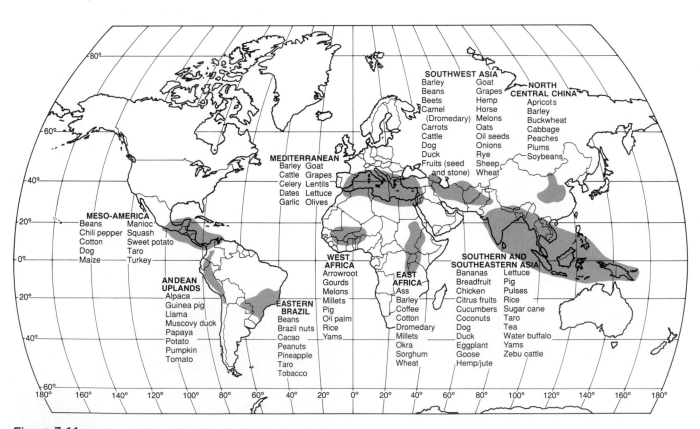

Figure 7.11

Chief centers of plant and animal domestication. The southern and southeastern Asian center was characterized by the domestication of plants, such as taro, that are propagated by the division and replanting of existing plants (vegetative reproduction). Reproduction by the planting of seeds (e.g., maize and wheat) was more characteristic of Meso-America and the Near East. The African and Andean areas developed crops reproduced by both methods. The lists of crops and livestock associated with the separate origin areas are selective not exhaustive.

mortaring, and building construction. They discovered the skills of mining, smelting, and casting metals. Special local advantages in resources or products promoted the development of long-distance trading connections. On the foundation of such technical advancements a more complex exploitative culture appeared, including a stratified society to replace the rough equality of hunting and gathering economies.

The source regions of such social and technical revolutions were initially spatially confined. The term **culture hearth** is used to describe those restricted areas of innovation from which key culture elements diffused to exert an influence on surrounding regions. The hearth may be viewed as the "cradle" of a culture group whose developed systems of livelihood and life created distinctive cultural landscapes. All hearth areas developed the trappings of *civilization,* which are usually assumed to include writing (or other form of record keeping), metallurgy, long-distance trade connections, astronomy and mathematics, social stratification and labor specialization, formalized governmental systems, and a structured urban society. Several major culture hearths emerged, some as early as 5500 years ago, following the initial revolution in food production. Prominent centers of early creativity were located in Egypt, Mesopotamia, the Indus Valley of the Indian subcontinent, northern China, Southeastern Asia, and several locations in Africa, the Americas, and elsewhere (Figure 7.12).

In most modern societies, innovative change has become common, expected, and inevitable. The rate of invention, at least as measured by the number of patents granted, has steadily increased, and the period between idea conception and product availability has been decreasing. A general axiom is that the more ideas available and the more minds able to exploit and combine them, the greater the rate of innovation. The spatial implication is that larger urban centers of advanced technologies tend to be centers of innovation, not just because of their size but because of the number of ideas interchanged. Indeed, ideas not only stimulate new ideas, but also create circumstances in which new solutions must be developed to maintain the forward momentum of the society (Figure 7.13).

Spatial diffusion is the process by which a concept, practice, innovation, or substance spreads from its point of origin to new territories. Diffusion may assume a variety of forms, but basically two processes are involved. Either people move to a new area and take their culture with them (as the immigrants to the American colonies did), or information about an innovation (like barbed wire or hybrid corn) may spread throughout a culture. In either case, new ideas are transferred from their source region to new areas and different culture groups. Spatial diffusion will be discussed in more detail in Chapter 9.

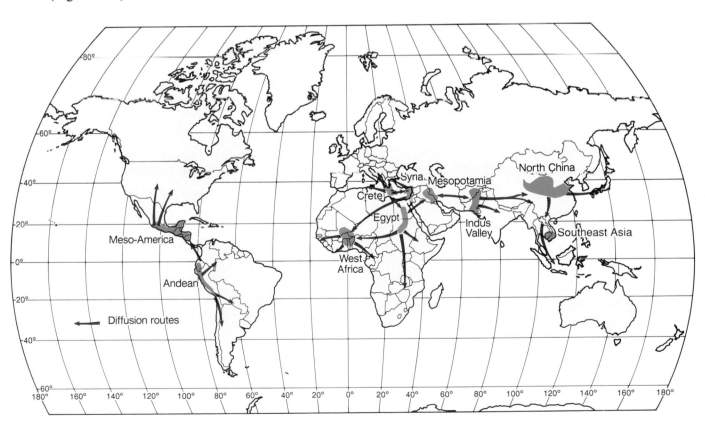

Figure 7.12
Early culture hearths of the Old World and the Americas.

It is not always possible to determine whether the existence of a culture trait in two different areas is the result of diffusion or of independent (or *parallel*) invention. Cultural similarities do not necessarily prove that spatial diffusion has occurred. The pyramids of Egypt and Central America, pictured in Figure 7.14, most likely were

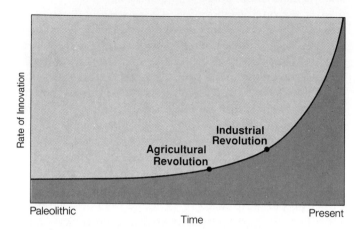

Figure 7.13

The rate of innovation through human history. Paleolithic hunter-gatherers, living in easy equilibrium with their environment and their resource base, had little need for innovation and no necessity for cultural change. Increased population pressures led to the Agricultural Revolution and the diffusion of the ideas and techniques of domestication, urbanization, and trade. With the Industrial Revolution, dramatic increases in innovation began to alter cultures throughout the world.

separately conceived and are not evidence, as some have proposed, of pre-Columbian voyages from the Mediterranean to the Americas. A monument-building culture, after all, has only a limited number of shapes from which to choose.

Historical examples of independent, parallel invention are numerous: logarithms by Napier (1614) and Bürgi (1620), the calculus by Newton (1672) and Leibnitz (1675), the telephone by Elisha Gray and Alexander Graham Bell (1876) are commonly cited examples. It appears beyond doubt that agriculture was independently invented not only in the New World and in the Old, but also in more than one culture hearth in each of the hemispheres.

Acculturation

Acculturation is the process by which one culture group undergoes a major modification by adopting many of the characteristics of another, usually dominant culture group. In practice, acculturation may involve changes in the original cultural patterns of either or both of two groups involved in prolonged firsthand contact. Such contact and subsequent cultural alteration may occur in a conquered or colonized region. Very often the subordinate or subject population is forced to acculturate or does so voluntarily, overwhelmed by the superiority in numbers or the technical level of the conqueror.

(a)

(b)

Figure 7.14

Outward appearance is not the only similarity between Egyptian (a) and Mayan (b) pyramids. In both societies, they were royal burial sites, goods were placed inside to accompany the corpse on its journey to eternity, and pains were taken to hide the entrance to the tomb.

The tribal Europeans in areas of Roman conquest, native populations in the wake of Slavic occupation of Siberia, and Native Americans following European settlement of North America experienced this kind of acculturation. In a different fashion, it is evident in the changes in Japanese political organization and philosophy imposed by occupying Americans after World War II or in the Japanese adoption of some aspects of American life (Figure 7.15). In turn, American life was enriched by awareness of Japanese food, architecture, and philosophy, which demonstrates the two-way nature of acculturation.

On occasion, the invading group is assimilated into the conquered society as, for example, when the older, richer Chinese culture prevailed over that of the conquering tribes of invading Mongols during the 13th and 14th centuries. The relationship of a mother country to its colony may also result in permanent changes in the culture of the colonizer even though little direct population contact is involved. The early European spread of tobacco addiction (see "Documenting Diffusion," in Chapter 9) may serve as an example, as can the impact on Old World diets and agriculture of potatoes, maize, and turkeys introduced from America.

All cultures are amalgams of innumerable innovations spread spatially from their points of origin and integrated into the structure of the receiving societies. It has been estimated that no more than 10% of the cultural items of any society are traceable to innovations created by its members, and that the other 90% come to the society through diffusion. (See "A Homemade Culture.") Since, as we have seen, the pace of innovation is affected strongly by the mixing of ideas among alert, responsive people and is increased by exposure to a variety of cultures, the most active and innovative historical hearths of culture were those at crossroad locations and those deeply involved in distant trade and colonization. Ancient Mesopotamia or classical Greece and Rome had such locations and involvements, as did the West African culture hearth after the 5th century and, much later, England during the Industrial Revolution and during the spread of its empire.

Barriers to diffusion do exist, of course, as Chapter 9 explains. Generally, the closer and the more similar two cultural areas are to one another, the lower those barriers are and the greater is the likelihood of the adoption of an innovation, for diffusion is a selective process. Of course, the receiver culture may selectively adopt some goods or

Figure 7.15
Baseball, an import from America, is one of the most popular sports in Japan, attracting millions of spectators annually.

ideas from the donor society and reject others. The decision to adopt is governed by the receiving group's own culture. Political restrictions, religious taboos, and other social customs are cultural barriers to diffusion. The French Canadians, although close geographically to many centers of diffusion, such as Toronto, New York, and Boston, are only minimally influenced by such centers. Both their language and culture complex govern their selective acceptance of Anglo influences. Traditional groups, perhaps controlled by firm religious conviction, may very largely reject culture traits and technologies of the larger society in whose midst they live (Figure 7.16).

Adopting cultures do not usually accept intact items originating from the outside. Diffused ideas and artifacts commonly undergo some alteration of meaning or form that makes them acceptable to a borrowing group. The process of the fusion of old and new, called **syncretism**, is a major feature of culture change. It can be seen in alterations to religious ritual and dogma made by convert societies seeking acceptable conformity between old and new beliefs. The mixture of Catholic rites and voodooism in Haiti is an example. On a more familiar level, syncretism is reflected in subtle or blatant alterations of imported cuisines to make them conform to the demands of America's fast-food franchises.

Cultural Diversity

We began our discussion of culture with its subsystems of technological, sociological, and ideological content. We have learned that the distinctive makeup of those subsystems—the combinations and interactions of traits and complexes characteristic of particular cultures—is subjected to, and the product of, change through innovation, spatial diffusion, adoption, and acculturation. Those processes of cultural development and alteration have not, however, led to a homogenized world culture even after thousands of years of cultural contact and exchange since the origins of agriculture. It is true, as we earlier observed, that in an increasingly integrated world, access to the material trappings and technologies of modern life and economy is widely available to all peoples and societies. As a result, important cultural commonalities have developed.

Nevertheless, all of our experience and observation indicates a world still divided, not unified, in culture. Our concern as geographers is to identify the traits of a culture that both have spatial expression and indicate significant differences from other culture complexes. We may reject as superficial and meaningless generalizations derived from trivialities: the foods people eat for breakfast, for example, or the kinds of eating implements they use.

Figure 7.16
Motivated by religious conviction that the "good life" must be reduced to its simplest forms, the Amish community of east central Illinois shuns all modern luxuries of the secular community around it. Children use horse and buggy, not school bus or automobile, on the daily trip to their rural school.

This rejection is a reflection of the kinds of understanding and the level of generalization we desire. There is no single most appropriate way to designate or recognize a culture or to delimit a culture region. As geographers concerned with world systems, we are interested in those aspects of culture that vary over extensive regions of the world and differentiate societies in a broad, summary fashion.

Language, religion, and ethnicity meet our criteria and are among the most prominent of the differentiating cultural traits of societies and regions. Language and religion are basic components of culture, helping to identify who and what we are as individuals and clearly placing us within larger communities of persons with similar characteristics. In our earlier terminology, they are mentifacts, components of the ideological subsystem of culture that help shape the belief system of a society and transmit it to succeeding generations. Ethnicity is a cultural summary rather than a single trait. It is based on the firm understanding by members of a group that they are in some fundamental ways different from others who do not share their distinguishing composite characteristics, which may include language, religion, national origin, unique customs, or other identifiers. Like language and religion, ethnicity has spatial identification. Like them, too, it may serve as an element of diversity and division within culturally complex societies and states.

A Homemade Culture

Reflecting on an average morning in the life of a "100% American," Ralph Linton noted:

Our solid American citizen awakens in a bed built on a pattern which originated in the Near East but which was modified in Northern Europe before it was transmitted to America. He throws back covers made from cotton, domesticated in India, or linen, domesticated in the Near East, or wool from sheep, also domesticated in the Near East, or silk, the use of which was discovered in China. All of these materials have been spun and woven by processes invented in the Near East. . . . He takes off his pajamas, a garment invented in India, and washes with soap invented by the ancient Gauls. . . .

Returning to the bedroom, . . . he puts on garments whose form originally derived from the skin clothing of the nomads of the Asiatic steppes [and] puts on shoes made from skins tanned by a process invented in ancient Egypt and cut to a pattern derived from the classical civilizations of the Mediterranean. . . . Before going out for breakfast he glances through the window, made of glass invented in Egypt, and if it is raining puts on overshoes made of rubber discovered by the Central American Indians and takes an umbrella invented in southeastern Asia. . . .

[At breakfast] a whole new series of borrowed elements confronts him. His plate is made of a form of pottery invented in China. His knife is of steel, an alloy first made in southern India, his fork a medieval Italian invention, and his spoon a derivative of a Roman original. He begins breakfast with an orange, from the eastern Mediterranean, a canteloupe from Persia, or perhaps a piece of African watermelon. With this he has coffee, an Abyssinian plant. . . . [H]e may have the egg of a species of bird domesticated in Indo-China, or thin strips of flesh of an animal domesticated in Eastern Asia which have been salted and smoked by a process developed in northern Europe.

When our friend has finished eating . . . he reads the news of the day, imprinted in characters invented by the ancient Semites upon a material invented in China by a process invented in Germany. As he absorbs the accounts of foreign troubles he will, if he is a good conservative citizen, thank a Hebrew deity in an Indo-European language that he is 100 per cent American.

Ralph Linton, *The Study of Man*, © 1936, renewed 1964, pp. 326–27. Reprinted by permission of Prentice Hall, Inc., Englewood Cliffs, New Jersey.

Language

Forever changing and evolving, language in spoken or written form makes possible the cooperative efforts, the group understandings, and shared behavior patterns that distinguish culture groups. *Language,* defined simply as an organized system of speech by which people communicate with each other with mutual comprehension, is the most important medium by which culture is transmitted. It is what enables parents to teach their children what the world they live in is like and what they must do to become functioning members of society. Some argue that the language of a society structures the perceptions of its speakers. By the words that it contains and the concepts that it can formulate, language is said to determine the attitudes, understandings, and responses of the society. Language therefore may be both a cause and a symbol of cultural differentiation (Figure 7.17).

If that conclusion is true, one aspect of cultural heterogeneity may be easily understood. The more than 5 billion people on earth speak many thousands of different languages. Knowing that as many as 1000 languages are spoken in Africa gives us a clearer appreciation of the political and social divisions in that continent. Europe alone has more than 100 languages and dialects, and language differences were a partial basis for drawing boundaries in the political restructuring of Europe after World War I. Language, then, is a hallmark of cultural diversity, and the present world distribution of major languages (Figure

Figure 7.17
In their mountainous homeland, the Basques have maintained a linguistic uniqueness despite more than 2000 years of encirclement by dominant lowland speakers of Latin or Romance languages. This sign of friendly farewell gives its message in both Spanish and the Basque language, Euskara.

7.18) records not only the migrations and conquests of our linguistic ancestors but also the continuing dynamic pattern of human movements, settlements, and colonizations of more recent centuries.

Figure 7.18

World language families. Language families are assumed to group individual tongues that had a common but remote ancestor. By suggesting that the area assigned to a language or language family uses that language exclusively, the map pattern conceals important linguistic detail. Many countries and regions have local languages spoken in territories too small to be recorded at this scale. The map also fails to report that the population in many regions is fluent in more than one language or that a second language serves as the necessary vehicle of commerce, education, or government. Nor is important information given about the number of speakers of different languages. The fact that there are more speakers of English in India or Africa than in Australia is not even hinted at by a map at this scale.

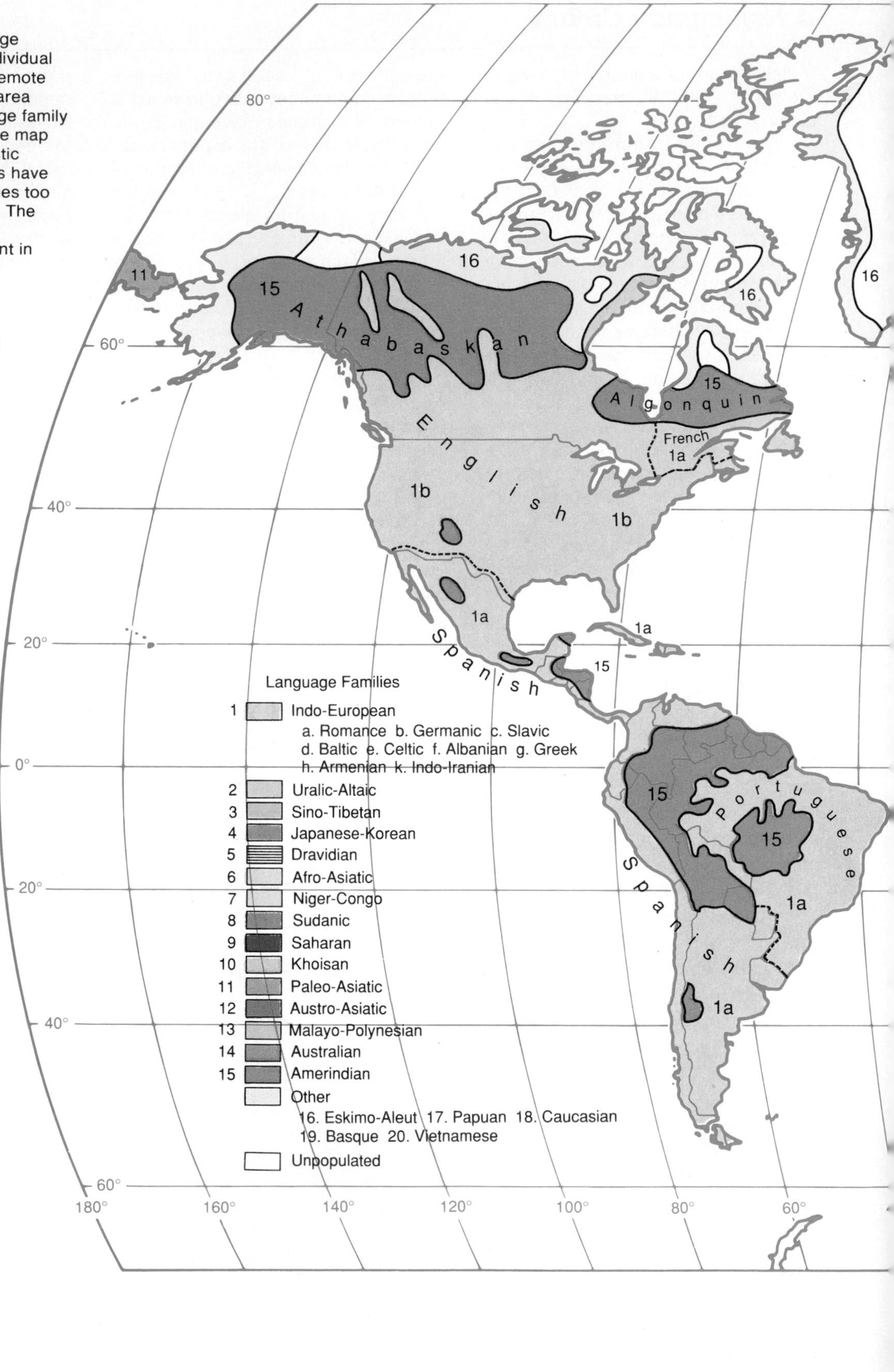

Language Families
1 Indo-European
a. Romance b. Germanic c. Slavic
d. Baltic e. Celtic f. Albanian g. Greek
h. Armenian k. Indo-Iranian
2 Uralic-Altaic
3 Sino-Tibetan
4 Japanese-Korean
5 Dravidian
6 Afro-Asiatic
7 Niger-Congo
8 Sudanic
9 Saharan
10 Khoisan
11 Paleo-Asiatic
12 Austro-Asiatic
13 Malayo-Polynesian
14 Australian
15 Amerindian
Other
16. Eskimo-Aleut 17. Papuan 18. Caucasian
19. Basque 20. Vietnamese
Unpopulated
Athabaskan
Algonquin
English
French
Spanish
Portuguese
1a
1b
11
15
16
80°
60°
40°
20°
0°
180°
160°
140°
120°
100°
80°
60°

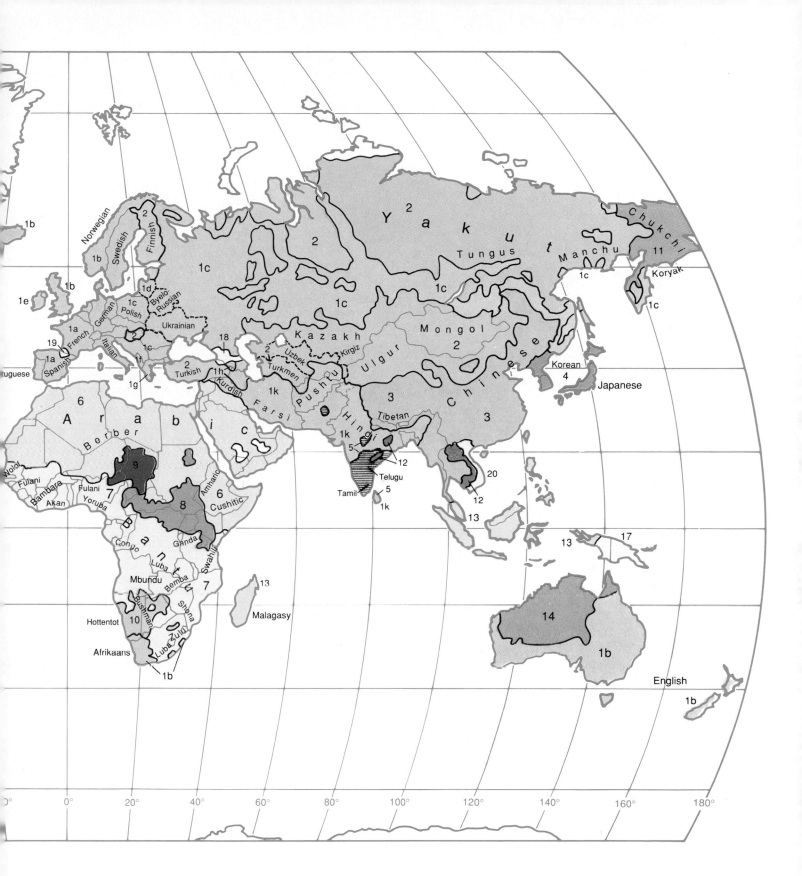

1b

Norwegian
Swedish
Finnish
2
Y ² a k u t
Chukchi

1b
1c
2
Tungus
Manchu
11
Koryak

1e
1b
1d
German
Byelo-
Russian
1c
1c
1c

1c
Polish
Ukrainian
1c
1c

1a
French
German
18
K a z a k h
Mongol
Korean

19
Italian
2
2
Uzbek
Kirgiz
U l g u r
2
C
h
i
4
Japanese

1a
Spanish
Turkish
Turkmen
P
u
s
h
t
u
n
e
s
e

ruguese
1f
1g
1h
F a r s i
3
Tibetan
3

6
Kurdish
1k
Hindi

A r a b i c
1k
5
12

Berber
9
Telugu
20

Wolof
Tamil
5
12

Fulani
Fulani
7
8
Amharic
1k
13

Bambara
Yoruba
Cushitic
6

Akan
Congo
Ganda
13
17

Luba
Swahili

Mbundu
Bemba
B
a
n
t
u
7
13

Hottentot
Bushman
Shona
Malagasy
14

Afrikaans
Luba
Zulu
1b

1b

English

1b

0°   0°   20°   40°   60°   80°   100°   120°   140°   160°   180°

Languages differ greatly in their relative importance, if "importance" can be taken to mean the number of people using them. More than half of the world's inhabitants are speakers of just 8 of its thousands of tongues. Table 7.1 lists those languages spoken by more than 40 million people, a list that includes four-fifths of the world's population. At the other end of the scale are a number of rapidly declining languages whose speakers number in the hundreds or, at most, a few thousand.

Table 7.1

Languages Spoken by More Than 40 Million People

Language	Millions of speakers
Mandarin (China)	802
English	425
Hindi[a] (India, Pakistan)	315
Spanish	310
Russian	225
Arabic	185
Bengali (Bangladesh, India)	175
Portuguese	163
Malay-Indonesian	131
Japanese	123
German	117
French	114
Urdu[a] (Pakistan, India)	86
Punjabi (India, Pakistan)	74
Korean	66
Telugu (India)	64
Tamil (India, Sri Lanka)	63
Marathi (India)	61
Cantonese (China)	60
Italian	59
Wu (China)	59
Javanese	52
Vietnamese	52
Turkish	51
Min (China)	46
Thai	45
Ukrainian	45
Polish	41
Swahili (East Africa)	40

[a] Hindi and Urdu are basically the same language, Hindustani. Written in the Devanagari script, it is called *Hindi,* the official language of India; in the Arabic script it is called *Urdu,* the official language of Pakistan.

The diversity of languages is simplified when we recognize among them related families. A **language** (or *linguistic*) **family** is a group of languages thought to have a common origin in a single, earlier tongue. The Indo-European family is among the most prominent of such groupings, embracing most of the languages of Europe and a large part of Asia, and the introduced—not the native—languages of the Americas (Figure 7.19). All told, languages in the Indo–European family are spoken by about half the world's peoples.

By recognizing similar words in most Indo–European languages, linguists deduce that these languages derived from a common ancestor tongue called *proto–Indo-European,* which was spoken by people living somewhere in eastern Europe about 5000 years ago (though some conclude that the southern Russian plains eastward to the Caspian Sea were the more likely site of origin). About 2500 B.C. their society apparently fragmented. The homeland was left and segments of the parent culture migrated in different directions. Some moved into Greece, others settled in Italy, still others crossed central and western Europe, ultimately reaching the British Isles. Another group headed into the Russian forest lands, and still another branch crossed Iran and Afghanistan, eventually

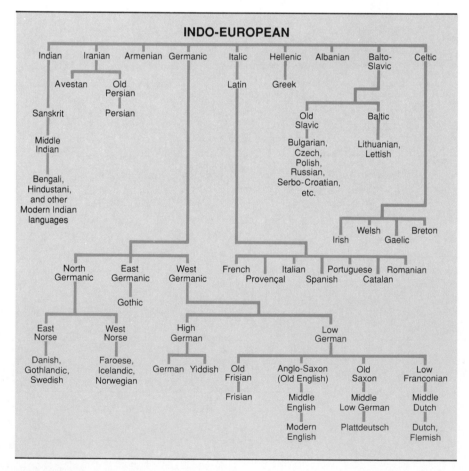

Figure 7.19
The Indo-European linguistic family tree.

to reach India. Wherever this remarkable people settled, they appear to have dominated local populations and imposed their language upon them.

Within a language family we can distinguish *subfamilies*. The Romance languages (including French, Spanish, and Italian)—offsprings of Latin—and the Germanic languages (such as English, German, and Dutch) are subfamilies of Indo-European. The languages in a subfamily often show similarities in sounds, grammatical structure, and vocabulary even though they are mutually unintelligible. English *daughter,* German *Tochter,* and Swedish *dotter* are Germanic examples. *Panis,* the Latin root word for "bread," appears as *pane* (Italian), *pain* (French), *pan* (Spanish), *pão* (Portuguese), and *pîine* (Romanian) among the Romance languages.

Language Spread and Change

Language spread as a geographical event represents the increase or relocation through time in the area over which a language is spoken. The more than 300 Bantu languages found south of the "Bantu line" in sub-Saharan Africa, for example, are variants of a proto-Bantu carried by an expanding, culturally advanced population who displaced linguistically different preexisting populations (Figure 7.20). More recently, the languages of European colonists similarly replaced native tongues in their areas of settlement in North and South America, Australasia, and Siberia. That is, languages may spread because their speakers occupy new territory.

Latin, however, replaced earlier Celtic languages in Western Europe not by force of numbers—Roman legionnaires, administrators, and settlers never represented a majority population—but by the gradual abandonment of their former tongues by native populations brought under the influence and control of the Roman Empire. Adoption rather than eviction of language appears the rule followed in the majority of historical and contemporary instances of language spread. That is, languages may expand spatially through the acquisition of new speakers.

Language spread of either form may, through segregation and isolation, give rise to separate mutually incomprehensible tongues because the society speaking the parent protolanguage no longer remains unitary. Comparable changes occur normally and naturally within a single language in word meaning, pronunciation, vocabulary, and *syntax* (the way words are put together in phrases and sentences). Because they are gradual, such changes tend to go unremarked. Yet, cumulatively, they can result in language change so great that in the course of centuries an essentially new language has been created.

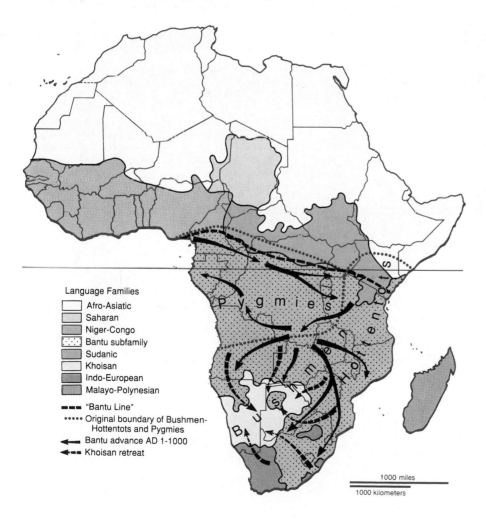

Figure 7.20

Bantu advance, Khoisan retreat in Africa. Linguistic evidence suggests that proto-Bantu speakers originated in the region of the Cameroon-Nigeria border, spread eastward across the southern Sudan, then turned southward to Central Africa. From there they dispersed slowly eastward, westward and, against slight resistance, southward. The earlier *Khoisan*-speaking hunter-gatherers of sub-Saharan Africa were no match against the advancing metal-using Bantu agriculturalists. Pygmies, adopting a Bantu tongue, retreated deep into the forests. Bushmen and Hottentots retained their distinctive Khoisan "click" language but were forced out of forest and grasslands into the dry steppes and deserts of the southwest.

Language Families
- Afro-Asiatic
- Saharan
- Niger-Congo
- Bantu subfamily
- Sudanic
- Khoisan
- Indo-European
- Malayo-Polynesian

- – – – "Bantu Line"
- • • • • Original boundary of Bushmen-Hottentots and Pygmies
- ◄— Bantu advance AD 1–1000
- ◄– – Khoisan retreat

1000 miles
1000 kilometers

The English of 17th-century Shakespearean writings or the King James Bible (1611) sounds stilted to our ears. Few of us can easily read Chaucer's 14th-century *Canterbury Tales* in their original Middle English, and 8th century *Beowulf* is practically unintelligible.

Language evolution may be gradual and cumulative, with each generation deviating in small degree from the speech patterns and vocabulary of its parents, or it may be massive and abrupt—reflecting conquests, migrations, new trade contacts, and other disruptions of cultural isolation. English owes its form to the Celts, the original inhabitants of the British Isles, and to successive waves of invaders, including the Latin-speaking Romans and the Teutonic Angles, Saxons, and Danes. The French-speaking Norman conquerors of the 11th century added about 10,000 new words to the evolving English tongue.

The English language continued to change in the centuries following the Norman conquest, as did the Gaelic of Britons who had been pushed into the outlying, relatively isolated areas of Ireland, Wales, and Scotland. A worldwide diffusion of the language resulted as English colonists carried it to the Western Hemisphere and Australasia and as trade, conquest, and territorial claim took it to Africa and Asia. In that areal spread, English was further enriched by its contacts with other languages. By becoming the accepted language of commerce and science, it contributed in turn to the common vocabularies of other tongues. (See "Language Exchange.") Within roughly 400 years, spreading and changing English has developed from a localized language of 7 million islanders off the European coast to become the native tongue of over 400 million speakers and the official language of more than 40 countries (Figure 7.21).

Standard and Variant Languages

People who speak a common language such as English are members of a *speech community,* but membership does not necessarily imply linguistic uniformity. A speech community usually possesses both a *standard language* comprising the accepted community norms of syntax, vocabulary, and pronunciation and a number of more or less distinctive **dialects**, the ordinary speech of areal, social, professional, or other subdivisions of the general population.

An official or unofficial standard language is the form carrying governmental, educational, or societal sanction. In Arab countries, for example, classical Arabic is the language of the mosque, of education, and of the newspapers

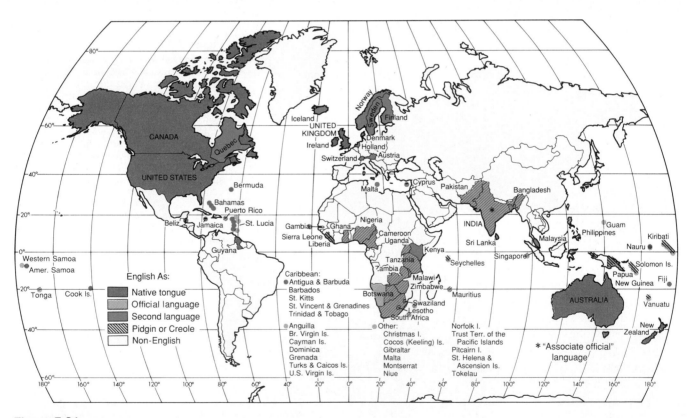

Figure 7.21

International English. In worldwide diffusion and acceptance, English has no past or present rivals. Along with French, it is one of the two working languages of the United Nations, and some two-thirds of all scientific papers are published in it. English is the sole or joint official language of more countries and territories, some too small to be shown here, than any other tongue. It also serves as the effective unofficial language of administration in other multilingual countries with different formal official languages. "English as a second language" is indicated for countries with near-universal or mandatory instruction in it in public schools.

and is standardized throughout the Arabic speaking world. Colloquial Arabic is used at home, in the street, and at the market—and in its regional variants may be as widely different as are, for example, Portuguese and Italian. On the other hand, the United States, English-speaking Canada, Australia, and the United Kingdom all have only slightly different forms of standard English. In England, *British Received Pronunciation*—the speech of educated people of London and southeastern England and used by the British Broadcasting System—is the accepted standard.

Just as no two individuals talk exactly the same, all but the smallest and most closely knit speech communities display recognizable speech variants called *dialects*. Vocabulary, pronunciation, rhythm, and the speed at which the language is spoken may clearly set groups of speakers apart from one another. Dialects may coexist in space. Cockney and cultured English share the streets of London; black English and Standard American are heard in the same school yards throughout the United States. In many societies, *social dialects* denote social class and educational levels, with speakers of higher socioeconomic status or educational achievement most likely to follow the norms of their standard language. Less educated or lower status people are more apt to use the *vernacular*—nonstandard language or dialect adopted by the social group.

More commonly, we think of dialects in spatial terms. Speech is a geographic variable. Each locale is likely to have its own, perhaps slight, language differences from neighboring places. Such differences in pronunciation, vocabulary, word meanings, and other language characteristics help define the *linguistic geography*—the study of the character and spatial pattern of *geographic* or *regional dialects*—of a generalized speech community. Figure 7.22 records the variation in usage associated with just one phrase. In the United States, Southern English and New England speech are among the regional dialects that are most easily recognized by their distinctive accents. In some instances there may be so much variation among geographic dialects that some are almost foreign tongues to other speakers of the same language. Effort is required for Americans to understand Australian English or that spoken in Liverpool, England, or in Glasgow, Scotland. An interesting U.S. example is discussed in the regional study, "Gullah as Language," in Chapter 13.

Language is rarely a total barrier in communication between different peoples. Bilingualism or multilingualism may permit skilled linguists to communicate in a jointly understood third language, but long-term contact between less-able populations may require the creation of new language—a pidgin—learned by both parties. A **pidgin** is an amalgamation of languages, usually a simplified form of one, such as English or French, with borrowings from another, perhaps non-European, native language. In its original form, a pidgin is not the mother tongue of any of its speakers; it is a second language for everyone who uses it, one generally restricted to such specific functions as commerce, administration, or work supervision.

Pidgins are characterized by a highly simplified grammatical structure and a sharply reduced vocabulary adequate to express basic ideas but not complex concepts. If a pidgin becomes the first language of a group of speakers—who may have lost their former native tongue through disuse—a **creole** has evolved. Creoles invariably acquire a more complex grammatical structure and enhanced vocabulary.

Creole languages have proved useful integrative tools in linguistically diverse areas, and several have become symbols of nationhood. Swahili, a pidgin formed from a number of Bantu dialects, originated in the coastal areas of East Africa and spread by trade during the period of English and German colonial rules. When Kenya and Tanzania gained independence, they made Swahili the national language of administration and education. Other examples of creolization are Afrikaans (a pidginized form

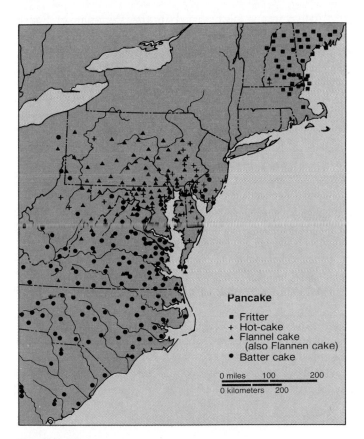

Pancake

- ■ Fritter
- + Hot-cake
- ▲ Flannel cake
  (also Flannen cake)
- ● Batter cake

0 miles   100   200
0 kilometers   200

Figure 7.22

Maps such as this are used to record variations over space and among social classes in word usage, accent, and pronunciation. The differences are due not only to initial settlement patterns but also to more recent large-scale movements of people, for example, from rural to urban areas and from the South to the North. Despite the presumed influence of national radio and television programs in promoting a "general" or "standard" American accent and usage, regional and ethnic language variations persist.

Language Exchange

English has a happily eclectic vocabulary. Its foundations are Anglo-Saxon (*was, that, eat, cow*) reinforced by Norse (*sky, get, bath, husband, skill*); its superstructure is Norman-French (*soldier, Parliament, prayer, beef*). The Norman aristocracy used their words for the food, but the Saxon serfs kept theirs for the animals. Its decor comes from Renaissance and Enlightenment Europe: 16th century France yielded *etiquette, naive, reprimand,* and *police.*

Italy provided *umbrella, duet, bandit,* and *dilettante;* Holland gave *cruise, yacht, trigger, landscape,* and *decoy.* Its elaborations come from Latin and Greek: *misanthrope, meditate,* and *parenthesis* all first appeared during the 1560s. In this century, English adopted *penicillin* from Latin, *polystyrene* from Greek, and *sociology* and *television* from both. And English's ornaments come from all round the world: *slogan* and *spree* from Gaelic, *hammock* and *hurricane* from Caribbean languages, *caviar* and *kiosk* from Turkish, *dinghy* and *dungarees* from Hindi, *caravan* and *candy* from Persian, *mattress* and *masquerade* from Arabic.

Redressing the balance of trade, English is sharply stepping up its linguistic exports. Not just the necessary *imotokali* (motor car) and *izingilazi* (glasses) to Zulu; or *motokaa* and *shillingi* (shilling) to Swahili; but also *der Bestseller, der Kommunikations Manager, das Teeshirt,* and *der Babysitter* to German; and to Italian, *la pop art, il popcorn* and *la spray.* In some Spanish-speaking countries you might wear *un sueter* to *el beisbol,* or witness *un nocaut* at *el boxeo.* Indeed, a sort of global English wordlist can be drawn up: *airport, passport, hotel, telephone; bar, soda, cigarette; sport, golf, tennis; stop, OK,* and increasingly, *weekend, jeans, know-how, sex appeal,* and *no problem.*

---

of 17th century Dutch used in the Republic of South Africa); Haitian Creole (the language of Haiti, derived from the pidginized French used in the slave trade); and Bazaar Malay (a pidginized form of the Malay language, a version of which is the official national language of Indonesia).

A **lingua franca** is an established language used habitually for communication by people whose native tongues are mutually incomprehensible. For them it is a *second language,* one learned in addition to the native tongue. Lingua franca (literally "Frankish tongue") was named from the French dialect adopted as a common tongue by the Crusaders at war in the Holy Land. Later, Latin became the lingua franca of the Mediterranean world until, finally, it was displaced by vernacular European tongues. Outside of the European sphere, Aramaic served the role from the 5th century B.C. to the 4th century A.D. in the Near East and Egypt. Arabic followed Muslim conquest as the unifying language of that international religion after the 7th century. Mandarin Chinese and Hindi in India have traditionally had a lingua franca role in their linguistically diverse countries. The immense linguistic diversity of Africa has made regional lingua francas there necessary and inevitable.

Language and Culture

Language embodies the culture complex of a people, reflecting both environment and technology. Arabic has 80 words related to camels, an animal on which a regional culture relied for food, transport, and labor, and Japanese contains over 20 words for various types of rice. Russian is rich in terms for ice and snow, indicative of the prevailing climate of its linguistic cradle, and the 15,000 tributaries and subtributaries of the Amazon River have obliged the Brazilians to enrich Portuguese with words that go beyond "river." Among them are *paraná* (a stream that leaves and reenters the same river), *igarapé* (an offshoot that runs until it dries up), and *furo* (a waterway that connects two rivers).

A common language fosters unity among people. It promotes a feeling for a region; if it is spoken throughout a country, it fosters nationalism. For this reason, languages often gain political significance and serve as a focus of opposition to what is perceived as foreign domination. Although nearly all people in Wales speak English, many also want to preserve Welsh because they consider it an important aspect of their culture. They think that if the language is forgotten, their entire culture may also be threatened. French Canadians have received government recognition of their language and now have established it as the official language of Quebec Province; Canada itself is officially bilingual. In India, with 15 constitutional languages and 1652 other tongues, serious riots were caused in 1965 by people expressing opposition to the imposition of Hindi as the single official national language. In the Soviet Union, protest marches and demonstrations in the Georgian Republic of the Caucasus area during 1981 marked the exclusion of reference to Georgian as the official language in the republic's revised constitution. Other of the culturally based divisions of the USSR have similarly demanded linguistic identity in the years since.

Bilingualism or multilingualism complicates national linguistic structure. Areas are considered bilingual if more than one language is spoken by a significant proportion of the population. In some countries—Belgium or Switzerland, for example—there is more than one official language. In many others, such as the United States, only one language may have implicit or official government sanction, although several others are spoken. Speakers of one of these may be concentrated in restricted areas (e.g., most speakers of French in Canada live in Quebec Province). Less often, they may be distributed fairly evenly throughout the country. There may be roughly equal numbers of speakers of each language, or one group may be significantly in the majority. When this latter is the case, the minority group may have to learn the language of the majority to receive an education, to obtain certain types of employment, or to fill a position in the government. In some countries, the language in which instruction, commercial transactions, and government business take place is neither a native language nor the majority language, but yet another tongue. In linguistically complex sub-Saharan Africa, all but two countries have selected a European language—usually that of their former colonial governors—as an official language (Figure 7.23).

Toponyms—place-names—are language on the land, the record of past and present cultures whose namings endure as reminders of their passing and their existence. **Toponymy**, the study of place-names, therefore is a revealing tool of historical cultural geography, for place-names become a part of the cultural landscape that remains long after the name givers have passed from the scene.

In England, for example, place-names ending in *chester* (as in Winchester and Manchester) evolved from the Latin *castra,* meaning "camp." Common Anglo-Saxon suffixes for tribal and family settlements were *ing* (people or family) and *ham* (hamlet or, perhaps, meadow) as in Birmingham or Gillingham. Norse and Danish settlers contributed place-names ending in *thwaite* (meadow) and others denoting such landscape features as *fell* (an uncultivated hill) and *beck* (a small brook). The Arabs, sweeping out from Arabia across North Africa and into Iberia, left their imprint in place-names to mark their conquest and control. *Cairo* means "victorious," *Sudan* is "the land of the blacks," and *Sahara* is "wasteland" or "wilderness." In Spain, a corrupted version of the Arabic *wadi,* "watercourse," is found in *Guadalajara* and *Guadalquiver.*

In the New World, not one people but many placed names on landscape features and new settlements. In doing so they remembered their homes and homelands, honored their monarchs and heroes, borrowed and mispronounced from rivals, adopted and distorted Amerindian names, followed fads, and recalled the Bible. Homelands were recalled in New England, New France, or New Holland; settlers' hometown memories brought Boston, New Bern, and New Rochelle from England, Switzerland, and France. Monarchs were remembered in Virginia for the Virgin Queen Elizabeth, Carolina for one English king, Georgia for another, and Louisiana for a king of France. Washington, D.C.; Jackson, Mississippi and Michigan; Austin, Texas; and Lincoln, Illinois, memorialized heroes and leaders.

Figure 7.23

Europe in Africa through official languages. Both the linguistic complexity of sub-Saharan Africa and the colonial histories of its present political units are implicit in the designation of a European language as the sole or joint official language of the different countries.

Names given by the Dutch in New York were often distorted by the English; Breukelyn, Vlissingen, and Haarlem became Brooklyn, Flushing, and Harlem. French names underwent similar twisting or translation, and Spanish names were adopted, altered, or, later, put into such bilingual combinations as Hermosa Beach. Amerindian tribal names—the Yenrish, Maha, Kansa were modified, first by French and later by English speakers, to Erie, Omaha, and Kansas. A faddish classical revival after the American Revolution gave us Troy, Athens, Rome, Sparta, and other ancient town names and later spread them across the country. Bethlehem, Ephrata, Nazareth, and Salem came from the Bible.

Religion

Enduring place-names are only one measure of the importance of language as a powerful unifying thread in the culture complex of people and as a fiercely defended symbol of the history and individuality of a distinctive social group. But language is not alone in that role. In some ways it yields to religion as a cultural rallying point. French Catholics and French Huguenots (Protestants) freely slaughtered each other in the name of religion in the 16th century. English Roman Catholics were hounded from the country after the establishment of the Anglican church.

Figure 7.24

Worshipers at al-Hassan mosque in Cairo, Egypt. Many rules concerning daily life are given in the Koran, the holy book of the Muslims. All Muslims are expected to observe the five pillars of the faith: (1) repeated saying of the basic creed; (2) prayers five times daily, facing Mecca; (3) a month of daytime fasting (Ramadan); (4) almsgiving; and (5) if possible, a pilgrimage to Mecca.

Religious enmity between Muslims and Hindus forced the partition of the Indian subcontinent after the departure of the British in 1947. And the 1980s witnessed continuing religious confrontations including those of Catholic and Protestant Christian groups in Northern Ireland; Muslim sects in Lebanon, Iran, and Iraq; Muslims and Jews in Palestine; Christians and Muslims in the Philippines and Lebanon; and Buddhists and Hindus in Sri Lanka.

However, unlike language, which is an attribute of all people, religion varies in its cultural role—dominating among some societies, unimportant or denied totally in others. All societies have value systems—common beliefs, understandings, expectations, and controls—that unite their members and set them off from other, different culture groups. Such a value system is termed a *religion* when it involves systems of formal or informal worship and faith in the sacred and divine.

Religion may intimately affect all facets of a culture. Religious belief is by definition an element of the ideological subsystem; formalized and organized religion is an institutional expression of the sociological subsystem. And religious beliefs strongly influence attitudes toward the tools and rewards of the technological subsystem.

Nonreligious value systems can exist—humanism or Marxism, for example—that are just as binding upon the societies that espouse them as are more traditional religious beliefs. Indeed, in a few countries—the Soviet Union during its period of Communist Party domination, for example—political ideologies have a quasi-religious role. They have many of the elements of a religion, including a set of beliefs, ethical standards, revered leaders, an or-

ganization, and a body of literature akin to scripture. In addition, the adherents may display an almost religious fervor in their desire to proselytize (convert nonbelievers) and to root out heretical beliefs and practices. Even societies that are officially atheistic, however, are strongly influenced by traditional values and customs set by predecessor religions—in days of work and rest or in legal principles for example.

Since religions are formalized views about the relation of the individual to this world and to the hereafter, each carries a distinct conception of the meaning and value of this life, and most contain strictures about what must be done to achieve salvation (Figure 7.24). These rules become interwoven with the traditions of a culture. One cannot understand India without a knowledge of Hinduism or Israel without an appreciation of Judaism.

Economic patterns may be intertwined with past or present religious beliefs. Traditional restrictions on food and drink may affect the kinds of animals that are raised or avoided, the crops that are grown, and the importance of those crops in the daily diet. Occupational assignment in the Hindu caste system is in part religiously supported. In many countries, there is a state religion, that is, religious and political structures are intertwined. Buddhism, for example, has been the state religion in Burma, Laos, and Thailand. By their official names, the Islamic Republic of Pakistan and the Islamic Republic of Iran proclaim their identity of church and government. Despite the country's overwhelming Muslim majority, Indonesia seeks domestic harmony by recognizing five official religions and a state ideology—*pancasila*—whose first tenet is belief in one god.

Classification and Distribution of Religions

Religions are cultural innovations. They may be unique to a single culture group, closely related to the faiths professed in nearby areas, or derived from or identical to belief systems spatially far removed. Although interconnections and derivations among religions can frequently be discerned—as Christianity and Islam can trace descent from Judaism—family groupings are not as useful to us in classifying religions as they were in studying languages. A distinction between *monotheism,* belief in a single deity, and *polytheism,* belief in many gods, is frequent, but not particularly spatially relevant. It is more useful for the spatial interests of geographers to categorize religions as *universalizing, ethnic,* or *tribal* (*traditional*).

Christianity, Islam, and Buddhism are the major world **universalizing religions**, faiths that claim applicability to all humans and that seek to transmit their beliefs to all lands through missionary work and conversion. Membership in universalizing religions is open to anyone who chooses to make some sort of symbolic commitment, such as baptism in Christianity. No one is excluded because of nationality, ethnicity, or previous religious belief.

Ethnic religions have strong territorial and cultural group identification. One becomes a member of an ethnic religion by birth or by adoption of a complex life-style and cultural identity, not by a simple declaration of faith. These religions do not usually proselytize, and their members form distinctive closed communities identified with a particular ethnic group, region, or political unit. An ethnic religion—for example, Judaism, Indian Hinduism, or Japanese Shinto—is an integral element of a specific culture. To be part of the religion is to be immersed in the totality of the culture.

Tribal (or *traditional*) **religions** are special forms of ethnic religions distinguished by their small size, their unique identity with localized culture groups not yet fully absorbed into modern society, and their close ties to nature. *Animism* is the name given to their belief that life exists in all objects, from rocks and trees to lakes and mountains, or that such objects are the abode of the dead, of spirits, and of gods. *Shamanism* is a form of tribal religion that involves community acceptance of a *shaman* who, through special powers, can intercede with and interpret the spirit world.

The nature of the different classes of religions is reflected in their distribution over the world (Figure 7.25) and in their number of adherents (Table 7.2). Universalizing religions tend to be expansionary, carrying their message to new peoples and areas. Ethnic religions, unless their adherents are dispersed, tend to be regionally confined or to expand only slowly and over long periods. Tribal religions tend to contract spatially as their adherents are incorporated increasingly into modern society and converted by proselytizing faiths.

As we expect in cultural geography, both map and table record only the latest stage of a constantly changing reality. While established religious institutions tend to be conservative and resistant to change, religion as a culture trait is dynamic. Personal and collective beliefs may alter in response to developing individual and societal needs and challenges. Religions may be imposed by conquest, adopted by conversion, or defended and preserved in the face of surrounding hostility or indifference.

Neither map nor table, however, presents a full picture even of current religious regionalization or affiliation. Few societies are homogeneous, and most modern ones contain a variety of different faiths or, at least, variants of the dominant professed religion. Frequently, members of a particular religion show areal concentration within a country. Thus, in urban Northern Ireland, Protestants and Catholics reside in separate areas whose boundaries are clearly understood and respected. The "Green Line" in Beirut, Lebanon, marks a guarded border between the Christian east and the Muslim west sides of the city, while within the country as a whole regional concentrations of adherents of different faiths and sects are clearly recognized. Religious diversity within countries may reflect the degree of toleration a majority culture affords minority religions. In dominantly (90%) Muslim Indonesia, Christian Bataks, Hindu Balinese, and Muslim Javanese live in easy coexistence. By contrast, the fundamentalist Islamic regime in Iran has persecuted and executed those of the Baha'i faith.

Data on numbers of adherents must be considered suspect and tentative. One cannot assume that all people within a mapped religious region are adherents of the designated faith, nor can it be assumed that membership in a religious community means active participation in its belief system. *Secularism,* an indifference to or rejection of religion and religious belief, is an increasing part of many modern societies, particularly of the industrialized countries and those now or recently under Communist regimes. The incidence of secularism at the end of the 1980s in atheistic Communist societies is suggested on the map by letter symbol. Its widespread occurrence in other, largely Christian, countries should be understood though it is not mapped. In England, for example, the state Church of England claims only 5% of the British population as communicants.

Figure 7.25
The pattern of principal world religions.

Christianity
- Mainly Roman Catholic
- Mainly Protestant
- Mainly Eastern Orthodox

Islam
- Sunni
- Shiah

- Judaism
- Buddhism
- Hindu
- Chinese faiths
- Shinto
- Tribal religions
- S Secularism

The Principal Religions

Each of the major religions has its own unique mix of cultural values and expressions, each has had its own pattern of innovation and spatial diffusion (Figure 7.26), and each has had its own impact upon the cultural landscape. Together they contribute importantly to the worldwide pattern of human diversity.

Judaism

We may logically begin our review of world faiths with *Judaism,* whose belief in a single God laid the foundation for both Christianity and Islam. Unlike its universalizing offspring, Judaism is closely identified with a single ethnic group and with a complex and restrictive set of beliefs and laws. It emerged some 3000 to 4000 years ago in the Near East, one of the ancient cultural hearth regions (Figure 7.12).

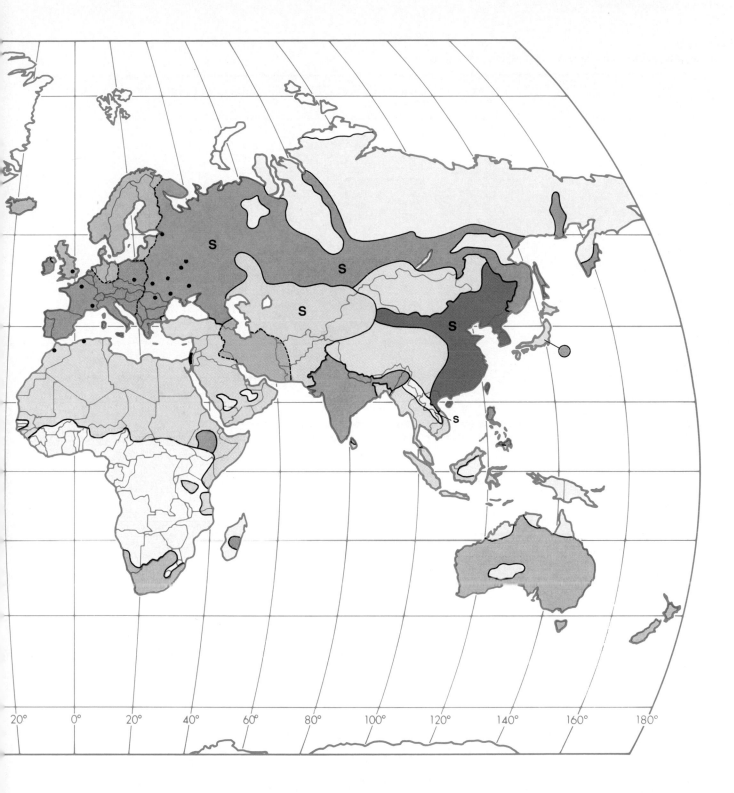

The Israelites' conviction that they were a chosen people, bound with God through a covenant of mutual loyalty and guided by complex formal rules of behavior, set them apart from other peoples of the Near East. Theirs is a distinctively *ethnic* religion, the determining factors of which are descent from Israel (the patriarch Jacob), the Torah (law and scripture), and the traditions of the culture and the faith. Early military success gave the Jews a sense of territorial and political identity to supplement their religious self-awareness. Later conquest by nonbelievers led to their dispersion (*diaspora*) to much of the Mediterranean world and farther east into Asia by A.D. 500 (Figure 7.27).

Between the 13th and 14th centuries, many Jews sought refuge in Poland and Russia from persecution in western and central Europe. During the later 19th and early 20th centuries Jews were important elements of the European immigrant stream to the Western Hemisphere.

Table 7.2

Estimated Adherents of Principal World Religions, 1989 (millions)

Religion or belief	Anglo America	Latin America[a]	Europe[b]	Asia[c]	Africa	Oceania[d]	World	% of World
Universalizing								
Christianity	233.5	404.6	506.7	198.3	261.7	21.5	1626.3	31.1
Roman Catholic	93.6	376.8	199.3	82.2	86.7	6.7	845.3	16.2
Protestant	95.2	17.7	112.7	53.4	89.6	8.4	377.0	7.2
Eastern Orthodox	6.0	0.7	148.3	3.3	24.9	0.6	183.8	3.5
Other Christian[e]	38.7	9.4	46.4	59.4	60.5	5.8	220.2	4.2
Islam	3.1	0.8	44.8	612.2	266.6	0.2	927.7[f]	17.7
Buddhism	0.4	0.6	0.5	317.6	—	—	319.1	6.1
Ethnic								
Hindu	1.0	0.8	0.6	651.5	1.4	0.4	655.7	12.5
Chinese faiths	0.3	—	0.2	359.5	—	—	360.0	6.9
Shinto	0.6	—	0.3	29.7	—	—	30.6	0.6
Judaism	8.3	1.2	4.7	3.8	0.3	—	18.3	0.3
Tribal	0.2	1.7	—	28.8	79.3	—	110.0	2.1
Secular[g]	24.7	20.5	222.5	835.1	2.1	3.3	1108.2	21.2
Other religions and unassigned	2.9	7.8	7.7	24.5	34.6	0.6	78.1	1.5
Total	275.0	438.0	788.0	3061.0	646.0	26.0	5234.0	100.0

[a]Includes Central and Caribbean America.
[b]Includes USSR and East European Communist nations where religious affiliation is difficult to determine.
[c]Includes areas in which multiple affiliation is common; also includes the People's Republic of China where affiliations can be only crudely estimated.
[d]Includes Australia, New Zealand, and South Pacific island groups.
[e]Includes Anglicans.
[f]According to the Islamic Center, Washington, D.C., there are over 1 billion Muslims worldwide.
[g]Includes both the nonreligious and atheists.

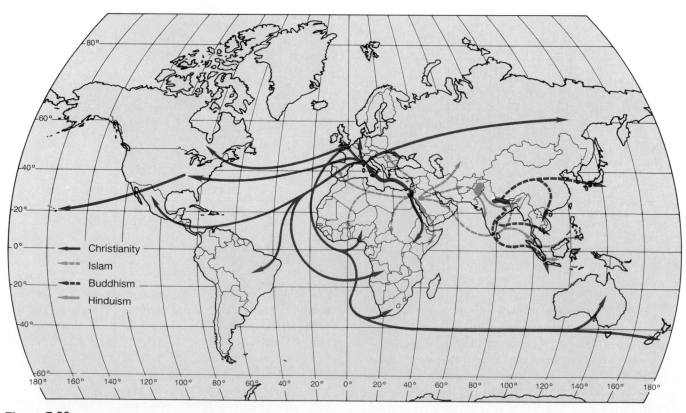

Figure 7.26

Innovation areas and diffusion routes of major world religions.
The monotheistic (single deity) faiths of Judaism, Christianity, and Islam arose in southwestern Asia, the first two in Palestine in the eastern Mediterranean region and the latter in western Arabia near the Red Sea. Hinduism and Buddhism originated within a confined hearth region in the northern part of the Indian subcontinent. Their rates, extent, and directions of spread are suggested here and detailed on later maps.

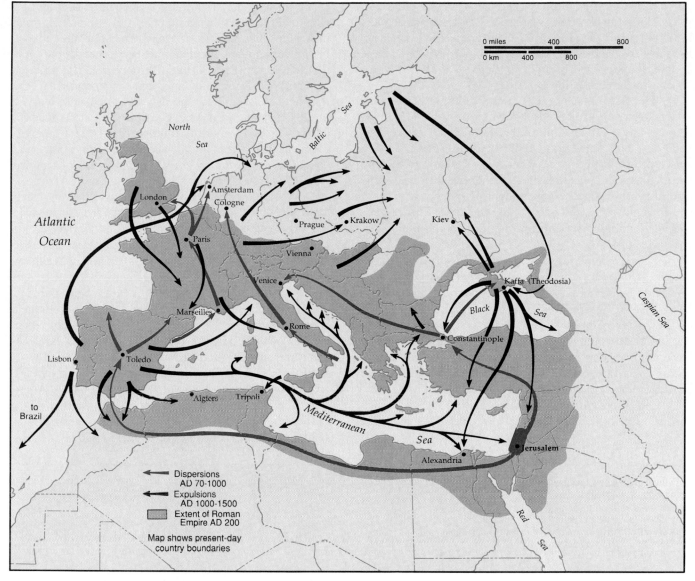

Figure 7.27

Jewish dispersions, A.D. 70–1500. A revolt against Roman rule in A.D. 66 was followed by the destruction of the Jewish Temple four years later and an imperial decision to Romanize the city of Jerusalem. Judaism spread from the hearth region, carried by its adherents dispersing from their homeland to Europe, Africa, and eventually in great numbers to the Western Hemisphere. Although Jews established themselves and their religion in new lands, they did not lose their sense of cultural identity nor did they seek to attract converts to their faith.

The establishment of the state of Israel in 1948 was a fulfillment of the goal of *Zionism,* the belief in the need to create an autonomous Jewish state in Palestine. It demonstrated a determination that Jews not lose their identity by absorption into alien cultures and societies.

Judaism's imprint upon the cultural landscape has been subtle and unobtrusive. The Jewish community reserves space for the practice of communal burial; the spread of the cultivated citron in the Mediterranean area during Roman times has been traced to Jewish ritual needs; and the religious use of grape wine assured the cultivation of the vine in their areas of settlement. The synagogue as place of worship has tended to be less elaborate than its Christian counterpart. The essential for religious service is a community of at least 10 adult males, not a specific structure.

Christianity

Christianity had its origin in the life and teachings of Jesus, a Jewish preacher of the 1st century of the modern era, whom his followers believed was the messiah promised by God. The new covenant he preached was not a rejection of traditional Judaism but a promise of salvation to all humankind rather than to just a chosen people.

Christianity's mission was conversion. As a universal religion of salvation and hope it spread quickly among the underclasses of both the eastern and western parts of the Roman Empire, carried to major cities and ports along the excellent system of Roman roads and sea lanes (Figure 7.28). In A.D. 313, the emperor Constantine proclaimed Christianity the state religion. Much later, of course, the faith was brought to the New World with European settlement (Figure 7.25).

The dissolution of the Roman Empire into a western and eastern half after the fall of Rome also divided Christianity. The Western Church, based in Rome, was one of the very few stabilizing and civilizing forces uniting western Europe during the Dark Ages. Its bishops became the civil as well as ecclesiastical authorities over vast areas devoid of other effective government. Parish churches were the focus of rural and urban life, and the cathedrals replaced Roman monuments and temples as the symbols of the social order.

Secular imperial control endured in the eastern empire, whose capital was Constantinople. Thriving under its protection, the Eastern Church expanded into the Balkans, eastern Europe, Russia, and the Near East. The fall of the eastern empire to the Turks in the 15th century opened eastern Europe temporarily to Islam, though the Eastern Orthodox Church (the direct descendant of the Byzantine State Church) remains in its various ethnic branches a major component of Christianity (Table 7.2).

The Protestant Reformation of the 15th and 16th centuries split the church in the west, leaving Roman Catholicism supreme in southern Europe but installing a variety of Protestant denominations and national churches in western and northern Europe. The split was reflected in the subsequent worldwide dispersion of Christianity. Catholic Spain and Portugal colonized Latin America, bringing both their languages and the Roman church to that area (Figure 7.25), as they did to colonial outposts in the Philippines, India, and Africa. Catholic France colonized Quebec in North America. Protestants, many of them fleeing Catholic or repressive Protestant state churches, were primary early settlers of Anglo-America, Australia, New Zealand, Oceania, and South Africa.

Although religious intermingling rather than rigid territorial division is characteristic of the American scene, the beliefs and practices of various immigrant groups and the innovations of domestic congregations have created a particularly varied spatial patterning of "religious regions" in the United States (Figure 7.29). Strongly French-, Irish-, and Portuguese-Catholic New England, the Hispanic-Catholic Southwest, and the French-Catholic vicinity of New Orleans are commonly recognized regional subdivisions of the United States. Each has a cultural identity that includes but is not limited to its dominant religion. The western area of Mormon cultural and religious dominance is prominent and purely American. The Baptist presence in the South and that of the Lutherans in the Upper Midwest help determine the boundaries of other distinctive cultural regions.

Figure 7.28

Diffusion paths of Christianity, A.D. 100–1500. Routes and dates are for Christianity as a composite faith. No distinction is made between the Western Church and the various subdivisions of the Eastern Orthodox denominations.

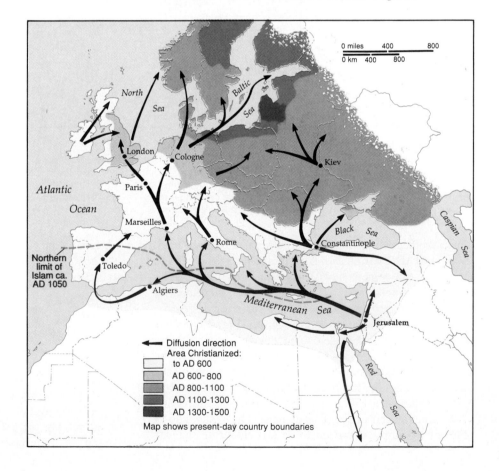

The mark of Christianity upon the cultural landscape has been conspicuous and enduring. In pre-Reformation Catholic Europe, the parish church formed the center of life for small neighborhoods of every town, the village church was the centerpiece of every rural community, and in larger cities the central cathedral served simultaneously as a glorification of God, a symbol of piety, and the focus of religious and secular life (Figure 7.30a).

Protestantism placed less importance upon the church as a monument and symbol, although in many communities—colonial New England, for example—the churches of the principal denominations were at the village center (Figure 7.30b). Frequently they were adjoined by a cemetery, for Christians—in common with Muslims and Jews—practice burial in areas reserved for the dead. In Christian countries, particularly, the cemetery—whether connected to the church, separate from it, or unrelated to a specific denomination—has traditionally been a significant land use within urban areas. Often the separate cemetery, originally on the outskirts of the community, becomes with urban expansion a more central land use and often one that distorts or blocks the growth of the city.

Islam

Islam springs from the same Judaic roots as Christianity and embodies many of the same beliefs: there is only one God, who may be revealed to humans through prophets; Adam was the first human; Abraham was one of his descendants. Mohammed is revered as the prophet of *Allah* (God), succeeding and completing the work of earlier prophets of Judaism and Christianity, including Moses, David, and Jesus. The Koran, the word of Allah revealed to Mohammed, contains not only rules of worship and details of doctrine but also instructions on the conduct of human affairs. For fundamentalists, it thus becomes the unquestioned guide to matters both religious and secular. Observance of the "five pillars" (Figure 7.24) and surrender to the will of Allah unite the faithful into a brotherhood that has no concern with race, color, or caste.

It was that law of brotherhood that served to unify an Arab world sorely divided by tribes, social ranks, and multiple local deities. Mohammed was a resident of Mecca, but fled in A.D. 622 to Medina, where the Prophet proclaimed a constitution and announced the universal mission of the Islamic community. That flight—*Hegira*—marks the starting point of the Islamic (lunar) calendar.

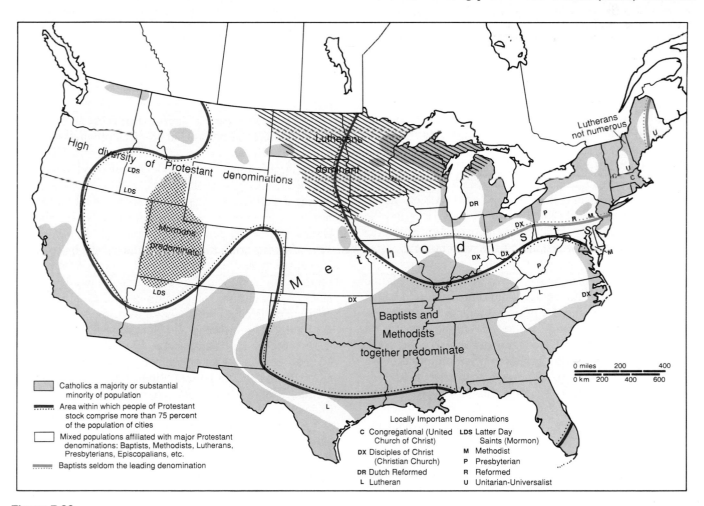

Figure 7.29
Religious affiliation in the conterminous United States.

(a)

(b)

Figure 7.30

In Christian societies the church assumes a prominent circumflex central position in the cultural landscape. (a) The building of Notre Dame Cathedral of Paris, France, begun in 1163, took more than 100 years to complete. Between 1170 and 1270, some 80 cathedrals were constructed in France alone. The cathedrals in all of Catholic Europe were located in the center of major cities. Their plazas were the sites of markets, public meetings, and religious ceremonies. (b) Individually less imposing than the central cathedral of Catholic areas, the several Protestant churches common in small and large American towns collectively constitute an important land use that is frequently sited in the center of the community.

By the time of Mohammed's death in A.H. 11 (*Anno*–in the year of—*Hegira,* or A.D. 632), all of Arabia had joined Islam. The new religion swept quickly outward from that source region over most of central Asia into parts of present-day USSR and, at the expense of Hinduism, into northern India (Figure 7.31). Its advance westward was particularly rapid and inclusive in North Africa. Later, Islam dispersed into Indonesia, southern Africa, and the Western Hemisphere. It continues its spatial spread as the fastest growing major religion at the present time.

The mosque—place of worship, community clubhouse, meeting hall, school—is the focal point of Islamic communal life and the primary imprint of the religion on the cultural landscape. Its principal purpose is to accommodate the Friday communal service, mandatory for all male Muslims. It is the congregation rather than the structure that is important; small or poor communities are as well served by a bare whitewashed room as are larger cities by architecturally splendid mosques. With its perfectly proportioned, frequently gilded or tiled domes, its graceful, soaring towers and minarets (from which the faithful are called to prayer), and its delicately wrought parapets and cupolas, the carefully tended mosque is frequently the most elaborate and imposing structure of the town (Figure 7.32).

Hinduism

Hinduism is the world's oldest major religion. Though it has no datable founding event or initial prophet, some evidence traces its origin back 4000 or more years. Hinduism is not just a religion but an intricate web of religious, philosophical, social, economic, and artistic elements comprising a distinctive Indian civilization. Its estimated 600 million adherents are largely confined to India, where it claims 75% of the population.

From its cradle area in the valley of the Indus River, Hinduism spread eastward down the Ganges River and southward throughout the subcontinent and adjacent regions by amalgamating, absorbing, and eventually supplanting earlier native religions and customs. Its practice eventually spread throughout Southeast Asia, into Indonesia, Malaysia, Kampuchea, Thailand, Laos, and Vietnam as well as into neighboring Myanmar (Burma) and Sri Lanka. The largest Hindu temple complex is in Kampuchea, not India, and Bali remains a Hindu pocket in dominantly Islamic Indonesia.

There is no common creed, single doctrine, or central ecclesiastical organization defining the Hindu. A Hindu is one born into a caste, a member of a complex social and economic—as well as religious—community.

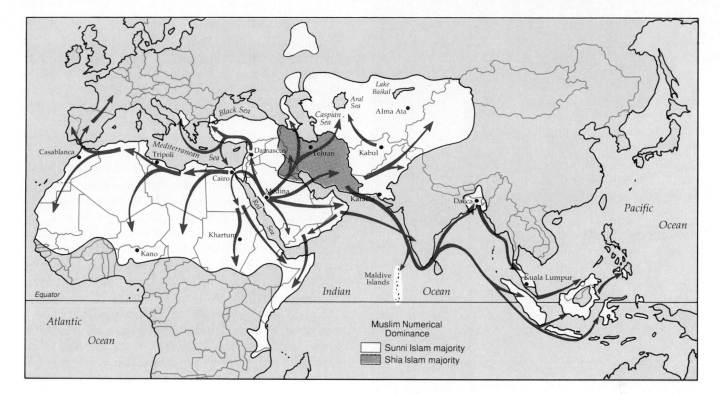

Figure 7.31

Spread and extent of Islam. Islam predominates in over 30 countries along a band across northern Africa to the southern borders of the USSR and the northern part of the Indian subcontinent. Still farther east, Indonesia has the largest Muslim population of any country. Islam's greatest development is in Asia, where it is second only to Hinduism, and in Africa, where it is the leading religion. Current Islamic expansion is particularly rapid in the Southern Hemisphere.

Hinduism accepts and incorporates all forms of belief; adherents may believe in one god or many or none. It emphasizes the divinity of the soul and is based on the concepts of reincarnation and passage from one state of existence to another in an unending cycle of birth and death in which all living things are caught. All humans are ranked, and one's position in this life is determined by one's *karma,* or deeds and conduct in previous lives. The social caste into which an individual is born is an indication of that person's spiritual status. The goal of existence is to move up the hierarchy, eventually to be liberated from the cycle of rebirth and redeath, and to achieve salvation and eternal peace through union with the *Brahman,* the universal soul.

The *caste* (meaning "birth") structure of society is an expression of the eternal transmigration of souls. For the Hindu, the primary aim of this life is to conform to prescribed social and ritual duties and to the rules of conduct for the assigned caste and profession. Those requirements comprise that individual's *dharma*—law and duties. Traditionally, each craft or profession is the property of a particular caste.

The practice of Hinduism is rich with rites and ceremonies, festivals and feasts, processions and ritual gatherings of literally millions of celebrants. It involves careful observance of food and marriage rules and the performance of duties within the framework of the caste system. Pilgrimages to holy rivers and sacred places are thought to secure deliverance from sin or pollution and to preserve religious worth. Worship in the temples and shrines that are found in every village (Figure 7.33) and the leaving of offerings to secure merit from the gods are required. (See "Religion in Nanpur.") The temples, shrines, daily rituals and worship, numerous specially garbed or marked holy men and ascetics, and the ever-present sacred animals mark the cultural landscape of Hindu societies, a landscape infused with religious symbols and sights that are part of a total cultural experience.

Buddhism

Numerous reform movements have derived from Hinduism over the centuries, some of which have endured to the present day as major religions on a regional or world scale. For example, *Sikhism* developed in the Punjab area of northwestern India in the late 15th century A.D., rejecting the formalism of both Hinduism and Islam and proclaiming a gospel of universal toleration. The great majority of some 16 million Sikhs still live in India, mostly in the Punjab, though others have settled in Malaysia, Singapore, East Africa, the United Kingdom, and North America.

Figure 7.32

The common architectural features of the mosque make it an unmistakable landscape evidence of the presence of Islam in any local culture. Here, the Jame mosque fits comfortably among the modern office buildings of Kuala Lumpur, Malaysia.

The largest and most influential of the dissident movements has been *Buddhism,* a universalizing faith founded in the 6th century B.C. in northern India by Siddhartha Gautama, the Buddha ("Enlightened One "). The Buddha's teachings were more a moral philosophy that offered an explanation for evil and human suffering than a formal religion. He viewed the road to enlightenment and salvation to lie in understanding the "four noble truths": existence involves suffering; suffering is the result of desire; pain ceases when desire is destroyed; the destruction of desire comes through knowledge of correct behavior and correct thoughts. The Buddha instructed his followers to carry his message as missionaries of a doctrine open to all castes, for no distinction among people was recognized. In that message all could aspire to ultimate enlightenment, a promise of salvation that raised the Buddha in popular imagination from teacher to saviour and Buddhism from philosophy to universalizing religion.

The belief system spread throughout India, where it was made the state religion in the 3d century B.C. It was carried elsewhere into Asia by missionaries, monks, and merchants. While expanding abroad, Buddhism began to decline at home as early as the 4th century A.D., slowly but irreversibly reabsorbed into a revived Hinduism. By the 8th century its dominance in northern India was broken by conversions to Islam, and by the 15th century it had essentially disappeared from all of the subcontinent.

Present-day spatial patterns of Buddhist adherence reflect the schools of thought, or *vehicles,* that were dominant during different periods of dispersion of the basic belief system (Figure 7.34). In all of its many variants, Buddhism imprints its presence vividly upon the cultural

Religion in Nanpur

The villagers of Nanpur are Hindus. They are religious. They believe in God and his many incarnations. For them He is everywhere, in a man, in a tree, in a stone. According to . . . the village Brahmin, God is light and energy, like the electric current. To him there is no difference between the gods of the Hindus, Muslims and Christians. Only the names are different.

Every village has a local deity. In Nanpur it is a piece of stone called Mahlia Buddha. He sits under the ancient *varuna* tree protecting the village. Kanhai Barik, the village barber, is the attendant to the deity. Kanhai, before starting his daily work, washes the deity, decorates it with vermilion and flowers and offers food given by the villagers. Clay animals are presented. It is believed that the deity rides them

during the night and goes from place to place guarding the village. . . . In the old days Mahlia Buddha had a special power to cure smallpox and cholera. Now, although modern medicines have brought the epidemics under control, the power of the deity has not diminished. People believe in him and worship him for everything, even for modern medicines to be effective.

Religious festivals provide entertainment. There is one almost every month. The most enjoyable is the Spring festival of Holi when people throw colored powder and water on each other as an expression of love. As the cuckoo sings, hidden among the mango blossoms, the villagers carry Gopinath (Krishna) in a palanquin [a chair with carrying poles] around the

village accompanied by musicians. . . .

The women in Nanpur worship Satyapir, a Hindu–Muslim god, to bless them with sons. "Satyka" is the Hindu part meaning "truth," and "pir" in Islam means "prophet." It was a deliberate attempt to bring the two communities together through religion. There is a large Muslim settlement three kilometers from Nanpur and in a village on the other side of the river a single Muslim family lives surrounded by Brahmins. In spite of Hindu–Muslim tensions in other parts of India, the atmosphere around the village has remained peaceful.

Excerpted with permission from: Prafulla Mohanti, "A Village called Nanpur," *Unesco Courier,* June 1983, pp. 11–12.

Figure 7.33
The Hindu temple complex at Khajuraho in central India. The creation of temples and the images they house has been a principal outlet of Indian artistry for more than 2000 years. At the village level, the structure may be simple, containing only the windowless central cell housing the divine image, a surmounting spire, and the temple porch or stoop to protect the doorway of the cell. The great temples, of immense size, are ornate extensions of the same basic design.

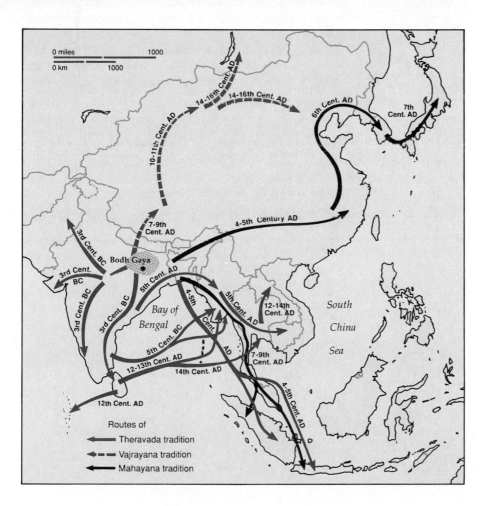

Figure 7.34
Diffusion paths, times, and "vehicles" of Buddhism.

landscape. Buddha images in stylized human form began to appear in the 1st century A.D. and are common in painting and sculpture throughout the Buddhist world. Equally widespread are the three main types of buildings and monuments: the *stupa,* a commemorative shrine; the temple or pagoda enshrining an image or relic of the Buddha (Figure 7.35); and the monastery, some of them the size of small cities.

East Asian Ethnic Religions

When Buddhism reached China from the south 1500 to 2000 years ago and was carried to Japan from Korea in the 7th century, it encountered and later amalgamated with already well established ethical belief systems. The Far Eastern ethnic religions are *syncretisms,* combinations of different forms of belief and practice. In China the union was with Confucianism and Taoism, themselves becoming intermingled by the time of Buddhism's arrival, and in Japan it was with Shinto, a polytheistic animism and shamanism.

Chinese belief systems address not so much the hereafter as the achievement of the best possible way of life in the present existence. They are more ethical or philosophical than religious in the pure sense. Confucius (K'ung Fu-tzu), a compiler of traditional wisdom who lived about the same time as Gautama Buddha, emphasized the importance of proper conduct between ruler and subjects and between family members. The family was extolled as the nucleus of the state, and filial piety was the loftiest of virtues. There are no churches or clergy in *Confucianism,* though its founder believed in a heaven seen in naturalistic terms, and the Chinese custom of ancestor worship as a mark of gratitude and respect was encouraged.

Confucianism was joined by, or blended with, *Taoism,* an ideology that according to legend was first taught by Lao Tsu in the 6th century B.C. Its central theme is *Tao,* the Way, a philosophy teaching that eternal happiness lies in total identification with nature and deploring passion, unnecessary invention, unneeded knowledge, and government interference in the simple life of individuals. Buddhism, stripped by Chinese pragmatism of much of its Indian otherworldliness and defining a *nirvana* achievable in this life, was easily accepted as a companion to these traditional Chinese belief systems. Along with Confucianism and Taoism, Buddhism became one of the honored Three Teachings, and to the average person there was no distinction in meaning or importance between a Confucian temple, Taoist shrine, or Buddhist stupa.

Buddhism also joined and influenced Japanese Shinto, the traditional religion of Japan that developed out of nature and ancestor worship. *Shinto*—The Way of the Gods—is basically a structure of customs and rituals rather than an ethical or moral system. It observes a complex set of deities, including deified emperors, family spirits, and the divinities residing in rivers, trees, certain animals, mountains and, particularly, the sun and moon. At first resisted, Buddhism was later amalgamated with traditional Shinto. Buddhist deities were seen as Japanese gods in a different form, and Buddhist priests assumed control of most of the numerous Shinto shrines in which the gods are believed to dwell and which are approached through ceremonial *torii,* or gateway arches (Figure 7.36).

Figure 7.35
The golden stupas of the Swedagon pagoda, Rangoon, Myanmar (Burma).

Figure 7.36
A Shinto shrine in Kyoto, Japan.

Ethnicity

Any discussion of cultural diversity would be incomplete without mention of **ethnicity.** Based on the root word *ethnos*, meaning "people" or "nation," the term is usually used to refer to the ancestry of a particular people who have in common distinguishing characteristics associated with their heritage. No single trait denotes ethnicity. Recognition of ethnic communities may be based on language, religion, national origin, unique customs, or an ill-defined concept of "race." (See "The Question of Race.") Whatever the unifying thread, ethnic groups may strive to preserve their special shared ancestry and cultural heritage through the collective retention of language, religion, festivals, cuisines, traditions, and in-group work relationships, friendships, and marriages. Those preserved relationships and associations are fostered by and support *ethnocentrism,* the feeling that one's own ethnic group is superior.

Normally, reference to ethnic communities is recognition of their minority status within a country or region dominated by a different, majority culture group. We do not identify Koreans living in Korea as an ethnic group because theirs is the dominant culture in their own land. Koreans living in Japan, however, constitute a discerned and segregated group in that foreign country. Ethnicity, therefore, is an evidence of areal cultural diversity and a reminder that culture regions are rarely homogeneous in the characteristics displayed by all of their occupants.

Territorial segregation is a strong and sustaining trait of ethnic identity, and one that assists groups to retain their distinction. On the world scene, indigenous ethnic groups have developed over time in specific locations and have established themselves in their own and others' eyes as distinctive peoples with defined homeland areas. The boundaries of most countries of the world encompass a number of racial or ethnic minorities (Figure 7.37). Their demands for special territorial recognition have sometimes increased with advances in economic development and self-awareness, as Chapter 8, "Political Geography," points out.

Increasingly in a world of movement, ethnicity is less a matter of indigenous populations and more one of outsiders in an alien culture. Immigrants, legal and illegal,

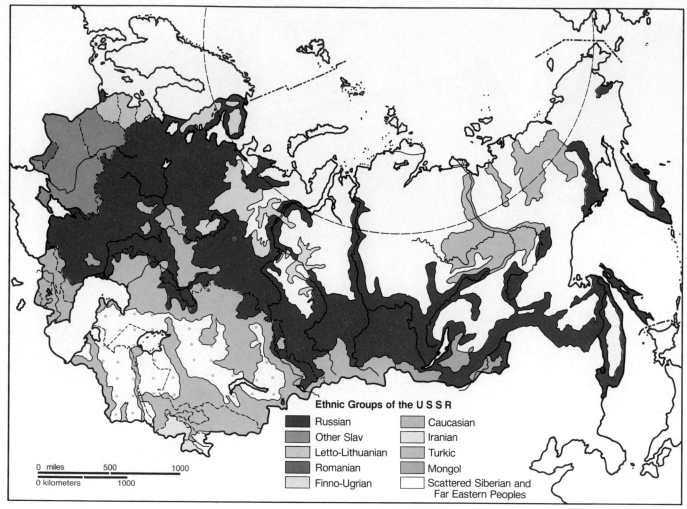

Ethnic Groups of the U S S R

◼ Russian	◼ Caucasian
◼ Other Slav	◼ Iranian
◻ Letto-Lithuanian	◼ Turkic
◼ Romanian	◼ Mongol
◻ Finno-Ugrian	◻ Scattered Siberian and Far Eastern Peoples

0 miles 500 1000
0 kilometers 1000

Figure 7.37

Ethnic regionalism in the USSR. More than 100 non-Slavic "nationalities" and ethnic groups have homelands within the boundaries of the USSR. Made part of the czarist empire or forcibly annexed to the Soviet Union, these increasingly restive ethnocentric populations pose challenges to established Slavic, particularly Russian, domination.

and refugees from war, famine, or persecution are a growing presence in countries throughout the world. Immigrants to a country typically have one of two choices. They may hope for **assimilation** by giving up many of their past cultural traits, losing their distinguishing characteristics and merging into the mainstream of the dominant culture. Or, they may try to retain their distinctive cultural heritage. In either case, they usually settle initially in an area where other members of their ethnic group live, as a place of refuge and learning (Figure 7.38). With the passage of time they may leave their protected community and move out among the general population.

The Chinatowns and Little Italys of North American cities have provided the support systems essential to new immigrants in an alien culture region. Japanese, Italians, Germans, and other ethnics have formed agricultural colonies in Brazil in much the same spirit. Such ethnic enclaves may provide an entry station, allowing both individuals and the groups to which they belong to undergo cultural and social modifications sufficient to enable them to operate effectively in the new, majority society. Sometimes, of course, settlers have no desire to assimilate or are not allowed to assimilate, so that they and their descendants form a more or less permanent subculture in the larger society. The Chinese in Malaysia belong to this category. Ethnicity in the context of nationality is discussed more fully in Chapter 8.

Other Aspects of Diversity

Culture is the sum total of the way of life of a society. It is misleading to isolate, as we have done, only a few elements of the technological, sociological, and ideological subsystems and imply that they are identifying characteristics differentiating culture groups. Economic developmental levels, language, religion, ethnicity all are

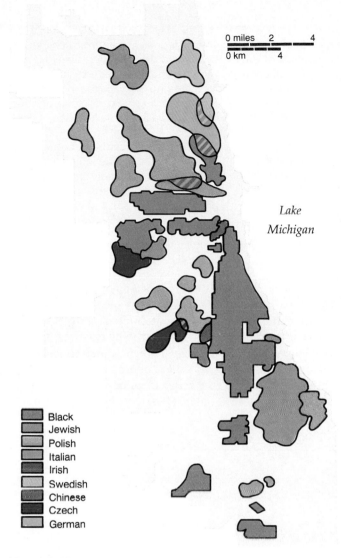

Black
Jewish
Polish
Italian
Irish
Swedish
Chinese
Czech
German

*Lake
Michigan*

0 miles 2 4
0 km 4

Figure 7.38

The distribution of ethnic groups in Chicago, 1957. The rapidly industrializing society of 19th- and early 20th-century America became a mosaic of ethnic enclaves, areas of initial refuge and dominance from which diffusion or absorption into the host society later occurred. The increasing subdivision of the U.S. immigrant stream and the consequent reduction in the size of identified enclaves make it nearly impossible to produce comparable maps for the last decade of the 20th century.

important and common distinguishing cultural traits, but they tell only a partial story. Other suggestive, though perhaps less pervasive, basic elements exist.

Architectural styles in public and private buildings are evocative of region of origin even when they are encountered in indiscriminate juxtaposition in American cities. The Gothic and New England churches, the neo-classical bank, and the skyscraper office building suggest not only the functions they house but the culturally and regionally variant design solutions that gave them form. The Spanish, Tudor, French Provincial, or ranch-style residence may not reveal the ethnic background of its

American occupant, but it does constitute a culture statement of the area and the society from which it diffused.

Music, food, games, and other evidences of the joys of life, too, are cultural indicators associated with particular world or national areas. Music is an emotional form of communication found in all societies, but, being culturally patterned, it varies among them. Instruments, scales, and types of compositions are technical forms of variants; the emotions aroused and the responses evoked among peoples to musical cues are learned behaviors. The Christian hymn means nothing emotionally to a pagan New Guinea clan. The music of a Chinese opera may be simply noise to the European ear. Where there is sufficient similarity between musical styles and instrumentation, blending (syncretism) and transferal may occur. American jazz represents a blend; calypso and flamenco music have been transferred to the North American scene. Foods identified with other culture regions have similarly been transferred to become part of the culinary environment of the American "melting pot."

These are but a few additional minor statements of the variety and the intricate interrelationships of that human mosaic called culture. Individually and collectively they are, in their areal expressions and variations, the subject matter of the cultural geographer.

Culture Realms

Our discussion of culture has had one consistent and recurrent message: culture has spatial expression in all of its details and composites. The individual culture traits we examined and mapped show the subdivision of earth space into special-purpose regions. Of course, the same trait—the Christian religion, perhaps, or the Spanish language—may be part of more than one culture, but each separate culture will be marked by a distinctive complex of such individual traits, setting it off spatially from adjacent cultures with their own identifying composites of traits.

If two or more culture complexes have a number of such traits in common, a **culture system** may be recognized as a larger spatial reality and generalization. Multiethnic societies, perhaps further subdivided by linguistic differences, varied food preferences, and a host of other internal differentiations, may nonetheless share enough common characteristics of the subsystems of culture to be recognizably distinctive cultural entities to themselves and others. Certainly, citizens of "melting-pot" United States would identify themselves as *Americans,* together comprising a unique culture system on the world scene.

Culture regions and complexes are elements in the spatial hierarchy of cultural geography. They may, at a still higher level of generalization, be combined into composite world regions, into **culture realms.** At that level of generalization, cultural specifics become obscured and perceptions of world regional differentiation come into

The Question of Race

Human populations may be differentiated on the basis of acquired cultural or inherent physical attributes. Culture is learned behavior transmitted to the successive generations of a social group through imitation and through that distinctively human capability, speech. Culture summarizes the way of life of a group of people and may be adopted by members of the group irrespective of their individual genetic heritage, or race.

The spread of human beings over the earth and their occupation of different environments were accompanied by the development of those physical variations in the basic human stock that are commonly called *racial*. In very general terms, a **race** may be defined as a population subset whose members have in common some hereditary biological characteristics that set them apart physically from other human groups. Since the reference is

solely to genes, the term *race* cannot be applied meaningfully to any human attribute that is acquired. Race, therefore, has no significance in reference to national, linguistic, religious, or other culturally based classifications of populations.

Although it has been argued that race is a misleading concept, there does exist a common understanding that populations are recognizably grouped on the basis of differences in pigmentation, hair characteristics,

play. Culture realms attempt to document those perceptions of world-scale cultural contrasts by identifying groups of culture systems with enough distinctive characteristics in common to set them apart from other realms with differing sets of identifying generalizations.

Clearly, our present database is inadequate to the task of definitive world cultural regionalization. Political structure, economic orientation, patterns of behavior, levels

of urbanization—all aspects of contemporary culture—are yet to be considered. A preliminary recognition of composite culture realms may be attempted, however, on the basis of the fundamental differentiating characteristics of development level, language, religion, and ethnicity already discussed. Figure 7.39 is offered as a world subdivision into such culture realms, regionally discrete areas that are more alike internally than they are like other realms.

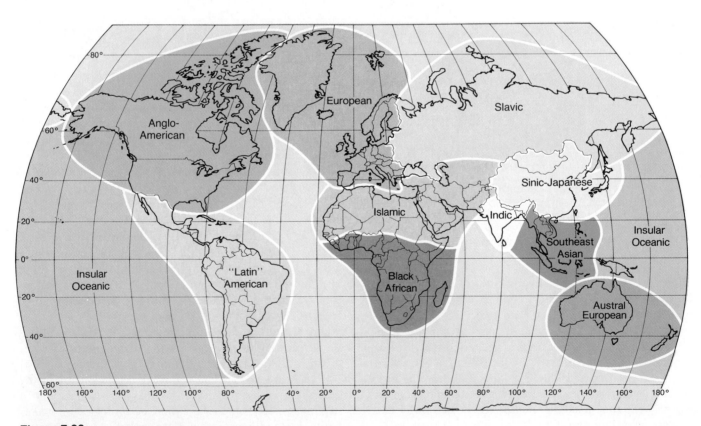

Figure 7.39
Culture realms of the modern world. This is just one of many possible subdivisions of the world into multifactor cultural regions.

facial features, and other traits largely related to variations in soft tissue. Some subtle skeletal differences among peoples also exist. Such differences form the basis for the division of humanity into racial groups by some anthropologists. Racial differentiation as commonly understood is old and can reasonably be dated at least to the Paleolithic (100,000 to about 11,000 years ago) spread and isolation of population groups.

If all humankind belongs to a single species that can freely interbreed and produce fertile offspring, how did any apparent differentiation occur? Among several causative factors, two appear to be most important: *genetic drift* and *adaptation*. Genetic drift refers to a heritable trait (such as a distinctive shape of the nose) that appears by chance in one group and becomes accentuated through inbreeding. If two populations are spatially too far apart for much interaction to occur (*isola-tion*), a trait may develop in one but not in the other. Adaptation to particular environments is thought to account for the rest of racial differentiation, although it should be recognized that we are only guessing here. One cannot say with certainty that any racial characteristic is the result of such evolutionary changes. Genetic studies, however, have suggested a relationship between solar radiation and blood type and between temperature and body size, for example.

Summary

Culture is the learned behaviors and beliefs of distinctive groups of people. Culture traits, the smallest distinctive items of culture, are the building blocks of integrated culture complexes. Together, traits and complexes in their spatial patterns create human—"cultural"—landscapes, define culture regions, and distinguish culture groups. Those landscapes, regions, and group characteristics change through time as human societies interact with their environment, develop new solutions to collective needs, or are altered through innovations adopted from outside the group.

The detailed complexities of culture can be simplified by recognition of its component subsystems. The technological subsystem is composed of the material objects (artifacts) and techniques of livelihood; the sociological subsystem comprises the formal and informal institutions (sociofacts) that control the social organization of a culture group. The ideological subsystem consists of the ideas and beliefs (mentifacts) a culture expresses in speech and through belief systems.

The presumed cultural uniformity of a preagricultural world was lost as domestication of plants and animals in many world areas led to the emergence of culture hearths of wide-ranging innovation and to a cultural divergence between different groups. While modern-day shared technologies contribute to cultural convergence throughout the world, many elements of cultural distinction remain to identify and separate social groups. Among the most prominent of the differentiating cultural traits are language, religion, and ethnicity.

Language and religion are both transmitters of culture and identifying traits of culture groups. Both have distinctive spatial patterns, reflecting past and present processes of interaction and change. Although languages may be grouped by origins and historical development, their world distributions depend as much on the movements of peoples and histories of conquest and colonization as they do on linguistic evolution. Toponymy, the study of place-names, helps document that history of movement. Linguistic geography studies spatial variations in languages, variations that may be minimized by encouragement of standard languages or overcome by pidgins, creoles, and lingua francas.

Religion is a less-pronounced identifier or transmitter of culture than is language, but even in secular societies religion may influence economic activities, legal systems, holiday observances, and the like. Although religions do not lend themselves to easy classification, their spatial patterns are distinct and reveal past and present histories of migration, conquest, and diffusion. Those patterns are also important components in the spatially distinctive cultural landscapes created in response to different religious belief systems.

Ethnicity, affiliation in a group sharing common identifying cultural traits, is fostered by territorial separation or isolation and preserved in ethnically complex societies by a feeling that one's own ethnic group is superior to others. Ethnic diversity is a reality in most countries of the world and is increasing in many of them. Many ethnic minority groups may seek absorption into their surrounding majority culture through acculturation and assimilation, but other groups choose to preserve their identifying distinctions through spatial separation or overt rejection of the majority cultural traits.

Cultural differentiation and regionalization are not limited to subsystems, traits, and complexes. Of nearly comparable importance as a determinant of societal identity at all levels of generalization are the political affiliations entered into or imposed upon individuals and groups. Formal political organization, the allocation and exercise of power it implies, and the role of nationality as a prime element of cultural variation are topics dealt with by political geography, to which we next turn our attention.

Key Words

acculturation	ethnic religion
assimilation	ideological subsystem
creole	innovation
cultural integration	language family
cultural landscape	lingua franca
culture	pidgin
culture complex	possibilism
culture hearth	race
culture realm	sociological subsystem
culture region	spatial diffusion
culture system	syncretism
culture trait	technological subsystem
dialect	toponymy
environmental determinism	tribal religion
ethnicity	universalizing religion

For Review

1. What is included in the concept of *culture?* How is culture transmitted? What personal characteristics affect the aspects of culture that any single individual acquires or fully masters?

2. What is a *culture hearth?* What new traits of culture characterized the early hearths? In the cultural geographic sense, what is meant by *innovation?*

3. Differentiate between *culture traits* and *culture complexes,* between *environmental determinism* and *possibilism.*

4. What are the components or subsystems of the three-part system of culture? What characteristics—aspects of culture—are included in each of the subsystems?

5. Why might one consider language the dominant differentiating element of culture separating societies?

6. In what ways may religion affect other cultural traits of a society?

7. How does the classification of religions as *universalizing, ethnic,* or *tribal* help us understand their patterns of distribution and spread?

8. How are the concepts of *ethnicity, race,* and *culture* related, if at all?

9. How does *acculturation* occur? Is *ethnocentrism* likely to be an obstacle in the acculturation process? How do acculturation and *assimilation* differ?

Selected References

Agnew, John A. "Bringing Culture Back in: Overcoming the Economic–Cultural Split in Development Studies." *Journal of Geography* 86, no. 6 (November-December 1987): 276–81.

al-Faruqi, Isma'il R., and David E. Sopher, eds. *Historical Atlas of the Religions of the World.* New York: Macmillan, 1974.

Allen, James Paul, and Eugene James Turner. *We the People: An Atlas of America's Ethnic Diversity.* New York: Macmillan, 1988.

Clarke, Colin, David Ley, and Ceri Peach, eds. *Geography and Ethnic Pluralism.* London: Allen & Unwin, 1984.

Cole, John P. *Development and Underdevelopment: A Profile of the Third World.* London and New York: Methuen, 1987.

Cooper, Robert L., ed. *Language Spread: Studies in Diffusion and Social Change.* Bloomington: Indiana University Press, 1982.

Fellmann, Jerome D., Arthur Getis, and Judith Getis. *Human Geography.* Dubuque, Ia.: Wm. C. Brown, 1990.

Gastil, Raymond D. *Cultural Regions of the United States.* Seattle: University of Washington Press, 1975.

Gaustad, Edwin S. *Historical Atlas of Religion in America.* New York: Harper and Row, 1976.

Gowlett, John. *Ascent to Civilization.* London: Collins, 1984.

International Bank for Reconstruction and Development/The World Bank. *World Development Report.* New York: Oxford University Press, Annual.

Isaac, Erich. "Religion, Landscape, and Space." *Landscape* 9, no. 2 (Winter 1959–60): 14–18.

Journal of Cultural Geography 7, no. 1 (Fall/Winter 1986). Special issue devoted to geography and religion.

Katzner, Kenneth. *The Languages of the World.* Rev. ed. London: Routledge & Kegan Paul, 1986.

Kurath, Hans. *A Word Geography of the Eastern United States.* Ann Arbor: University of Michigan Press, 1949.

Lieberson, Stanley, and Mary C. Waters. *From Many Strands: Ethnic and Racial Groups in Contemporary America.* Census Monograph Series. New York: Russell Sage Foundation, 1988.

McCrum, Robert, William Cran, and Robert MacNeil. *The Story of English.* New York: Elizabeth Sifton Books/ Viking, 1986.

Nielsen, Niels C., Jr., et al. *Religions of the World.* New York: St. Martin's Press, 1983.

Reed, Charles A., ed. *Origins of Agriculture.* The Hague: Mouton, 1977.

Shortridge, James R. "Patterns of Religion in the United States." *Geographical Review* 66 (1976): 420–34.

Sopher, David E. *The Geography of Religions.* Englewood Cliffs, N.J.: Prentice-Hall, 1967.

Stewart, George R. *Names on the Globe.* New York: Oxford University Press, 1975.

———. *Names on the Land.* 4th ed. San Francisco: Lexikos, 1982.

Thomas, William L., Jr., ed. *Man's Role in Changing the Face of the Earth.* Chicago: University of Chicago Press, 1956.

Trudgill, Peter. *On Dialect: Social and Geographical Perspectives.* Oxford, England: Basil Blackwell Ltd., 1983.

Wagner, Philip L., and Marvin W. Mikesell. *Readings in Cultural Geography.* Chicago: University of Chicago Press, 1962.

Zelinsky, Wilbur, "An Approach to the Religious Geography of the United States: Patterns of Church Membership in 1952." *Annals of the Association of American Geographers* 51 (1961): 139–93.

———. *The Cultural Geography of the United States.* Englewood Cliffs, N.J.: Prentice-Hall, 1973.

8

Political Geography

United States-Mexico international bridge, Laredo, Texas.

THEY met together in the cabin of the little ship on the day of the landfall. The journey from England had been long and stormy. Provisions ran out, a man had died, a boy had been born. Although they were grateful to have reached the calm waters off Cape Cod that November day of 1620, their gathering in the cramped cabin was not to offer prayers of thanksgiving but to create a political structure to govern the settlement they had come to establish. The Mayflower Compact was an agreement among themselves to "covenant and combine our selves togeather into a civill body politick . . . to enact, constitute, and frame such just and equall lawes, ordinances, acts, constitutions, and offices . . . convenient for ye generall good of ye Colonie. . . ." They elected one of their company governor, and only after those political acts did they launch a boat and put a party ashore.

The land they sought to colonize had for more than 100 years been claimed by the England they had left. The New World voyage of John Cabot in 1497 had invested their sovereign with title to all of the land of North America and a recognized legal right to govern his subjects dwelling there. That right was delegated by royal patent to colonizers and their sponsors, conferring upon them title to a defined tract and the right to govern it. Although the Mayflower settlers were originally without a charter or patent, they recognized themselves as part of an established political system. They chose their governor and his executive department annually by vote of the General Court, a legislature composed of all freemen of the settlement.

As the population grew, new towns were established too distant for their voters to attend the General Court. By 1636 the larger towns were sending representatives to cooperate with the executive branch in making laws. Each town became a legal entity, with election of local officials and enactment of local ordinances the prime purpose of the town meetings that are still common in New England today.

The Mayflower Compact, signed by 41 freemen as their first act in a New World, was the first step in a continuing journey of political development for the settlement and for the larger territory of which it became a part. From company patent to crown colony to rebellious commonwealth under the Continental Congress to state in a new country, Massachusetts (and Plymouth Plantation) were part of a continuing process of the political organization of space.

That process is as old as human history. From clans to kingdoms, human groups have laid claim to territory and have organized themselves and administered their affairs within it. Indeed, the political organizations of society are as fundamental an expression of culture and cultural differences as are forms of economy or religious beliefs. Geographers are interested in that structuring because it is both an expression of the human organization of space and is closely related to other spatial evidences of culture, such as religion, language, and ethnicity.

Political geography is the study of the organization and distribution of political phenomena in their areal expression. Nationality is a basic element in cultural variation among people, and political geography traditionally has had a primary interest in country units, or *states* (Figure 8.1). Of central concern have been spatial patterns that reflect the exercise of central governmental control, such as questions of boundary delimitation and effect. Increasingly, however, attention has shifted both upward and downward on the political scale. International alliances, regional compacts, and producer cartels have increased in prominence since World War II, representing new forms of spatial interaction. At the local level, voting patterns, constituency boundaries and districting rules, and political fragmentation have directed public attention to the significance of area in the domestic political process.

In this chapter, we discuss some of the characteristics of political entities, examine the problems involved in defining jurisdictions, seek the elements that lend cohesion to a political entity, explore the implications of partial surrender of sovereignty, and consider the significance of the fragmentation of political power. We begin with states and end with local political systems.

National Political Systems

One of the most significant elements in cultural geography is the nearly complete division of the earth's land surface into separate national units, as shown on the world political map inside the front cover. Even Antarctica is subject to the rival territorial claims of seven countries, although these claims have not been pressed because of the Antarctic Treaty of 1961 (Figure 8.2). A second element is that this division into country units is relatively recent. Although countries and empires have existed since the days of early Egypt and Mesopotamia, only in the last century has the world been almost completely divided into independent governing entities. Now people everywhere accept the idea of the state and its claim to sovereignty within its borders as normal.

States, Nations, and Nation-States

Before we begin our consideration of political systems, we need to clarify some terminology. Geographers use the words *state* and *nation* somewhat differently than the way they are used in everyday speech; the confusion arises because each word has more than one meaning. A state can be defined as either (1) any of the political units forming a federal government (e.g., one of the United States) or as (2) an independent political unit holding sovereignty over a territory (e.g., the United States). In this latter sense, *state* is synonymous with *country* or *nation*. That

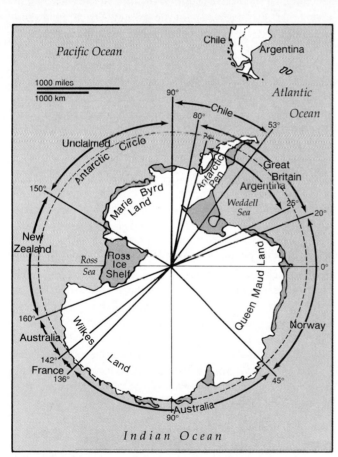

Figure 8.2

Territorial claims in Antarctica. Although seven countries claim sovereignty over portions of Antarctica, and three of the claims overlap, the continent has no permanent inhabitants or established local government.

is, a nation can also be defined as (1) an independent political unit holding sovereignty over a territory (e.g., a member of the United Nations). But it can also be used to describe (2) a community of people with a common culture and territory (e.g., the Kurdish nation). The second definition is *not* synonymous with state or country.

To avoid confusion, we shall define a **state** on the international level as an independent political unit occupying a defined, permanently populated territory and having full sovereign control over its internal and foreign affairs. We will use *country* as a synonym for the territorial and political concept of "state." Not all recognized territorial entities are states. Antarctica, for example, has neither established government nor permanent population, and it is, therefore, not a state. Nor are *colonies* or *protectorates* recognized as states. Although they have defined extent, permanent inhabitants, and some degree of separate governmental structure, they lack full control over all of their internal and external affairs.

We use nation in its second sense, as a reference to people, not to political structure. A **nation** is a group of people with a common culture occupying a particular territory, bound together by a strong sense of unity arising from shared beliefs and customs. Language and religion may be unifying elements, but even more important are an emotional conviction of cultural distinctiveness and a sense of ethnocentrism.

Used thus, it follows that a state can have more than one nation, and, alternatively, that a single nation may be divided among two or more states. In the constitutional structure of the Soviet Union before the creation of a single-chamber Congress of People's Deputies in December 1988, one division of the legislative branch of the government was termed the *Soviet of Nationalities*. It was composed of representatives from civil divisions of the Soviet Union populated by groups of officially recognized "nations": Ukrainians, Kazakhs, Estonians, and others. The new Congress retains representatives of the 15 ethnically based Soviet republics and autonomous territories and regions that are official homelands of more than 100 ethnic groups. In the Soviet instance, the concept of nationality is territorially less than the extent of the state.

Conversely, the Kurds are a nation of some 20 million people divided among five states: Turkey, Iran, Iraq, Syria, and the Soviet Union (Figure 8.3). Kurdish nationalism has survived over the centuries, and many Kurds nurture a vision of an independent Kurdistan. Here, a people's sense of nationality exceeds the areal limits of a single state.

The composite term **nation-state** properly refers to a state whose territorial extent coincides with that occupied by a distinct nation or people or, at least, whose population shares a general sense of cohesion and adherence to a set of common values. Although all countries strive for consensus values and loyalty to the state, few can claim to be ethnic nation-states. France, Denmark, and Poland are often cited as acceptable European examples, though France has several dissident ethnic minorities within its borders. Japan is an Asian illustration.

The Evolution of the Modern State

The idea of the modern state was developed by political philosophers in the 18th century. Their views gave rise to the concept that people owe allegiance to a state and the people it represents rather than to its leader, such as a king or a feudal lord. The new concept coincided in France with the French Revolution and spread over Western Europe, to England, Spain, and Germany.

Many states are the result of European expansion during the 17th, 18th, and 19th centuries, when much of Africa, Asia, and the Americas was divided into colonies. Usually these colonial claims were given fixed and described boundaries where none had earlier been formally defined. Of course, precolonial native populations had relatively fixed home areas of control within which there was recognized dominance and border defense and from which there were, perhaps, raids of plunder or conquest of neighboring "foreign" territories. Beyond tribal territories, great empires arose, again with recognized outer limits of influence or control: Mogul and Chinese; Benin and Zulu; Incan and Aztec. Upon them where they still existed, and upon the less formally organized spatial patterns of effective tribal control, European colonizers imposed their arbitrary new administrative divisions of the land. In fact, tribes that had little in common were often joined in the same colony (Figure 8.4). The new divisions, therefore,

Figure 8.3
Traditional homelands of the Kurdish nation. An ancient group with a distinctive language, Kurdish, the Kurds are concentrated in Turkey, Iran, and Iraq. Smaller numbers live in Syria and the Soviet Union.

were not usually based on meaningful cultural or physical lines. Instead, the boundaries simply represented the limits of the colonizing empire's power.

As these former colonies have gained political independence, they have retained the idea of the state. They have generally accepted—in the case of Africa, by a conscious decision to avoid precolonial territorial or ethnic claims that could lead to war—the borders established by their former European rulers. The problem that many of the new countries face is to develop feelings of loyalty to the state among their arbitrarily associated citizens. Zaire, the former Belgian Congo, contains more than 250 frequently antagonistic tribes. Only if past tribal animosities can be converted into an overriding spirit of national cohesion will countries like Zaire truly be nation-states.

The end of the colonial era brought a rapid increase in the number of sovereign states. At the beginning of World War II in 1939, there were about 70 independent countries. By 1970, the number had more than doubled, and by 1990, 159 independent states had membership in the United Nations. Ten or more others were not UN members. From the former British Empire and Commonwealth, there have come, just in the period since World War II, the independent countries of India, Pakistan, Bangladesh, Malaysia and Singapore in Asia, and Ghana, Nigeria, Kenya, Uganda, Tanzania, Malawi, Botswana, Zimbabwe, and Zambia in Africa. Even this extensive list is not complete. A similar process has occurred in most of the former overseas possessions of the Netherlands, Spain, Portugal, and France.

Geographic Characteristics of States

Every state has certain geographic characteristics by which it can be described and that set it apart from all other states. A look at the world political map inside the front cover of this book confirms that every state is unique. The size, shape, and location of any one state combine to

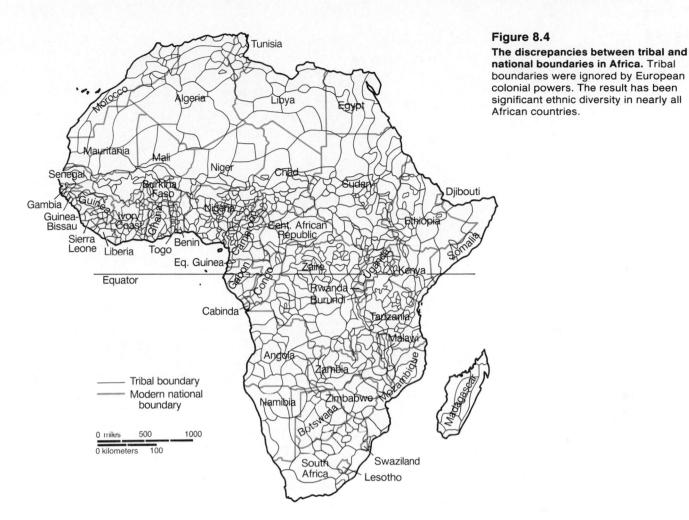

Figure 8.4
The discrepancies between tribal and national boundaries in Africa. Tribal boundaries were ignored by European colonial powers. The result has been significant ethnic diversity in nearly all African countries.

differentiate it from all others. These characteristics are of more than academic interest, because they also affect the power and stability of states.

Size

The area that a state occupies may be large, as is true of China, or small, as is true of Liechtenstein. The world's largest country, the USSR, occupies over 8.5 million square miles (22.4 million km²), or some 15% of the land surface of the world. It is more than a million times as large as Nauru. (See "The Ministates.")

An easy assumption would be that the larger a state's area, the greater is the chance that it will have resources, such as fertile soil and minerals, from which it can benefit. In general, that assumption is valid, but much depends upon accidents of location. Mineral resources are unevenly distributed, and size alone does not guarantee their presence within a state. And Australia, Canada, and the USSR, though large, have relatively small areas capable of supporting productive agriculture. In addition, a very large country may have vast areas that are inaccessible, sparsely populated, and hard to govern. Small states are more apt than large ones to have a culturally homogeneous population. They find it easier to develop transportation and communication systems to link the sections of

the country, and, of course, they have shorter boundaries to defend against invasion. Size alone, then, is not critical in determining a country's stability and strength, but it is a contributing factor.

Shape

Like size, a country's shape can affect its well-being as a state by fostering or hindering effective organization. Assuming no major topographical barriers, the most efficient form would be a circle, with the capital located in the center. In such a country, all places could be reached from the center in a minimal amount of time and with the least expenditure for roads, railway lines, and so on. It would also have the shortest possible borders to defend. Uruguay and Poland have roughly circular shapes, forming a **compact state** (Figure 8.5).

Prorupt states are nearly compact but possess one or sometimes two narrow extensions of territory. Proruption may simply reflect peninsular elongations of land area, as in the case of Myanmar (Burma) and Thailand. In other cases, the extensions have an economic or strategic significance, having been designed to secure state access to resources or to establish a buffer zone between states that would otherwise adjoin. The proruptions of Afghanistan, Zaire, and Namibia fall into this category. The Caprivi

Figure 8.5

Shapes of states. The sizes of the countries should not be compared. Each is drawn on a different scale.

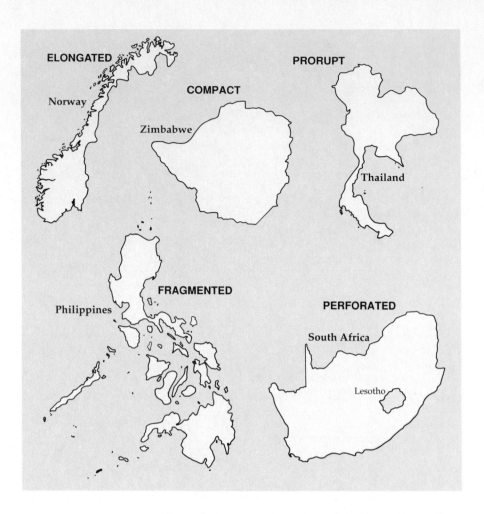

Strip of Namibia, for example, which extends eastward from the main part of the country, was designed by the Germans to give what was then their colony of Southwest Africa access to the Zambezi River. In recent years, the strip has had strategic importance, containing military bases of the Republic of South Africa. Whatever their origin, proruptions tend to isolate a portion of a state.

The least efficient shape is represented by countries like Norway and Chile, which are long and narrow. In such **elongated states,** the parts of the country far from the capital are likely to be isolated because great expenditures are required to link them to the core. These countries are also likely to encompass more diversity of climate, resources, and peoples than compact states, perhaps to the detriment of national cohesion or, perhaps, to the promotion of economic strength.

A fourth class of shapes, characterizing the **fragmented state,** includes countries composed entirely of islands (e.g., the Philippines and Indonesia), countries that are partly on islands and partly on the mainland (Italy and Malaysia), and those that are chiefly on the mainland but whose territory is separated by another state (the United States). Fragmentation makes it harder for the state to impose centralized control over its territory, particularly when the parts of the state are far from one another. This is a problem in the Philippines and Indonesia,

the latter made up of over 13,000 islands, and helped lead to the disintegration of Pakistan. It was created in 1947 as a fragmented state, but East and West Pakistan were 1000 miles from one another. That distance exacerbated economic and cultural differences between the two, and when the eastern part of the country seceded in 1971 and declared itself the independent state of Bangladesh, West Pakistan was unable to impose its control.

The type of fragmentation that occurs when the parts of one state are separated by the territory of a second state can give rise to what are called exclaves. An **exclave** is a territorial outlier located inside another state. Perhaps the best known example is West Berlin, an exclave of West Germany, but Europe has many such outlying bits of one country inside another. Kleinwalsertal is a piece of Austria accessible only from West Germany; Baarle-Hertog is a fragment of Belgium inside Holland; and Llivia is a Spanish town of 930 residents three miles inside France (Figure 8.6). Exclaves are not limited to Europe, of course. African examples include Cabinda, an exclave of Angola, and Melilla and Ceuta, two Spanish exclaves in Morocco.

The counterpart of an exclave, an **enclave,** helps to define the fifth class of shapes, the **perforated state.** A perforated state completely surrounds a territory that it does not rule, as, for example, the Republic of South Africa surrounds Lesotho. The enclave, the surrounded territory,

The Ministates

Totally or partially autonomous political units that are small in area and population pose some intriguing questions. Should size be a criterion for statehood? What is the potential of ministates to cause friction among the major powers? Under what conditions are they entitled to representation in international assemblies like the United Nations?

About half the world's independent countries contain fewer than 5 million people. Of these, 37 have under 1 million, the population size adopted by the United Nations as the upper limit defining "small states," though not too small to be members of that organization. Nauru has about 8000 inhabitants on its 8.2 square miles (21 km²). Other areally small states like Singapore (224 square miles, 580 km²) have populations (2.6 million) well above the UN criterion. Many are island territories located in the West Indies and the Pacific Ocean (such as Grenada and Tonga islands), but Europe (Vatican City and Andorra), Asia (Bahrain and Macao), and Africa (Djibouti and Equatorial Guinea) have their share.

Many statelets are vestiges of colonial systems that no longer exist. Some of the small states of West Africa and the Arabian peninsula fall into this category. Others, such as Mauritius, served primarily as refueling stops on

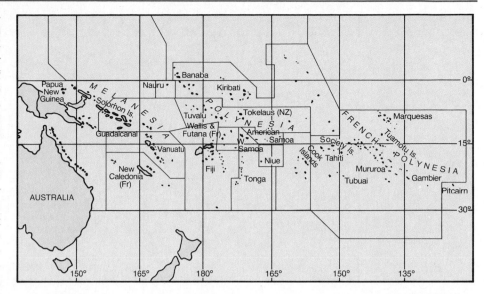

transoceanic voyages. However, some occupy strategic locations (such as Bahrain, Malta, and Singapore) and others contain valuable minerals (Kuwait, Nauru, and Trinidad). The possibility of claiming 200-mile-wide zones of adjacent seas adds to the attraction of yet others.

The proliferation of tiny countries raises the question of their representation and their voting weight in international assemblies. Should there be a minimum size necessary for participation in such bodies? Should countries receive a vote proportional to their population? The influence of the United States and other major powers in the United Nations has already been eroded by the small states. Although the United States pays 25% of the UN budget and has about one and one-half times the population of all the small countries combined, its vote can be balanced by that of any of them. The fact that as many as 50 additional territories may gain independence in the next few years underscores the international interest in them.

---

may be independent or may be part of another state. Two of Europe's smallest independent states, San Marino and Vatican City, are enclaves that perforate Italy. As an *exclave* of West Germany, West Berlin perforated the national territory of East Germany and was an *enclave*. The stability of the perforated state can be weakened if the enclave is occupied by people whose value systems differ from those of the surrounding country.

Location

The significance of size and shape as factors in national well-being can be modified by a state's location, both absolute and relative. Although both Canada and the USSR are extremely large, their *absolute* location in the upper-middle latitudes reduces their size advantages when agricultural potential is considered. To take another example, Iceland has a reasonably compact shape, but its location in the North Atlantic Ocean, just south of the Arctic Circle, means that most of the country is barren. Settlement is confined to the rim of the island.

As important as absolute location is a state's *relative* location, its position compared to that of other countries. *Landlocked* states, those lacking ocean frontage and surrounded by other states, are at a geographical disadvantage. They lack easy access to both maritime (seaborne) trade and the resources found in coastal waters and submerged lands. Bolivia gained 300 miles (480 km) of sea frontier along with its independence in 1825, but lost its ocean frontage by conquest to Chile in 1879. Its annual Day of the Sea ceremony reminds Bolivians of their loss and of continuing diplomatic efforts to secure an alternate outlet.

In a few instances, a favorable relative location constitutes the primary resource of a state. Singapore, a state of only 224 square miles (580 km²) and 2.7 million population, is located at a crossroads of world shipping and commerce. Based on its port and commercial activities and buttressed by its more recent industrial development, Singapore has become a notable Southeast Asia economic

Figure 8.6

Spanish exclaves in North Africa and France. Spanish troops seized the garrison towns of Melilla and Ceuta almost 500 years ago after evicting the Moors from Spain proper. Llivia became an exclave in 1660 when Spain ceded the surrounding area to France in the Treaty of the Pyrenees. Gibraltar is a British colony, and Andorra is an independent ministate.

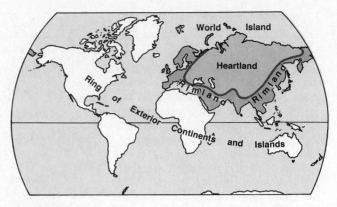

Figure 8.7

Both Mackinder and Spykman believed Eurasia possessed strategic advantages, but they disagreed on whether its heartland or rimland provided the most likely base for world domination.

success. In general, history has shown that countries benefit from a location on major trade routes, not only from the economic advantages such a location carries, but also because they are exposed to the diffusion of new ideas and technologies.

Many analysts have generalized about the relationship between location, territorial extent, and national power. These students of **geostrategy** examine the economic, political, and military value of space in order to present recommendations about courses of action in international relations best designed to advance the cause of national security, projection of power and influence, or territorial aggrandizement. Two of the best known geostrategic theories were those developed by Mackinder and Spykman.

An eminent British professor and a member of Parliament, Halford Mackinder was concerned with the balance of power in the world at the beginning of the 20th century. Believing that the major powers would be those that controlled the land, not the seas, he developed what came to be known as the **heartland theory.** The greatest land power, he argued, would be sited in Eurasia, the "world island" containing the world's largest landmass in terms of both size and population. Its interior or heartland, he warned, would provide a base for world conquest, and Eastern Europe was the core of that heartland (Figure 8.7). Mackinder warned, "Who rules East Europe com-

mands the Heartland, who rules the Heartland commands the World-Island, who rules the World-Island commands the World."[1]

Developed in a century that saw first Germany and then the Soviet Union dominate East Europe, and the decline of Britain as a superpower, Mackinder's theory impressed many. Near the end of the Second World War, the theory was modified by Nicholas Spykman, who agreed with Mackinder that Eurasia was the likely base for world conquest but argued that the coastal fringes of the landmass, not its heartland, were the key. The coastal fringes, or rimland, contained dense populations, abundant resources, and had access both to the seas and to the continental interior. Spykman's **rimland theory,** published in 1944, stated, "Who controls the Rimland rules Eurasia, who rules Eurasia controls the destinies of the world."[2] The rimland has tended throughout history to be politically fragmented, and Spykman believed that it would be to the advantage of both the United States and USSR if it remained that way.

The importance of location changes with developments in technology, resource use, political ideologies, and alliances. The relatively recent rise of air power and the possibility of missile warfare render all states vulnerable to attack. New economic, military, and political alliances, and the economic ascendancy of Southeast Asia, suggest that there are no hard and fast correlations between location and national power.

Cores and Capitals

Many states have come to assume their present shape, and thus the location they occupy, as a result of growth over centuries of time. They grew outward from a central region, gradually expanding into surrounding territory.

1. Halford J. MacKinder, *Democratic Ideals and Reality* (London: Constable, 1919), p. 150.
2. Nicholas J. Spykman, *The Geography of the Peace* (New York: Harcourt Brace, 1944), p. 43.

The original nucleus, or **core area,** of a state usually contains its densest population and largest cities, the most highly developed transportation system, the most developed economies. All of these elements become less intense away from the national core. Urbanization ratios and city sizes decline, transport networks are less well developed, and economic development is less intensive.

Easily recognized and unmistakably dominant national cores include the Paris Basin of France, London and southeastern England, Moscow and the major cities of western European USSR, northeastern United States, and the Buenos Aires megalopolis in Argentina. Not all countries have such clearly defined cores—Chad, Mongolia, or Saudi Arabia, for example—and some may have two or more rival core areas.

The capital city of a state is usually within its core region, and frequently is the very focus of it, dominant not only because it is the seat of central authority but because of the concentration of population and economic functions as well. That is, in many countries the capital city is also the largest or *primate* city, dominating the structure of the entire country. Paris in France, London in the United Kingdom, and Mexico City are all examples of that kind of political, cultural, and economic primacy.

This association of capital with core is common in what have been called the *unitary states*, countries with highly centralized governments, relatively few internal cultural contrasts, a strong sense of national identity, and borders that are clearly cultural as well as political boundaries. Most European cores and capitals are of this type. It is also found in many newly independent countries whose former colonial occupiers established a primary center of exploitation and administration and developed a functioning core in a region that lacked an urban structure or organized government. With independence, the new states retained the established infrastructure, added new functions to the capital, and, through lavish expenditures on governmental, public, and commercial buildings, sought to create prestigious symbols of nationhood.

In *federal states*, associations of more or less equal provinces or states with strong regional governmental responsibilities, the capital city may have been newly created to serve as the administrative center. Although part of a generalized core region of the country, the capital was not its largest city and acquired few of the additional functions to make it so. Ottawa, Canada; Washington, D.C.; and Canberra, Australia, are examples (Figure 8.8).

Figure 8.8
Canberra, the planned capital of Australia, was deliberately sited away from the country's largest cities. Planned capitals are often architectural showcases, providing a focus for national pride.

All other things being equal, a capital located in the center of the country provides equal access to the government, facilitates communication to and from the political hub, and enables the government to exert its authority easily. Many capital cities, such as Washington, D.C., were centrally located when they were designated as seats of government but lost their centrality as the state expanded. Some capital cities have been *relocated* outside of peripheral national core regions, at least in part to achieve the presumed advantages of centrality.

A particular type of relocated capital is the *forward-thrust capital* city, one that has been deliberately sited in a state's frontier zone to signal the government's awareness of regions away from the core and its interest in encouraging more uniform development. In the late 1950s, Brazil moved its capital from Rio de Janeiro to the new city of Brasília to demonstrate its intent to develop the vast interior of the country. Nigeria has been building its new capital of Abuja near the geographic center of that West African country since the late 1970s, with relocation there of government offices and foreign embassies scheduled for the early 1990s. In 1986 Argentina's president proposed moving the capital from Buenos Aires to a more central location, both to redistribute population away from the present capital and to promote economic growth by exploiting the natural resources of Patagonia (Figure 8.9).

In a few states, two cities share governmental functions. Countries with *divided capitals* include the Republic of South Africa, the Netherlands, and Bolivia.

It should be noted that the location of capital cities has been a concern of political units other than nation-states. Capital cities in compromise locations exist all over the United States. In Pennsylvania, Ohio, Michigan, and Illinois, the capitals were moved from previous sites to more central locations. For over 20 years, Alaskans have debated proposals to move the capital from isolated and often inaccessible Juneau closer to Anchorage, the center of population and commerce.

Boundaries: The Limits of the State

We noted earlier that no portion of the earth's land surface is outside the claimed control of a national unit, that even uninhabited Antarctica has had territorial claims imposed upon it (Figure 8.2). Each of the world's states is separated from its neighbors by *international boundaries*, or lines that establish the limit of each state's jurisdiction and authority. Boundaries indicate where the sovereignty of one state ends and that of another begins.

Within its own bounded territory, a state administers laws, collects taxes, provides for defense, and performs other such governmental functions. Thus, the location of the boundary determines the kind of money people in a given area use, the legal code to which they

Figure 8.9

The relocation of Argentina's capital city from Buenos Aires to Viedma was proposed in 1986 as a way of reducing the primacy of Buenos Aires and of directing national attention to the less developed sections of the country.

are subject, the army they may be called upon to join, and the language and perhaps the religion children are taught in school. These examples suggest how boundaries serve as powerful reinforcers of cultural variation over the earth's surface.

Territorial claims of sovereignty, it should be noted, are three-dimensional. International boundaries mark not only the outer limits of a state's claim to land (or water) surface, but are also projected downward to the center of the earth in accordance with international consensus allocating rights to subsurface resources. States also project their sovereignty upward, but with less certainty because of a lack of agreement on the upper limits of territorial airspace. Properly viewed, then, an international boundary is a line without breadth; it is a vertical interface between adjacent state sovereignties.

Before boundaries were delimited, nations or empires were likely to be separated by *frontier zones*, ill-defined and fluctuating areas marking the effective end of a state's authority. Such zones were often uninhabited or only sparsely populated and were liable to change with shifting settlement patterns. Many present-day international boundaries lie in former frontier zones, and in that sense the boundary line has replaced the broader frontier as a marker of a state's authority.

Transferring the National Capital

President Raúl Alfonsín of Argentina announced in 1986 a plan to move the nation's capital from Buenos Aires to the small city of Viedma, about 500 miles to the south. . . . The proposed transfer highlights the recurrent puzzle of why governments throughout history have felt compelled to uproot themselves and move elsewhere.

Argentina's national government is presently entrenched in Buenos Aires, an urban agglomeration of over 10 million people. Viedma, in contrast, has a population of about 34,000 (less than the number of government employees in just one ministry). . . . The move to Viedma, according to Alfonsín, will create a more even national distribution of resources and government attention.

Alfonsín's proposal has a number of historical precedents, including capital transfers in Brazil, Australia, Pakistan, and the United States. Each of the precedents sheds some light on Alfonsín's decision. . . .

Thirty years ago President Kubitschek of Brazil acted upon a constitutional provision for relocating the capital to the country's uninhabited interior and away from the urban sprawl of Rio de Janeiro and São Paulo. Brasília . . . is now the center of national government activity and . . . has established itself as an urban center for Brazil's vast interior.

Australia's capital was established at its present site as a result of rivalry between the country's two largest cities, Sydney and Melbourne. Canberra, located roughly halfway between these two cities, was intended as a demonstration of the national government's commitment to unify Australia's independent-minded states.

Pakistan shifted its capital from Karachi to Islamabad in 1965. . . . Islamabad . . . was to be the capital of a new nation formed from continuous Muslim-majority districts. . . . The capital transfer away from the hot, humid climate of the coast and to the country's cultural heartland also served to emphasize Pakistan's interest in the disputed northern region.

In the early years of the United States, the founding fathers decided to move their capital from New York City to a more centralized location to serve better all of the 13 states and to separate the capital from the administration of individual state governments. . . .

A general look at motives for capital transfers reveals several common themes. First, a capital transfer represents a "fresh start" for a national government concerned with establishing its own mark on a country's future rather than simply continuing previous institutions. For some developing countries, a new, interior capital represents a physical break from a colonial heritage which was centered around the port cities. . . . The capital is often moved to a more centralized hinterland location to demonstrate the government's commitment to develop all parts of the country equally.

Abridged from William B. Wood, "Transferring the National Capital," *Focus*, Summer 1986, 31–32.

Classification of Boundaries

Geographers have traditionally distinguished between "natural" and "artificial" boundaries. **Natural** (or *physical*) **boundaries** are those based on recognizable physiographic features, such as mountains, rivers, and lakes. Although they might seem to be attractive as borders because they actually exist in the landscape and are visible dividing elements, many natural boundaries have proved to be unsatisfactory. That is, they do not effectively separate states.

Many international boundaries lie along mountain ranges, for example in the Alps, Himalayas, and Andes, but while some have proved to be stable, others have not. Mountains are rarely total barriers to interaction. Although they do not invite movement, they are crossed by passes, roads, and tunnels. High pastures may be used for seasonal grazing, and the mountain region may be the source of water for hydroelectric power. Nor is the definition of a boundary along a mountain range a simple matter. Should it follow the crests of the mountains or the *water divide* (the line dividing two drainage areas)? The two are not always the same. Border disputes between China and India are in part the result of the failure of mountain crests and headwaters of major streams to coincide.

Rivers can be even less satisfactory as boundaries. In contrast to mountains, rivers foster interaction. River valleys are likely to be agriculturally or industrially productive, and to be densely populated. For example, for hundreds of miles the Rhine River serves as an international boundary in Western Europe. It is also a primary traffic route lined by chemical plants, factories, blast furnaces, and power stations, and is dotted by the castles and cathedrals that make it one of Europe's major tourist attractions. It is more a common intensively used resource than a barrier in the lives of the states it borders.

With any river, it is not clear precisely where the boundary line should lie: along the right or left bank, along the center of the river, or perhaps along the middle of the navigable channel? Soviet insistence that its sovereignty extended to the Manchurian (Dongbei) bank of the Amur and Ussuri rivers was a long-standing matter of dispute and border conflict between the USSR and the People's Republic of China, only resolved with Russian agreement in 1987 that the boundary should pass along the main channel of the rivers. Any decision about a river boundary tends to complicate the use of the waterway by people of the states involved. Even an agreement in accordance with international custom that the boundary be drawn along the main channel may be impermanent if the river changes its course, floods, or dries up.

The alternative to natural boundaries are artificial or **geometric boundaries.** Frequently delimited as sections of parallels of latitude or meridians of longitude, they are found chiefly in Africa, Asia, and the Americas. The western portion of the United States-Canada border, which follows the 49th parallel, is an example of a geometric boundary. Many such boundaries were established when the areas in question were colonies, the land was only sparsely settled, and detailed geographic knowledge of the frontier region was lacking.

Boundaries can also be classified according to whether they were laid out before or after the principal features of the cultural landscape developed. An **antecedent boundary** is one drawn across an area before it is well populated, that is, before most of the cultural landscape features developed. To continue our earlier example, the western portion of the United States-Canada boundary is such an antecedent line, having been established by a treaty between the United States and Great Britain in 1846.

Boundaries drawn after the development of the cultural landscape are termed **subsequent.** One type of subsequent boundary is **consequent** (also called *ethnographic*), a border drawn to accommodate existing religious, linguistic, ethnic, or economic differences between countries. An example is the boundary drawn between Northern Ireland and Eire (Ireland). Subsequent **superimposed boundaries** may also be forced upon existing cultural landscapes, a country, or a people by a conquering or colonizing power that is unconcerned about preexisting cultural patterns. The colonial powers in 19th century Africa superimposed boundaries upon established African cultures without regard to the tradition, language, religion, or tribal affiliation of those whom they divided (Figure 8.4).

When Great Britain prepared to leave the Indian subcontinent after World War II, it was decided that two independent states would be established in the region: India and Pakistan. The boundary between the two countries, defined in the partition settlement of 1947, was thus both a *subsequent* and a *superimposed* line. As millions of Hindus migrated from the northwestern portion of the subcontinent to seek homes in India, millions of Muslims left what would become India for Pakistan. In a sense, they were attempting to insure that the boundary would be *consequent*, that is, that it would coincide with a division based on religion. This boundary example is more fully discussed in Political Regions in the Indian Subcontinent in Chapter 13.

Stages in the Development of Boundaries

There are three distinct stages in boundary line development, but any individual boundary need not have pro-

ceeded through all three stages, and decades or even centuries can elapse between the stages. **Boundary definition** is a general agreement between two states about the allocation of territory, a verbal description of the boundary and the area through which it passes. If the area is only sparsely settled or otherwise lacks value, there is usually little pressure to proceed to the second stage, **delimitation** of the boundary. This involves actually plotting, as precisely as possible, the boundary line on maps or aerial photographs. In Africa and Asia, delimitation became important when former colonies became independent.

Finally, the boundary line may actually be marked on the ground in a process of **demarcation.** The markers may be intermittent, like poles or pillars, or be continuous fences or walls. Most international boundaries are not demarcated—for example, that between the United States and Canada for most of its length. In contrast, the former "Iron Curtain" separating Soviet-dominated Eastern Europe from Western European countries was marked along much of its length by fences, watchtowers, and mine fields (Figure 8.10).

If a former boundary line that no longer functions as such is still marked by some landscape features or differences on the two sides, it is termed a *relict* boundary. The abandoned castles dotting the former frontier zone between Wales and England are examples of a relict boundary. They are also evidence of the disputes that sometimes attend the process of boundary making.

Boundary Disputes

Boundaries create many possibilities and provocations for conflict. Since World War II, almost half of the world's sovereign states have been involved in border disputes with neighboring countries. Just like householders, states are far more likely to have disputes with their neighbors than with more distant parties. It follows that the more neighbors a state has, the greater the likelihood of conflict.

Although the causes of boundary disputes and open conflict are many and varied, they can reasonably be placed into four categories.

1. **Positional disputes** occur when states disagree about the interpretation of documents that define a boundary and/or the way the boundary was delimited. Such disputes typically arise when the boundary is antecedent, preceding effective human settlement in the border region. Once the area becomes populated and gains value, the exact location of the boundary becomes important.

The boundary between Argentina and Chile, originally defined during Spanish colonial rule, was to follow the highest peaks of the southern Andes and the watershed divides between east- and west-flowing rivers. Because the terrain had not been adequately explored, it wasn't apparent that the two do not always coincide. In some places, the water divide is many miles east of the highest peaks, leaving a long, narrow area of several hundred square miles in dispute (Figure 8.11). During the late 1970s, Argentina and Chile nearly went to war over

Figure 8.10
Like Hadrian's Wall in the north of England or the Great Wall of China, the Berlin Wall was a demarcated boundary. Unlike them, it cut across a large city and disrupted established cultural patterns. The Berlin Wall, therefore, was a *subsequent, superimposed, demarcated* boundary. Openings were made in the wall in late 1989. Should it cease to function as a barrier, but portions of it remain standing, the wall will become a *relict* boundary.

Figure 8.11
Areas of international dispute in Latin America. Among the countries disputing the precise location of their boundaries are Argentina and Chile, Argentina and Uruguay, Venezuela and Guyana, and Honduras and El Salvador.

the disputed territory, whose significance had been increased by the discovery of oil and natural gas deposits. Argentina has also contested its border with Uruguay in the Rio de la Plata estuary. Argentina claims that the international boundary is in the deepest part of the river, while Uruguay contends that it lies in the middle of the estuary.

2. **Territorial disputes** over the ownership of a region often, though not always, arise when a boundary that has been superimposed on the landscape divides an ethnically homogeneous population. Each of the two states then has some justification for claiming the territory inhabited by the ethnic group in question. The Balkan countries of Eastern Europe offer numerous examples of such territorial disputes. Regional tensions provided the sparks that helped ignite both world wars, and the area is far from stable today.

Even land that might seem to be without value can become the subject of a territorial conflict. Since the early 1970s, thousands of people have been killed in a series of battles between Chad and Libya over ownership of the Aozou Strip, a 36,000-square-mile (100,000 km²) piece of desert (Figure 8.12). The boundary between what are now Libya and Chad was originally set by France and Britain in 1899. In 1935, at the request of Italy, which had seized Libya, France agreed to move the boundary 60 miles (100 km) south. Italy did not ratify the agreement, however, and Chad gained its independence with the original boundary intact. Libya disagrees, claiming the strip belongs to it.

3. Closely related to territorial conflicts are **resource disputes.** Neighboring states are likely to covet the resources—whether they be valuable mineral deposits, fertile farmland, or rich fishing grounds—lying in border areas and to disagree over their use. In recent years, the United States has been involved in disputes with both its immediate neighbors, Mexico and Canada, over the shared resources of the Colorado River and Gulf of Mexico in the south and the Georges Bank fishing grounds in the northeast.

For over 35 years, India and Bangladesh have disputed the shared water resources of the Ganges and Brahmaputra rivers, two of the world's largest waterways (Figure 8.13). At present, the river basin supports a population of about 500 million, a figure that is expected to double by the middle of the next century. During the monsoon season, the rivers can become torrents of destruction, but in dry months they slow and the region suffers from water shortages and drought. The two countries have been unable to agree on a long-term water management plan that would permit irrigation of more of the arable land, improve flood control, help stem deforestation, and allow development of the basin's hydroelectric potential. Some 90% of the water flow originates in India, which has proposed building a 200-mile-long (320 km) canal to channel water from the Brahmaputra to the Ganges. Fearing the loss of any of its water, and believing that the impact of the canal would be disastrous, Bangladesh has countered with a proposal to build a series of dams at the headwaters of tributaries of the Ganges, dams that would help control floods and store water during the wet season.

4. **Functional disputes** arise when neighboring states disagree over policies to be applied along a boundary. Such policies may concern immigration, the movement of traditionally nomadic groups, customs regulations, or land use. U.S. relations with Mexico, for example, have been affected by the increasing number of illegal aliens and the flow of drugs entering the United States from Mexico. One of the world's most contentious frontiers, that between

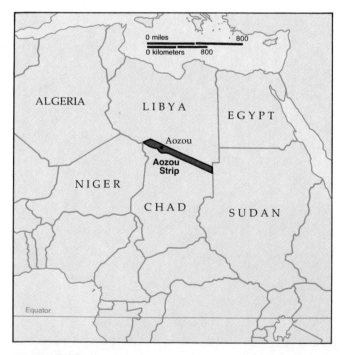

Figure 8.12
Both Chad and Libya claim ownership of the Aozou Strip, an example of a territorial dispute.

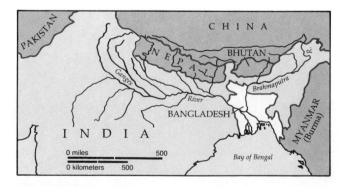

Figure 8.13
India and Bangladesh have been unable to agree on a long-term plan for managing the waters of the Ganges and Brahmaputra rivers. Their dispute helps block development of one of the world's poorest and most densely populated areas.

northern India and Pakistan, has strained relations between the two countries for decades. Insurgent Sikh extremists demand a separate state in the Punjab, and India claims that Pakistan harbors guerrillas and allows weapons to be smuggled across the border.

Maritime Boundaries

Boundaries define political jurisdictions and thus areas in which certain activities may or may not be allowed to take place. However, we have considered only boundaries on land. As water covers about two-thirds of the earth's surface and people have used the seas since ancient times, it is not surprising that boundaries across water are also important. A basic question involves the right of states to use water and the resources that it contains. The inland waters of a country, such as rivers and lakes, have traditionally been regarded as being within the sovereignty of that country. Oceans, however, are not within any country's borders. Are they, then, to be open to all states to use, or may a single country claim sovereignty and limit access and use by other countries?

For most of human history, the oceans remained effectively outside individual national control or international jurisdiction. The seas were a common highway for those daring enough to venture on them, an inexhaustible larder for fishermen, and a vast refuse pit for the muck of civilization. By the end of the 19th century, however, most coastal countries claimed sovereignty over a continuous belt 3 or 4 nautical miles wide (a *nautical mile*, or *nm*, equals 1.15 statute miles, or 1.85 km). At the time, the 3-mile limit represented the farthest range of artillery and thus the effective limit of control by the coastal state. Though recognizing the rights of others to innocent passage, such sovereignty permitted the enforcement of quarantine and customs regulations, allowed national protection of coastal fisheries, and made claims of neutrality effective during other people's wars. The primary concern was with security and unrestricted commerce. No separately codified law of the sea existed, however, and none seemed to be needed until after World War I.

A League of Nations Conference for the Codification of International Law, convened in 1930, inconclusively discussed maritime legal matters and served to identify areas of concern that were to become increasingly pressing after World War II. Important among these was an emerging shift from interest in commerce and national security to a preoccupation with the resources of the seas, an interest fanned by the *Truman Proclamation* of 1945. Motivated by a desire to exploit offshore oil deposits, the federal government under this doctrine laid claim to all resources on the continental shelf contiguous to its coasts. Other states, many claiming even broader areas of control, hurried to annex marine resources. Within a few years, a quarter of the earth's surface was appropriated by individual coastal countries.

Unrestricted extensions of jurisdiction and territorial disputes over proliferating claims to maritime space and resources led to a series of United Nations conferences on the Law of the Sea. Meeting over a period of years, delegates from over 150 countries attempted to achieve consensus on a treaty that would establish an internationally agreed-upon "convention dealing with all matters relating to the Law of the Sea." The meetings culminated in a draft treaty in 1982, the **United Nations Convention on the Law of the Sea.**

An International Law of the Sea

The convention delimits territorial boundaries and rights by defining four zones of diminishing control (Figure 8.14).

1. The treaty allows for the establishment of a *territorial sea* of up to 12 nm (19 km) in breadth, providing various measures for distinguishing between internal and territorial waters. Coastal states have sovereignty over the territorial sea, including the exclusive right to fish in it. Vessels of all types have the right of innocent passage through the territorial sea, although in certain instances coastal states can challenge noncommercial vessels (primarily military and research).

2. A *contiguous zone* is permitted out to 24 nm (38 km). Although a coastal state does not have complete sovereignty in this zone, it can enforce its customs, immigration, and sanitation laws and has the right of hot pursuit out of its territorial waters.

3. The convention allows the creation of an **exclusive economic zone (EEZ)** of up to 200 nm (370 km)(Figure

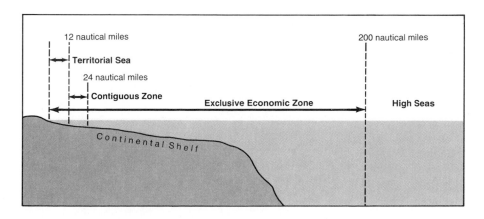

Figure 8.14
Territorial claims permitted by the 1982 United Nations Convention on the Law of the Sea.

8.15). In this zone, the state has certain rights for the purpose of economic advantage, notably sovereign rights to explore, exploit, conserve, and manage the natural resources, both living and nonliving, of the seabed and waters. Countries have exclusive rights to the resources lying within the continental shelf when this extends farther, up to 350 nm (560 km), beyond their coasts. The traditional freedoms of the high seas are to be maintained in this zone.

4. All parts of the sea that lie beyond the EEZ constitute the fourth zone, that of the *high seas*. Outside any national jurisdiction, they are open to all states, whether coastal or landlocked. Freedom of the high seas includes the right to sail ships, fish, fly over, lay submarine cables and pipelines, construct artificial platforms and other installations, and pursue scientific research. Mineral resources in the international deep seabed area beyond national jurisdiction are declared the common heritage of humankind, to be managed for the benefit of all the peoples of the earth.

The 1982 convention will not formally take effect until one year after 60 states have ratified it. Although the requisite number have not yet done so, by the end of the decade most coastal countries, including the United States, had used its provisions to proclaim and reciprocally recognize jurisdiction over 12-nm territorial seas and 200-nm economic zones. The treaty is generally recognized as a masterly accomplishment and viewed with favor by most countries as fairly balancing the interests of all states. Except for reservations about the deep seabed mining provisions, the convention is now so widely accepted as to be, for all practical purposes, international law.

A New Maritime Map

General acceptance of the Law of the Sea Convention has in effect changed the maritime map of the world. Three important consequences flow from the 200-nm EEZ concept: (1) islands have gained a new significance, (2) countries have a host of new neighbors, and (3) the EEZ lines result in overlapping claims.

EEZ lines are drawn around a country's possessions as well as around the country itself. Every island that can sustain human habitation or has an economic life has its own 200-nm EEZ. This means that while the United States shares continental borders only with Canada and Mexico, it has maritime boundaries with countries in Asia, South America, and Europe. Thus the United States and the USSR share a maritime boundary, yet to be precisely determined, in the waters of the Bering and Chukchi seas. All told, the United States may have to negotiate some 30 maritime boundaries, 10 of them situated off one or more of the 50 states, and the remainder located off the shores of such American possessions as Guam, Wake, and American Samoa. Delimitation of these boundaries is likely to take decades. Only after years of bitter dispute was the boundary resolved between the United States and Canada in the Atlantic Ocean. (See "The Georges Bank.") Other countries, particularly those with many possessions, will have to engage in similar lengthy negotiations.

Islands have gained a new significance because the Law of the Sea Convention gives their owners claims over immense areas of the surrounding sea, and to the fisheries and mineral resources the water contains. Tiny specks of land formerly too small or too distant to arouse the emotion of any state are now avidly sought and claimed. One reason British forces retook the Falkland (Malvinas) Islands after Argentina seized them in 1982 was because their ownership enables Britain to claim a sea area three times as large as Britain. Japan is spending hundreds of millions of dollars to keep a tiny islet over 1000 miles (1600 km) from Tokyo from vanishing beneath the Pacific. Only 15 feet long, Okinotorishima will be shielded against further erosion by being encased in concrete. Ownership of the island may entitle Japan to claim fishing rights in the 120,000 square miles (310,000 km²) of sea surrounding it.

When several countries claim sovereignty over an island or group of islands, the possibilities of armed dispute are ever present. The Spratly Islands, spread over some 600 miles (960 km) in the South China Sea, have attracted recent attention. Mere dots in the sea, with the largest island considerably smaller than New York's Central Park, the islands straddle trade routes between the Pacific and Indian oceans and are thought to lie atop large reserves of oil and gas. The Philippines, Malaysia, Vietnam, and Taiwan each occupy several of the islands, though each country claims sovereignty over all of them. In 1988, China sent destroyers and submarines to islands near the Vietnamese holdings, sank two Vietnamese naval ships, and staked its claim to the islands.

State Cohesiveness

At any moment in time, a state is characterized by forces that promote unity and national stability and by others that disrupt them. Political geographers refer to the former as **centripetal forces.** These are factors that bind together the people of a state, that enable it to function and give it strength. **Centrifugal forces,** on the other hand, destabilize and weaken a state. If centrifugal forces are stronger than those promoting unity, the very existence of the state will be threatened. In the sections that follow we examine four centripetal forces: nationalism, unifying institutions, effective organization and administration of government, and systems of transportation and communication.

Nationalism

One of the most powerful of the centripetal forces is **nationalism,** an identification with the state and the acceptance of national goals. Nationalism is based on the concept of allegiance to a single country; it thus fosters a feeling

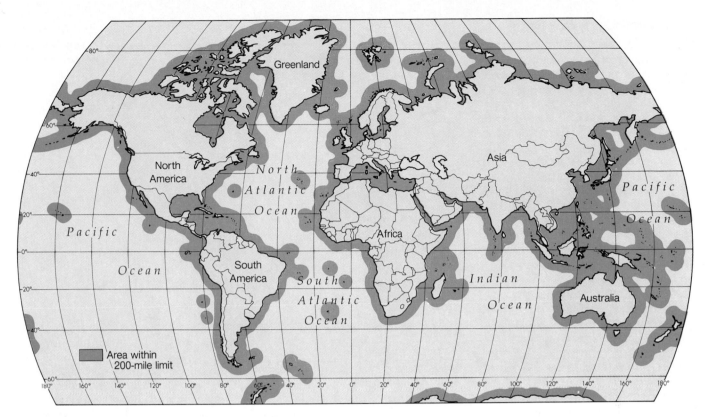

Figure 8.15

The 200-nautical mile exclusive economic zone (EEZ) claims of coastal states. Overlapping claims may take decades to negotiate.

of collective distinction from all other peoples and lands. It is an emotion that provides a sense of identity and loyalty.

States purposely try to instill feelings of allegiance in their constituents, for such feelings give the political system strength. People who have such allegiance are likely to accept the rules governing behavior in the area and to contribute to the decision-making process. In addition, a sense of unity binding the people of a state together is necessary to overcome the divisive forces present in most societies. Not everyone, of course, will feel the same degree of commitment or loyalty. The important consideration is that the majority of a state's population accept its ideologies, adhere to its laws, and participate in its effective operation. For many countries, such acceptance and adherence has come only recently and partially; in some, it is frail and endangered.

We noted earlier that true nation-states are rare; in only a few countries do the territory occupied by the people of a particular nation and the territorial limits of the state coincide. Most countries have more than one culture group that considers itself separate in some important way from other citizens. In a multicultural society, nationalism helps integrate different groups into a unified population. This has occurred in countries such as the United States and Canada, where different culture groups, few or none with

defined North American homelands, have joined together to create political entities commanding the loyalties of all of their citizens.

States promote nationalism in a number of ways. *Iconography* is the study of the symbols that help unite people. National anthems and other patriotic songs, flags, national flowers and animals, colors, and rituals are all developed as symbols of a state in order to attract allegiance (Figure 8.16). They ensure that all citizens, no matter how diverse the population may be, will have at least these symbols in common. They impart a sense of belonging to a political entity called, for example, Japan or Canada. In some countries, certain documents, such as the Magna Charta in England or the Declaration of Independence in the United States, serve the same purpose. Symbols such as these are significant insofar as ideologies and beliefs are an important aspect of culture. When a culture is very heterogeneous—composed of people with different customs, religions, and languages—belief in the national unit can help weld them together.

Unifying Institutions

A number of institutions help to develop the sense of commitment and cohesiveness essential to the state. Schools, particularly elementary schools, are among the most important of these. Children learn the history of their own

The Georges Bank

One of the richest fishing grounds in the world for scallops, haddock, cod, and flounder is the Georges Bank, a 9000-square-mile (23,300 km²) expanse east of Cape Cod in the Gulf of Maine. For centuries, Canadians and New Englanders shared the grounds with little conflict. Then the United States in 1976, and Canada a year later, extended their claims over fishing rights in waters up to 200 (320 km) miles offshore, creating the basis for a bitter boundary dispute. The claims overlapped in the eastern portion of the Georges Bank around a section called the Northeast Peak, generally agreed to be the most productive grounds. Approximately half the Georges Bank haddock and pollock, a third of the flounder and scallops, a quarter of the cod, and the best swordfish and lobster were harvested in the Northeast Peak, where deep waters provide rich harvests even in winter months.

Canada and the United States tried without success to settle the dispute with a treaty providing for joint fishing and management. They eventually agreed to refer the conflict to the International Court of Justice at The Hague, Netherlands. Canada claimed slightly less than half of the disputed area; the United States claimed the whole bank.

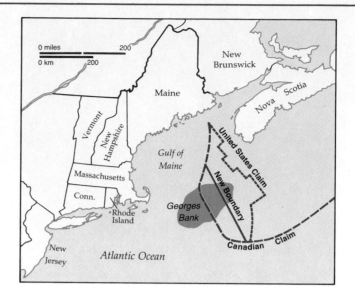

Perhaps predictably, the decision issued by the World Court in October 1984 satisfied neither country entirely. It awarded the United States about 80% of the Georges Bank. Canada received the remainder, but that portion included the Northeast Peak. Fishermen from both countries are confined to their side of the new line. U.S. fishermen contend they are now crowded into more restricted, less-productive waters, while Canadians deplore their loss of access to most of the bank.

The Georges Bank issue reflects many political-geographic concerns: the significance of boundaries to nation-states and their potential to cause disputes, the growing significance of maritime boundaries as states seek both to protect and to exploit the resources of the sea, and the use of international authorities to resolve disputes peacefully.

---

country and relatively little about other countries. Schools are expected to inculcate the society's goals, values, and traditions; allegiance to the state is accepted as the norm. As a rule, schools teach youngsters to identify with their country rather than with the world or with humanity as a whole.

Other institutions that promote nationalism are the armed forces and, sometimes, a state church. The military organization fulfills a primary state goal: the provision of security, both internal and external. A high percentage of most states' budgets is spent to secure such protection. The armed forces are of necessity taught to identify with the state. They see themselves as protecting the state's welfare from what are perceived to be its enemies.

In some countries, the religion of the majority of the people may be designated a state church. In such cases the church sometimes becomes a force for cohesion, helping to unify the population. This is true of the Roman Catholic church in the Republic of Ireland, Islam in Pakistan, and Judaism in Israel. In countries like these, the religion and the church are so identified with the state that belief in one is transferred to allegiance to the other.

The schools, the armed forces, and the church are just three of the institutions that teach people what it is like to be members of a state. These institutions operate primarily on the level of the ideological subsystem of culture. By themselves, they are not enough to give cohesion, and thus strength, to a state.

Organization and Administration

A further bonding force is public confidence in the effective organization of the state. Can it provide security from external aggression and internal conflict? Are its resources distributed and allocated in such a way as to be perceived to promote the economic welfare of all its citizens? Are there institutions that encourage consultation

Figure 8.16
The ritual of the pledge of allegiance is just one way in which schools in the United States seek to instill a sense of national identity in students.

and the peaceful settlement of disputes? How firmly established are the rule of law and the power of the courts? Is the system of decision making responsive to the people's needs?

These are not questions of democracy or dictatorship but of citizen perception of the propriety and legitimacy of governmental control. Thus, while the degree to which the state is involved in the economy varies from country to country, from the public ownership of the means of production to private ownership, in every country there is some governmental role. People expect the government to promote their economic well-being. Tariff barriers restrict trade, protecting home industries from foreign competition. Direct aid is given to certain industries in many countries, and some states strictly control or discourage foreign investment. Governments extend subsidies to those engaged in certain kinds of agriculture. They may sponsor programs aimed at increasing agricultural yields or at supporting basic industry.

The answers to the questions posed above, and the relative importance of the answers, will vary from country to country, but they and similar ones are implicit in the expectation that the state will, in the words of the Constitution of the United States, "establish justice, insure domestic tranquility, provide for the common defense, (and) promote the general welfare. . . ." If those expectations are not fulfilled, the loyalties promoted by national symbols and unifying institutions may be weakened or lost.

Transportation and Communication

A state's transportation network fosters political integration by promoting interaction between areas and by joining them economically and socially. The role of a transportation network in uniting a country has been recognized since ancient times. The saying that all roads lead to Rome had its origin in the impressive system of roads that linked Rome to the rest of the empire. Centuries later, a similar network was built in France, linking Paris to the various departments of the country. Often the capital city is better connected to other cities than the outlying cities are to one another. In France, for example, it can take less time to travel from one city to another by way of Paris than by direct route.

Roads and railroads have played a historically significant role in promoting political integration. In the United States and Canada, they not only opened up new areas for settlement but increased interaction between rural and urban areas. The Soviet Union anticipates comparable benefits of economic, social, and political interaction and territorial development from the recently completed Baikal-Amur Railroad (BAM) through Siberia (Figure 8.17). The cost—some $38 billion—of the 2000-mile (3200-km) project is expected to yield great returns by tapping a development area of 600,000 square miles (1.5 million km²) with some of the world's richest deposits of minerals and timber, by providing access to 13 planned major new industrial centers, and by establishing a militarily secure link between the western and extreme far-eastern sections of the USSR.

Because transportation systems play a major role in a state's economic development, it follows that the more economically advanced a country is, the more extensive its transport network is likely to be. By fostering interdependence between regions, a transport network keeps

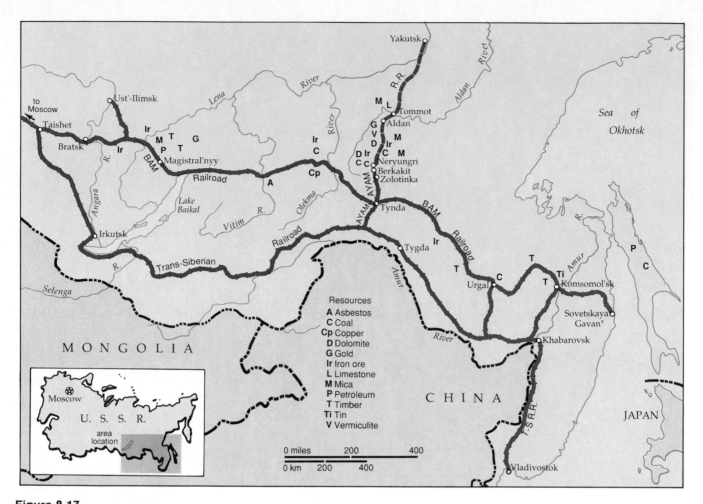

Figure 8.17

The Baikal-Amur Mainline Railroad is expected to be a vital link in the longer-term development of Siberian resources undertaken according to Soviet developmental plans. A ceremonial golden spike marked the road's completion on October 1, 1984. Work still continues on tunnels and structures to make it operational to planned capacity. Other projects, including the Amur-Yakutsk Mainline leading northward from the southern rail lines into interior Siberia, are also planned or under construction.

regions from having to be self-sufficient. Each can specialize in the production of the goods and services for which it is best suited. Specialization results in an overall increase in productivity. At the same time, the higher the level of development, the more money there is to be spent on building transport routes. In other words, the two feed on one another.

Transportation and communication are encouraged within a state and are curtailed or at least controlled between states. The mechanisms of control include restrictions on trade through tariffs or embargoes, legal barriers to immigration and emigration, and limitations on travel through passports and visa requirements. A pointed illustration is the Australian railroad system (Figure 8.18). Until federation in 1901, the states of Australia were independent of one another. As competitors for British markets, they had no desire to foster mutual interaction. Different rail gauges, rather than a single, standard gauge, were used, so that people and goods traveling between states had to change trains at the border.

Separatism and Intrastate Conflict

The cohesion that the spirit of nationalism implies is not easily achieved or, once gained, invariably retained. Destabilizing *centrifugal forces* are ever-present, sowing internal discord and challenges to the state's authority. The most serious manifestation of such forces is a civil war, and in a typical week in 1989, civil wars raged in over 15 countries, or some 10% of the world's total.

Some rebellions are truly revolutionary, seeking to replace the existing government and/or ideology with another. Latin American examples have included the Shining Path movement in Peru and the contras of Nicaragua. These wars lie more in the domain of the historian or political scientist than of the geographer. Other rebel groups, however, have total or partial secession from the state as their primary goal. The Western Sahara, Ethiopia, Myanmar (Burma), and Indonesia are a few of the countries currently characterized by such insurgencies. These wars have attracted the attention of political geographers

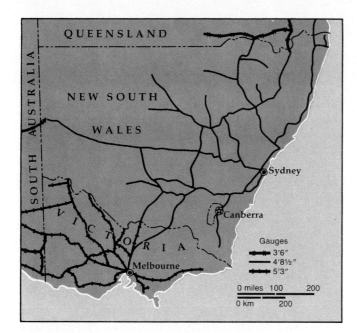

Figure 8.18
Railway gauges in eastern Australia. Gauges of different widths were established to hinder interaction between the separate states of Australia, which until federation in 1901 owed primary allegiance to Great Britain rather than to one another.

Figure 8.19
Regions in Western Europe demanding autonomy. Each was once an independent entity incorporated within a larger state, and each has recently sought or is currently seeking a degree of separatism that recognizes its separate identity.

because they are an expression of **regionalism,** minority-group self-awareness and identification with a region rather than with the state. The minority group may be a tribe, an ethnic group, or even a nation.

In recent years, regionalism has created currents of unrest within many countries, even long-established ones. In Western Europe, for example, five countries (the United Kingdom, France, Belgium, Italy, and Spain) house separatist political movements whose members reject the existing sovereign state and who claim to be the core of a separate national group (Figure 8.19). Their basic demand is for *regional autonomy*, usually in the form of self-government or "home rule" rather than complete independence. Both the Welsh and the Scottish nationalist parties would like their regions to have commonwealth status within the United Kingdom, for example.

Separatist movements affect many states outside of Western Europe and indeed are more characteristic of Second and Third World countries, especially those formed since the end of the Second World War and containing disparate groups more motivated by enmity than affinity. The Basques of Spain and the Bretons of France have their counterparts in the Palestinians in Israel, the Sikhs in India, the Tigre in Ethiopia, the Moros in the Philippines, and many others. For over a decade, for example, the island state of Sri Lanka has been wracked by a bitter and bloody ethnic conflict between a minority Tamil community and a government dominated by the majority Sinhalese population. The Sinhalese are predominantly Buddhist, while the Tamils are mostly Hindu and Christian. The goal of the Tamil guerrillas is to create an independent state in the northeast section of the country. More moderate Tamils desire autonomy for a Tamil province in the north and some sort of power sharing in the ethnically mixed eastern areas (Figure 8.20).

The Soviet Union and Communist states along its western periphery, from the Baltic to the Balkans, have recently seen an explosion of long-suppressed ethnic and nationalist feelings (Figure 8.21). "Popular fronts" demanding greater economic, cultural, and political independence have emerged in Estonia, Latvia, and Lithuania almost a half-century after those Baltic republics were annexed by the Soviet Union in 1940. Ukrainians demand official recognition of their language and religion; the Armenian National Movement seeks a measure of economic sovereignty as well as priority for the Armenian language; and Georgian nationalists demand greater political and economic autonomy. In Czechoslovakia, Slovak separatists protest union with the Czechs. Indeed, with the exception of Albania, territorial and ethnic tensions beset all the Balkan countries. The situation may be most serious in Yugoslavia, where warring nationalities threaten the stability of the state.

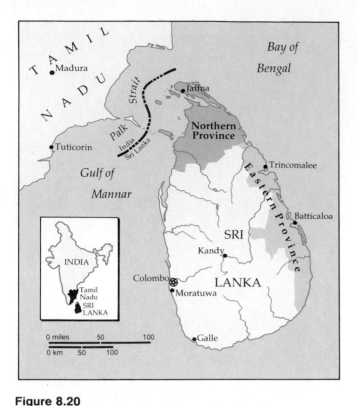

Figure 8.20
Traditional Tamil areas in Sri Lanka. Although Tamils are not a majority in the eastern province, militant Tamils demand a "homeland" embracing the northern and eastern provinces. These were merged into a single administrative unit in late 1988 by executive order of the Sri Lankan president, meeting a major Tamil demand.

The two preconditions common to all regional autonomist movements are *territory* and *nationality*. First, the group must be concentrated in a core region that it claims as a national homeland. It seeks to regain control of land and power that it believes were unjustly taken by the ruling party. Second, certain cultural characteristics must provide a basis for the group's perception of separateness and cultural unity. These might be language, religion, or distinctive group customs, which promote feelings of group identity at the same time that they foster exclusivity. Normally, these cultural differences have persisted over several generations and have survived despite strong pressures toward assimilation.

Other characteristics common to many separatist movements are a *peripheral location* and *social and economic inequality*. Troubled regions tend to be peripheral, often isolated in rural pockets, and their location away from the seat of central government engenders feelings of alienation and neglect. Second, the dominant culture group is often seen as an exploiting class that has suppressed the local language, controlled access to the civil service, and taken more than its share of wealth and power. Poorer regions complain that they have lower incomes and greater unemployment than prevail in the sovereign state, and that "outsiders" control key resources and industry. Separatists in relatively rich regions believe that they could exploit their resources for themselves and do better without the constraints imposed by the central state.

Figure 8.21
A blossoming of ethnic and nationalist feeling in these Eastern European countries and Soviet republics, in part the result of the increased political tolerance characterizing the policy of *glasnost*, poses a challenge to state control.

International Political Systems

The strivings of groups with distinct territory-based identities and a desire for fuller control of that territory are a reminder of the fragility of the modern state. In many ways, countries are now weaker than ever before. Despite their independence, many are economically frail, others are politically unstable, and some are both. Strategically, no state is safe from military attack, for technology now enables countries to shoot weapons halfway around the world. Some people believe that no national security is possible in the atomic age.

The recognition that a country cannot by itself guarantee either its prosperity or its own security has led to increased cooperation among states. In a sense, these cooperative ventures are replacing the empires of yesterday. They are proliferating quickly, and they involve countries everywhere.

The United Nations and Its Agencies

The United Nations (UN) is the only organization that tries to be universal, and even it is not all-inclusive. The UN is the most ambitious attempt ever undertaken to bring together the world's countries in international assembly and to promote world peace. It is stronger and more representative than its predecessor, the League of Nations. It provides a forum where countries may meet to discuss international problems and a mechanism, admittedly weak but still significant, for forestalling disputes or, when necessary, for ending wars (Figure 8.22). The United Nations also sponsors 40 programs and agencies aimed at fostering international cooperation with respect to specific goals. Among these are the World Health Organization (WHO), the Food and Agriculture Organization (FAO), and the United Nations Educational, Scientific, and Cultural Organization (UNESCO).

Member states have not surrendered sovereignty to the UN, and the world body is legally and effectively unable to make or enforce a world law. Nor is there a world police force. Although there is recognized international law adjudicated by the International Court of Justice, rulings by this body are sought only by countries agreeing beforehand to abide by its arbitration. The United Nations has no authority over the military forces of individual countries, and it cannot take prompt and effective action to counter aggression.

Countries have shown themselves to be more willing to relinquish some of their independence to participate in smaller multinational systems. These systems may be economic, military, or political, and many have been formed since 1945.

Figure 8.22

United Nations peacekeeping forces on duty in Lebanon. Under the auspices of the United Nations, soldiers from many different countries staff peacekeeping forces and military observer groups in an effort to halt or mitigate conflicts. Demand for peacekeeping operations is indicated by the recent or current deployment of UN forces in Cyprus, Lebanon, Israel, Syria, Iran, Iraq, India, Pakistan, Afghanistan, Angola, and Namibia.

Regional Alliances

Cooperation in the economic sphere seems to come more easily to states than does political or military cooperation. Among the most powerful and far-reaching of the *economic alliances* are those that have evolved in Europe, particularly the Common Market and its several forerunners.

Shortly after the end of World War II, the Benelux countries (Belgium, the Netherlands, and Luxembourg) formed an economic union to create a common set of tariffs and to eliminate import licenses and quotas. Formed at about the same time were the Organization for European Cooperation (1948), which coordinated the distribution and use of Marshall Plan funds, and the European Coal and Steel Community (1952), which integrated the development of that industry in the member countries. A few years later, in 1957, the **European Economic Community (EEC)**, or Common Market, was created, composed at first of only six states: France, Italy, West Germany, and the Benelux countries.

To counteract these Inner Six, as they were called, other countries joined in the European Free Trade Association (EFTA). Known as the Outer Seven, they were the United Kingdom, Norway, Denmark, Sweden, Switzerland, Austria, and Portugal (Figure 8.23). In 1973, the United Kingdom, Denmark, and Ireland applied for and were granted membership in the Common Market. They were joined by Greece in 1981; Spain and Portugal entered in 1986. After 23 years of associate status, Turkey applied for full membership in 1987, but will not achieve it for some years.

The members of the European Economic Community have taken many steps to integrate their economies and coordinate their policies in such areas as transportation and agriculture. A council of ministers, a commission, a European Parliament, and a Court of Justice give the EEC supranational institutions with effective ability to make and enforce laws. The community objective is to abolish—on December 31, 1992—all barriers to trade, migration, and capital and labor movement among its members (Figure 8.24). Political, military, and cultural integration are scheduled to follow.

We have traced this European development process not because it is important to remember all the forerunners or the present structure of the Common Market, but to illustrate the fluid process by which regional alliances are made. Countries come together in an association, some drop out, and others join. New treaties are made, and new coalitions emerge. It seems safe to predict that although the alliances themselves will change, the idea of economic associations has been permanently added to that of political and military leagues, which are as old as nation-states themselves.

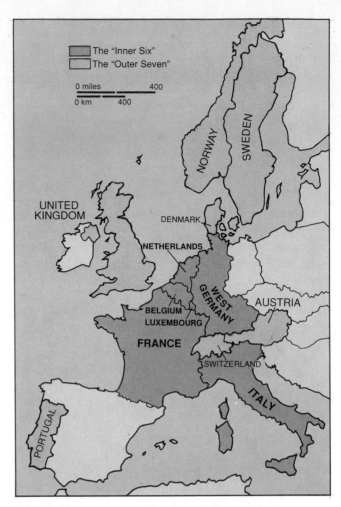

Figure 8.23
The original Inner Six and Outer Seven of Europe.

Three further points about economic unions are worth noting. The first, which also applies to both military and political alliances, is that the formation of a coalition in one area often stimulates the creation of another alliance by countries left out of the first. Thus, the union of the Inner Six gave rise to the treaty among the Outer Seven. Similarly, a counterpart of the Common Market was the Council of Mutual Economic Assistance (CMEA), also known as Comecon, which linked the former Communist countries of Eastern Europe and the USSR through trade agreements.

Second, the new economic unions tend to be composed of contiguous states (Figure 8.25). This was not the case with the recently dissolved empires, which included far-flung territories. Contiguity facilitates the movement of people and goods. Communication and transportation are simpler and more effective among adjoining countries than among those far removed from one another.

Finally, it does not seem to matter whether countries are alike or different in their economies, as far as joining economic unions is concerned. There are examples of both.

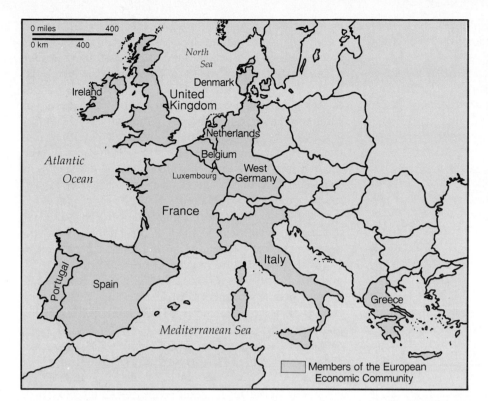

Figure 8.24

The European Economic Community. By the end of 1992, the 12 members of the EEC are to have dismantled any remnant trade barriers among themselves, and the world's largest free trade region will be a reality. Its unified market of more than 325 million will increasingly be controlled from Community headquarters in Brussels, Belgium, as individual countries relinquish their sovereign control over industrial standards, residency laws, labor agreements, and the like. In addition to the 12 Community members, some 60 states in Africa, the Caribbean, and the Pacific have been affiliated with the EEC by the Lomé Convention, which provides for development aid and access to EEC markets.

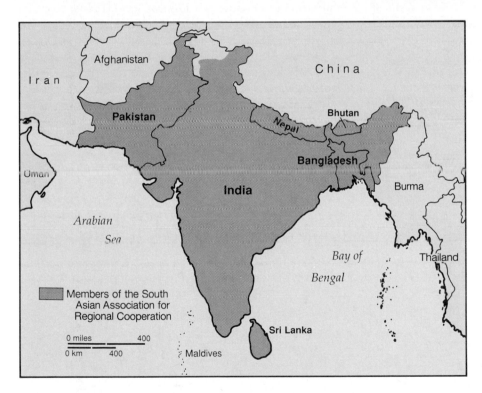

Figure 8.25

Members of the South Asian Association for Regional Cooperation. Formed in December 1985 to promote cooperation in such areas as agriculture and rural development, transportation, and telecommunications, the Association is composed of seven neighboring states that contain about one-fifth of the world's population. Until the formation of the Association, South Asia was the only major part of the world with no organization for regional cooperation.

If the countries are dissimilar, they may complement each other. This was one basis for the European Common Market. Dairy products and furniture from Denmark are sold in France, freeing that country to specialize in the production of machinery and clothing. On the other hand, countries that produce the same raw materials hope that by joining together in an economic alliance, they might be able to enhance their control of markets and prices for their products (Figure 8.26). The Organization of Petroleum Exporting Countries (OPEC) is a case in point. Other attempts to form commodity cartels and price agreements between producing and consuming countries are represented by the International Tin Agreement, the International Coffee Agreement, and others.

Countries form alliances for other than economic reasons. Strategic, political, and cultural considerations may also foster cooperation. *Military alliances* are based on the principle that in unity there is strength. Such pacts usually provide for mutual assistance in the case of aggression. Once again, action breeds reaction when such an association is created. The formation of the North Atlantic Treaty Organization (NATO), a defensive alliance of many European countries and the United States, was countered by the establishment of the Warsaw Treaty Organization, which joined the USSR and its satellite countries of Eastern Europe (Figure 8.27). Both pacts allowed the member states to base armed forces in one another's territories, a relinquishment of a certain degree of sovereignty unique to this century.

Military alliances depend on the perceived common interests and political goodwill of the countries involved. As political realities change, so do the strategic alliances. NATO was altered in the 1960s as a result of policy disagreements between France and the United States. The conflict over Cyprus between Greece and Turkey, both NATO members, altered the original cohesiveness of that alliance. The organization similar to NATO in Asia, SEATO (Southeast Asia Treaty Organization), was disbanded in response to changes in the policies of its member countries.

All international alliances recognize communities of interest. In economic and military associations, common objectives are clearly seen and described, and joint actions are agreed on with respect to the achievement of those objectives. More generalized common concerns or appeals to historical interest may be the basis for primarily *political alliances*. Such associations tend to be rather loose, not requiring their members to yield much power to the

Figure 8.26
Members of the Union of Banana Exporting Countries (UPEB). Founded in 1974, the Union represents producers of about half of the world's banana exports. The cartel has established a multinational trading corporation in an attempt to challenge the control of three large U.S. corporations that dominate the banana trade.

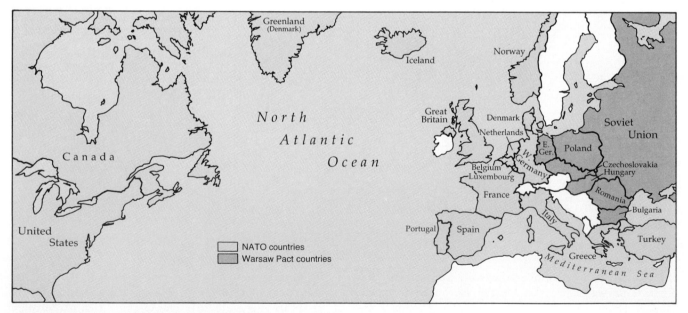

Figure 8.27
Countries belonging to NATO and the Warsaw Pact. Military alliances change to fit perceived political realities. The division of Europe evolved from the rivalries and tensions of the Cold War.

union. Examples are the Commonwealth of Nations (formerly the British Commonwealth), composed of many former British colonies and dominions, and the Organization of American States, both of which offer economic as well as political benefits.

There are many examples of abortive political unions that have foundered precisely because the individual countries could not agree on questions of policy and were unwilling to subordinate individual interests to make the union succeed. The United Arab Republic, the Central African Federation, the Federation of Malaysia and Singapore, and the Federation of the West Indies fall within this category.

Although many such political associations have failed, observers of the world scene speculate about the possibility that "superstates" will emerge from one or more of the international alliances that now exist. Will a "United States of Europe," for example, under a single common government be the logical outcome of the successes of the EEC? No one knows, but as long as the individual state is regarded as the highest form of political and social organization (as it is now) and as the body in which sovereignty rests, such total unification presents difficulties.

Local and Regional Political Organization

The most profound contrasts in cultures tend to occur between, rather than within, states, one reason political geographers traditionally have been primarily interested in country units. The emphasis on the state, however, should not obscure the fact that in most countries the actions of subnational political organizations have a direct impact on people's behavior. In the United States, for example, an individual is subject to the decisions made by the local school board, the municipality, the county, the state, and many other governing bodies. Among other things, local political entities determine where children go to school, the minimum size lot on which a person can build a house, and where one may legally park a car. Adjacent states of the United States may be characterized by sharply differing personal and business tax rates, differing controls on the sale of firearms, alcohol, and tobacco, variant administrative systems for public services, and different levels of expenditures for them (Figure 8.28).

Because all of these governmental entities operate within defined geographic areas, and because they make behavior-governing decisions, they have attracted the attention of political geographers. In the concluding sections of this chapter we examine two aspects of political organization at the local and regional level. Our emphasis will be on the United States and Canada because those are the areas with which we are most familiar.

The Districting Problem

There are more than 85,000 local governmental units in the United States. Slightly more than half of these are municipalities, townships, and counties. The remainder are school districts, water-control districts, airport authorities, sanitary districts, and other special-purpose bodies. Around each of these districts, boundaries have been drawn. Although the number of districts does not change greatly from year to year, many boundary lines are redrawn in any single year.

For example, the ruling of the U.S. Supreme Court in 1954 in *Brown v. Board of Education of Topeka, Kansas*, that the doctrine of "separate but equal" school systems for blacks and whites was unconstitutional, led to the redrawing of thousands of attendance boundaries of school districts. Likewise, the court's "one person, one vote" ruling in *Baker v. Carr* (1962) signified the end of overrepresentation of sparsely populated rural districts in state legislatures and led to the frequent adjustment of electoral districts within states and cities to attain roughly equal numbers of voters. Such *redistricting* or *reapportionment* is made necessary by shifts in population, as areas gain or lose people. It may also be mandated when an electoral district is seen to be *gerrymandered*, that is, structured in such a way as to give one political party an unfair advantage in elections, to fragment voting blocs, or to achieve other nondemocratic objectives. However, the drawing of boundaries is a tremendously complicated exercise in spatial decision making, and it is not always clear what is "unfair."

Figure 8.28
The Four Corners Monument, marking the meeting of Utah, Colorado, Arizona, and New Mexico. Jurisdictional boundaries within countries may be precisely located but usually are not highly visible on the landscape. At the same time, those boundaries may be very significant in citizens' personal affairs and in the conduct of economic activities.

The analysis of how boundaries are drawn around *voting* districts is one aspect of **electoral geography,** which also studies the spatial patterns yielded by election results and their relationship to the socioeconomic characteristics of voters. In a democracy, it might be assumed that election districts should contain roughly equal numbers of voters, that electoral districts should be reasonably compact, and that the proportion of elected representatives should correspond to the share of votes cast by members of a given political party. Problems arise because the way in which the boundary lines are drawn can maximize, minimize, or effectively nullify the power of a group of people.

Assume that X and Y represent two groups with different policy preferences. In Figure 8.29a, the Xs are concentrated in one district and have one representative of four. In Figure 8.29b, the Xs are dispersed, are not a majority in any district, and stand the chance of not electing any representative at all. The power of the Xs is maximized in Figure 8.29c, where they may control two of the four districts. Which of the districting arrangements is "fair" or "unfair" may be taken to the courts to decide.

Figure 8.29 depicts an idealized district, square in shape with a uniform population distribution and only two groups competing for representation. In actuality, American municipal voting districts may be oddly shaped to consider such factors as the city limits, current population distribution, historic settlement patterns, and transportation routes. In any large city, many groups vie for power. A U.S. Justice Department lawsuit filed successfully against the Los Angeles City Council claimed that the council's 1982 district boundary lines, shown in Figure 8.30, violated the Voting Rights Act because they diluted the voting strength of the city's Hispanic residents by dispersing them among several council districts. The council agreed to draft a redistricting plan, to take effect in 1987, although each subsequent census promises other ethnic political conflicts and redistricting demands. Each electoral interest group promotes its version of fairness in the way boundaries are delimited. Minorities seek representation in proportion to their numbers, so that they will be able to elect representatives who are concerned about and responsive to ethnic community issues. But ethnically homogeneous districts are impossible when—as in the case of Los Angeles—there are only 15 council members representing 3 million people and over 80 ethnic groups.

In Canada, the possibilities for gerrymandering in municipal elections have been reduced since the 1960s by assigning responsibility for redrawing constituency boundaries to independent boundary commissions. That task is made unnecessary in the many Canadian cities that do not observe neighborhood or interest-group election districts ("wards"). Many smaller cities have at-large elections, with the municipality as a whole the single constituency. In the United States, at-large election has been

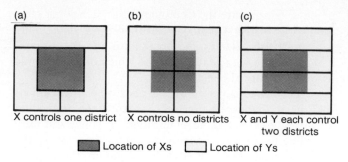

X controls one district   X controls no districts   X and Y each control two districts

Location of Xs   Location of Ys

Figure 8.29

The problem of drawing electoral boundaries. Xs and Ys might represent Republicans and Democrats, urban and rural voters, blacks and whites, or any other distinctive groups.

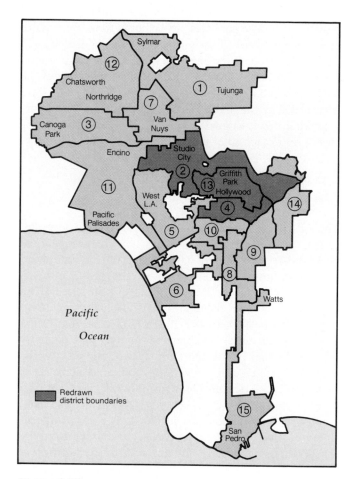

Figure 8.30

City council districts in Los Angeles created by the 1982 redistricting plan had to be redrawn when it was ruled that they violated the Voting Rights Act by splitting the city's Hispanic population. Although Hispanics constituted more than 30% of the city's population, they were a majority and elected a council member in only one district, the 14th. In 1986 the city council agreed to redraw the district boundary lines to create a second predominantly Hispanic council district and a third district in which Hispanics are likely to be a majority in a few years.

successfully challenged as a device to dilute or prevent minority representation. Other Canadian cities have adopted division into "strip" wards that are unrelated to natural or community boundary lines. Some cities, however, including Montreal, Toronto, and Winnipeg, do retain wards based on city blocks to respect distinctive neighborhood units.

Boundary drawing at any electoral level is never easy, particularly when political groups want to maximize their representation and minimize that of opposition groups. Furthermore, the boundaries that we may want for one set of districts may *not* be those that we want for another. For example, sewage districts must take natural drainage features into account, whereas police districts may be based on the distribution of the population or the number of miles of street to be patrolled, and school attendance zones must consider the numbers of school-age children and the capacities of individual schools.

The Fragmentation of Political Power

The United States is subdivided into great numbers of political administrative units whose areas of control are spatially limited. The 50 states are partitioned into more than 3000 counties ("parishes" in Louisiana), most of which are further subdivided into townships, each with a still lower level of governing power. This political fragmentation is further increased by the existence of nearly innumerable special-purpose districts whose boundaries rarely coincide with the standard major and minor civil divisions of the country or even with each other (Figure 8.31). Each district represents a form of political allocation of territory to achieve a specific aim of local need or legislative intent. (See "Too Many Governments.")

Canada, a federation of ten provinces and two territories, has a similar pattern of political subdivision. Each of the provinces contains minor civil divisions—municipalities—under provincial control, and all (cities, towns, villages, and rural municipalities) are governed by elected councils. Ontario and Quebec also have counties that group smaller municipal units for certain purposes. In general, municipalities are responsible for police and fire protection, local jails, roads and hospitals, water supply and sanitation, and schools, duties that are discharged either by elected agencies or appointed commissions.

Most North Americans live in large and small cities. In the United States, these, too, are subdivided, not only into wards or precincts for voting purposes but also into special districts for such functions as fire and police protection, water and electricity supply, education, recreation, and sanitation. These districts almost never coincide with one another, and the larger the urban area, the greater the proliferation of small, special-purpose governing and taxing units. Although no Canadian community has quite the multiplication of governmental entities as, for example, Chicago, Illinois, with well over 1000 special- and general-purpose governments, major Canadian cities may find themselves with complex and growing systems of similar nature. Metropolitan Toronto, for example, has more than 100 authorities that can be classified as "local governments."

The existence of such a great number of districts in metropolitan areas may cause inefficiency in public services and hinder the orderly use of space. *Zoning ordinances,* for example, are determined by each municipality. They are intended to allow the citizens to decide how land is to be used and, thus, are a clear example of the effect of political decisions on the division and development of space. Zoning policies dictate where light and heavy industries may be located; the sites of parks and other recreational areas; the location of business districts; and the types and locations of housing. Unfortunately, in large urban areas, the efforts of one community may be hindered by the practices of neighboring communities. Thus, land zoned for an industrial park in one city may abut land zoned for single-family residences in an adjoining municipality. Each community pursues its own interests, which may not coincide with those of its neighbors or the larger region.

Inefficiency and duplication of effort characterize not just zoning but many of the services provided by local governments. The efforts of one community to avert air and water pollution may be, and often are, counteracted by the rules and practices of other towns in the region. Social as well as physical problems spread beyond city boundaries. Thus, nearby suburban communities are affected when a central city lacks the resources to maintain high-quality schools or to attack social ills. The provision of health care facilities, electricity and water, transportation, and recreational space affects the whole region and, many professionals think, should be under the control of a single unified metropolitan government.

The growth in the number and size of metropolitan areas has increased awareness of the problems of their administrative fragmentation. In response, new approaches to the integration of those areas have been proposed and adopted. The aims of all plans of metropolitan government are to create or preserve coherence in the management of areawide concerns and to assure that both the problems and the benefits of growth are shared without regard to their jurisdictional locations. Two approaches—*unified government* and *predevelopment annexation*—have been followed to secure the efficient administration of extensive urbanized areas.

Unified Government
The modern concept of metropolitan government was developed early in the 20th century in the United States as a means of coping with urban growth. The applied North American model of **unified government**—sometimes called *Unigov* or *Metro*—is found, however, in the Province of

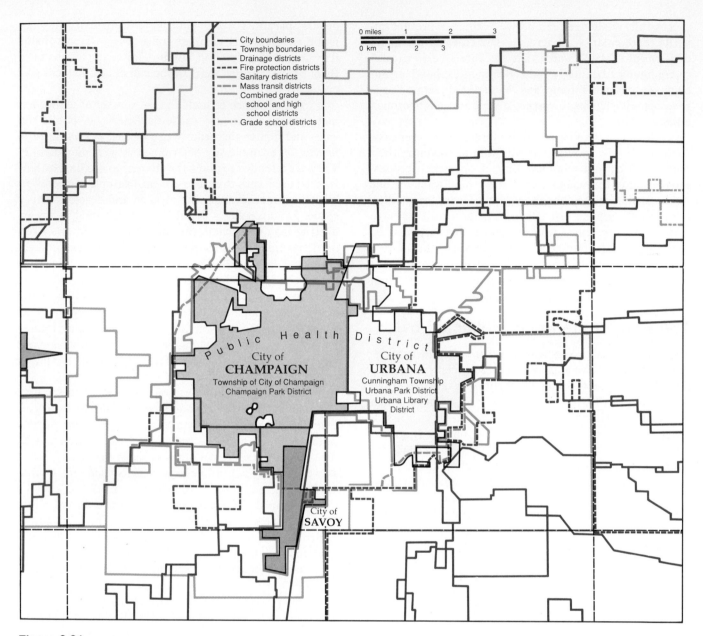

Figure 8.31

Political fragmentation in Champaign County, Illinois. The map shows a few of the independent administrative agencies with separate jurisdictions, responsibilities, and taxing powers in a portion of a single Illinois county. Among the other such agencies forming the fragmented political landscape are Champaign County itself, a forest preserve district, a public health district, and a community college district.

Ontario, Canada, and, particularly, in metropolitan Toronto. The provincial government undertook the reorganization of Ontario's 900-plus cities, towns, and villages into some 30 two-tiered systems of metropolitan government. The upper tier provided broad-scale direction and fiscal control, the lower tier preserved some measure of local autonomy.

The Metropolitan Government of Toronto, established in 1954, was the prototype. The province federated the city of Toronto and its suburbs as a metropolitan corporation and assigned it responsibility for public transportation, water and sewerage, police protection, regional parks, and centralized land use planning. To preserve a sense of identity and to free Metro government from concern with problems best handled locally, the area municipalities retained their separate names and continued to perform such tasks as maintaining local roads and parks, providing fire protection, and collecting garbage. Metropolitanwide school boards, independent of both the metropolitan corporation and area municipalities, are responsible for the administration and financing of the public schools. Metro Toronto served as a model for other Unigov introductions in Canada. The Montreal, Calgary, and Vancouver regions (among others) have Metro governments, though not all are structured identically to that of Toronto.

Too Many Governments

If you are a property owner in Wheeling Township in the city of Arlington Heights in Cook County, Illinois, here's who divvies up your taxes: the city, the county, the township, an elementary school district, a high school district, a junior college district, a fire protection district, a park district, a sanitary district, a forest preserve district, a library district, a tuberculosis sanitarium district, and a mosquito abatement district.

Lest you attribute this to population density or the Byzantine ways of Cook County politics, it's not that much different elsewhere in Illinois—home to more governmental units than any other state in the United States. According to late-1980s figures from the U.S. Bureau of the Census, there were 6627 local government units in Illinois. Second-place Pennsylvania had 4956 governments—1670 fewer than Illinois—and the average for all states was 1663.

Along with its 102 counties, Illinois has nearly 1300 municipalities, more than 1400 townships, and over 1000 school districts; but the biggest factor in the governmental unit total are single-function special districts. These were up from 2600 in 1982 to nearly 2800 in 1987, with no end to their increase in sight. Special districts range from Chicago's Metropolitan Sanitary District to the Caseyville Township Street Lighting District. Most of these governments have property-taxing power. Some also impose sales or utility taxes.

This proliferation is in part a historical by-product of good intentions. The framers of the state's 1870 constitution, wanting to prevent overtaxation, limited the borrowing and taxing power of local governments to 5% of the assessed value of properties in their jurisdictions. When this limit was reached and the need for government services continued to grow with population, voters and officials circumvented the constitutional proscription by creating new taxing bodies—special districts. Illinois' special districts grew because they could be fitted to service users without regard to city or county boundaries.

Critics say all these governments result in duplication of effort, inefficiencies, higher costs, and higher taxes. Supporters of special districts, townships, and small school districts argue that such units fulfill the ideal of a government close and responsive to its constituents.

Reprinted with permission from the Winter, 1985 issue of *Crain's Illinois Business*. Copyright 1985 by Crain Communications Inc.

Unigov has won many adherents and some successes in the United States as well, where a variety of approaches to regionwide administration and planning have been tried. Nashville-Davidson County, Tennessee, and Indianapolis-Marion County, Indiana, are among the areas where city-county consolidation has occurred. A different system operates in Metropolitan Dade County, Florida, which includes Miami and 26 other municipalities. There, local governments have retained some powers (such as police and fire services), but others (land use planning and tax collection) have been transferred to the county. Other Metro government operations are found in Jacksonville, Florida; Portland, Oregon; Seattle, Washington, and elsewhere.

Yet another approach to regionalism is represented by the Metropolitan Council of the Twin Cities area of Minneapolis/St. Paul. Here, Minnesota's two largest cities, some 130 smaller municipalities, seven counties, and great numbers of special purpose agencies were partially subordinated to a metropolitan council responsible for activities that demand centralized, areawide administration (Figure 8.32). These include sewerage, water supply, airport location, highway routes, and the preservation of open space. Local communities kept their own forms of government and the control of fire and police services, schools, street maintenance, and other local functions.

Predevelopment Annexation

The preservation of the tax base and the retention of expansion room for the central city are the driving forces

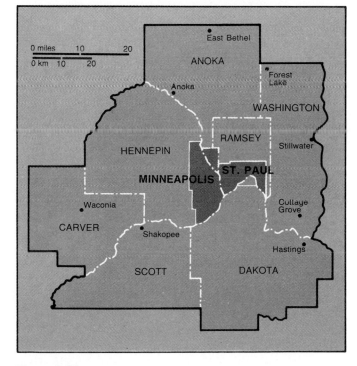

Figure 8.32

The Metropolitan Council of the Twin Cities Area, created in 1967, is responsible for the orderly development of a seven-county area that contains over 300 local units of government, including Minneapolis and St. Paul. The council coordinates such services as sewers and solid-waste disposal and the development of highways and parks. It reviews all land use plans of individual governments and can delay their implementation for 60 days while it attempts to coordinate them with regional goals.

behind the **predevelopment annexations** permitted by the laws of some states, particularly in the West. Cities such as Houston (510 square miles; 1320 km²) and Oklahoma City (650 square miles; 1680 km²) have taken advantage of such permissive regulation to assure that they will not be constricted by a ring of incorporated suburbs that would siphon off growth, population, jobs, and income generated by the presence of the central city itself. However, because all incorporated communities of a specific minimum size possess such annexation rights over contiguous unincorporated areas, rival communities have sometimes initiated unseemly "land grabs," to the detriment of both undeveloped rural areas and the rational control of logically single metropolitan areas.

Summary

The sovereign state is the dominant entity in the political subdivision of the world. It constitutes an expression of cultural separation and identity as pervasive as that inherent in language, religion, or ethnicity. A product of 18th century political philosophy, the idea of the state was diffused globally by colonizing European powers. In most instances, the colonial boundaries they established have been retained as their international boundaries by newly independent countries.

The greatly varying geographic characteristics of states contribute to national strength and stability. Size, shape, and location influence countries' economies and international roles, while national cores and capitals are the heartlands of states. Boundaries, the legal definition of a state's size and shape, determine the limits of its sovereignty. They may or may not reflect preexisting cultural landscapes and in any given case may or may not prove to be viable. Whatever their nature, boundaries are at the root of many international disputes. Maritime boundary claims, particularly as reflected in the Convention on the Law of the Sea, add a new dimension to traditional claims of territorial sovereignty.

State cohesiveness is promoted by a number of centripetal forces. Among these are national symbols, a variety of institutions, and confidence in the aims, organization, and administration of government. Also helping to foster political and economic integration are transportation and communication connections. Destabilizing centrifugal forces, particularly ethnically based separatist movements, threaten the cohesion and stability of many states.

Although the state remains central to the partitioning of the world, a broadening array of political entities affects people individually and collectively. Recent decades have seen a significant increase in the number and variety of global and regional alliances to which states have surrendered some sovereign powers. At the other end of the spectrum, expanding North American urban areas and governmental responsibilities raise questions of fairness in districting procedures and of effectiveness when political power is fragmented. Metropolitan government and predevelopment annexation are two approaches to region-wide administration and planning.

Key Words

antecedent boundary	geometric boundary
boundary definition	geostrategy
boundary delimitation	heartland theory
boundary demarcation	nation
centrifugal force	nationalism
centripetal force	nation-state
compact state	natural boundary
consequent boundary	perforated state
core area	positional dispute
delimitation	predevelopment annexation
demarcation	prorupt state
electoral geography	regionalism
elongated state	resource dispute
enclave	rimland theory
European Economic	state
Community (EEC)	subsequent boundary
exclave	superimposed boundary
exclusive economic zone	territorial dispute
(EEZ)	unified government
fragmented state	United Nations Convention
functional dispute	on the Law of the Sea

For Review

1. What are the differences between a *state*, a *nation*, and a *nation-state?* Why is a colony not a state? How can one account for the rapid increase in the number of states since World War II?

2. What attributes differentiate states from one another? How do a country's size and shape affect its power and stability? Can a piece of land be both an *enclave* and an *exclave?*

3. How did Mackinder and Spykman differ in their assessment of Eurasia as a likely base for world conquest? What post-1945 developments suggest that there might be no enduring correlation between location and national power?

4. How may boundaries be classified? How do they create opportunities for conflict? Describe and give examples of three types of border disputes.

5. How does the *United Nations Convention on the Law of the Sea* define zones of diminishing national control? What are the consequences of the concept of the 200-nm *exclusive economic zone?*

6. Distinguish between *centripetal* and *centrifugal* political forces. What are some of the ways national cohesion and identity are achieved?

7. What characteristics are common to all or most regional autonomist movements? Where are some of these movements active? Why do they tend to be located on the periphery rather than at the national core?

8. What types of international organizations and alliances can you name? What were the purposes of their establishment? What generalizations can you make regarding economic alliances?

9. Why does it matter how boundaries are drawn around electoral districts? Theoretically, is it always possible to delimit boundaries "fairly"? Support your answer.

10. What reasons can you suggest for the great political fragmentation of the United States? What problems stem from such fragmentation? Describe two approaches to insuring the more efficient administration of large urban areas.

Selected References

Archer, J. Clark, and Fred M. Shelley. *American Electoral Mosaics*. Washington, D.C.: Association of American Geographers, 1986.

Bennett, D. Gordon. *Tension Areas of the World*. Delray Beach, Fla.: Park Press, 1982.

Bergman, Edward F. *Modern Political Geography*. Dubuque, Ia.: Wm. C. Brown, 1975.

Boyd, Andrew. *An Atlas of World Affairs*. 7th ed. New York: Methuen, 1983.

Brass, Paul, ed. *Ethnic Groups and the State*. London: Croom Helm, 1985.

Burnett, Alan D., and Peter J. Taylor, eds. *Political Studies from Spatial Perspectives*. Chichester, England: Wiley, 1981.

Chaliand, Gerard, and Jean-Pierre Rageau. *A Strategic Atlas: Comparative Geopolitics of the World's Powers*. 2d ed. New York: Harper and Row, 1985.

Christopher, A.J. "Continuity and Change of African Capitals." *Geographical Review* 75 (1985): 44–57.

Clarke, C., D. Ley, and C. Peach, eds. *Geography and Ethnic Pluralism*. London: Allen & Unwin, 1984.

Gellner, Ernest. *Nations and Nationalism*. Ithaca, N.Y.: Cornell University Press, 1983.

Glassner, M. I., and Harm J. de Blij. *Systematic Political Geography*. 4th ed. New York: Wiley, 1989.

Hartshorne, Richard. "The Functional Approach in Political Geography." *Annals of the Association of American Geographers* 40 (1950): 95–130.

Johnston, Douglas M., and Phillip M. Saunders, eds. *Ocean Boundary Making: Regional Issues and Developments*. London and New York: Croom Helm, 1988.

Johnston, R. J. *Geography and the State: An Essay in Political Geography*. New York: St. Martin's Press, 1982.

Johnston, R. J., and P. J. Taylor, eds. *A World in Crisis?: Geographical Perspectives*. 2d ed. Oxford, England: Basil Blackwell Ltd., 1989.

Keagan, John, and Andrew Wheatcroft. *Zones of Conflict*. London: Jonathan Cape, 1986.

Knight, David B. "Identity and Territory: Geographical Perspectives on Nationalism and Regionalism." *Annals of the Association of American Geographers* 72 (1982): 514–31.

The Law of the Sea. New York: United Nations, 1983.

Mackinder, Halford J. *Democratic Ideals and Reality*. London: Constable, 1919.

———. "The Geographical Pivot of History." *Geographical Journal* 23 (1904): 421–37.

Mikesell, W. Marvin. "The Myth of the Nation State." *Journal of Geography* 82 (1983): 257–60.

Morrill, Richard L. *Political Redistricting and Geographic Theory*. Washington, D.C.: Association of American Geographers, 1981.

Norris, Robert, and Lloyd Haring. *Political Geography*. Columbus, Ohio: Charles E. Merrill, 1980.

O'Loughlin, John. "The Identification and Evaluation of Racial Gerrymandering." *Annals of the Association of American Geographers* 72 (1982): 165–84.

———. "Political Geography: Tilling the Fallow Field." *Progress in Human Geography* 10 (1986): 69–83.

Pacione, Michael. *Progress in Political Geography*. London: Croom Helm, 1985.

Paddison, Ronan. *The Fragmented State: The Political Geography of Power*. New York: St. Martin's Press, 1983.

Prescott, J. R. V. *The Maritime Political Boundaries of the World*. London: Methuen, 1985.

———. *Political Frontiers and Boundaries*. London: Allen & Unwin, 1987.

Rokkan, Stein, and Derek W. Urwin, eds. *The Politics of Territorial Identity: Studies in European Regionalism*. Beverly Hills, Calif.: Sage, 1982.

Sanger, Clyde. *Ordering the Oceans: The Making of the Law of the Sea*. Toronto: University of Toronto Press, 1987.

Smith, Anthony D. *The Ethnic Origins of Nations*. Oxford, England: Basil Blackwell Ltd., 1986.

Spykman, Nicholas J. *The Geography of the Peace*. New York: Harcourt Brace, 1944.

Taylor, Peter J. *Political Geography: World-Economy, Nation-State and Locality*. London and New York: Longman, 1986.

Wheatcroft, Andrew. *The World Atlas of Revolutions*. London: Hamish Hamilton, 1983.

Williams, C. H., ed. *National Separatism*. Cardiff: University of Wales Press, 1982.

C H A P T E R

9

Behavioral Geography

Commuters in Tokyo, Japan.

EARLY in January of 1849 we first thought of migrating to California. It was a period of National hard times . . . and we longed to go to the new El Dorado and "pick up" gold enough with which to return and pay off our debts. Our discontent and restlessness were enhanced by the fact that my health was not good. . . . The physician advised an entire change of climate thus to avoid the intense cold of Iowa, and recommended a sea voyage, but finally approved of our contemplated trip across the plains in a "prairie schooner." Full of the energy and enthusiasm of youth, the prospects of so hazardous an undertaking had no terror for us, indeed, as we had been married but a few months, it appealed to us as a romantic wedding tour.*

So begins Catherine Haun's account of her journey from Clinton, Iowa, to California in 1849, a trip that was to last 9 months and cover 2400 miles (3900 km). The Hauns were just two of the 250,000 people who traveled across the continent on the Overland Trail in one of the world's great migrations. The dangers inherent in such a trip were numerous; thousands were to die en route, and many others stopped or turned back short of their goal. The migrants faced at least 6 months, and often more, of grueling travel over badly marked routes that crossed swollen rivers, deserts, and mountains. The weather was often foul, with hailstorms, drenching rains, and summer temperatures that could exceed 110°F (43°C) inside the covered wagons. Wagon breakdowns were frequent. Graves along the route were a silent testimony to the lives claimed by buffalo stampedes, Indian skirmishes, cholera epidemics, and other disasters.

What inducements were so great as to make emigrants leave behind all that was familiar and risk their lives on an uncertain venture? Catherine Haun's account is unusual in that it gives the reasons for their trip. She alludes to economic hard times; the depression that swept the United States in 1837 inaugurated a prolonged period of bank closures, depressed prices for agricultural goods, and high rates of unemployment. The Hauns hoped to strike it rich by mining gold. Other migrants were attracted by reports of free land and rich soil in the Oregon and California territories, of productive fishing grounds, and of ample furs for trapping. Like other migrants, the Hauns were also attracted by the climate in the West, which was said to be always sunny and free of disease. Finally, like most who undertook the trip along the Overland Trail, the Hauns were young, moved by restlessness and a sense of adventure.

*Source: Catherine Haun, "A Woman's Trip Across the Plains in 1849," as quoted in Lillian Schlissel, *Women's Diaries of the Westward Journey*, Schocken Books, New York, 1982.

Catherine Haun's story is unique only in its particulars. As did her predecessors back to the beginnings of humankind, she and her family acted in space and over space on the basis of acquired information and awareness of opportunity (Figure 9.1). Her story summarizes the content of this chapter, a survey of how individuals make spatial behavioral decisions and how those separate decisions may be summarized by models and generalizations to explain collective actions.

Behavioral geography is a relatively new field of the discipline of geography. In the last 30 years, geographers have become more interested in studying culture from the viewpoint of the individual. In previous chapters in this section we outlined what might be called the general patterns of cultural geography. In this chapter we pose a question basic to the themes that we have been exploring in preceding chapters: What considerations influence how

Figure 9.1

Cross-country movement was slow, arduous, and dangerous early in the 19th century, and the price of long-distance spatial interaction was far higher in time and risks than a comparable journey today.

individual human beings use space and act within it? Implicit in this question are subsidiary analytical concerns: How do individuals view (perceive) their environment? How is information transmitted through space and acted on? How might all the separate decisions of many individuals be summarized so that we may understand the order that underlies the seeming randomness of individual action?

We will begin with a discussion of human environmental perception. This leads to a consideration of how people define the space within which they carry out their activities. Then we turn to the way ideas are transmitted over space. Finally, we bring together all of these concepts in a section devoted to the migration process.

Perception of the Environment

The term **environmental perception** refers to our awareness, as individuals, of home and distant places and the beliefs we have about them. It involves our feelings, reasoned or irrational, about the complex of natural and cultural characteristics of an area. Whether our view accords with that of others or truly reflects the "real" world seen in abstract descriptive terms is not the major concern. Our perceptions are our reality. The decisions people make about the use of their lives are based not necessarily on reality but on their perceptions of reality.

Geographers interested in determining how we arrive at our environmental understanding employ the concept taken from psychology termed **cognition**. Cognition refers to the way individuals give mental meaning to the information they receive. Perceptions of the environment are sent to the brain where they are stored together with our previously accumulated environmental knowledge. The mental or cognitive structures influence the way we process and recall our perceptions. We might perceive the college classroom by physically being there, but our later recollections of the classroom depend on the way we mentally have organized our perceptions.

The fact that people have had different experiences has much to do with the way they perceive their environment. As people order new information, they store it within their own developing mental structures. Thus, it is not unusual for people to interpret similar phenomena in different ways. The more uniform the life experiences of a group of people, the greater the chance that they will have similar perceptions.

In technologically advanced societies, television and radio, magazines and newspapers, books and lectures, travel brochures and hearsay all combine to help develop a mental picture of unfamiliar places. The most effectively transmitted information seems to come from word-of-mouth reports. These may be in the form of letters or conversations with relatives, friends, and associates. Probably the strongest lines of attachment to relatively un-known regions, whether nearby or far away, develop through the information supplied by family members and friends.

Of course, our knowledge of close places is greater than our knowledge of far places. But barriers to information flow give rise to *directional biases*. Not having friends or relatives in one part of a country may represent a barrier to individuals, so that interest in and knowledge of the area beyond the "unknown" region are sketchy. In the United States, both northerners and southerners tend to be less well informed about each other's areas than about the western part of the country. Traditional communication lines in the United States follow an east-west rather than a north-south direction, which is the result of early migration patterns, business connections, and the pattern of the development of principal cities.

Mental Maps

When information about a place is sketchy, blurred pictures develop. These influence the impression we have of places and cannot be discounted. We might say that each individual has a **mental map** of the world. No single person, of course, has a true and complete image of the world; therefore there can be no completely accurate mental map. In fact the best mental map that most individuals have is that of their own residential neighborhood, the place where they spend the most time.

No one can reproduce on paper an exact replica of the mental image that he or she might have of an area. The study of mental maps must by necessity be indirect. If we want to know how particular people envisage their town, we must either ask them questions about the town or ask them to draw sketch maps. Although the result will not be a completely accurate picture of what they have in mind, it will be suggestive of the mental map.

Whenever individuals think about a place or how to get to a place, they produce a mental map. What are believed to be unnecessary details are left out, and only the important elements are incorporated (Figure 9.2). Those elements usually include awareness that the object or the destination does indeed exist, some conception of the distances separating the starting point and the named object(s), and a feeling for the directional relationships between points. A mental route map might also include reference points to be encountered on the chosen path of connection or on alternate lines of travel. Although mental maps are highly personalized, people with similar experiences tend to give similar answers to questions about the environment and to produce roughly comparable sketch maps.

Awareness of places is usually accompanied by opinions about them, but there is no necessary relationship between the depth of knowledge and the perceptions held. In general, the more familiar we are with a locale, the more sound will be the factual basis of our mental

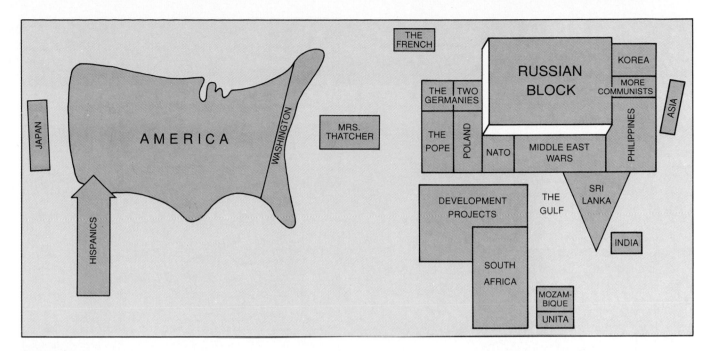

Figure 9.2

A mental map of the world as fashioned from newscasts heard by a Peace Corps volunteer working in the Central African Republic. The map is based on several months of listening to the Africa service of the Voice of America during 1987. Both areas and topics shown were those in the news and selected for emphasis by the broadcasters, whose choices helped form the mental maps and awareness patterns of their listeners. The "Russian Block" took on a three-dimensional reality not intended by the VOA. (Courtesy of Hilary Hope Getis.)

image of it. But individuals form firm impressions of places totally unknown by them personally, and these may color travel or migration decisions.

One way to ascertain how individuals envisage the environment is to ask them what they think of various places. For instance, they may be asked to rate places according to desirability—perhaps residential desirability— or to make a list of the best and worst places in a region such as the United States. Certain regularities appear in such studies. Figure 9.3 presents some residential desirability data as elicited from college students in three provinces of Canada. These and comparable mental maps suggest that near places are preferred to far places unless much information is available about the far places. Places with similar cultural forms are preferred, as are places with high standards of living. Individuals tend to be indifferent to unfamiliar places and to dislike unfamiliar areas that have competing cultural interests (such as disliked political and military activities) or a physical environment known to be unpleasant.

People tend mentally to increase the size of the familiar and to decrease the size of all else. They tend to place their own location in a central position and to increase the size of things nearby. The regional study, "The Yurok World View" (Chapter 13), demonstrates these conclusions in the collective mental map of a primitive society. Figure 9.4 gives two examples of the way people are affected by the familiar. Also, as we grow older our perspectives change, as shown in Figure 9.5.

Perception of Natural Hazards

Mental maps of home areas do not generally include as an overriding concern an acknowledgment of potential natural dangers. An intriguing area of geographical research deals with how people perceive *natural hazards*, defined as processes or events in the physical environment that are not caused by humans but that have consequences harmful to them. Most climatic (hurricanes, tornados, blizzards) and geological (earthquakes, volcanic eruptions) hazards cannot be prevented, and their consequences may be disastrous. The hurricanes that struck the delta area of Bangladesh in 1970 and again in 1985 left dead at least 500,000 people; the 1976 earthquake in the Tangshan area of China devastated a major urban-industrial complex with casualties reported between 700,000 and 1 million. These were major and exceptional natural hazards, but more common occurrences are experienced and are apparently discounted by those affected. Johnstown, Pennsylvania, has suffered recurrent floods, and yet its residents rebuild; violent storms strike the Gulf and East coasts of the United States, and people remain or return.

Why do people choose to settle in high-hazard areas, in spite of the potential threat to their lives and property? Why do hundreds of thousands of people live along the San Andreas fault in California, build houses in Pacific coastal areas known to experience severe erosion during storms (Figure 9.6), or farm in flood-prone areas adjacent

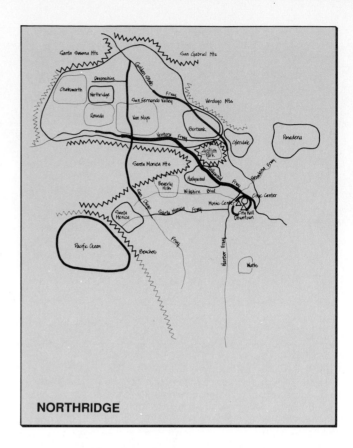

NORTHRIDGE

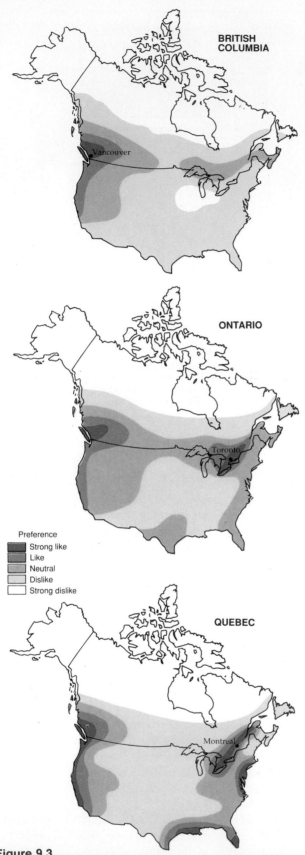

BRITISH COLUMBIA

ONTARIO

Preference

	Strong like
	Like
	Neutral
	Dislike
	Strong dislike

QUEBEC

Figure 9.3
Each of these maps shows the residential preference of a sampled group of Canadians from the provinces of British Columbia, Ontario, and Quebec, respectively. Note that each group of respondents prefers its own area but that all like the Canadian and U.S. west coasts.

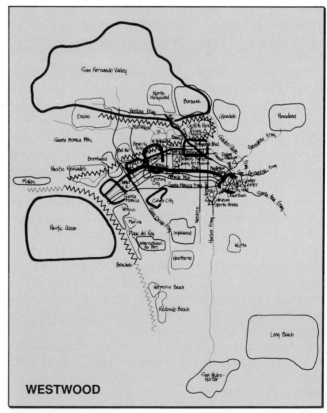

WESTWOOD

Figure 9.4
Four mental maps of Los Angeles. The upper–middle-income residents of Northridge and Westwood have expansive views of the metropolis, reflecting their mobility and area of travel.

BOYLE HEIGHTS

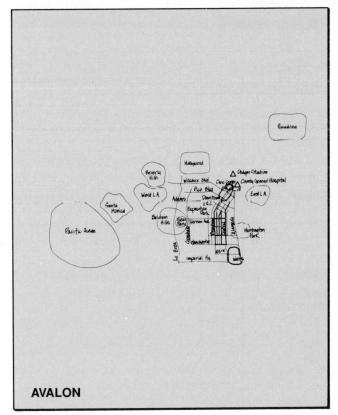

AVALON

Figure 9.4 continued. Residents of Boyle Heights and Avalon, both minority districts, have a much more restricted and incomplete mental image of the city. Their limited mental maps reflect and reinforce their spatial isolation within the metropolitan area.

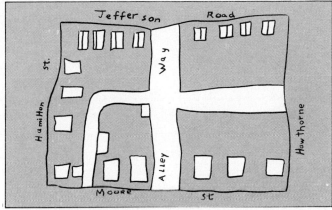

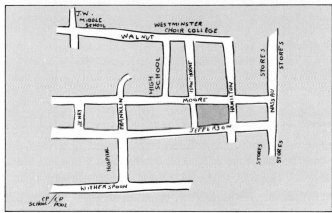

Figure 9.5

Three children, aged 6, 10, and 13, who lived in the same house, were asked to draw maps of their neighborhood. No further instructions were given. Notice how perspectives broaden and neighborhoods expand with age. For the 6-year-old, the neighborhood consisted of the houses on either side of her own. The square block on which she lived was the neighborhood for the 10-year-old. The wider activity space of the 13-year-old is also evident. The square block that the 10-year-old drew is shaded in the 13-year-old's sketch.

Figure 9.6

The aftermath of Hurricane Hugo, which struck the coast of
South Carolina in September 1989.

to the Mississippi River? What is it that makes the risk
worth taking? There are many reasons that hazardous
areas are perceived differently by people and thus that
some people choose to settle in them. Of major importance is the fact that specific hazards are relatively rare
occurrences. Many people think that the likelihood of an
earthquake, flood, or other natural calamity is sufficiently
remote that it is not economically feasible to protect
themselves against it. They are also influenced by the fact
that the scientists who study such hazards may themselves differ on the probability of an event or on the damage
that it may inflict. And, in fact, the prediction of hazards
is not an exact science, being based on the calculation of
the probability of occurrence of what are uncommon
events.

People are also influenced by their past experiences
in high-hazard areas. If they have not suffered much
damage in the past, they may be optimistic about the
future. If, on the other hand, past damage has been great,
they may think that the probability of similar occurrences
in the future is low (Table 9.1). People's memories can be
short. In the years following an earthquake, for example,

a sense of security grows, building codes or their interpretation are relaxed, and population in the area increases.

High-hazard areas are often sought out not because
they pose risks but because they possess desirable topography or scenic views, as do, for instance, the Atlantic and
Pacific coasts. Once people have purchased property in a
known hazard area, they may be unable to sell it for a
reasonable price even if they so desire. They think that
they have no choice but to remain and protect their investment. The cultural hazard—loss of livelihood and investment—appears to be more serious than whatever
natural hazards there may be.

Individual Activity Space

We saw in Chapter 8 that groups and countries draw
boundaries around themselves and divide space into territories that are defended, if necessary. The concept of **territoriality**—the emotional attachment to and the defense
of home ground—has been seen by some as a root explanation of much of human action and response. It is true
that some collective activity appears to be governed by

Table 9.1
Common Responses to the Uncertainty of Natural Hazards

Eliminate the Hazard	
Deny or denigrate its existence	*Deny or denigrate its recurrence*
"We have no floods here, only high water."	"Lightning never strikes twice in the same place."
"It can't happen here."	"It's a freak of nature."

Eliminate the Uncertainty	
Make it determinate and knowable	*Transfer uncertainty to a higher power*
"Seven years of great plenty. . . . After them seven years of famine."	"It's in the hands of God."
"Floods come every five years."	"The government is taking care of it."

territorial defense responses: the conflict between street groups in claiming and protecting their "turf" (and their fear for their lives when venturing beyond it) and the sometimes violent rejection by ethnic urban neighborhoods of an encroaching black, Hispanic, or other population group.

But for most, our personal sense of territoriality is a tempered one. Homes and property are regarded as defensible private domains but are opened to innocent visitors, known or unknown, or to those on private or official business. Nor do we confine our activities so exclusively within controlled home territories as street-gang members do within theirs. Rather, we have a more or less extended home range, an **activity space** within which we move freely on our rounds of regular activity, sharing that space with others who are also about their daily affairs. Figure 9.7 suggests a probable activity space for a suburban family of five for one day. Note that the activity space for each individual for one day is rather limited, even though two members of the family use automobiles. If one week's activity were shown, more paths would have to be added to the map, and in a year's time, several long trips would probably have to be noted. Because long trips are taken irregularly, we will confine our idea of activity space to often-visited places.

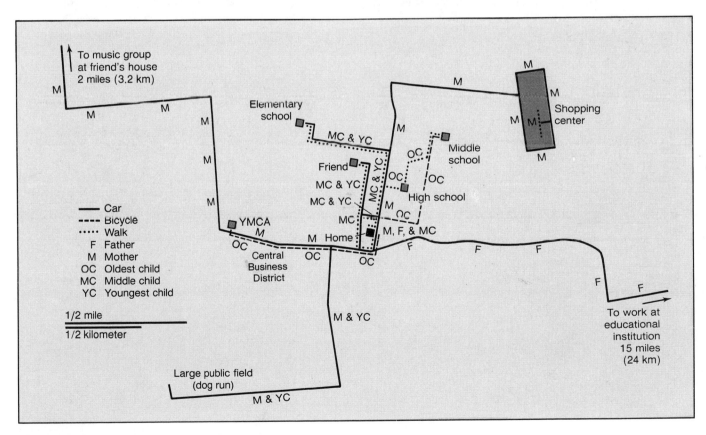

Figure 9.7
Activity space for each member of a family of five for a typical weekday. Routes of regular movement and areas recurrently visited help to foster a sense of territoriality and to color one's perceptions of space.

The kind of activities individuals engage in can be classified according to type of trip: journeys to work, to school, to shops, for recreation, and so on. People in nearly all parts of the world make these same types of journeys, though the spatially variable requirements of culture and economy dictate their frequency, duration, and significance in the time budget of an individual. Figures 9.8, 9.9, and 9.10 illustrate this point. Figure 9.8 depicts variations in travel patterns in two different culture groups in rural midwestern Canada. It suggests that "modern" rural Canadians, who want to take advantage of the variety of goods offered in the regional capital, are willing to travel longer distances than are people of a traditionalist culture who have different tastes in clothing and consumer goods and whose demands are satisfied in local settlements. Also, in this case the traditionalists do not own cars, thus limiting their spatial range. Figures 9.9 and 9.10 suggest the importance of the journey to work among urban populations in two different cities.

The types of trips that individuals make and thus the extent of their activity space depend on at least three variables: their stage in the life cycle, the means of mobility at their command, and the demands or opportunities implicit in their daily activities. The first, *stage in the life cycle*, refers to membership in specific age groups. Stages include preschooler, school-age, young adult, adult, and elderly. Preschoolers stay close to home unless they accompany their parents. School-age children usually travel short distances to lower schools and longer distances to upper-level schools. After-school activities tend to be limited to walking or bicycle trips to nearby locations. High-school students are usually more mobile and take part in more activities than do younger children. Adults responsible for household duties make shopping trips and trips related to child care as well as journeys away from home for social, cultural, or recreational purposes. Wage-earning adults usually travel farther from home than other family members. Elderly people normally do not find it feasible or desirable to have extended activity spaces.

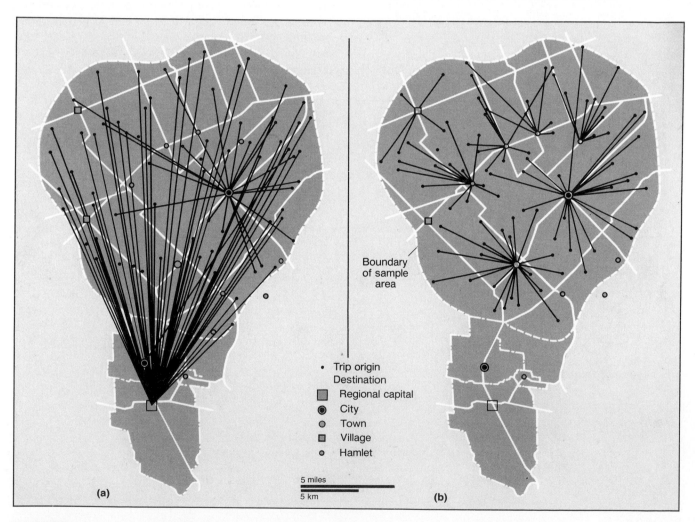

Figure 9.8

Travel patterns for purchase of clothing and yard goods of (a) rural cash-economy Canadians and (b) Canadians of the old-order Mennonite sect. These strikingly different travel behaviors demonstrate the great differences that may exist in the action spaces of different culture groups occupying the same territory.

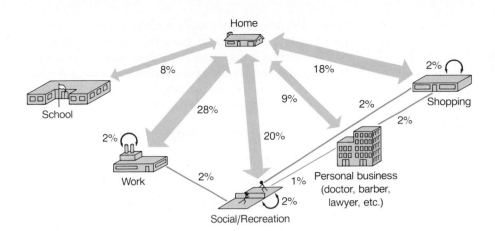

Home

School

8%

28%

2%

Work

2%

2%

Social/Recreation

18%

9%

20%

2%

1%

2%

Shopping

2%

2%

Personal business
(doctor, barber,
lawyer, etc.)

Figure 9.9
Chicago travel patterns. The numbers are the percentages of all urban trips taken in Chicago. The greatest single movement is the journey to and from work. Over 96% of all trips are represented on the diagram.

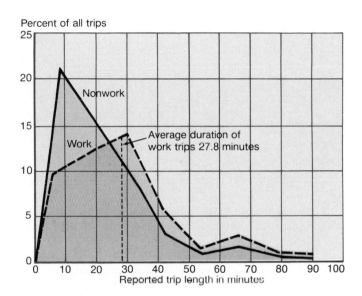

Percent of all trips

Nonwork

Work

Average duration of work trips 27.8 minutes

Reported trip length in minutes

Figure 9.10
The frequency distribution of work and nonwork trip lengths in minutes in Toronto. Work trips are usually longer than other internal city journeys. For North America, 20 minutes seems to be the upper limit of commuting time before people begin to consider time seriously in their choice of homesite. Most nonwork trips are very short.

The second variable that affects the extent of activity space is *mobility*, or the ability to travel. An informal consideration of the cost and effort required to overcome the friction of distance is implicit. Where incomes are high, automobiles are available, and the cost of fuel is a minor item in the family budget, mobility may be great and individual action space large. In societies where cars are not a standard means of personal conveyance, the daily activity space may be limited to the shorter range afforded by the bicycle or by walking. Obviously, both intensity of purpose and the condition of the roadway affect the execution of movement decisions.

The mobility of individuals in countries or in sections of countries with high incomes is relatively great; their activity space horizons are broad. These horizons, however, are not limitless. There are a fixed number of

hours in a day, most of them consumed in performing work, preparing and eating food, and sleeping. In addition, there are a fixed number of road, rail, and air routes, so that even the most mobile individuals are constrained in the amount of activity space that they can use. No one can easily claim the world as his or her activity space.

A third factor limiting activity space is the individual assessment of the availability of possible activities or *opportunities*. In the subsistence economies discussed in Chapter 10, the needs of daily life are satisfied at home, so the impetus for journeys away from the residence is minimal. If there are no stores, schools, factories, or roads, one's expectations and opportunities are limited, and one's activity space is therefore reduced. In impoverished nations or neighborhoods, low incomes limit the inducements, opportunities, destinations, and necessity of travel.

Distance and Spatial Interaction

People make many more short-distance trips than long ones. In the words of the behavioral geographer, there is greater human interaction over short distances than long distances. If we drew a boundary line around our activity space, it would be evident that trips to the boundary are taken much less often than short-distance trips near the home. Think of activity space as more intensively used (greater spatial interaction) near one's home place or base and as declining in use with increasing distance from the base. This is the principle of **distance decay** (introduced briefly in Chapter 1), the exponential decline of an activity, function, or amount of interaction with increasing distance from the point of origin. The tendency is for the frequency of trips to fall off very rapidly beyond an individual's **critical distance**. Figure 9.11 illustrates this principle with regard to journeys from the homesite.

The critical distance is the distance beyond which cost, effort, and perception play an overriding role in our willingness to travel. A small child, for example, will make many trips up and down its block but is inhibited by parental admonitions from crossing the street. Different but

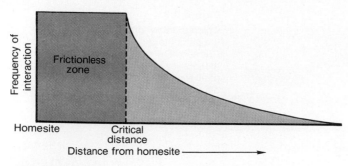

Frequency of interaction

Frictionless zone

Homesite

Critical distance

Distance from homesite ——————→

Figure 9.11

This general diagram indicates how distance is observed by most people. For each activity, there is a distance beyond which the intensity of contact declines. This is called the *critical distance*, if distance alone is being considered, or the *critical isochrone*, if time is the measuring rod. For the distance up to the critical distance, a frictionless zone is identified in which time or distance considerations do not effectively figure in the trip decision.

equally effective constraints control adult behavior. Daily or weekly shopping may be within the critical distance of an individual, and little thought may be given to the cost or the effort involved, but shopping for special goods is relegated to infrequent trips, and cost and effort are considered.

Effort may be measured in terms of *time-distance*, that is, the time required to complete the trip. For the journey to work, often time rather than cost plays the major role in the determination of the critical distance. When there are significant differences between our cognition of distance and real distance, we use the term *psychological distance* to describe our perception of distance. A number of studies show that people tend to psychologically consider known places as nearer than they really are and little-known places as farther than true distance.

Spatial Interaction and the Accumulation of Information

The critical distance is different for each person. The variables of life-cycle stage, mobility, and opportunity, together with an individual's interests and demands, help to define how much and how far a person will travel. On the basis of these variables, we can make inferences about the likelihood that a person will gain more or less information about his or her activity space and the space beyond.

We gain information about the world from many sources. Although information obtained from radio, television, and newspapers is important to us, face-to-face contact is assumed to be the most effective means of communication. If we combine the ideas of activity space and distance decay, we see that as the distance away from the home place increases, the number of possible face-to-face contacts usually decreases. We expect more human interactions at short distances than at long distances. Where

population densities are high, such as in cities (particularly central business districts during business hours), the spatial interaction between individuals can be at a very high level, which is one reason that these centers of commerce and entertainment are often also centers for the development of new ideas.

Barriers to Interaction

Recent changes in technology permit us to travel farther than ever before, with greater safety and speed, and to communicate without physical contact more easily and completely than previously possible. This intensification of contact has resulted in an acceleration of innovation and in rapid spread of goods and ideas. Several millennia ago, innovations such as smelting of metals took hundreds of years to spread. Today, worldwide diffusion may be almost instantaneous.

The fact that the possible number of interactions is high, however, does not necessarily mean that the effective occurrence of interactions will be high. That is, there are a number of barriers to interaction. Such barriers are any conditions that hinder either the flow of information or the movement of people, and thus retard or prevent the acceptance of an innovation.

Distance itself is a barrier to interaction. Generally, the farther two areas are from one another, the less likely is interaction. The concept of *distance decay* says that, all else being equal, the amount of interaction decreases as the distance between two areas increases.

Cost represents another barrier to interaction. Relatives, friends, and associates living long distances apart may find it difficult to afford to interact. The frequency and time allocated to telephone communication, a relatively inexpensive form of interaction, is very much a function of the rate structure—which, of course, favors short-distance interaction.

Interregional contact can also be hindered by the physical environment and by the cultural barriers of differing religions, languages, ideologies, and political systems. Mountains and deserts, oceans and rivers can, and have, acted as physical barriers slowing or preventing diffusion. Cultural barriers may be equally impenetrable. For at least 1500 years, for example, most American Indians in California were in contact with cultures utilizing both maize and pottery, yet they failed to accept either innovation. Should such nonadopters intervene between hearths and receptive cultures, the spread of an innovation can be delayed. It can also be delayed when cultural contact is overtly impeded by governments that interfere with radio reception, control the flow of foreign literature, and discourage contact between their citizens and foreign nationals.

Even when some members of a given population are receptive to change, there will be others who will not innovate, that is, who will not accept new ideas. In some societies—particularly traditional, dispersed agricultural ones—the number of nonadopters for any innovation may be very high indeed. In other societies, where innovation and change are given high status, and where the innovations are not risky, ideas spread much more rapidly.

Finally, in crowded areas, people commonly set psychological barriers around themselves so that only a limited number of interactions take place. The barriers are raised in defense against information overload and for psychological well-being. We must have a sense of privacy in order to filter out the information that does not directly concern us. As a result, individuals tend to reduce their interests to a narrow range when they find themselves in crowded situations, allowing their wider interests to be satisfied by use of the communications media.

Spatial Interaction and Innovation

The probability that new ideas will be generated out of old ideas is a function of the number of available old ideas in contact with one another. People who specialize in a particular field of interest seek out others with whom they wish to interact. Crowded central cities are characteristically composed of specialists in very narrow fields of interest. Consequently, under short-distance, high-density circumstances, the old ideas are given a hearing and new ideas are generated by the interaction. New inventions and new social movements usually arise in circumstances of high spatial interaction. An exception, of course, is the case of intensely traditional societies—Japan in the 17th and 18th centuries, for example—where the culture rejects innovation and clings steadfastly to customary ideas and methods.

The culture hearths (discussed in Chapter 7) of an earlier day were the most densely settled, high-interaction centers of the world. At the present time, the great national and regional capital cities attract people who want or need to interact with others in special-interest fields. The association of population concentrations and the expression of human ingenuity has long been noted. The home addresses recorded for patent applicants by the U.S. Patent Office over the last century indicate that the inventors were typically residents of major urban centers, presumably closely in contact and exchanging ideas with those in shared fields of interest. It still appears that the metropolitan centers of the world attract those who are young and ambitious, and that face-to-face or word-of-mouth contact is important in the creation of new ideas and products. The recent revolution in communications that now allows for inexpensive interaction by a variety of telephone services and computer equipment have suggested to some that the traditional importance of cities as collectors of creative talent may decline in the future.

Diffusion and Innovation

As we noted in Chapter 7, **diffusion** is the process by which a concept, practice, or substance spreads from its point of origin to new territories. This concept will be discussed in more detail here since it is at the conceptual heart of behavioral geography. When such aspects of a culture as ideas, behaviors, or substances are transferred from one place to another through the migration of people, the process is called *relocation diffusion*. It is discussed later in this chapter under the heading "Migration."

Ideas generated in a center of activity will remain there unless some process is available for their spread. *Innovations*, the changes to a culture that result from the adoption of new ideas, spread in various ways. Some new inventions are so obviously advantageous that they are quickly put to use by those who can afford and profit from them. A new development in petroleum extraction may promise such material reward as to lead to its quick adoption by all of the major petroleum companies, irrespective of their distance from the point of introduction. The new strains of wheat and rice that were part of the Green Revolution were quickly made known to agronomists in all cereal-producing countries, but they were more slowly taken up in poor countries, which could benefit most from them, partly because the farmers had difficulty paying for them.

Many innovations are of little consequence by themselves, but sometimes the widespread adoption of seemingly inconsequential innovations brings about large changes when viewed over a period of time. A new musical tune, "adopted" by a few people, may lead many individuals to fancy that tune plus others of a similar sound, which in turn may have a bearing on dance routines, which in turn may bear on clothes selection, which in turn may affect retailers' advertising campaigns and consumers' patterns of expenditure. Eventually a new cultural form will be identified that may have an important impact on the thinking processes of the adopters and on those who come into contact with the adopters. Notice that a broad definition of innovation is used, but notice also that what is important is whether or not innovations are adopted.

In spatial terms, we may identify a number of processes for the diffusion of innovations. Each is based on the manner in which innovations spread from person to person and, therefore, from place to place. These are discussed below.

Contagious Diffusion

Let us suppose that a scientist develops a gasoline additive that noticeably improves the performance of his or her car. Suppose further that the person shows friends and associates the invention and that they in turn tell others. This process is similar to the spread of a contagious disease. The innovation will continue to diffuse until barriers are met (that is, people not interested in adopting the new idea) or until the area is saturated (that is, all available people have adopted the innovation). This **contagious diffusion** process follows the rules of distance decay at each step. Short-distance contacts are more likely than long-distance contacts, but over time the idea may have spread far from the original site. Figure 9.12 illustrates how a theoretical contagious diffusion process differs from relocation diffusion.

A number of characteristics of this kind of diffusion are worth noting. If the idea has merit in the eyes of potential adopters, the number of contacts of adopters with potential adopters will compound. Consequently, the innovation will spread slowly at first and then more and more rapidly, until there is saturation or a barrier is reached. The incidence of adoption is represented by the S-shaped curve in Figure 9.13 and shown for an actual case in Figure 9.14. In a similar manner, the area in which the adopters are located will at first be small, and then will enlarge at a faster and faster rate. The spreading process will slow as the available areas and/or people decrease.

If an inventor's idea fell into the hands of a commercial distributor, the diffusion process might follow a somewhat different course than that discussed above. The distributor might "force" the idea into the minds of individuals by using the mass media. If the media were local

in impact, such as newspapers, then the pattern of adoptions would be similar to that described earlier (Figure 9.15). If, however, a nationwide television, newspaper, or magazine advertising campaign were undertaken, the innovation would become known in numbers roughly corresponding to the population density. Where more people live, there would, of course, be more potential adopters. Economic or other barriers may also affect the diffusion. One immediately sees, however, why large TV markets are so valuable and why national advertising is so expensive.

This type of contagious diffusion process may act together with the distance-decay process. Many of those who accept the innovation after learning of it in the mass media will tell others, so that a locally contagious effect will begin to take over soon after the original contact is made. Each

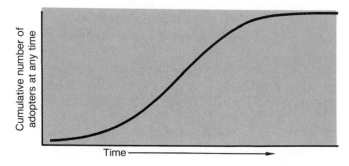

Figure 9.13

The diffusion of innovations over time. The number of adopters of an innovation rises at an increasing rate until the point at which about one-half of the total who ultimately decide to adopt the innovation have made the decision. At this point, the number of adopters increases at a decreasing rate.

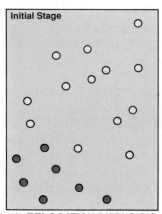
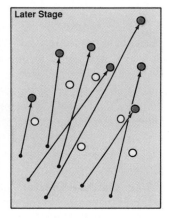

(a) RELOCATION DIFFUSION

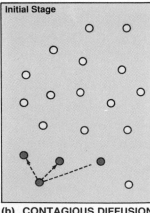
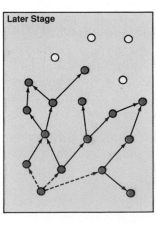

(b) CONTAGIOUS DIFFUSION

Figure 9.12

Patterns of diffusion. (a) In *relocation diffusion*, innovations or ideas are transported to new areas by carriers who permanently leave the home locale. The process is spatially selective. (b) In *contagious diffusion*, a phenomenon spreads from one place to neighboring locations, but in the process it remains and is often intensified in the place of origin.

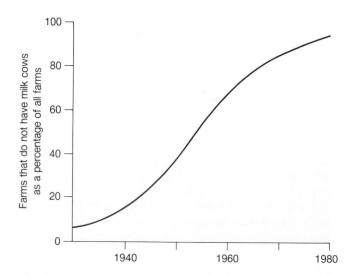

Figure 9.14

Farmers in Illinois found it disadvantageous to herd dairy cattle during the period 1930 to 1978. The idea of discontinuing having milk cows was adopted slowly but increased rapidly from 1940 to 1960. Compare this real example with the theoretical diagram shown in Figure 9.13.

type of medium has its own level of effectiveness. Advertisers have found that they must repeat messages time and again before they are accepted as important information. This fact says something about the effectiveness of the mass media as opposed to, say, face-to-face contact.

Hierarchical Diffusion

A second way innovations are spread combines some aspects of contagious diffusion with a new element: the hierarchy. A *hierarchy* is a classification of objects into categories so that each category is increasingly complex or has increasingly higher status. Hierarchies are found in many systems of organization, such as government offices (the organization chart), universities (instructors, professors, deans, president), and cities (villages, towns, regional centers, metropolises). By **hierarchical diffusion** we mean the spread of innovation up or down a hierarchy of places.

As an example, let us suppose that a new way of processing passengers is adopted at an airport in a major city. Information on the innovation is spread, but only officials at comparably-sized airports are in a position to accept the idea at first. It may be that the quality of information diffused to the larger airports is better or that the larger airports are more financially able to adopt the idea than smaller airports. Eventually, the innovation is adopted at the airports of somewhat smaller cities, and so on down the hierarchy as it becomes better known or as it becomes financially feasible. A hypothetical scheme showing how a four-level hierarchy might be connected in the flow of information is presented in Figure 9.16. Note that the lowest-level centers are connected to higher-level centers but not to each other. Note, too, that connections might bypass intermediate levels and link only with the highest-level center.

Many times, hierarchical diffusion takes place simultaneously with contagious diffusion. One might expect variations when the density of high-level centers is great and when the distances between centers are short. A quick and inexpensive way to spread an idea is to communicate information about it at high-order hierarchical levels. Then the three types of diffusion processes may be used most effectively; even while an idea is diffusing through a high level in the hierarchy, it is also spreading outward from the high-level centers. Consequently, low-level centers that are a short distance from high-level centers may be apprised of the innovation before more distant medium-level centers. People living in suburbs and small towns near a large city are privy to much that is new in the large city, as are individuals in other large cities half a continent away. Figures 9.17 and 9.18 show these patterns for a case taken from Japan.

These forms of diffusion operate in the spread of culture. The consequences are the spatial interaction and innovation discussed earlier in this chapter. We should also recall from Chapter 7 that migration, invasions, selective cultural adoptions, and cultural transference aid the diffusion of innovation. These broader movements and exchanges represent interactions of people beyond their usual activity spaces.

Migration

An important aspect of human history has been the migration of peoples, the evolution of their separate cultures, and the relocation diffusion of those cultures. Portions of that story have been touched on in Chapter 6. The settlement of North America, Australia, and New Zealand involved great long-distance movements of peoples. The flight of refugees from past and recent wars, the settlement of Jews in Israel, the movement south in Africa from the drought-stricken Sahel, the current migration of workers to the United States from overpopulated Mexico,

Figure 9.15
A street scene in Guatemala City. In modern society advertising is a potent force for diffusion. Advertisements over radio and television, in newspapers and magazines, and on billboards and signs communicate information about many different products and innovations.

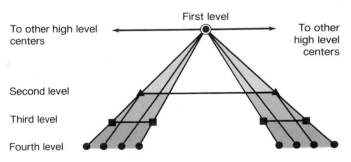

Figure 9.16
A four-level communication hierarchy.

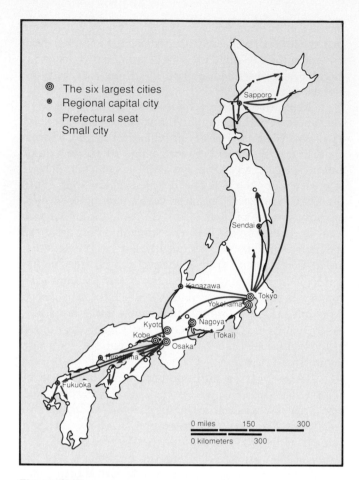

Figure 9.17
Rotary Clubs, members of the international service association, were established in the large cities of Japan during the 1920s. New clubs were established under the sponsorship of the original clubs. This map shows the pattern of diffusion. Note the regional and spatial effects of the diffusion process.

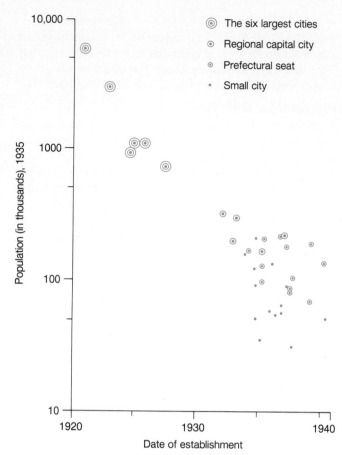

Figure 9.18
This diagram shows the hierarchical diffusion of Rotary Clubs in Japan. The largest cities were the first centers of Rotary Club activity, followed by cities at lower and lower levels of urban population and city function.

and innumerable other examples of mass movement come quickly to mind. In all cases, societies transplanted their cultures to the new areas, their cultures therefore diffused and intermixed, and history was altered. The vast numbers who suffered the hardships of movement to the American West, like Catherine Haun, were the predecessors of the regionally relocating job seekers and retirees of today.

The *planned two-way trip* is a long-distance movement that is not considered to be a migration. It includes the exciting, perhaps historically decisive, daring journey of exploration about which we read in history, or the more common contemporary vacation, business, or social trip in which modern societies indulge so freely. The latter sort of trip, of course, enhances the mental maps and enlarges the awareness space of the participants and may be the prelude to migration. It may also contribute to the diffusion of information through cultural contact, but its individual impact is small because it is so transitory.

Much more important is **migration**: a relocation of both residential environment and activity space. Naturally, the length of the move and its degree of disruption of normal household activities raise distinctions important in the study of migration. A change of residence from the central city to the suburbs certainly changes the activity space of schoolchildren and adults in many of their nonworking activities, but the workers may still retain the city—indeed, the same place of employment there—as an action space. On the other hand, immigration from Europe to the United States and the massive farm-to-city movements of rural Americans around the turn of the century meant a total change of all aspects of behavioral patterns.

The Decision to Migrate

The decision to move is a cultural and temporal variable. Nomads fleeing the famine and the spreading deserts in the Sahel obviously are motivated by different considerations from those of the executive receiving a job transfer to Chicago, the resident of Appalachia seeking factory employment in the city, or the retired couple searching for sun and sand. In general, people who decide to migrate are seeking better economic, political, or cultural conditions or certain amenities. Of course, as in the case of the Hauns, the reasons for migration are frequently a combination of several of these.

Economic causes have impelled more migrations than any other single category. If migrants face unsatisfactory conditions at home (unemployment or famine) and believe that the economic opportunities are better elsewhere, they will be attracted to the thought of a move. To a nomadic herdsman, better opportunities might be abundant grass and water; to a western emigrant, rich soil; to a modern worker, the promise of a high-paying job in a Los Angeles suburb. Attractions such as these are called **pull factors**, and those that contribute to dissatisfaction at home are called **push factors**. Very often migration is a result of both push and pull factors (Figure 9.19).

The desire to escape war and persecution at home and to pursue the promise of freedom in a new location is a political incentive for migration (Figure 9.20). Americans are familiar with the history of settlers who emigrated to North America seeking religious and political freedom. In more recent times, the United States has received hundreds of thousands of refugees from countries such as Hungary (following the uprising of 1956), Cuba

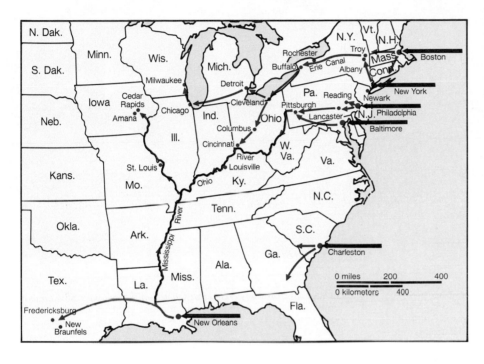

Figure 9.19
The major paths of the early migration of Germans to America constitute a relocation diffusion process. Most emigrants left Germany because of religious and political persecution. They chose the United States not only because immigrants were made welcome, but also because labor was in demand and farmland was available. The first immigrants landed, and many settled, in Boston, New York, Philadelphia, Baltimore, and Charleston. As is characteristic of a relocation diffusion process, the migrants carried with them such aspects of their culture as religion, language, and food preferences.

Documenting Diffusion

The places of origin of many ideas, items, and technologies important in contemporary cultures are only dimly known or supposed, and their routes of diffusion are at best speculative. Gunpowder, printing, and spaghetti are presumed to be the products of Chinese inventiveness; the lateen sail has been traced to the Near Eastern culture world. The moldboard plow is ascribed to 6th century Slavs of northeastern Europe. The sequence and routes of the diffusion of these innovations have not been documented. In other cases, such documentation exists, and the process of diffusion is open to analysis. For example, hybrid corn was originally adopted by imaginative farmers of northern Illinois and eastern Iowa in the mid-1930s. By the late 1930s and early 1940s the new seeds were being planted as far east as Ohio and north to Minnesota, Wisconsin, and northern Michigan. By the late 1940s, all commercial corn-growing districts of the United States and southern Canada were cultivating hybrid varieties.

Also clearly marked is the diffusion path of the custom of smoking tobacco, a practice that originated with Amerindians. Sir Walter Raleigh's Virginia colonists, returning home in 1586, introduced smoking in English court circles, and the habit very quickly spread among the general populace. England became the source region of the new custom for northern Europe; smoking was introduced to Holland by English medical students in 1590. Dutch and English together spread the habit to the Baltic and Scandinavian areas and through Germany to Russia. The innovation continued its eastward diffusion, and within a hundred years, tobacco had spread across Siberia and was, in the 1740s, reintroduced to the American continent at Alaska by Russian fur traders. A second route of diffusion for tobacco smoking can be traced from Spain through the Mediterranean area into Africa, the Near East, and Southeast Asia.

Figure 9.20

Some of the approximately 350,000 Vietnamese "boat people" who fled their country after 1975 to seek a new life in a new location. The United Nations High Commissioner for Refugees in 1988 estimated that worldwide there were over 13 million refugees, people who had left their country because of the well-founded fear of being persecuted for their political views, economic activities, or membership in certain social or religious groups.

(after its takeover by Fidel Castro), and Vietnam (after the fall of South Vietnam). The massive movements of Hindus and Muslims across the Indian subcontinent in 1947, when Pakistan and India were established as governing entities, and the exodus of Jews fleeing persecution in Nazi Germany in the 1930s are other examples of politically inspired moves. Figure 9.21 identifies recent major migrations.

Migration normally involves a hierarchy of decisions. Once people have decided to move and have selected a general destination (e.g., America or the West), they must still choose a particular site at which to settle. At this scale, cultural variables can be important pull factors. Migrants tend to be attracted to areas where the language, the religion, and the racial or ethnic background of the inhabitants is similar to their own. This similarity can help the migrants feel at home when they arrive at their destination and may make it easier to find a job and become assimilated into the new culture. The Chinatowns and Little Italys of large cities attest to the drawing power of cultural factors, as we saw in Chapter 7.

Another set of inducements is grouped under the heading *amenities*, the particularly attractive or agreeable features characteristic of a place. Amenities may be natural (mountains, oceans, climate, and the like) or cultural (the arts and music opportunities available in large cities). They are particularly important to relatively affluent people seeking "the good life." They help account for the attractiveness of the U.S. Sunbelt states for retirees; a similar movement to the south coast has also been observed in countries such as the United Kingdom and France.

The significance of the various factors also varies according to the age, sex, education, and economic status of the migrants. For the modern American, reasons to migrate have been summarized into a limited number of categories that are not mutually exclusive. They include:

1. changes in life cycle (e.g., getting married, having children, getting a divorce, or needing less dwelling space when the children leave home)

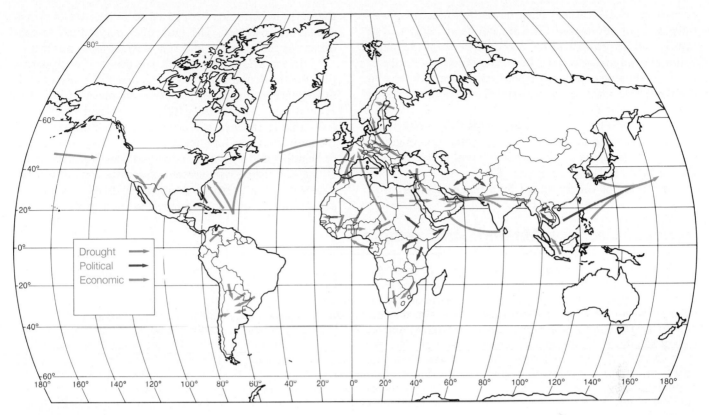

Figure 9.21

Major recent international migration streams. Most of those shown are economic, that is, people moving to the location of available jobs. In the case of the Vietnamese, Afghan, and Sahel migrations, however, politics or drought is the fundamental cause.

2. changes in the career cycle (e.g., leaving college, getting a first job or a promotion, receiving a career transfer, or retiring)
3. forced migrations associated with urban development, construction projects, and the like
4. neighborhood changes from which there is flight, perhaps pressures from new and unwelcome ethnic groups, building deterioration, street gangs, and similar rejected alterations in activity space
5. changes of residence associated with individual personality

Some people simply seem to move often for no easily discernible reason, whereas others, *stayers*, settle into a community permanently. Of course, for a country such as the Soviet Union, with its limitations on emigration, a tightly controlled system of internal passports, restrictions on job changes, and severe housing shortages, a totally different set of summary migration factors would be present.

The factors that contribute to mobility tend to change over time. There is a group, however, that in most societies has always been the most mobile: young adults (Figure 9.22). They are the members of society who are launching careers and making their first decisions about

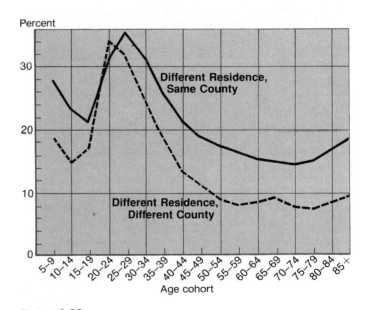

Figure 9.22

Percentage of 1970 population more than 5 years of age with different residence in 1965. Young adults figured most prominently in both short- and long-distance moves in the United States between 1965 and 1970. Movement data through the middle 1980s recorded the same age-related pattern of mobility.

occupation and location. They have the fewest responsibilities of all adults and thus are not as strongly tied to family and institutions as older people. Most of the major voluntary migrations have been composed primarily of young people who suffered from lack of opportunities in the home area and who were easily able to take advantage of opportunities elsewhere.

The concept of **place utility** helps us to understand the decision-making process that potential migrants undergo. It is the value that an individual places on each known, potential migration site. The decision to migrate is a reflection of the appraisal by a potential migrant of the current homesite as opposed to other sites of which something is known. The individual may adjust to conditions at the homesite and thus decide not to migrate.

In the evaluation of the place utility of each potential site, the decision maker considers not only present relative place utility but also expected place utility. The evaluations are matched with the individual's aspiration level, that is, the level of accomplishment or ambition that the individual sees for himself or herself. Aspirations tend to be adjusted to what an individual considers attainable. If one is satisfied with present circumstances, search behavior is not initiated. If, on the other hand, there is dissatisfaction, a utility is assigned to each of the possible

Rational Decision Making: An Example

An example of a structure developed for making a rational decision whether or not to migrate can be found in the case of the selection by a hypothetical unmarried male who is a recent college graduate. This person partitioned his consideration into four major aspects: monetary compensation, geographic location, travel requirements, and the nature of the work. Each of these aspects was subdivided, and some were subdivided again, as illustrated in the table.

Each category was given a value on a scale from 0 to 1.0, according to the importance that the potential migrant placed on it. The restriction that the total of the values must add up to

1.0 enabled the graduate to multiply across any path to find the true relative worth of the 15 categories. In the table, we see that the graduate placed the greatest emphasis on starting salary and the kind of work to be done at the outset. Retirement and insurance benefits were least important.

Once the graduate had a clearer idea of his values, he could take each of his job opportunities and evaluate them according to the best information available about each subcategory. For each possible job, he assigned a value to every category according to the system 90–100 is excellent, 80–89 is good, 70–79 is fair, and so on. By multiplying each of these by the relative

worth of each category and summing for each column, the graduate was able to make an objective decision. For example, for job opportunity No. 1, 95 x 0.040 = 3.80; 80 x 0.059 = 4.72; and so on. For job opportunity No. 1, the score was 71.8; for No. 2, the value was 71.2; and for No. 3, it was 77.3.

This decision was based on the subject's own value system. Each person's categories and weighting systems would be different. In the case illustrated, the graduate finally ranked the local opportunity ahead of the national and regional opportunities. The ranking led to the decision not to migrate at the present time.

			Relative worth placed on each category	Ratings of job opportunities		
				1 Large national company	2 Regional office	3 Local job
Nature of work	Current and future work features	Management training program	0.040	× 95 = 3.80	80 = 3.20	50 = 2.00
		Variety of work	0.059	× 80 = 4.72	75 = 4.43	70 = 4.13
		Technical challenge	0.049	× 80 = 3.92	80 = 3.92	95 = 4.66
	Immediate work features		0.132	× 70 = 9.24	80 = 10.56	90 = 11.88
Travel requirements	Long trips away from office	Trip lengths	0.082	× 90 = 7.38	70 = 5.74	60 = 4.92
		Proportion time away	0.054	× 80 = 4.32	70 = 3.78	60 = 3.24
	Daily commuting characteristics		0.034	× 50 = 1.70	70 = 2.38	100 = 3.40
Geographic location		Climate	0.034	× 80 = 2.72	70 = 2.38	60 = 2.04
		Proximity to leisure-time activities	0.068	× 90 = 6.12	80 = 5.44	60 = 4.08
		Proximity to relatives	0.068	× 60 = 4.08	70 = 4.76	100 = 6.80
Monetary compensation	Future salary prospects	10-year increase	0.035	× 90 = 3.15	80 = 2.80	80 = 2.80
		3-year increase	0.064	× 60 = 3.84	70 = 4.48	80 = 5.12
	Immediate prospects	Fringe benefits	Retirement	0.009	× 95 = 0.86	70 = 0.63	60 = 0.54
			Insurance	0.014	× 95 = 1.33	70 = 0.98	60 = 0.84
		Starting salary	0.209	× 70 = 14.63	75 = 15.68	100 = 20.90
			1.000	71.81	71.16	77.35

new sites. The utility is based on past or expected future rewards at the various sites. Because the new places are unfamiliar to the individual, the information received about them acts as a substitute for the personal experience of the homesite. The decision maker can do no more than sample information about the new sites, and, of course, there may be sampling errors.

One goal of the potential migrant is to minimize uncertainty. Most decision makers either elect not to migrate or postpone the decision unless uncertainty can be lowered sufficiently. Most migrants reduce uncertainty by imitating the successful procedures followed by others. We see that the decision to migrate is not a perfunctory, spur-of-the-moment reaction to information. It is usually a long, drawn-out process based on a great deal of sifting and evaluation of data.

An example of some of these observations and generalizations in action can be seen in the case of the large numbers of young Mexicans who have migrated both legally and illegally to the United States over the last 20 years. Faced with rural poverty and overpopulation at home, they regard the place utility there as minimal. Their space-searching ability is, however, limited by lack of money and by lack of alternatives in the land of their birth. With a willingness to work and with aspirations for success—perhaps wealth—in the United States, they are eager listeners to friends and relatives who tell of numerous job opportunities in the United States. Hundreds of thousands, presented with the glittering prospect of a certain job, low-paying though it might be, quickly place high utility on perhaps a temporary relocation (maybe 5 or 10 years) to the United States. Many know that dangerous risks are involved if they attempt to enter the country illegally, but the rewards are worth the risk. Their arrival indicates their assignment of higher utility to the new site than to the old.

In the 20th century, nearly all countries have experienced a movement of population to the cities from their agricultural areas. The migration has presumably paralleled the number of perceived opportunities within the cities and convictions of absence of place utility in the rural districts. Perceptions, of course, do not necessarily accord with reality. The enormous influx of rural folk to major urban areas in the developing countries and the economic destitution of many of those in-migrants suggest recurrent gross misperceptions, faulty information, and sampling error. In Chapter 12, we discuss the reasons for this phenomenon.

Barriers to Migration

Paralleling the incentives to migration is a set of disincentives, or barriers, to migration. They help to account for the fact that many people do not choose to move even when conditions are bad at home and are known to be better elsewhere. Migration depends on a knowledge of the opportunities in other areas. People with a limited

Problems in the Sunbelt

Since 1970, there has been a net in-migration of over 12 million people to the region known as the Sunbelt, the tier of southern states stretching from the Atlantic to the Pacific Coast. This region captures the bulk of the country's migrants and immigrants from other countries. Arizona's population grew to 3.4 million in 1987 from 1.8 million in 1970. Florida's growth rate for the period was over 50%. Ironically, the rapid growth of the Sunbelt causes problems that threaten some of the very qualities that make the area attractive.

Low energy costs, low wages, and low taxes attracted businesses and industries from the North to the Sunbelt. Both blue- and white-collar workers from the North followed the companies, drawn by the promise of jobs and the particular attractions of the region: warmth and sunshine, unspoiled land, recreational opportunities, and other amenities. Although the Sunbelt continues to attract retirees, companies, and workers, its advantages are eroding rapidly, and growth itself is the reason.

With economic growth have come higher wages. In some parts of the region, taxes have risen already, and they will rise more as the states and cities try to cope with the services that population and economic growth demand. Across the Sunbelt, the rapid influx of people has strained the capacity of sewage-treatment plants. Depending on which city is considered, poor roads, inadequate fire and police protection, and limited mass transit are problems. Florida and the Southwest face water shortages as well.

Many Sunbelt cities, although prosperous, exhibit the kinds of problems that we associate with northern and eastern cities: crowded freeways and traffic jams, noise, urban sprawl, rising crime rates, unemployment, poor inner cities, and air pollution. The clean air that attracted people with allergies, asthma, and other respiratory problems to Arizona is being polluted not only by cars and mining operations but also by the foliage that the migrants planted to remind them of home. For example, the mulberry, a fast-growing tree that provides good shade, produces pollen levels four times as great as those of other plants. Overall, Arizona has experienced a tenfold increase in airborne pollen since 1960. The problems caused by growth tend to inhibit further growth. The high crime rate in Florida, for example, hurts the huge tourist industry, and higher wages in the Southeast deter new companies from moving there.

A city does not create urban problems independently of the people who live there. When a substantial number of those people move to another city, the potential for the reappearance in the new city of the problems that they hoped to leave behind is considerable.

knowledge of the opportunities elsewhere are less likely to migrate than are better-informed individuals. Other barriers include physical features, the costs of moving, ties to individuals and institutions in the original activity space, and government regulations.

Physical barriers to travel include seas, mountains, swamps, deserts, and other natural features. In prehistoric times, they played an especially significant role in limiting movement. Thus, the spread of the ice sheets across most of Europe in Pleistocene times was a barrier to both migration and human habitation. Physical barriers to movement have probably assumed less importance only within the last 400 years. The developments that made possible the great age of exploration, beginning about A.D. 1500, and the technological advances associated with industrialization have enabled people to conquer space more easily. With industrialization came improved forms of transportation that made travel faster, easier, and cheaper. Still, as the account of conditions on the Overland Trail illustrated, only a century and a half ago travel could be arduous, and in some parts of the world, it remains so today.

Economic barriers to migration include the cost of travel and establishing a residence elsewhere. Frequently the additional expense of maintaining contact with those left behind is also a cost factor that must be considered. Normally all these costs increase with the distance traveled and are a more significant barrier to travel for the poor than for the rich. Many immigrants to this country were married men who came alone; when they had acquired enough money, they sent for their families to join them. This phenomenon is still evident among recent immigrants from the Caribbean area to the United States, and among Turks, Yugoslavs, and West Indians who have settled in Northern and Western European countries.

The *cost factor* limits long-distance movement, but the larger the differential between present circumstances and perceived opportunities, the more individuals are willing to spend on moving. Figure 9.23 is an example of the effect of distance on relocation. For many people, especially older people, the differentials must be extraordinarily high for movement to take place.

Cultural factors also contribute to decisions not to migrate. Family, religious, ethnic, and community relationships defy the principle of differential opportunities. Many people will not migrate under any but the most pressing of circumstances. The fear of change and human inertia—the fact that it is easier not to move than to do so—may be so great that people consider but reject a move. Ties to one's own country, cultural group, neighborhood, or family may be so strong as to compensate for the disadvantages of the home location. Returning migrants may convince potential leavers that the opportunities are not in fact better elsewhere—or, even if better, that they are not worth entering an alien culture or sacrificing home and family.

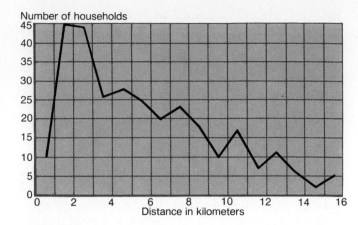

Figure 9.23
Distance between old and new residences in the Asby area of Sweden. Notice how the number of movers decreased with increasing distance.

Restrictions on immigration and emigration constitute *political barriers* to migration. Many governments frown on movements into or outside their own borders. Recognizing that immigration might be economically or politically disadvantageous, many nations restrict outmigration. These restrictions may make it impossible for potential migrants to leave, and they certainly limit the number who can do so.

On the other hand, countries suffering from an excess of workers often encourage emigration. The huge migration of people to the Americas in the late 19th and early 20th centuries is a good example of perceived opportunities for economic gain far greater than in the home country. Many European countries were overpopulated, and their political and economic systems stifled domestic economic opportunity at a time when people were needed by American entrepreneurs hoping to increase their wealth in untapped resource-rich areas.

Countries where per capita incomes are high or are perceived as high are generally the most desired international destinations. In order to protect themselves against overwhelming migration streams, such countries have restrictive policies on immigration. In addition to absolute quotas on the number of immigrants (usually classified by country of origin), a country may impose other requirements, such as the possession of a labor permit or sponsorship by a recognized association.

Patterns of Migration

Several geographic concepts deal with patterns of migration. The first of these is the **migration field**. For any single place, the origin of its in-migrants and the destination of its out-migrants remain fairly stable spatially over time. Areas that dominate a locale's in- and out-migration patterns constitute the migration fields for the place in question, as shown in Figure 9.24. As we would expect, areas

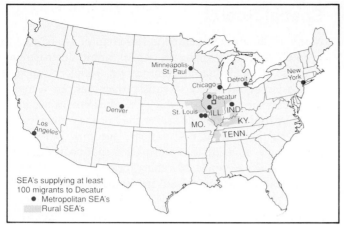

Figure 9.24

The in-migration fields of Decatur, Illinois, and Fresno, California, by State Economic Areas (SEA). For Decatur, as expected, most of the in-migrants originate in nearby rural areas and cities. For Fresno, the pattern is similar with an important addition—rural areas in the Southwest and Kansas. This secondary pattern represents the constant westward movement of population in the United States.

near the point of origin make up the largest part of the migration field. However, places far away, especially large cities, may also be prominent. These characteristics of migration fields are functions of the hierarchical movement to larger places (as we shall see) and the fact that there are so many people in large metropolitan areas that one might expect some migration into and out of them from most areas within a country.

Migration fields do not conform exactly to the diffusion concepts mentioned earlier. As shown by Figure 9.25, some migration fields reveal a distinctly **channelized** pattern of flow. The channels link areas that are socially and economically tied to one another by past migration patterns, by economic trade considerations, or by some other affinity. As a result, flows of migration along these channels are greater than would otherwise be the case. The movements of blacks from the southern United States to the North, of Scandinavians to Minnesota and Wisconsin, of Mexicans to such border states as California, Texas, and New Mexico, and of retirees to Florida and Arizona are all examples of channelized flows.

Sometimes the channelized migration is specific to occupational as well as to ethnic groups. For example, nearly all newspaper vendors in New Delhi, in the north of India, come from one small district in Tamil Nadu, in the south of India. Most construction workers in New Delhi come either from Orissa, in the east of India, or Rajasthan, in the northwest, and taxicab drivers originate in the Punjab. The diamond trade of Bombay, India, is dominated by a network of about 250 related families who come from a small town several hundred miles to the north.

Return migration is a term used to refer to those who, soon after migrating, decide to return to their point of origin. If freedom of movement is not restricted, it is not unusual for as many as 25% of all migrants to return to

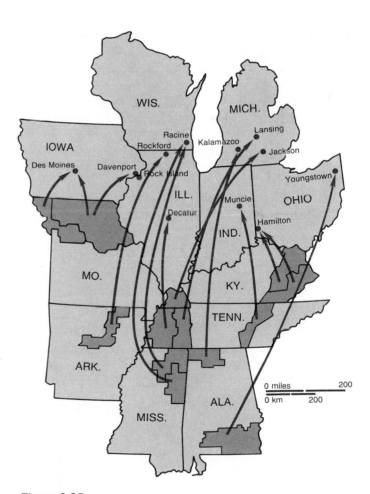

Figure 9.25

Channelized migration flows from the rural South to midwestern cities of medium size. Distance is not necessarily the main determinant of flow direction. Perhaps through family and friendship links, the rural southern areas are tied to particular midwestern destinations.

their place of origin. Unsuccessful migration is sometimes due to an inability to adjust to the new environment. More often, it is a result of false expectations based on distorted mental images of the destination at the time of the move. Myths, secondhand and false information, and people's own exaggerations contribute to what turn out to be mistaken decisions to move. Although return migration often represents the unsuccessful adjustment of individuals to a new environment, it does not necessarily mean that negative information about a place returns with the migrant. It usually means a reinforcement of the channel, as communication lines between the unsuccessful migrant and would-be migrants take on added meaning and understanding.

In addition to channelization, the influence of large cities causes migration fields to deviate from the distance-decay pattern. The concept of **hierarchical migration** assists in understanding the nature of migration fields. Earlier we noted that sometimes information diffuses according to a hierarchical rule, that is, from city to city at the highest level in the hierarchy and then to lower levels. Hierarchical migration, in a sense, is a response to that flow. The tendency is for individuals in domestic relocations to move up the level in the hierarchy from small places to larger ones. Very often, levels are skipped on the way up; only in periods of general economic decline is there considerable movement down the hierarchy. The suburbs of large cities are considered part of the metropolitan area, so that movement from a town to a small suburb would be considered moving up the hierarchy. From this pattern, we can envisage information flowing to a place via a hierarchical routing. Once the information is digested, some respond by migrating to the area from which the information came.

Spatial Search

Many of the ideas presented in this chapter are related to people's decision making in a geographic environment. That is, after evaluating alternatives, people come to a conclusion about where to satisfy their desires. Geographers call the process by which the alternatives are evaluated **spatial search**. The seeker may or may not actually travel to the various destinations; one may pursue the search by evaluating the information that comes to one's residence. For example, searching for a new residence may entail reading newspaper advertisements.

The quality of the decision is very much a function of the quality of the available information. Some decisions, like the decision to buy a new home, are so important that individuals take months or even years to gather information. And, of course, in such circumstances, one is usually obliged to visit each of the possible homesites (Figure 9.26).

Decision making is based on a relationship between information availability and the preferences and motivations of the individual decision maker. Unfortunately, many decisions are made with inadequate or distorted information. Gathering information is a time-consuming activity. Many people are unable or unwilling to use their time for information gathering, and many disappointments are based on insufficient information. Of course, even if the time were available and people were willing to gather information, the needed information might just be unavailable. Many of the return migrants discussed above were responding to an inadequate understanding of the migration location.

Decision makers evaluate information by assessing the utility of each of the possible outcomes of the various courses of action. They depend on their image (perhaps a

Figure 9.26

An example of spatial search in the San Fernando Valley area of Los Angeles. The dots represent the house vacancies in the price range of a sample family. Note (1) the relationship of the new house location to the workplaces of the married couple, (2) the relationship of the old house location to the new house location, and (3) the limited spatial search space. This example is typical of intraurban moves.

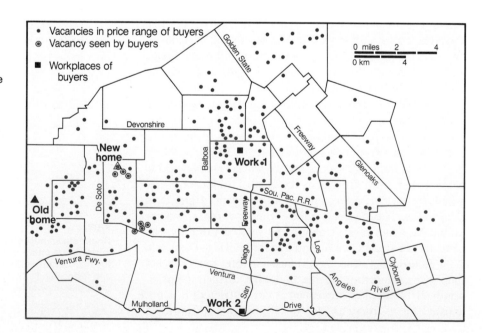

The young merchant assessed his options. If he remained at home, there was little chance that his life would improve materially. But if he were willing to take the risk, he could travel surreptiously to the border, slip across it at night, and enter one of the richest countries in the region . . . a country whose high rate of economic growth was marked by modern office buildings, highways, and well-stocked stores. If he weren't deported, he could expect to earn many times the annual wages he received at home.

The story of a Mexican immigrant to the United States? Not at all. Marcel Amao is Malian, and his destination was the Ivory Coast (Côte d'Ivoire), which along with Nigeria is the richest of the West African countries. In recent years, immigrants have flooded into the Ivory Coast at such a rate that they now constitute about one-third of the country's population of 12 million (1989). By the thousands, they come not only from Mali but also from Benin, Burkina Faso, Ghana, and Senegal.

That migration echoes many of the themes explored in this chapter. Most migrants leave their home for economic reasons; "pushed" by poverty, they are "pulled" by the chance to better their lot in the Ivory Coast. The migration is not strictly distance-

affected; emigrants from Benin and Senegal bypass neighboring countries and cross two or three national borders before entering the Ivory Coast. The movement is somewhat channelized. In the capital city of Abidjan, people from Benin tend to be cooks and cabinetmakers, while immigrants from the impoverished Sahelian na-

tions of Burkina Faso and Mali find work as street sweepers and night-watchmen, laundrymen, and sellers of cloth. Finally, migrants are both "stayers" and "leavers." In the words of its president, the Ivory Coast "is a country without a passport: one comes, one leaves, one stays, but more often one stays."

mental map) of the thing or place being considered, and they use some type of scaling or preference criteria, keeping in mind the uncertainties that exist. Their motivations are based on their belief and value systems. For example, does the decision maker expect instant gratification or is she or he more inclined to expect gratification in several weeks, months, or years? Does one think that family satisfaction is more important than individual satisfaction? And so on.

In a study of the search for housing and residential choice in Toronto, many of the variables and factors involved in spatial search were evaluated. Among other things, it was found that those looking for apartments spent much less time looking than did potential home owners. The difference was ascribed to the importance people place on buying as opposed to renting. It was found, however, that apartment dwellers, even though they may complain about noise, inadequate space, and so on, are not much less satisfied with their decision than home owners. Most young apartment dwellers think of their quarters as tem-

porary and therefore set lower standards of comfort in order to take advantage of lower-cost housing, more convenience, and less responsibility.

In terms of the search itself, people spend longer in deciding to look for new housing than they do in actually inspecting and choosing it. Most home buyers look at seven or more housing units, compared to only about three for those who eventually choose apartments. A Los Angeles study concluded that those home buyers not leaving the metropolitan area altogether search for homes within a short distance of both their present home place and their workplace. The limiting factor, it was determined, is the income level of the decision maker.

Summary

In this discussion of behavioral geography, we have emphasized the factors that influence how individuals view and use space. At the beginning of the chapter, we posed a number of questions used as our theme. In considering how individuals view their environment, the concepts of spatial perception and spatial cognition were helpful. We spoke of the nature of the information available and of the age of people, their past experiences, and their values. The question of differences in the extent of space used led us to the concept of activity space. The age of an individual, his or her degree of mobility, and the availability of opportunities all play a role in defining the limits of individual activity space. We saw, however, that as distance increases, familiarity with the environment decreases. The concept of critical distance identifies the distance beyond which the decrease in familiarity with places begins to be significant.

How space is used is a function of all of the factors mentioned, but certain factors having to do with the diffusion of innovations indicate what opportunities exist for individuals living in various places. Contagious and hierarchical diffusion influence the geographic direction that cultural change will take. Of course, there are barriers to diffusion, such as effort and cost. A special type of human interaction is migration. When strong enough the various push and pull forces motivate a long-distance, permanent move. Migration fosters the spread of culture by means of a relocation diffusion process.

Finally, we examined how the ability of many to make well-reasoned, meaningful decisions is a function of the utility that they assign to places and the opportunities at those places.

Behavioral geographers view individual action in a special way. There is strong emphasis on information flow and the effect of distance decay. In many respects, it is an interdisciplinary field in that the combined ideas of geographers, psychologists, and other social scientists are crucial to an understanding of spatial behavior. Such behavior has an economic component, and to that we next turn our attention.

Key Words

activity space	hierarchical migration
channelized migration	mental map
cognition	migration
contagious diffusion	migration field
critical distance	place utility
diffusion	push and pull factors
distance decay	return migration
environmental perception	spatial search
hierarchical diffusion	territoriality

For Review

1. What is the difference in meaning between *environmental perception* and *cognition?* Give an example of each.

2. On a blank piece of paper, and without any maps to guide you, draw a map of the United States, putting in state boundaries wherever possible; this is your mental map of the country. Compare it with a standard atlas map. What conclusions can you reach?

3. What is meant by *activity space?* What factors affect the areal extent of the activity space of an individual?

4. Recall the places that you have visited in the past week. In your movements, were the distance-decay and critical-distance rules operative? What variables affect an individual's critical distance?

5. Briefly distinguish between *contagious diffusion* and *hierarchical diffusion.* In what ways, if any, were these forms of diffusion in operation in the culture hearths discussed in Chapter 7?

6. What considerations affect a decision to migrate? What is *place utility,* and how does its perception induce or inhibit migration?

7. What common barriers to migration exist? Why do most people migrate within their own country?

8. Define the term *migration field.* Some migration fields show a channelized flow of people. Select a particular channelized migration flow (such as the movement of Scandinavians to the United States, or people from the Great Plains to California, or southern blacks to the North) and explain why a channelized flow developed.

Selected References

Brown, Lawrence A. *Innovation Diffusion: A New Perspective.* New York: Methuen, 1981.

Cox, Kevin R., and Reginald G. Golledge. *Behavioral Problems in Geography Revisited.* New York: Methuen, 1982.

Downs, Roger M., and David Stea. *Cognitive Mapping and Spatial Behavior.* Chicago: Aldine, 1974.

Fligstein, Neil. *Going North: Migration of Whites and Blacks from the South, 1900–1950.* New York: Academic Press, 1981.

Gold, John R. *An Introduction to Behavioral Geography.* New York: Oxford University Press, 1980.

Golledge, Reginald G., and Robert J. Stimson. *Analytical Behavioural Geography.* New York: Croom Helm, 1987.

Gould, P. R., and Rodney White. *Mental Maps,* 2d ed. Boston: Allen & Unwin, 1986.

Hägerstrand, Torsten. *Innovation Diffusion as a Spatial Process.* Chicago: University of Chicago Press, 1967.

Leis, G. J. *Human Migration: A Geographical Perspective.* London: Croom Helm, 1982.

Lynch, Kevin. *The Image of the City.* Cambridge: MIT Press, 1960.

Newman, Oscar. *Defensible Space*. New York: Macmillan, 1972.

Porteous, J. Douglas. *Environment and Behavior: Planning and Everyday Urban Life*. Reading, Mass.: Addison-Wesley, 1977.

Roseman, Curtis C. *Changing Migration Patterns within the United States*. Resource Papers for College Geography, No. 77–2. Washington, D.C.: Association of American Geographers, 1977.

Rushton, Gerard, and R. G. Golledge, eds. *Spatial Choice and Spatial Behavior*. Columbus: Ohio State University Press, 1976.

Saarinen, Thomas F. *Environmental Planning: Perception and Behavior*. Boston: Houghton Mifflin, 1976.

P A R T

3

The Locational Tradition

Given, then, our population-map, what has it to show us? Starting from the most generally known before proceeding towards the less familiar, observe first the mapping of London—here plainly shown, as it is properly known, as Greater London—with its vast population streaming out in all directions—east, west, north, south—flooding all the levels, flowing up the main Thames valley and all the minor ones, filling them up, crowded and dark, and leaving only the intervening patches of high ground pale. . . . This octopus of London, polypus rather, is something curious exceedingly, a vast irregular growth without previous parallel in the world of life—perhaps likest to the spreading of a great coral reef. Like this, it has a stony skeleton, and living polypes—call it, then, a "man-reef" if you will. Onward it grows, thinly at first, the pale tints spreading further and faster than the others, but the deeper tints of thicker population at every point steadily following on. Within lies a dark crowded area; of which, however, the daily pulsating centre calls on us to seek some fresh comparison to higher than corraline life.[1]

Thus, in the lavish prose of the early 20th century, did Patrick Geddes, "spokesman for man and the environment," capsulize the spread of people in the London region. The physical environment, the vibrating human life of cities, and change in human settlement patterns are all implicit in his remarks, as they are in the locational tradition of geography.

A theme that has run like a thread through the first two parts of this book may aptly be termed the *theme of location*. Distributions of climates, landforms, culture hearths, and religions were points of interest in our examinations of physical and cultural landscapes. In Part III of our study, the analysis of location becomes our central concern rather than just one strand among many, and the locational tradition of geography is brought to the fore.

In the study of economic, resource, or urban geography, a central question has to do with *where* certain types of human activities take place, not just in absolute terms but also in relation to one another. In studying the distribution of a given activity, such as commercial grain farming, the concern is with identifying the locational pattern, analyzing it to see why it is arranged the way it is, and searching for underlying principles of location. Because the elements of culture are integrated, such a study often must take into consideration the ways the location of one element affects the locations of others. For example, certain characteristics

1. From *Patrick Geddes: Spokesman for Man and the Environment,* ed. by Marshall Stalley (New Brunswick, N.J.: Rutgers University Press, 1972), p. 123.

of the labor pool have helped to alter the distribution of the textile industry in America, and this alteration has had economic consequences for both New England and the South, and has affected the location of other economic activities as well.

In Chapter 10, "Economic Geography," our attention is directed to the location of economic activities as we seek to answer the question of why they are distributed as they are. What forces operate to make some regions extremely productive and others less so or some enterprises successful and others not? The stages of economic activities from primary production to sophisticated technical services are examined as are the similarities and contrasts between broad types of economic systems, such as commercial, subsistence, and planned.

The resources upon which humankind draws and depends for sustenance and development are not distributed uniformly in quantity or quality over the earth. Chapter 11, "The Geography of Natural Resources," centers on the relationship between the demands people place upon the spatially varying environment—demands that are culture-bound—and the ability of the environment to sustain those demands. In an increasingly integrated economic and social world of growing consumption demands, both access to resources (particularly energy and mineral resources) and their wise and productive use loom ever more important in human affairs. The understanding of resources in both their distributive and their exploitative characteristics is a theme of growing interest in the locational tradition of geography.

The increasing urbanization of the world's population represents another stage of the process of cultural development, economic change, and environmental control, aspects of which have been the topics of separate chapters. In Chapter 12, urban areas are viewed in two ways. First, we focus upon cities as points in space with patterns of distribution and specializations of function that invite analysis. Second, we consider cities as landscape entities with specialized arrangements of land uses resulting from recognizable processes of urban growth and development. Within their confines, and by means of the interconnected functional systems that they form, cities summarize a present stage of economic patterns and resource use of humankind. The consideration of urban geography thus logically concludes our examination of the locational tradition of geography.

Oil drill in Lloydminster, Saskatchewan, Canada.

Economic Geography

Open air market in Rwanda, Africa.

THE crop bloomed luxuriantly that summer of 1846. The disaster of the preceding year seemed over and the potato, the sole sustenance of some 8 million Irish peasants, would again yield in the bounty needed. Yet within a week, wrote Father Mathew, "I beheld one wide waste of putrefying vegetation. The wretched people were seated on the fences of their decaying gardens . . . bewailing bitterly the destruction that had left them foodless." Colonel Gore found that "every field was black," and an estate steward noted that "the fields . . . look as if fire has passed over them." The potato was irretrievably gone for a second year; famine and pestilence were inevitable.

Within five years, the settlement geography of the most densely populated country in Europe was forever altered. The United States received a million immigrants, who provided the cheap labor needed for the canals, railroads, and mines that it was creating in its rush to economic development. New patterns of commodity flows were initiated as American maize for the first time found an Anglo-Irish market—as part of Poor Relief—and then entered a wider European market that had also suffered general crop failure in that bitter year. Within days, a microscopic organism, the cause of the potato blight, had altered the economic and human geography of two continents.

That alteration resulted from a chain of interlocking causes and consequences dramatically demonstrating that all of those physical and cultural patterns making up that world view called geography are really parts of one whole. Central among those patterns are the ones the economic geographer isolates for special study.

Simply stated, *economic geography* is the study of how people earn their living, how livelihood systems vary by area, and how economic activities are interrelated and linked. In reality, of course, we cannot really comprehend the totality of the economic pursuits of over 5 billion human beings. We cannot examine the infinite variety of production and service activities found everywhere on the earth's surface; nor can we trace all their innumerable interrelationships, linkages, and flows. Even if that level of understanding were possible, it would be valid for only a fleeting instant of time, for economic activities are constantly undergoing change.

Economic geographers seek consistencies. They attempt to develop generalizations that will aid in the comprehension of the maze of economic variations characterizing human existence. From their studies emerges a deeper awareness of the dynamic, interlocking diversity of human enterprise, of the impact of economic activity upon all other facets of human life and culture, and of the increasing interdependence of differing national and regional economic systems. (See "Economic Regions" in Chapter 13.) The potato blight, although it struck only one small island, ultimately affected the economies of continents. In like fashion, the depletion of America's natural resources and the "deindustrialization" of its economy are altering the relative wealth of countries, flows of international trade, domestic employment and income patterns, and more (Figure 10.1).

The Classification of Economic Activity and Economies

The search for understanding of livelihood patterns is made more difficult by the complex environmental and cultural realities controlling the economic activities of humans. Many production patterns are rooted in the spatially variable circumstances of the *physical environment*. The staple crops of the humid tropics, for example, are not part of the agricultural systems of the midlatitudes; livestock types that thrive in American feedlots or on western ranges are not adapted to the Arctic tundra or the margins of the Sahara desert. The unequal distribution of useful mineral deposits gives some regions and countries economic prospects and employment opportunities that are denied to others. Forestry and fishing depend on still other natural resources unequal in occurrence, type, and value.

Within the bounds of the environmentally possible, economic or production decisions may be conditioned by *cultural considerations*. For example, culturally based food preferences rather than environmental limitations may dictate the choice of crops or livestock. Maize is a preferred grain in Africa and the Americas, wheat in North America, Australia, Argentina, and the Soviet Union, and rice in much of Asia. Pigs are not produced in Muslim areas. Level of *technological development* of a culture will affect its recognition of resources or its ability to exploit them. Preindustrial societies do not know of, or need, iron ore or coking coal underlying their hunting, gathering, or gardening grounds. *Political decisions* may encourage or discourage—through subsidies, protective tariffs, or production restrictions—patterns of economic activity. And, ultimately, production is controlled by *economic factors* of demand, whether that demand is expressed through a free-market mechanism, through government instruction, or through the consumption requirements of a single family producing for its own needs.

Categories of Activity

Such regionally varying environmental, cultural, technological, political, or market conditions add spatial details to more generalized ways of categorizing the world's productive work. One approach to that categorization is to view economic activity as ranged along a continuum of both increasing complexity of product or service and increasing distance from the natural environment. Seen from that

Figure 10.1
These Japanese cars are part of a continuing flow of imported goods capturing a share of the domestic market traditionally held by American manufacturers. Established patterns of production and exchange are constantly subject to change in a world of increasing economic and cultural interdependence.

perspective, a small number of distinctive stages of production and service activities may be distinguished (see Figure 10.2).

Primary activities are those that harvest or extract something from the earth. They are at the beginning of the production cycle, where humans are in closest contact with the resources and potentialities of the environment. Such primary activities encompass basic foodstuff and raw material production. Hunting and gathering, grazing, agriculture, fishing, forestry, and mining and quarrying are examples. **Secondary activities** are those that add value to materials by changing their form or combining them into more useful, therefore more valuable, commodities. That provision of *form utility* may range from simple handicraft production of pottery or woodenware to the delicate assembly of electronic goods or space vehicles (Figure 10.3). Iron smelting, steel-making, metalworking, automobile production, textile and chemical industries—indeed, the full array of *manufacturing* and *processing industries*—are included in this phase of the production process. Also included are the production of *energy* (the "power company") and the *construction* industry.

Tertiary activities consist of those business and labor specializations that provide *services* to the primary and secondary sectors and *goods* and *services* to the general community and to the individual. They include professional, clerical, and personal services. They constitute the

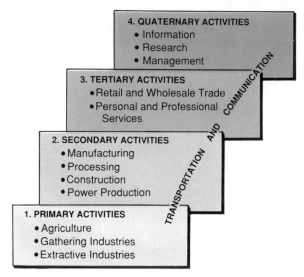

Figure 10.2

The categories of economic activity. The four main sectors of the economy do not stand alone. They are connected and integrated by transportation and communication services and facilities not assigned to any single sector but common to all.

vital link between producer and consumer, for tertiary occupations importantly include the wholesale and retail *trade* activities necessary in highly interdependent societies.

Figure 10.3
These logs entering a lumber mill are products of *primary production*. Processing them into boards or prefabricated houses is a *secondary activity* that increases their value by altering their form.

In economically advanced societies, many individuals and some entire organizations are engaged in the processing and dissemination of information and in the administration and control of their own or other enterprises. The term **quaternary** is applied to this fourth class of economic activities, which is composed entirely of services rendered by white collar professionals working in education, government, management, information processing, and research. The distinctions between tertiary and quaternary activities are further developed later under "Tertiary and Beyond."

These categories of production and service activities help us to see an underlying structure to the near-infinite variety of things people do to earn a living and to sustain themselves. But they tell us little about the organization of the larger economy of which the individual worker or enterprise is a part. For that sort of organizational understanding of world and regional economies we look to *systems* rather than *components* of economies.

Types of Economic Systems

Broadly speaking, there are three major types of economic systems: subsistence, commercial, and planned. None of them is "pure", that is, none exists in isolation in an increasingly interdependent world. Each, however, has certain underlying characteristics that mark it as a distinctive form of resource management and economic control.

In a **subsistence economy**, goods and services are created for the use of the producers and their kinship groups. Therefore, there is little exchange of goods and only limited need for markets. In a **commercial economy**

producers or their agents freely market their goods and services, the laws of supply and demand determine price and quantity, and market competition is the primary force shaping production decisions and distributions. A **planned economy** is one in which the producers or their agents dispose of goods and services, usually through a government agency that controls both supply and price. The quantities produced and the locational patterns of production are carefully programmed by central planning departments. In all economic systems, transportation is a key variable. No advanced economy can flourish without a well-connected transport system. All subsistence societies—or subsistence areas of developing countries—are characterized by their isolation from regional and world routeways (Figure 10.4).

In actuality, few people are members of only one of these systems, although one may be dominant. A farmer in India may produce rice and vegetables primarily for the family's consumption, but also save some of the produce to sell. In addition, members of the family may market cloth or other handicrafts they make. With the money derived from those sales, the Indian peasant is able to buy, among other things, clothes for the family, tools, or fuel. Thus, that Indian farmer is a member of at least two systems: subsistence and commercial. Similarly, in Eastern Europe before its dramatic economic and political changes of 1989 and 1990, an East German farmer might have produced a specified amount of grain for a government food agency and another amount for the family; any surplus might have been sold in a nearby marketplace. That East German was a member of all three types of systems.

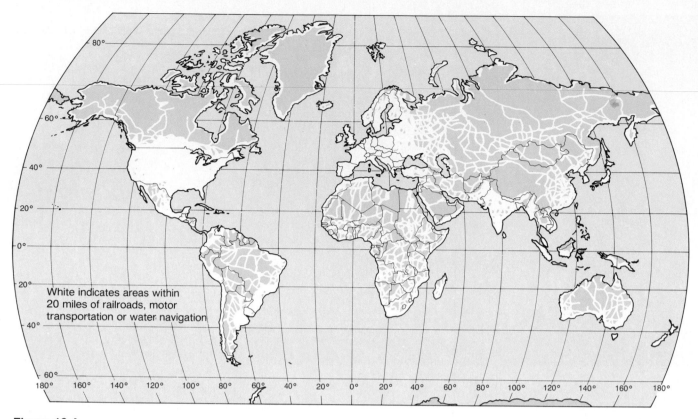

White indicates areas within 20 miles of railroads, motor transportation or water navigation

Figure 10.4

Accessibility is a key measure of economic development and of the degree to which a world region can participate in interconnected market activities. Isolated areas of countries with advanced economies suffer a price disadvantage because of high transportation costs. Lack of accessibility in subsistence economic areas slows their modernization and hinders their participation in the world market.

In the United States, government controls on the production of various types of goods and services (such as growing wheat or tobacco, producing alcohol, constructing and operating nuclear power plants, or engaging in licensed personal and professional services) mean that the country does not have a purely commercial economy. To a limited extent, it is also a controlled and planned one. Example after example would show that there are very few people in the world who are members of only one type of economic system.

Nonetheless, in a given country, one of the three economic systems tends to dominate, and it has in the past been relatively easy subjectively to classify countries by that dominance even while recognizing that some elements of the other two systems may exist within the controlling scheme. However, the one geographic certainty is that spatial patterns—including those of economic activities and systems—will change. For example, the commercial economies of Western European countries, some with sizable infusions of planned economy controls, will be affected by new, supranational direction and increased free-market competition as the Common Market of the European Community (see p. 262) takes effect in the 1990s. Many of the countries of Latin America, Africa,

Asia, and the Middle East that traditionally were dominated by subsistence economies are in the process of adopting advanced technologies and simultaneously become both more commercial and more planned.

Perhaps most striking has been the abrupt retreat from the extremes of planned control that formerly characterized the economies of Communist countries of Europe and Asia. In the late 1970s, China effectively abolished communes, reestablished a private farming system, and permitted privately owned small businesses and market competition (Figure 10.5). In 1990, as part of its *perestroika* (restructuring) policies, the Soviet Union instituted programs for long-term leasing of agricultural land to individual farmers, gave private citizens the right to own small-scale factories and other businesses, and introduced plans for conversion from governmental to market control of the economy.

Although the former sharp contrasts in economic organization are becoming blurred and national economic orientations are changing, the contrasts between subsistence, planned, and commercial systems still remain evident and still differently affect national patterns of livelihood, production, and economic decision making. Indeed, both approaches to economic classification—by

Figure 10.5

Independent food sellers in modern China. In early 1989 the country's 14.5 million registered private businesses far exceeded the total number of private enterprises operating in 1949 when the Communists took power. Since government price controls on most food items were removed in May 1985, free markets have multiplied. In Beijing, one-third of the city's sales of vegetables, eggs, and beef are sold on free markets. In rural districts up to two-thirds of fruits, vegetables, and meat are sold privately. As an immediate consequence of new government restrictions on the economy following the June 1989 student massacre in Tiananmen Square, the number of workers in the private sector fell from 23 million to 19.4 million. Politics and economy are closely intertwined in the People's Republic.

types of activities and by organization of economies—help us to visualize and understand world economic geographic patterns. In this chapter our path to that understanding leads through the successive categories of economic activity, from primary to quaternary, with emphasis on the technologies, spatial patterns, and organizational systems that are involved in each category.

Primary Activities: Agriculture

Before there was farming, *hunting and gathering* were the universal forms of primary production. These preagricultural pursuits are now practiced by at most a few thousand persons worldwide, primarily in isolated and remote pockets within the low latitudes and among the sparse populations of very high latitudes. The interior of New Guinea, rugged areas of interior Southeast Asia, diminishing segments of the Amazon rain forest, a few districts of tropical Africa and northern Australia, and parts of the Arctic regions still contain such preagricultural people.

Their numbers are few and declining, and, wherever they are brought into contact with more advanced cultures, their primitive way of life is eroded or lost.

Agriculture, defined as the growing of crops and the tending of livestock whether for the subsistence of the producers or for sale or exchange, has replaced hunting and gathering as economically the most significant and spatially the most widespread of the primary activities. It occupies a large majority of the world's workers. In many Third World subsistence economies, at least three-fourths of the labor force is directly involved in farming and herding. In some, such as Nepal in Asia or Burundi in Africa, the figure is more than 90%. In highly developed commercial economies, on the other hand, agriculture involves only a small fraction of the labor force: less than 10% in most of Western Europe, 5% in Canada, and only 3% or less in the United States (see Figure 7.5).

It has been customary to classify agricultural societies on the twin bases of the importance of off-farm sales (in either a free or a planned market economy) and the level of mechanization and technological advancement. "Subsistence," "traditional" (or "intermediate"), and "advanced" (or "modern") are usual terms employed to recognize both aspects. These are not mutually exclusive, but recognized stages along a continuum of farm economy variants. At one end lies production solely for family sustenance, using primitive tools and native plants. At the other is the specialized, highly capitalized, near-industrialized agriculture for off-farm delivery that marks advanced countries of both commercial and planned economies. Between these extremes is the middle ground of traditional agriculture, where farm production is in part destined for home consumption and in part oriented towards off-farm sale either locally or in national and international markets. The variety of agricultural activities and the diversity of controls upon their spatial patterns are most clearly seen by examining the "subsistence" and "advanced" ends of the agricultural continuum.

Subsistence Agriculture

By definition, a *subsistence* economic system involves nearly total self-sufficiency on the part of its members. Production for exchange is minimal, and each family or close-knit social group relies upon itself for its food and other most essential requirements. Farming for the immediate needs of the family is, even today, the predominant occupation of humankind. In most of Africa, much of Latin America, and most of Asia outside of the Soviet Union, the majority of people are primarily concerned with feeding themselves from their own land and livestock. That concern reflects the needs of people themselves and may bear no relation to the economic philosophy—whether favoring free markets or central planning—adopted by the country in which they reside.

Two chief types of subsistence agriculture may be recognized: *extensive* and *intensive*. Although each type has several variants, the essential contrast between the two types is realizable yield per acre utilized and, therefore, population-supporting potential. **Extensive subsistence agriculture** involves large areas of land and minimal labor input per acre. Both product per acre and population densities are low. **Intensive subsistence agriculture** involves the cultivation of small landholdings through the expenditure of great amounts of labor per acre. Yields per unit area and population densities are both high (Figure 10.6).

Extensive Subsistence Agriculture

Of the several types of *extensive subsistence* agriculture—varying one from another in their intensities of land use—two are of particular interest.

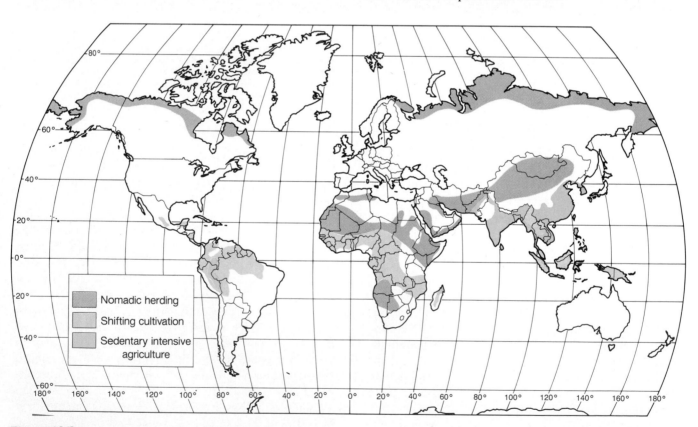

Figure 10.6

Subsistence agricultural areas of the world. Nomadic herding, supporting relatively few people, was the ages-old way of life in large parts of the dry and cold world. Shifting or swidden agriculture maintains soil fertility by tested native practices in tropical wet and wet-and-dry climates. Large parts of Asia support millions of people engaged in sedentary intensive cultivation, with rice and wheat the chief crops.

Nomadic herding, the wandering but controlled movement of livestock solely dependent upon natural forage, is the most extensive type of land use system (Figure 10.6). That is, it requires the greatest amount of land area per person sustained. Over large portions of the Asian semidesert and desert areas, in certain highland areas, and on the fringes of and within the Sahara, a relatively small number of people graze animals not for market sale but for consumption by the herder group. Sheep, goats, and camels are most common, while cattle, horses, and yaks are locally important. The reindeer of Lapland were formerly part of the same system.

Whatever the animals involved, their common characteristics are hardiness, mobility, and an ability to subsist on sparse forage. The animals provide a variety of products: milk, cheese, and meat for food; hair, wool, and skins for clothing; skins for shelter; and excrement for fuel. For the herder, they represent primary subsistence. Nomadic movement is tied to sparse and seasonal rainfall or to cold temperature regimes and to the areally varying appearance and exhaustion of forage. Extended stays in a given location are neither desirable nor possible.

As a type of economic system, nomadic herding is declining. Many economic, social, and cultural changes are causing nomadic groups to alter their way of life or to disappear entirely. On the Arctic fringe of the Soviet Union, herders are now members of state or collective herding enterprises. In the Scandinavian North, Lapps (Saami) are engaged in commercial more than in subsistence livestock farming. In the Sahel region of Africa on the margins of the Sahara desert, oases formerly controlled by nomads have been taken over by farmers, and the great droughts of recent decades have forever altered the formerly nomadic way of life of thousands.

A much differently based and distributed form of extensive subsistence agriculture is found in all of the warm, moist, low-latitude areas of the world. There, many people engage in a kind of nomadic farming. Through clearing and use, the soils of those areas lose many of their nutrients (as soil chemicals are dissolved and removed by surface and groundwater or nutrients are removed from the land in the vegetables picked and eaten), and farmers cultivating them need to move on after harvesting several crops. In a sense, they rotate fields rather than crops to maintain productivity. This type of **shifting cultivation** has a number of names, the most common of which are *swidden* (an English localism for "burned clearing") and *slash-and-burn.*

Characteristically, the farmers hack down the natural vegetation, burn the cuttings, and then plant such crops as maize (corn), millet (a cereal grain), rice, manioc or cassava, yams, and sugar (Figure 10.7). Increasingly included in many of the crop combinations are such high value, labor-intensive commercial crops as coffee, providing the cash income that is evidence of the growing integration of all peoples into exchange economies. Initial yields—the first and second crops—may be very high, but they quickly become lower with each successive planting on the same plot. As that occurs, cropping ceases, native vegetation is allowed to reclaim the clearing, and gardening shifts to another newly prepared site. The first clearing will ideally not be used again for crops until, after many years, natural fallowing replenishes its fertility. (See "Swidden Agriculture.")

Nearly 5% of the world's people are still predominantly engaged in tropical shifting cultivation on more than one-fifth of the world's land area (Figure 10.6). They are found on the islands of Kalimantan (Borneo), New Guinea, and Sumatera (Sumatra) in Indonesia. It is part of the economy of the uplands of South Asia in Vietnam, Thailand, Myanmar (Burma), and the Philippines. Nearly the whole of Central and West Africa away from the coasts, Brazil's Amazon Basin, and large portions of Central

Figure 10.7
A swidden plot in the Ivory Coast, West Africa. The second-growth slash in the background will soon be set afire; burning will kill, but not fell, the large tree in the foreground. Stumps and trees left in the clearing will remain after the burn.

Swidden Agriculture

The following account describes shifting cultivation among the Hanunóo people of the Philippines. Nearly identical procedures are followed in all swidden farming regions.

When a garden site has been selected, the swidden farmer begins to remove unwanted vegetation. The first phase of this process consists of slashing and cutting the undergrowth and smaller trees with bush knives. The principal aim is to cover the entire site with highly inflammable dead vegetation so that the later stage of burning will be most effective. Because of the threat of soil erosion the ground must not be exposed directly to the elements at any time during the cutting stage. During the first months of the agricultural year, activities connected with cutting take priority over all others.

Once most of the undergrowth has been slashed, the larger trees must be felled or killed by girdling (cutting a complete ring of bark) so that unwanted shade will be removed.

Some trees, however, are merely trimmed but not killed or cut, both to reduce labor and to leave trees to re-seed the swidden during the subsequent fallow period.

The crucial and most important single event in the agricultural cycle is swidden burning. The main firing of a swidden is the culmination of many weeks of preparation in spreading and leveling chopped vegetation, preparing firebreaks to prevent flames escaping into the jungle, and allowing time for the drying process. An ideal burn rapidly consumes every bit of litter; in no more than an hour or an hour and a half, only smouldering remains are left.

The Hanunóo, swidden farmers of the Philippines, note the following as the benefits of a good burn: 1) removal of unwanted vegetation, resulting in a cleared swidden; 2) extermination of many animal and some weed pests; 3) preparation of the soil for dibble (any small hand tool or stick to make a hole)

planting by making it softer and more friable; 4) provision of an evenly-distributed cover of wood ashes, good for young crop plants and protective of newly-planted grain seed. Within the first year of the swidden cycle, an average of between 40 and 50 different crops have been planted and harvested.

The most critical feature of swidden agriculture is the maintenance of soil fertility and structure. The solution is to pursue a system of rotation of one to three years in crop and ten to twenty in woody or bush fallow regeneration. When population pressures mandate a reduction in the length of fallow period, productivity of the region tends to drop as soil fertility is lowered, marginal land is utilized, and environmental degradation occurs. The balance is delicate.

Adapted from *Hanunóo Agriculture*, by Harold C. Conklin (FAO Forestry Development Paper No. 12), 1957.

America are all noted for this type of extensive subsistence agriculture. It may be argued that shifting cultivation is a highly efficient cultural adaptation where land is abundant in relation to population, and levels of technology and capital availability are low. As those conditions change, the system becomes less viable.

Intensive Subsistence Agriculture

Over one-half of the people of the world are engaged in intensive subsistence agriculture, which predominates in areas shown on Figure 10.6. As a descriptive term, *intensive subsistence* is no longer fully applicable to a changing way of life and economy in which the distinction between subsistence and commercial is decreasingly valid. While families may still be fed primarily with the produce of their individual plots, the exchange of farm commodities within the system is considerable. Production of foodstuffs for sale in rapidly growing urban markets is increasingly vital for the rural economies of "subsistence farming" areas and for the sustenance of the growing proportion of national and regional populations no longer themselves engaged in farming. Nevertheless, hundreds of millions of Indians, Chinese, Pakistanis, Bangladeshis, and Indonesians plus further millions in other Asian, African, and Latin American countries remain small-plot, mainly subsistence producers of rice, wheat, maize, millet, or pulses (peas, beans, and other legumes). Most live in monsoon Asia, and we will devote our attention to that area.

Intensive subsistence farmers are concentrated in such major river valleys and deltas as the Ganges and the Chang Jiang (Yangtze), and in smaller valleys close to coasts—level areas with fertile alluvial soils. These warm, moist districts are well suited to the production of rice, a crop that under ideal conditions can provide large amounts of food per unit of land. Rice also requires a great deal of time and attention, for planting rice shoots by hand in standing fresh water is a tedious art (Figure 10.8). In the cooler and drier portions of Asia, wheat is grown intensively, along with millet and, less commonly, upland rice.

Intensive subsistence farming is characterized by large inputs of labor per unit of land, by small plots, by the intensive use of fertilizers, mostly animal manure (see "The Economy of a Chinese Village"), and by the promise of high yields in good years. For food security and dietary custom, some other products are also grown. Vegetables and some livestock are part of the agricultural system, and fish may be reared in rice paddies and ponds. Cattle are a source of labor and food. Food animals include swine, ducks, and chickens, but since Muslims eat no pork, hogs are absent in their areas of settlement. Hindus generally eat no meat, but the large number of cattle in India are vital for labor, as a source of milk, and as producers of fertilizer and fuel.

Figure 10.8

Transplanting rice seedlings requires arduous hand labor by all members of the family. The newly flooded diked fields, previously plowed and fertilized, will have their water level maintained until the grain is ripe. This photograph was taken in India. The scene is repeated wherever subsistence wet-rice agriculture is practiced.

Population pressures and new agricultural technologies are forcing change upon traditional intensive subsistence farming practices and societies. **Green Revolution** is the shorthand reference to a complex of seed and management improvements adapted to the needs of intensive agriculture that have brought larger harvests from a given area of farmland. In the 1970s alone, the world's average annual grain production increased by 2.6% per year. Increases in per acre yields accounted for some 80% of that growth. These yield increases and the improved food supplies they represent have been particularly important in densely populated, subsistence farming areas heavily dependent upon rice and wheat cultivation (Figure 10.9).

The increases in food availability made possible through the Green Revolution have helped alleviate some of the shortages and famines predicted for intensive subsistence agricultural regions during the 1960s and 1970s. But a price has been paid. The Green Revolution is commercially oriented, and demands high inputs of costly hybrid seeds, mechanization, irrigation, fertilizers, and pesticides. As it is adopted, traditional and subsistence agriculture are being displaced. Lost, too, are the food security that distinctive locally adapted native crop varieties provided and the nutritional diversity and balance that multiple-crop intensive gardening assured. Subsistence farming, wherever practiced, was oriented toward risk minimization. Many differentially hardy varieties of a single crop guaranteed some yield whatever adverse weather, disease, or pest problems might occur. Commercial agriculture, however, aims at profit maximization, not minimal food security.

Commercial Agriculture

Few people or areas still retain the isolation and self-containment characteristic of pure subsistence economies. Nearly all have been touched by a modern world of trade and exchange and have adjusted their traditional economies in response. Farmers in advanced economies, whether commercial or planned, produce not for their own subsistence but primarily for a market off the farm itself. They are part of integrated exchange economies in which agriculture is but one element in a complex structure. In commercial economies, agricultural patterns presumably mark production responses to market demand expressed through price. In planned economies, those patterns of production may reflect administrative decisions influenced by noneconomic constraints but they are, nonetheless, related to the consumption requirements of the larger society.

Production Controls

Agriculture within modern, developed economies is characterized by *specialization*—by enterprise (farm), area, and even country; by *off-farm sale* rather than subsistence production; and by *interdependence* of producers and buyers linked through markets. Farmers in a free-market

The Economy of a Chinese Village

The village of Nanching is in subtropical southern China on the Zhu River delta near Guangzhou (Canton). Its pre-Communist subsistence agricultural system was described by a field investigator, whose account is here condensed. The system is found in its essentials in other rice-oriented societies.

In this double-crop region, rice was planted in March and August and harvested in late June or July and again in November. March to November was the major farming season. Early in March the earth was turned with an iron-tipped wooden plow pulled by a water buffalo. The very poor who could not afford a buffalo used a large iron-tipped wooden hoe for the same purpose.

The plowed soil was raked smooth, fertilizer was applied, and water was let into the field, which was then ready for the transplanting of rice seedlings. Seedlings were raised in a seedbed, a tiny patch fenced off on the side or corner of the field. Beginning from the middle of March, the transplanting of seedlings took place. The whole family was on the scene. Each took the seedlings by the bunch, ten to fifteen plants, and pushed them into the soft inundated soil. For the first thirty or forty days the emerald green crop demanded little attention except keeping the water at a proper level. But after this period came the first weeding; the second weeding followed a month later. This was done by hand, and everyone old enough for such work participated. With the second weeding went the job of adding fertilizer. The grain was now allowed to stand to "draw starch" to fill the hull of the kernels. When the kernels had "drawn enough starch," water was let out of the field, and both the soil and the stalks were allowed to dry under the hot sun.

Then came the harvest, when all the rice plants were cut off a few inches above the ground with a sickle. Threshing was done on a threshing board. Then the grain and the stalks and leaves were taken home with a carrying pole on the peasant's shoulder. The plant was used as fuel at home.

As soon as the exhausting harvest work was done, no time could be lost before starting the chores of plowing, fertilizing, pumping water into the fields, and transplanting seedlings for the second crop. The slack season of the rice crop was taken up by chores required for the vegetables which demanded continuous attention, since every peasant family devoted a part of the farm to vegetable gardening. In the hot and damp period of late spring and summer, eggplant and several varieties of squash and beans were grown. The green-leafed vegetables thrived in the cooler and drier period of fall, winter, and early spring. Leeks grew the year round.

When one crop of vegetables was harvested, the soil was turned and the clods broken up by a digging hoe and leveled with an iron rake. Fertilizer was applied, and seeds or seedlings of a new crop were planted. Hand weeding was a constant job; watering with the long-handled wooden dipper had to be done an average of three times a day, and in the very hot season when evaporation was rapid, as frequently as six times a day. The soil had to be cultivated with the hoe frequently as the heavy tropical rains packed the earth continuously. Instead of the two applications of fertilizer common with the rice crop, fertilizing was much more frequent for vegetables. Besides the heavy fertilizing of the soil at the beginning of a crop, usually with city garbage, additional fertilizer, usually diluted urine or a mixture of diluted urine and excreta, was given every ten days or so to most vegetables.

Adapted from C. K. Yang, *A Chinese Village in Early Communist Transition.* The MIT Press, Cambridge, MA. Copyright © 1959 by the Massachusetts Institute of Technology.

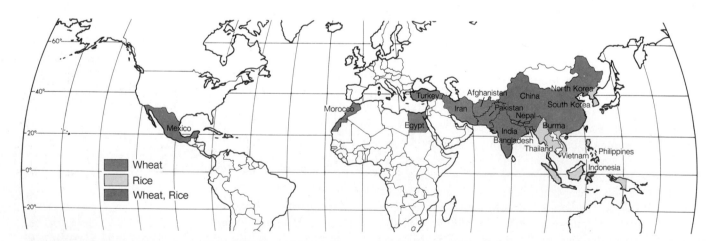

Figure 10.9

Chief beneficiaries of the Green Revolution. In the 11 countries adopting new rice varieties and cropping technologies, average yields increased by 52% between 1965 and 1983. In the rest of the world, they actually dropped by 4% during the same period. Wheat yields increased 66% in the 9 reporting countries (excluding Mexico). In the rest of the world they grew only 29%.

economy supposedly produce those crops that their estimates of market price and production cost indicate will yield the greatest return. Theoretically, farm products in short supply will command an increased market price. That, in turn, should induce increased production to meet the demand with a consequent reduction of market price to a level of equilibrium with production costs. In planned economies, both market demand and price are more prominently dictated by government directive than by purely economic forces. Frequently, production costs are not closely reflected in sale prices, for example when government policy requires uneconomically low food prices for urban workers.

In theory, supply, demand, and the market price mechanism are the controls on agricultural production in commercial economies. In actuality, they are joined by a number of nonmarket governmental influences that may be as decisive as market forces in shaping farmers' options and spatial production patterns. If there is a glut of wheat on the market, for example, the price per ton will come down and the area sown to it should diminish. It will also diminish regardless of supply if governments, responding to economic or political considerations, impose acreage controls. Agriculture in even the least controlled of commercial economies is strongly affected by a variety of governmental policies. In many, it deviates as much from theoretical production and marketing patterns as it does in the most tightly planned economies. The political power of farmers in the European Economic Community, for example, has secured for them generous product subsidies and for the EEC immense unsold stores of butter, wine, and grains. In the United States, programs of farm price supports, acreage controls, financial assistance, and other governmental involvements in agriculture are of long standing and have an equally distorting effect (Figure 10.10).

A Model of Agricultural Location

Early in the last century, before such governmental influences were the norm, Johann Heinrich von Thünen (1783–1850) observed that lands of apparently identical physical properties were used for different agricultural purposes. Around each major urban market center, he noted, there developed a set of concentric rings of different farm products (Figure 10.11). The ring closest to the market specialized in perishable commodities that were both expensive to ship and in high demand. Surrounding rings of farm lands farther away from the city were used for less perishable commodities with lower transport costs, reduced demand, and lower market prices. General farming and grain farming replaced the market gardening of the inner ring. At the outer margins of profitable agriculture, farthest from the single central market, livestock grazing and similar extensive land uses predominated.

To explain why this should be so, von Thünen constructed a formal spatial model—perhaps the first developed to analyze human activity patterns. He deduced that

Figure 10.10
Open storage of corn. In the world of commercial agriculture, supply and demand are not always in balance. Both the bounty of nature in favorable crop years and the intervention of governmental programs that distort production decisions can create surpluses for which no market is readily available.

the uses to which parcels were put was a function of the differing values (land rents) placed upon seemingly identical lands. Those differences, he determined, reflected the cost of overcoming the distance separating a given farm from a central market town. The greater the distance, the higher the operating cost to the farmer, since transport charges had to be added to other expenses. When a commodity's production costs plus its transport costs just equaled its value at the market, a farmer was at the economic margin of its cultivation. A simple exchange relationship ensued: the greater the transportation cost, the lower the rent that could be paid for land if the crop produced was to remain competitive in the market.

Since in the simplest form of the model, transport costs are the only variable, the relationship between land rent and distance from market can be easily calculated by reference to each competing crop's *transport gradient*. Perishable commodities, such as fruits and vegetables, would encounter high transport rates per unit of distance; other items such as grain would have lower rates. Land rent for any farm commodity decreases with increasing distance from the central market, and the rate of decline is determined by the transport gradient for that commodity. Crops that have both the highest market price and the highest transport costs will be grown nearest to the market. Less perishable crops with lower production and transport costs will be grown at greater distances away (Figure 10.12). Since in this model transport costs are uniform in all directions away from the center, the concentric zonal pattern of land use called the **von Thünen rings** results.

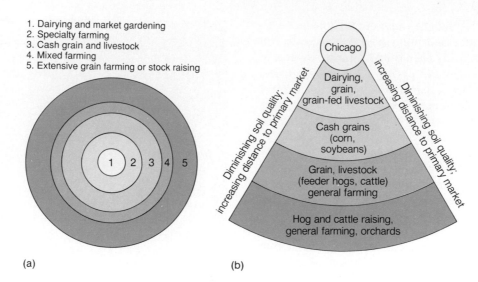

1. Dairying and market gardening
2. Specialty farming
3. Cash grain and livestock
4. Mixed farming
5. Extensive grain farming or stock raising

Chicago

Dairying, grain, grain-fed livestock

Cash grains (corn, soybeans)

Grain, livestock (feeder hogs, cattle) general farming

Hog and cattle raising, general farming, orchards

Diminishing soil quality; increasing distance to primary market

Diminishing soil quality; increasing distance to primary market

(a)   (b)

Figure 10.11

(a) **von Thünen's model.** Recognizing that as distance from the market increases, the value of land decreases, von Thünen developed a descriptive model of intensity of land use that holds up reasonably well in practice. The most intensively produced agricultural crops are found on the land close to the market; the less intensively produced commodities are located at more distant points. The numbered zones of the diagram represent modern equivalents of the theoretical land use sequence von Thünen suggested over 150 years ago. As the metropolitan area at the center increases in size, the agricultural speciality areas are displaced outward, but the relative position of each is retained. (b) A schematic view of the von Thünen zones in the sector south of Chicago. There, farmland quality decreases southward as the boundary of recent glaciation is passed and hill lands are encountered in southern Illinois. On the margins of the city near the market, dairying competes for space with livestock feeding and suburbanization. Southward into flat, fertile central Illinois, cash grains dominate. In southern Illinois, livestock rearing and fattening, general farming, and some orchard crops are the rule.

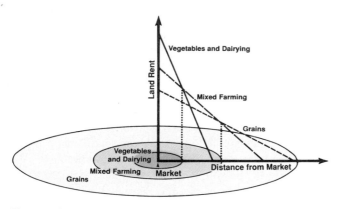

Figure 10.12
Transport gradients and agricultural zones.

The von Thünen model may be modified by introducing ideas of differential transport costs or variations in topography or soil fertility, but with or without such modifications von Thünen's analysis helps explain the changing crop patterns and farm sizes evident on the landscape at increasing distance from major cities. Farmland close to markets takes on high value, is used *intensively* for high-value crops, and is subdivided into relatively small units. Land far from markets is used *extensively* and in larger units.

Intensive Commercial Agriculture

Farmers who apply large amounts of capital (for machinery and fertilizers, for example) and/or labor per unit of land engage in **intensive commercial agriculture**. The crops that justify such costly inputs are characterized by high yields and high market value per unit of land; they include fruits, vegetables, and dairy products, all of which are highly perishable. Near most medium-size and large cities, are dairy farms and *truck farms* (horticultural or "market garden" farms) that produce a wide range of vegetables and fruits. Since the produce is perishable, transport costs increase because of the special handling that is needed, such as refrigerated trucks and custom packaging. This is another reason for locations close to market. Note the distribution of truck and fruit farming in Figure 10.13.

Livestock-grain farming involves the growing of grain to be fed on the producing farm to livestock, which constitute the farm's cash product. In Western Europe, three-fourths of cropland is devoted to production for animal consumption; in Denmark, 90% of all grains are fed to livestock for conversion into meat, butter, cheese, and milk. Although livestock-grain farmers work their land intensively, the value of their product per unit of land is usually less than that of the truck farm. Consequently, in the United States at least, livestock-grain farms are farther from the main markets than are horticultural and dairy farms. In general, the livestock-grain belts of the world are close to the great coastal and industrial zone markets. The Corn Belt of the United States and the livestock region of Western Europe are two examples.

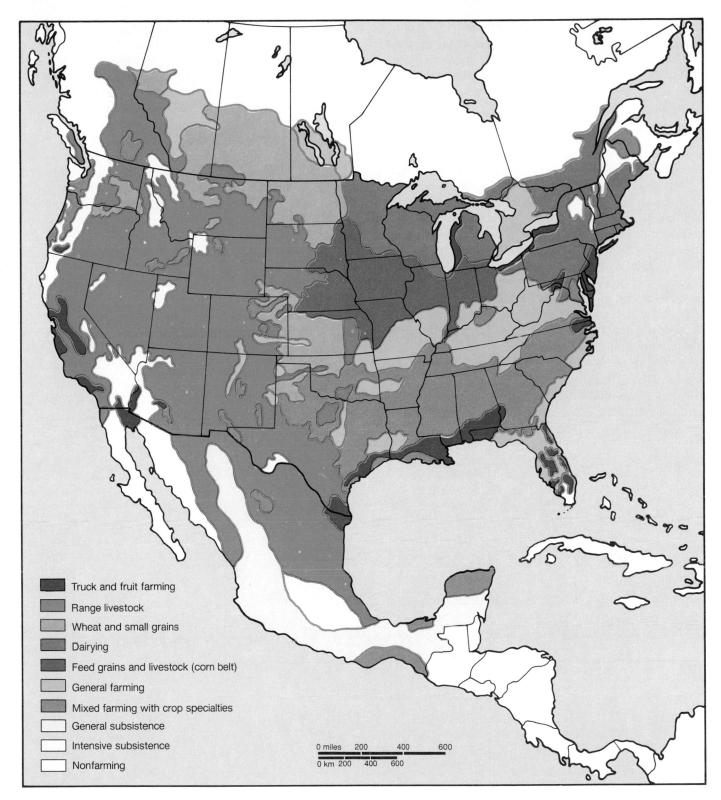

Figure 10.13
Generalized agricultural regions of North America.

Truck and fruit farming
Range livestock
Wheat and small grains
Dairying
Feed grains and livestock (corn belt)
General farming
Mixed farming with crop specialties
General subsistence
Intensive subsistence
Nonfarming

0 miles 200 400 600
0 km 200 400 600

Extensive Commercial Agriculture

Farther from the market, on less expensive land, there is less need to use the land intensively. Cheaper land gives rise to larger farm units. **Extensive commercial agriculture** is typified by large wheat farms and livestock ranching.

Large-scale wheat farming requires sizable capital inputs for planting and harvesting machinery, but the inputs per unit of land are low; wheat farms are very large. In North America, the spring wheat (planted in spring, harvested in autumn) region includes the Dakotas, eastern Montana, and the southern parts of the Prairie Provinces of Canada. The winter wheat (planted in fall, harvested in midsummer) belt focuses on Kansas and includes adjacent sections of neighboring states. Argentina is the only South American country to have comparable large-scale wheat farming. In the Eastern Hemisphere, the system is fully developed only in the Soviet Union on vast state farms east of the Volga River (in northern Kazakhstan and the southern part of Western Siberia) and in southeastern and western Australia.

Livestock ranching differs significantly from livestock-grain farming and, by its commercial orientation and distribution, from the nomadism it superficially resembles. A product of the 19th-century growth of urban markets for beef and wool in Western Europe and northeastern United States, ranching has been primarily confined to areas of European settlement. It is found in western United States and adjacent sections of Mexico and Canada (Figure 10.13); the grasslands of Argentina, Brazil, Uruguay, and Venezuela; the interior of Australia; the uplands of South Island, New Zealand; and the Karoo and adjacent areas of South Africa (Figure 10.14). All except New Zealand and the humid pampas of South America have semiarid climates. All, even the most remote from markets, were a product of improvements in transportation by land and sea, refrigeration of carriers, and meat-canning technology.

In all of the ranching regions, livestock range (and the area exclusively in ranching) has been reduced as crop farming has encroached on its more humid margins, as pasture improvement has replaced less nutritious native grasses, and as grain fattening has supplemented traditional grazing. Since ranching can be an economic activity only where alternative land uses are nonexistent and land quality is low, ranching regions of the world characteristically have low population densities, low capitalizations per land unit, and relatively low labor requirements.

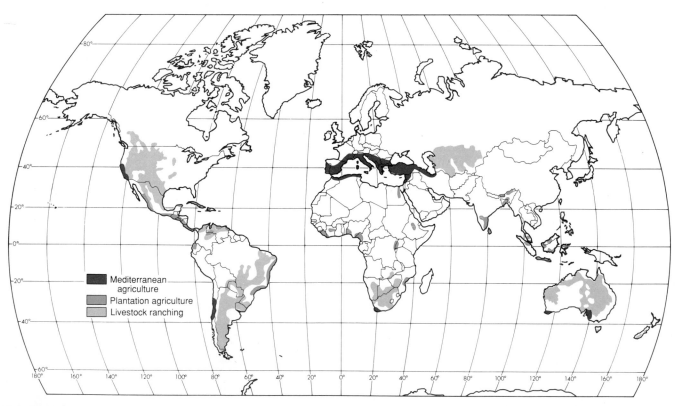

Figure 10.14

Livestock ranching and special crop agriculture. Livestock ranching is primarily a midlatitude enterprise catering to the urban markets of industrialized countries. Mediterranean and plantation agriculture are similarly oriented to the markets provided by advanced economies of Western Europe and North America. Areas of Mediterranean agriculture—all of roughly comparable climatic conditions—specialize in similar commodities, such as grapes, oranges, olives, peaches, and vegetables. The specialized crops of plantation agriculture are influenced by both physical geographic conditions and present or, particularly, former (colonial) control of area.

Special Crops

Proximity to the market does not guarantee the intensive production of high-value crops, should terrain or climatic circumstances prevent it. Nor does great distance from the market inevitably determine that extensive farming on low-priced land will be the sole agricultural option. Special circumstances, most often climatic, make some places far from markets intensively developed agricultural areas. Two special cases are agriculture in Mediterranean climates and in plantation areas (Figure 10.14).

Most of the arable land in the Mediterranean basin itself is planted to grains, and much of the agricultural area is used for grazing. *Mediterranean agriculture* as a specialized farming economy, however, is known for grapes, olives, oranges, figs, vegetables, and similar commodities. These crops need warm temperatures all year round and a great deal of sunshine in the summer. The distribution of Mediterranean agriculture is indicated in Figure 10.14. These are among the most productive agricultural lands in the world. The precipitation regime of Mediterranean climate areas—winter rain and summer drought—lends itself to the controlled use of water. Of course, much capital must be spent for the irrigation systems, another reason for the intensive use of the land for high-value crops.

Climate is also considered the vital element in the production of what are commonly but imprecisely known as *plantation crops*. The implication of **plantation** is the introduction of a foreign element—investment, management, and marketing—into an indigenous culture and economy, often employing an introduced nonnative labor force. Even the crops, although native to the tropics, were frequently foreign to the areas of plantation establishment: African coffee and Asian sugar in the Western Hemisphere; American cacao, tobacco, and rubber in Southeast Asia and Africa are examples (Figure 10.15). Entrepreneurs in Western countries, such as England, France, the Netherlands, and the United States, became interested in the tropics partly because they afforded them the opportunity to satisfy a demand in temperate lands for agricultural commodities not producible in the market areas. "Plantation" becomes an inappropriate term where the foreign element is lacking, as in cola nut production in Guinea, spice growing in India or Sri Lanka, or sisal production in the Yucatán, though custom and convenience frequently retain the term even where native producers of local crops dominate.

The major plantation crops and the areas where they are produced include tea (India and Sri Lanka); jute (India and Bangladesh); rubber (Malaysia and Indonesia); cacao (Ghana and Nigeria); cane sugar (Cuba and the Caribbean area, Brazil, Mexico, India, and the Philippines); coffee (Brazil and Colombia); and bananas (Central America). As Figure 10.14 suggests, for ease of access to shipping, most plantation crops are cultivated along or near coasts since production for export rather than for local consumption is the rule.

Agriculture in Planned Economies

As their name implies, planned economies have a degree of centrally directed control of resources and key sectors of the economy that permits the pursuit of government-determined objectives. When that control is extended to the agricultural sector, state enterprises replace private farms, crop production is divorced from market control, and prices are established by plan rather than by demand or production cost. Such extremes of agricultural control have in recent years been relaxed in most formerly planned economies, including those of the Soviet Union and the People's Republic of China where more local control, private initiative, and market orientation have become the rule. In both of those countries, however, past centralized control of agriculture altered the rural cultural landscape and the organization of rural society. Their experiences help clarify the contrasts between commercial and planned farm economies.

Soviet Agriculture

The collectivization of agriculture beginning in 1928 forcibly deprived 25 million peasant households of their privately operated (but state-owned) holdings, which

Figure 10.15
This Malaysian rubber plantation typifies classical plantation agriculture in general. It was established by foreign capital to produce a nonnative (American) commercial crop for a distant, midlatitude market using nonnative (Chinese) labor supervised by foreign (English) managers. Present-day ownership, management, and labor may have changed, but the nature and market orientation of the enterprise remain.

averaged some 40 acres (15 hectares) in size, and of their livestock, seed reserves, and farm implements. All "voluntarily" joined a local village-based group agricultural enterprise as unpaid members of a producers' cooperative—the **collective farm**. Large-field mechanized agriculture replaced small-plot peasant farming, and government agents rather than market forces established quotas for crop and livestock production based on national planning objectives. Projections of national consumption needs, not farm profitability, determined farm and regional crop specializations.

Under the collective farm system, individual farm families resided in traditional agricultural villages or in newly constructed centralized "agro-towns," jointly working the allotted communal land. Brigades of workers were assigned (or, recently, contract to perform) specific tasks during the crop year. Originally they received remuneration only as a share in farm profit, if any was achieved. In more recent years fixed wages were paid, supplemented by incentive bonuses based on achievement.

About one-half of the some 500 million acres (200 million hectares) of cultivated land in the "socialized" sector was operated by collective farms at the end of the 1980s. The other half was included in **state farms**, government enterprises operated by paid employees of the state, which provided all inputs and claimed all yields. State farms are rural "factories," equivalent in every administrative way to their urban counterparts. They tend to be larger than collectives and more specialized in output.

To provide the subsistence food stuffs no longer part of the larger collective farm operation, *private plots* up to a maximum of about 1.25 acres (0.5 hectare) were granted to collective farm families. State farm workers got about 0.75 acre (0.33 hectare) and urban workers much less. Collective farmers depended on the plots for home-produced food, including a limited number of animals, and for cash income through sale in urban farmers' markets where state price controls did not apply (Figure 10.16). These carefully tended personal plots, together amounting to less than 5% of the Soviet Union's farmland, yielded more than 25% of the value of the gross agricultural output of the country at the end of the 1980s. From them came some 60% of the potatoes, 50% of the fruit, 35% of the eggs and vegetables, and 30% of the meat and milk produced in the USSR. On them, too, were reared a fifth of the sheep, cattle, and hogs.

Programs set in motion by Stalin and his successors fundamentally restructured the geography of agriculture of the Soviet Union. Together, collective and state farms transformed the Soviet countryside from millions of small peasant holdings to a consolidated pattern of fewer than 50,000 centrally controlled operating units. Prerevolutionary Russian agriculture was very largely confined to the European part of the country. To increase the total area available for food and feed grain production, the Virgin and Idle Lands program, launched in 1954, had by

Figure 10.16

A farmers' market in the Soviet Union. For a modest daily fee, collective and state farmers can rent stall space in a government-built and regulated market building to sell commodities produced on their "private" plots. Between one-quarter and one-third of farm family income is earned from the operation of those small properties.

the early 1960s expanded total cultivated land in the USSR by 30% (Figure 10.17). The new farm lands permitted former grain-producing districts to be converted to production of other desired food and industrial crops and livestock.

Under the "restructuring" (*perestroika*) programs of Mikhail Gorbachev, who became national leader in 1985, a greater degree of private initiative and financial reward has been permitted and encouraged in the agricultural sector. "Collective contracts" allocate to groups of farm workers the resources needed to carry out all the farming activities on given parcels of land, paying them according to their output. With a program of "family contracts" introduced in late 1987, permitting state and collective farms to lease land to families or groups of farmers for periods of 12–15 years with the right to sell some of their output on free markets, Soviet agricultural policy once again allowed small-scale private agriculture for profit. More recently, permanent leaseholds with right of inheritance were proposed and private farming on leased lands was suggested as the appropriate standard in Soviet agriculture.

Agriculture in China

A similar but more sweeping progression from private and peasant agriculture, through collectivization, and back to

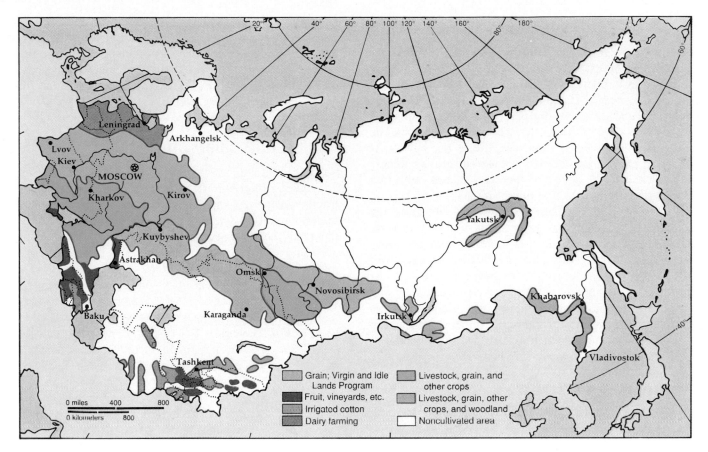

Figure 10.17

The Virgin and Idle Lands program extended grain production, primarily spring wheat, eastward onto marginal land. Wheat constitutes nearly 90% of total Soviet food-grain production and 50% of all grains grown. Total USSR sown area has decreased from its maximum of 538 million acres (218 million hectares) in 1978 as marginal land has been shifted to pasture and fodder crops, and larger areas have been put to fallow to store moisture during one year to use for a subsequent crop.

Map legend:
- Grain; Virgin and Idle Lands Program
- Fruit, vineyards, etc.
- Irrigated cotton
- Dairy farming
- Livestock, grain, and other crops
- Livestock, grain, other crops, and woodland
- Noncultivated area

what is virtually a private farming system has taken place in the planned economy of the People's Republic of China. Between its 1949 assumption of power and 1952, the Communist regime based its agricultural land reform program upon the ownership of land by individual farmers. Approximately 120 million acres (49 million hectares) of farmland were redistributed from "rich" peasants holding 5 to 8 acres (2 to 3 hectares) and "large" landlords with 8 or more acres (3 or more hectares) to about 350 million peasants. The resulting average farm holdings of less than one-half acre (0.2 hectares) created a highly inefficient, undercapitalized subsistence farming system totally inadequate for the growing agricultural needs of the country.

The government solution was, first, to establish cooperatives and, later, to collectivize the cooperatives into *communes* on the model of the Soviet collective farm. By the end of 1957, some 700,000 communes contained about 90% of the peasant households of the country. Except for garden plots, private ownership was abolished. In the 1970s all of China's farmland was again reorganized into 50,000 communes averaging some 13,000 persons, with production goals and directives issued by local governmental officials.

After the death of Chairman Mao in 1976, commune officials began to be stripped of their decision-making powers and once again what became effectively a private farming system was reintroduced. Farmers are not technically the owners of the land they work; they simply rent it for free for the term of a contract, which usually runs 15 years, planting what they choose. Operationally, the land is theirs. They have unrestricted use of the land during the contract period and may even bequeath it to their heirs.

Nearly overnight, 180 million new equal-size farms—all of them too small at an average 1.2 acres (0.5 hectare) to be more than family gardens—were created, and the problem of 1952 was recreated. Nonetheless, agricultural output increased by 4% annually between 1972 and 1978, by 7.5% from 1979–82, and 14% between 1983 and 1985. Productivity gains are attributed to the more efficient use of existing resources and the incentives of private gain. Most staple crops are sold to the government at fixed prices by enforced contract; increasingly, vegetables and meat are sold on the free market.

Other Primary Activities

In addition to agriculture, primary economic activities include fishing, forestry, and mining and quarrying of minerals. These industries involve the direct exploitation of natural resources that are unequally available in the environment and differentially perceived by different societies. Their development, therefore, depends upon the occurrence of perceived resources, the technology to exploit their natural availability, and the cultural awareness of their value. (The definition, perception, and utilization of resources are explored in depth in Chapter 11.)

Two of them—fishing and forestry—are **gathering industries** based upon harvesting the natural bounty of renewable resources, though ones in serious danger of depletion through overexploitation. Livelihoods based on both of these resources are areally widespread, and both involve subsistence and market-oriented components. Mining and quarrying are **extractive industries**, removing nonrenewable metallic and nonmetallic minerals, including mineral fuels, from the earth's crust. They are the initial raw material phase of modern industrial economies.

Fishing and Forestry

Fish as a food resource and forests as a source for building materials, cellulose, and fuel are heavily exploited renewable resources. Livelihoods based on both resources are areally widespread and parts of both subsistence and advanced economies. In both fishing and forestry, evidence is mounting that at least locally, their *maximum sustainable yield* is actually or potentially being exceeded. **Maximum sustainable yield** of a resource is the maximum volume or rate of use that will not impair its ability to be renewed or to maintain the same future productivity. For fishing or forestry, that level is marked by a catch or harvest equal to the net growth of the replacement stock. When sustainable yields are exceeded, stocks may be depleted below recovery levels and livelihoods and economies based upon the resource are endangered or destroyed. Both fishery and forestry resources and exploitations are discussed in Chapter 11.

Mining and Quarrying

Societies at all stages of economic development can and do engage in agriculture, fishing, and forestry. The extractive industries—mining and drilling for nonrenewable mineral wealth—emerged only when cultural advancement and economic necessity made possible a broader understanding of the earth's resources. Now those extractive industries provide the raw material and energy base for the way of life experienced by people in the advanced economies and are the basis for a major part of the international trade connecting the developed and developing countries of the world.

There is no meaningful distinction between the commercial and planned systems of advanced technology in importance and utilization of energy and minerals, though there are contrasts in their respective reactions to cost considerations. Commercial systems are, relatively, more sensitive to changes in the market price of these important components of production.

Transportation costs play a great role in determining where *low-value minerals* will be mined. Minerals such as gravel, limestone for cement, and aggregate are found in such abundance that they have value only when they are near the site where they are to be used (Figure 10.18). For example, gravel for road building has value if it is at the road-building site, not otherwise. Transporting gravel hundreds of miles is an unprofitable activity.

Figure 10.18
The Vancouver, British Columbia, municipal gravel quarry and storage yard. Proximity to market gives utility to low-value minerals unable to bear high transportation charges.

The production of other minerals, especially *metallic minerals,* such as copper, lead, and iron ore, is affected by a balance of three forces: the quantity available, the richness of the ore, and the distance to markets. Even if these conditions are favorable, mines may not be developed or even remain operating if supplies from competing sources are more cheaply available in the market. Between 1980 and 1985 more than 25 million tons of iron ore-producing capacity was permanently shut down in the United States and Canada. Similar declines occurred in North American copper, nickel, zinc, lead, and molybdenum mining as market prices fell below domestic production costs. Of course, increases in mineral prices may be reflected in opening or reopening mines that, at lower returns, were deemed unprofitable. However, the developed industrial countries of commercial economies, whatever their former or even present mineral endowment, frequently find themselves at a competitive disadvantage against Third World producers with lower cost labor and state-owned mines with abundant, rich reserves.

When the ore is rich in metallic content, it is profitable to ship it directly to the market for refining. But, of course, the highest-grade ores tend to be mined first. Consequently, the demand for low-grade ores has been increasing in recent years as richer deposits have been depleted. Low-grade ores are often upgraded by various types of separation treatments at the mine site to avoid the cost of transporting waste materials not wanted at the market. Concentration of copper is nearly always mine-oriented (Figure 10.19); refining takes place near areas of consumption.

The large amount of waste in copper (98–99% or more of the ore) and in most other industrially significant ores should not be considered the mark of an unattractive deposit. Indeed, the opposite may be true. Many higher-content ores are left unexploited—because of the cost of extraction or the smallness of the reserves—in favor of the utilization of large deposits of even very low-grade ore. The attraction of the latter is a size of reserve sufficient to justify the long-term commitment of development capital and, simultaneously, to assure a long-term source of supply. At one time, high-grade magnetite iron ore was mined and shipped from the Mesabi area of Minnesota. Those deposits are now exhausted. Yet immense amounts of capital have been invested in the mining and processing into high-grade iron ore pellets of the virtually unlimited supplies of low-grade, iron-bearing rock (taconite) still remaining.

Such investments do not assure the profitable exploitation of the resource. The metals market is highly volatile. Rapidly and widely fluctuating prices can quickly change profitable mining and refining ventures to losing undertakings. Marginal gold and silver deposits are opened or closed in reaction to trends in precious metals prices. Taconite beneficiation (waste material removal) in the Lake Superior region has virtually ceased in response to the decline of the U.S. steel industry and the price advantage of imported ores. In commercial economies, cost and market controls dominate economic decisions.

In planned economies, cost may be a less important consideration than are other controls on mining. In the Soviet Union, for example, a fundamental premise of Stalinist programs for the development of the national economy was that, whatever the cost, the country should fully utilize its domestic resources and should, to the extent

Figure 10.19
Copper ore concentrating and smelting facilities at the Phelps-Dodge mine in Morenci, Arizona. Concentrating mills crush the ore, separating copper-bearing material from the rocky mass containing it. The great volume of waste material removed assures that most concentrating operations are found near the ore bodies. Smelters separate concentrated copper from other, unwanted, minerals such as oxygen and sulfur. Because smelting is also a "weight-reducing" activity, it is frequently—though not invariably—located close to the mine as well.

possible, free itself from dependence upon foreign sources of supply of important industrial minerals. Following from that principle, massive investment has been made in transportation improvement and urban construction to exploit mineral deposits in formerly remote and undeveloped sections of the northern and eastern reaches of the country. Cost has been a less persuasive control.

The advanced economies, whether commercial or planned, have reached that status through their control and use of energy. Domestic supplies of mineral fuels, therefore, are often considered basic to national strength and independence. When those supplies are absent, developed countries are concerned participants in the availability and price of coal, oil, and natural gas in international trade. Mineral fuels and other energy sources are given extended discussion in Chapter 11.

Trade in Primary Products

Primary commodities—agricultural goods, minerals, and fuels—make up an important portion of total international trade. The world distribution of supply and demand for those items results in an understandable pattern of commodity flow: from the producers located within less developed countries to the processors, manufacturers, and consumers of the more developed ones (Figure 10.20). The trade benefits the developed countries by providing access to a continuing supply of industrial raw materials and food stuffs not available domestically. It provides less developed states with capital to invest in their own development or to expend on the importation of manufactured goods, food supplies, or commodities—such as petroleum—they do not themselves produce. The exchange has, however, been criticized as unequal and potentially damaging to commodity exporting countries.

Because commodity prices are volatile, rising sharply in periods of product shortage or international economic growth and, likely, falling abruptly with supply glut or international recession, commodity exporting countries risk disruptive fluctuations in levels of income. This, in turn, can create serious difficulties in economic planning and debt repayment.

Producers' cartels committed to production control and price maintenance have generally proved ineffective. This is so primarily because commodity selling countries are many, small, and weak in comparison to the industrialized market states. Often, there are many alternative sources of supply for a given commodity. Even OPEC (Organization of Petroleum Exporting Countries), once in near-total command of international petroleum supply and pricing, experienced a drop in share of world production from 57% in 1975 to only 30% a decade later and in income as non-OPEC countries expanded production and prices tumbled.

While prices paid for commodities tend to be low, prices charged for the manufactured goods offered in exchange by the developed countries tend to be high. To capture processing and manufacturing profits for themselves, some developing states have placed restrictions upon the export of unprocessed commodities. Malaysia, Philippines, and Cameroon, for example, have limited the export of logs in favor of increased domestic processing of sawlogs and exports of lumber; in 1985 Indonesia introduced a total ban on unprocessed log exports. They have also encouraged domestic manufacturing to reduce imports and to diversify their exports. Frequently, however, such exports meet with tariffs and quotas protecting the home markets of the industrialized countries, and the disparities in the economic roles and prospects of the developed and less developed countries (the "North–South" split seen on Figure 7.7) are continued in the established pattern of trade in primary commodities.

Secondary Activities: Manufacturing

Although primary industries are locationally tied to the natural resources that they gather or exploit, secondary and later stages of economic activity are less concerned with conditions of the physical environment. For them, enterprise location is more closely related to cultural and economic than to physical circumstances. They are movable rather than spatially tied, and are assumed to respond to recurring locational requirements and controls.

Those controls are rooted in observations about human spatial behavior in general and economic behavior in particular. We have already explored some of those assumptions in earlier discussions. We noted, for example, that the intensity of spatial interaction decreases with increasing separation of points—distance decay, we called it. Von Thünen's model of agricultural land use, you will recall, was rooted in conjectures about transportation cost and land value relationships.

Such simplifying assumptions help us to understand a presumed common set of controls and motivations guiding human economic behavior. We assume, for example, that people are *economically rational,* that is, given the information at their disposal, they make locational, production, or purchasing decisions in light of a perception of what is most cost-effective and advantageous. From the standpoint of producers or sellers of goods or services in commercial economies, it is assumed each is intent upon *maximizing profit*. To reach that objective, a host of production and marketing costs and political, competitive, and other limiting factors may be considered but the ultimate goal of profit-seeking remains clear. Finally, it is assumed that in commercial economies the best measure of the correctness of economic decisions is afforded by the market mechanism and the equilibrium between supply and demand that market prices establish (Figure 10.21).

Figure 10.20
Sugar being loaded for export at the port of Cebu in the Philippines. Much of the Third World depends on exports of mineral and agricultural products to the developed countries for the major portion of its income. Fluctuations in market demand and price of some of those commodities can have serious and unexpected consequences.

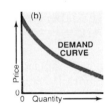

Figure 10.21
Supply, demand, and market equilibrium. The regulating mechanism of the market may be visualized graphically. (a) The *supply curve* tells us that as the price of a good increases, more of that good will be made available for sale. Countering any tendency for prices to rise to infinity is the market reality that the higher the price, the smaller the demand as potential customers find other purchases or products more cost-effective. (b) The *demand curve* shows how the market will expand as prices drop and the good becomes more affordable and attractive to more customers. (c) *Market equilibrium* is marked by the point of intersection of the supply and demand curves and determines the price of the good, the total demand, and the quantity bought and sold.

Industrial Locational Models

In commercial economies, entrepreneurs seek to maximize profits by locating manufacturing activities at sites of lowest total input costs (and high revenue yields). In order to assess the advantages of one location over another, industrialists must evaluate the most important **variable costs.** They subdivide their total costs into categories and note how each cost will vary from place to place. In different industries, transportation charges, labor rates, power costs, plant construction or operation expenses, the interest rate of money, or the price of raw materials may be the major variable cost. The industrialist must look at each of these and by a process of elimination eventually select the lowest-cost site. If the producer then determines that a large enough market can be reached cheaply enough, the location promises to be profitable.

In the economic world, nothing remains constant. Because of a changing mix of input costs, production techniques, and marketing activities, most initially profitable locations do not remain advantageous. Migrations of population, technological advances, and changes in the demand for products affect industrialists and industrial locations greatly. The abandoned mills and factories of New England or the steel towns of Pennsylvania, even the "deindustrialization" of America itself in the face of foreign competition, are testimony to the impermanence of the "best" locations.

The concern with variable costs as a determinant in industrial location decisions has inspired an extensive theoretical literature. Much of it is based upon and extends the **least cost theory** proposed by the German location economist Alfred Weber (1868–1958) and sometimes called *Weberian analysis.* Weber explained the optimum location of a manufacturing establishment in terms of minimization of three basic expenses: relative transport costs, labor costs, and agglomeration costs. **Agglomeration** refers to the clustering of productive activities and people for mutual advantage. Such clustering can produce *agglomeration economies* through shared facilities and services. Diseconomies, such as higher rents or wage levels resulting from competition for these resources, may also occur.

Weber concluded that transport costs were the major consideration determining location. That is, the optimum location would be found where the costs of transporting raw materials to the factory and finished goods to the market were at their lowest (Figure 10.22). He noted, however, if variations in labor or agglomeration costs were sufficiently great, a location determined solely on the basis of transportation costs might not in fact be the optimum one.

Assuming, however, transportation costs determine the "balance point," optimum location will depend on distances, the respective weights of the raw material inputs, and the final weight of the finished product. It may be either *material oriented* or *market oriented.* Material orientation reflects a sizable weight loss during the production process; market orientation indicates a weight gain (Figure 10.23).

For some theorists, Weber's least cost analysis is unnecessarily rigid and restrictive. They propose, for example, a *substitution principle* that recognizes that in many industrial processes it is possible to replace a declining amount of one input (e.g., labor) with an increase in another (e.g., capital for automated equipment) or to increase transportation costs while simultaneously reducing land rent. With substitution, a number of different points may be optimal manufacturing locations. Further, they suggest, a whole series of points may exist where total revenue of an enterprise just equals its total cost of producing a given output. These points, connected, mark the *spatial margin of profitability* and define the area within which profitable operation is possible (Figure 10.24). Location anywhere within the margin assures some profit and tolerates both imperfect knowledge and personal (rather than economic) considerations.

Other Locational Considerations

The behavior of individual firms seeking specific production sites under competitive commercial conditions forms the basis of most industrial location theory. But such theory does not fully explain world or regional patterns of industrial localization or specialization, nor does it account for locational behavior that is uncontrolled by objective "factors," or that is directed by noncapitalistic planning goals.

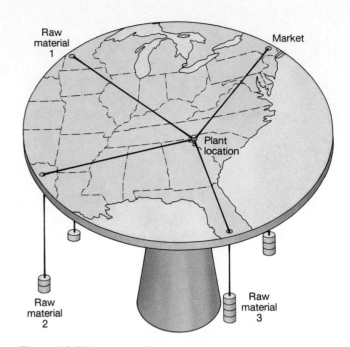

Figure 10.22

Plane table solution to a plant location problem. This mechanical model, suggested by Alfred Weber, uses weights to demonstrate the least transport cost point where there are several sources of raw materials and a single market. When a weight is allowed to represent the "pulls" of raw materials and market, an equilibrium point is found on the plane table. That point is the location at which all forces balance each other and represents the least-cost plant location.

Transport Characteristics

For example, within both national and international economies, the type and efficiency of transport media, as well as transportation costs, are central to the spatial patterning of production, explaining the location of a large variety of economic activities. Waterborne transportation is nearly always cheaper than any other mode of conveyance, and the enormous amount of commercial activity that takes place on coasts or on navigable rivers leading to coasts is an indication of that cost advantage. When railroads were developed and the commercial exploitation of inland areas could begin, coastal sites continued to be important as more and more goods were transferred there between low-cost water and land media. The advent of highway transportation vastly increased the number of potential "satisficing" manufacturing locations by freeing the locational decision from fixed-route production sites. Every change in carrier mode, efficiency, or cost structure has direct implications for locations of economic activity within commercial economies.

In the rare instance when transportation costs become a negligible factor in production and marketing, an economic activity is said to be *footloose.* Some manufacturing activities are located without reference to raw materials; for example, the raw materials for electronic products such as computers are so valuable, light, and compact that transportation costs have little bearing on

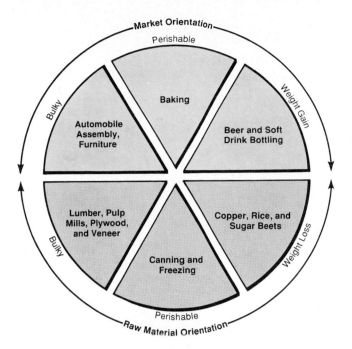

Figure 10.23

Spatial orientation tendencies. *Raw material orientation* is presumed to exist when there are limited alternative material sources, when the material is perishable, or when, in its natural state, it contains a large proportion of impurities or nonmarketable components. *Market orientation* represents the least-cost solution when manufacturing uses commonly available materials that add weight to the finished product, when the manufacturing process produces a commodity much bulkier or more expensive to ship than its separate components, or when the perishable nature of the product demands processing at individual market points.

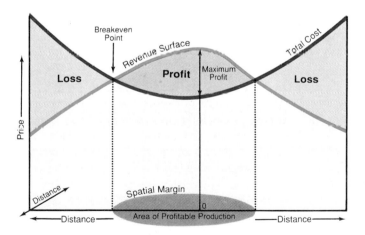

Figure 10.24

The spatial margin of profitability. In the diagram, O is the single optimal profit-maximizing location, but location anywhere within the area defined by the intersects of the total cost and total revenue surfaces will permit profitable operation. Some industries will have wide margins, others will be more spatially constricted. Skilled entrepreneurs may be able to expand the margins farther than less able industrialists. Importantly, a *satisficing* location may be selected by reasonable estimate even in the absence of the totality of information required for an *optimal* decision.

where production takes place. Others are inseparable from the markets they serve and are so widely distributed that they are known as *ubiquitous industries.* Examples of these are newspaper publishing, bakeries, and dairies, all of which produce a highly perishable commodity designed for immediate consumption.

Agglomeration Economies

The cumulative and reinforcing attractions of industrial concentration and urban growth are recognized locational factors, but ones not easily quantified. Both cost-minimizing and profit-maximizing theories make provision for *agglomeration,* the spatial concentration of people and activities for mutual benefit. That is, both recognize that areal grouping of industrial activities may produce benefits for individual firms that they could not experience in isolation. Those benefits—**external economies** or *agglomeration economies*—accrue in the form of savings from shared transport facilities, social services, public utilities, communication facilities, and the like. Collectively, these and other installations and services needed to facilitate industrial and other forms of economic development are called *infrastructure.* Areal concentration may also create pools of skilled and ordinary labor, of capital, ancillary business services, and, of course, a market built of other industries and urban populations. New firms, particularly, may find significant advantages in locating near other firms engaged in the same activity, for labor specializations and support services specific to that activity are already in place. Some may find profit in being near other firms with which they are linked either as customers or suppliers.

A concentration of capital, labor, management skills, customer base, and all that is implied by the term *infrastructure* will tend to attract still more industries from other locations to the agglomeration. In Weber's terms, economies of association distort or alter locational decisions that otherwise would be based solely upon transportation and labor costs, and once in existence agglomerations tend to grow (Figure 10.25). Through a *multiplier effect,* each new firm added to the agglomeration will lead to the further development of infrastructure and linkages. As we shall see in Chapter 12, the "multiplier effect" also implies total (urban) population growth and thus the expansion of the labor pool and the localized market that are part of agglomeration economies.

Comparative Advantage

A third consideration affecting industrial location and specialization is the principle of **comparative advantage.** It tells us that areas tend to specialize in the production of those items for which they have the greatest relative advantage over other areas, or for which they have the least relative disadvantage, as long as free trade exists. The principle, basic to the understanding of regional specializations, applies as long as areas have different relative advantages for two or more goods.

Figure 10.25

On a small scale, the planned industrial park furnishes its tenants external agglomeration economies similar to those offered by large urban concentrations to industry in general. An industrial park provides a subdivided tract of land developed according to a comprehensive plan for the use of (frequently) otherwise unconnected firms. Since the park developers, whether private companies or public agencies, supply the basic infrastructure of streets, water, sewage, power, transport facilities, and, perhaps, private police and fire protection, park tenants are spared the additional cost of providing these services themselves. In some instances factory buildings are available for rent, still further reducing firm development outlays. Counterparts of industrial parks for manufacturers are the office parks, research parks, science parks, and the like for "high-tech" firms and for enterprises in tertiary and quaternary services.

Assume that two countries both have a need for, and are able successfully to produce domestically, two commodities. Further assume that there is no transport cost consideration. No matter what its cost of production of either commodity, Country A will choose to specialize in only one of them if, by that specialization and through exchange with Country B for the other, A stands to gain more than it loses. The key to comparative advantage is the utilization of resources in such a fashion as to gain, by specialization, a volume of production and a selling price that permit exchange for a needed commodity at a cost level that is below that of the domestic production of both.

At first glance, the concept of comparative advantage may at times seem to defy logic. For example, Japan may be able to produce airplanes and home appliances more cheaply than the United States, thereby giving it an apparent advantage in both goods. But it benefits both countries if they specialize in the good in which they have a comparative advantage. In this instance, Japan's lower labor costs make it more profitable for Japan to specialize in the volume production of appliances and to buy airplanes from the United States, where large civilian and military markets encourage aircraft manufacturing specialization and efficiency.

When other countries' comparative advantages reflect lower labor, land, raw material, and capital costs, manufacturing activities may voluntarily relocate from higher-cost market locations to lower-cost foreign production sites. Such voluntary *out-sourcing*—producing parts or products abroad for domestic sale—by American manufacturers has employment and areal economic consequences no different from those resulting from successful competition by foreign companies or from industrial locational decisions favoring one section of the country over others.

A North American case in point is found along the northern border of Mexico. In the 1960s Mexico enacted legislation permitting foreign (specifically, American) companies to establish "sister" plants, called *maquiladoras,* within 12 miles of the U.S. border for the duty-free assembly of products destined for reexport. By the end of the 1980s, more than 1500 such assembly and manufacturing plants had been established to produce a diversity of goods including electronic products, textiles, furniture, leather goods, toys, and automotive parts. The plants generated direct employment for nearly half a million Mexican workers, many of them replacing U.S. employees (Figure 10.26).

Figure 10.26
American manufacturers, seeking lower labor costs, have increasingly established component manufacturing and assembly operations along the international border in Mexico. The finished or semifinished products are brought duty-free into the United States. In some cases, as here at a John Deere plant in Monterrey, Mexico, factories farther from the border are more traditional manufacturers, though the product is still destined for export. *Out-sourcing* has moved a large proportion of American electronics, small appliance, and garment industries to offshore subsidiaries or contractors in Asia and Latin America.

Imposed Considerations

Locational theories dictate that in a pure, competitive, commercial economy, the costs of material, transportation, labor, and plant should be dominant in locational decisions. Obviously, neither in the United States nor in any other commercial country do the idealized conditions exist. Other constraints—some representing cost considerations, others political or social impositions—also affect, perhaps decisively, the locational decision process. Land-use and zoning controls, environmental quality standards, governmental area-development inducements, local tax-abatement provisions or developmental bond authorizations, noneconomic pressures on quasi-governmental corporations, and other considerations constitute attractions or repulsions for industry outside of the context and consideration of pure theory. If these noneconomic forces become compelling, the assumptions of the commercial economy classification no longer obtain, and the transition to a controlled mixed or wholly planned economy has occurred.

Industrial Location in Planned Economies

The theoretical controls on plant location decisions that apply in commercial economies are not, by definition, determinant in centrally planned Marxist economies. In them, plant locational decisions are made by government planners rather than by individual firms. The shift in decision making from company to bureaucracy does not mean that economic logic is forgotten or that locational decisions based upon factor costs are unattainable. It does mean that central planners may, and do, incorporate into their deliberations and decisions other than the purely economic considerations that govern the locational process in commercial economies.

At the same time, under a totally planned Communist system locational decisions must be made without the informational guidance of accurate cost, comparative value, and demand information available in commercial economies through market prices. Although the same factors of production may be considered in establishing new plant locations, they cannot be weighted or evaluated with the same impersonal precision possible in commercial economies. In the USSR of the past, product prices were set, output targets established, and industrial inputs allocated by command of one of the more than 50 economic ministries created over the years to direct the national economy; they did not reflect judgment of the marketplace.

Recent administrations, particularly those of Yuri Andropov and Mikhail Gorbachev, have tried to decentralize planning control and introduce a modicum of market pricing to the industrial sector to reduce its rigidities and make more efficient an economy officially listing 20 million articles on its "index of products." Gorbachev's *perestroika* program aimed to transfer more decision-making power to locally elected officials and factory managers, and to transform state controlled industry into "self-financing enterprises" that must fund their own costs and investments from profits.

Even if fully successful, market pricing in centrally planned economies where it is now being introduced still would not be the sole determinant of industrial locational decisions. The reason lies in other developmental objectives influencing or determining the decisions of the central planners. Important in the Soviet Union, for example, has been a national policy of the *rationalization of industry* through full development of the resources of the country wherever they might be found and without regard to the cost or competitiveness of such development.

Rationalization reflects a Soviet application of the Marxist-Leninist call for *industrial diversification* and *regional self-sufficiency*. These developmental goals required each section of the Soviet Union to contain a variety of industrial types (and, if possible, a local agricultural base) to assure its independent operation even if its connections to other sections of the country should be severed.

The **territorial production complex** has been the planning mechanism created to achieve such economic development. It is an areally based organizational form assigned responsibility to develop necessary regional infrastructure, to facilitate specialization of individual enterprises, and to promote overall regional economic growth and integration. Inevitably, although the factors of industrial production are identical in capitalist and noncapitalist economies, the philosophies and patterns of industrial location and areal development differ between them.

Major Manufacturing Regions of the World

Whether locational decisions are made by private entrepreneurs or central planners, the results over many years have produced a distinctive world pattern of manufacturing that shows a striking prominence of a relatively small number of major industrial concentrations localized within a relatively few of the world's countries (Figure 10.27). These are not all countries of the developed world. Mexico, Brazil, China, and other countries of the developing world have created industrial regions of international significance, and the contribution to world manufacturing activity of the smaller newly industrializing countries (NICs) has been growing significantly.

Within the global industrial pattern outlined on Figure 10.27, the major manufacturing regions in North America; in Western, Central, and Eastern Europe; and in Eastern Asia are most prominent. However, also part of that globe-girdling industrial belt are a set of secondary centers, some of which are beginning to assume the status of major concentrations in their own right.

North America

The importance of manufacturing in North America, both as an employer of labor and as a contributor to the national income of Canada and the United States has been steadily declining. In 1960, the 25% of the labor force engaged in manufacturing generated nearly 30% of the region's wealth. By the late 1980s, manufacturing employment had dropped to well under 20% of a much larger labor force, and industry contributed only about a fifth of the gross developmental product of the Anglo-American realm.

Manufacturing is found throughout the urbanized sections of North America but is not uniformly distributed. Its primary concentration is in the northeastern United States and adjacent sections of southeastern

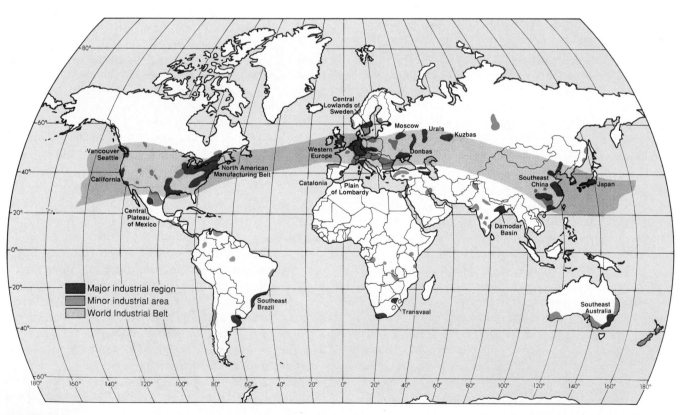

Figure 10.27

World industrial regions. Industrial districts are not as continuous or "solid" as the map suggests. Manufacturing is a relatively minor user of land even in the areas of its greatest concentration. A loose spatial association of major industrial districts extends, in an industrial belt, from Western Europe eastward to the Ural Mountains and, through outliers in Siberia, to the Far East. The belt picks up again on the west coast of North America, though its major American concentration lies east of the Mississippi River. More than 75% of the volume and 80% of the value of manufacturing production of the world is found within this global industrial belt.

Canada, the *North American manufacturing belt*. That belt extends from the St. Lawrence Valley on the north (though excluding northern New England) to the Ohio Valley, and from the Atlantic coast westward to just beyond the Mississippi River (Figure 10.28). Covering less than 5% of the land area of North America, the manufacturing belt contains the majority of the urban population of the continent, its densest and best developed transportation network, the largest number of its manufacturing establishments, and most of its heavy industry. It is, as well, part of the continent's agricultural heartland. These are all items of interrelated influence on the past history and present structure of industry within the belt.

North American manufacturing began early in the 19th century in southern New England, where water-powered textile mills, iron plants, and other small-scale industries began to free Canada and the new United States from total dependence upon European—particularly English—sources. The U.S. eastern seaboard still remains

an important producer of consumer goods, light industrial, and high-technology products on the basis of its market and developed labor skills. Important concentrations of both are found in *Megalopolis*[1], a 600-mile-long city-system stretching from southern Maine to Norfolk, Virginia—one of the longest, largest, most populous of the world's great metropolitan chains. As a major consumer concentration, it has attracted a tremendous array of market-oriented industries and thousands of individual industrial plants served by the largest and most varied labor force in the country.

The heart of the North American manufacturing belt developed across the Appalachians in the interior of the continent in the valleys of the Ohio River and its tributaries and along the shores of the Great Lakes. The rivers and the Lakes provided the early "highways" of the interior, supplemented by canals built throughout the eastern

1. A *megalopolis* or *conurbation* is an extended urbanized area formed by the gradual merger of several individual cities.

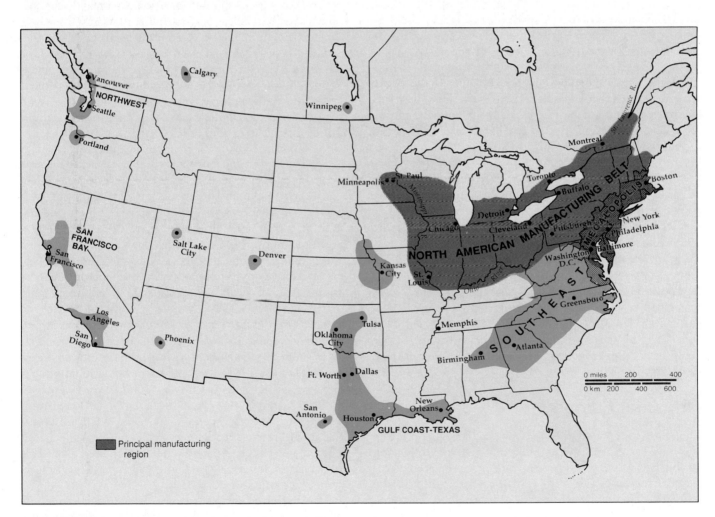

Figure 10.28

North American manufacturing districts. More precisely showing the Anglo-American manufacturing districts, the map (and text) excludes the emerging major manufacturing regions of the rapidly industrializing economy of Mexico.

Middle West of the United States during the 1830s and 1840s. They were followed after the 1850s by the railroads that tied together the agricultural and industrial raw materials, the growing urban centers, and the multiplying industrial plants of the interior with both the established eastern and the developing western reaches of the continent. The raw material endowment of the region was impressive: anthracite and bituminous coking and thermal coals, iron ore, petroleum, copper, lead and other nonferrous metals, timber resources, and agricultural land of richness and extent unknown elsewhere in the world.

The rule of life is change, and changing circumstances have altered the industrial strength and production mix of the interior manufacturing belt. Parts of it have been dubbed the "Rust Belt" to signify deterioration and abandonment of plants formerly dominating the metals industries but now outmoded and obsolete, unable to compete even in their home market with more efficient, cheaper, and more technologically advanced foreign producers. Raw material depletion, energy cost increases, water and air pollution control costs and restrictions, and direct and indirect labor costs that are high by the standards of competing countries have served to alter the industrial mix and patterns of success of the past.

Some of the same problems of change and competition have afflicted the Canadian portion of the North American manufacturing belt. Canadian industry is highly localized. About one-half of the country's manufacturing labor force is concentrated in southern Ontario, every bit the equivalent in industrial development and mix of its U.S. counterparts in Upstate New York and nearby Pennsylvania, Ohio, and eastern Michigan. Toronto forms the hub, but the industrial belt extends westward to Windsor, across from Detroit.

For much of the 19th century and as late as the late 1960s, the manufacturing belt contained from half to two-thirds of North American industry. By the late 1980s, however, its share had dropped to near 40%. Not only was the position of manufacturing declining in the North American economy, but what remained was continuing a pattern of relocation reflecting national population shifts and changing material and product orientations. Figure 10.28 indicates the location of some of the larger of these secondary industrial zones.

Industrial Europe

The Industrial Revolution that began in England in the 1730s and spread to the continent during the 19th century established western and central Europe as the premier industrial region of the world and the source area for the diffusion of industrialization across the globe. Although industry is part of the economic structure of every section and every metropolitan complex of Europe, the majority of industrial production is concentrated in a set of distinctive districts stretching from the Midlands of England in the west to the Ural Mountains in the east (Figure 10.29).

The textile industry of England was the start of the Industrial Revolution and of the areal concentrations and specializations that accompanied its spread. Water-powered mechanical spinning and weaving in Lancashire, Yorkshire, and the English Midlands established the model, but it was steam power, not water power, that provided the impetus for the industrialization of that country and of Europe as a whole. Consequently, coal fields, not rivers, were the sites of the new steel-making and diversified manufacturing districts in the English Midlands, Lancashire, Yorkshire, and the Scottish Lowlands. Although it was remote from coal deposits or other natural resources, London became the largest single manufacturing center of the United Kingdom. Already sizable at the time of the Industrial Revolution, its consumers and labor force were potent magnets for new industry.

Technologies developed in Britain spread to the continent. The coal fields distributed in a band across northern France, Belgium, central Germany, northern Czechoslovakia, southern Poland, and eastward to the southern Ukraine, as well as iron ore deposits, localize the metallurgical industries to the present day. Other pronounced industrial concentrations focus upon the major metropolitan districts and capital cities of the countries of Europe. In all districts, product specialization has been the tradition and the basis for regional exchange.

The largest and most important single industrial area of Europe today extends from the Sambre-Meuse area of the French-Belgium border to the Ruhr district of West Germany. Its core is the Ruhr, a compact, highly urbanized industrial concentration of more than 10 million inhabitants in over 50 major cities. Iron and steel, textiles, automobiles, chemicals, and all the metal-forming and metal-using industries of modern economies are found here in the hub of Germany's industrial structure (Figure 10.30). In France, heavy industry located near the iron ore of Nancy and the coal of Lille, which also specialized in textile production. Like London, Paris lacks raw materials, but as an early center of population and commerce with easy access to the sea and to the domestic market, it became and remains the major manufacturing center of France. Its population (including suburbs) of some 9 million make it the largest city in mainland Western Europe, and its full range of consumer goods, transport, electrical, and engineering equipment, and of such specialty items as perfumes, toiletries, and high-fashion garments reflects its market pull and labor skills.

The East German Saxony district began to industrialize as early as the 1600s, in part benefiting from labor skills brought by immigrant artisans from France and Holland. Those skills have been preserved in a district

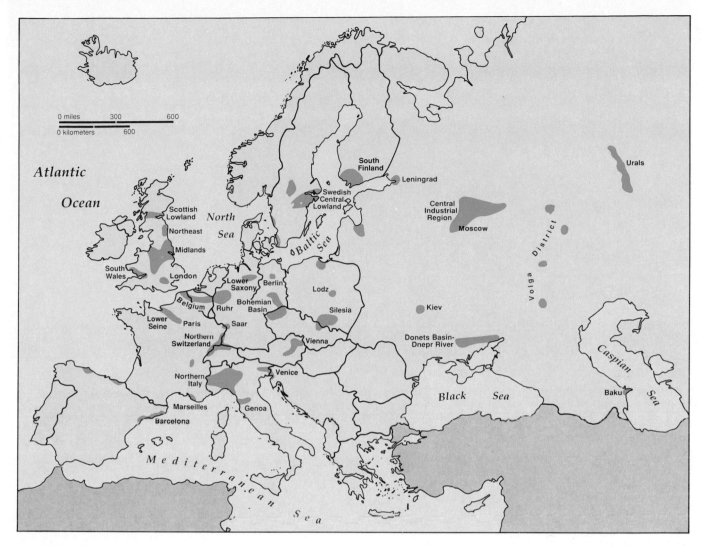

Figure 10.29
The industrial regions of Europe.

noted for the quality of its manufactured goods, including porcelain, textiles, chemicals, printing and publishing, and precision machinery and optical goods. Czechoslovak industry, too, is noted for diversity and quality. Heavy industry—iron and steel, transportation equipment, armaments, chemicals, and the like—characterizes the manufacturing pattern of much of the rest of the Eastern European region of Poland, Hungary, Bulgaria, Romania, and Yugoslavia. (See: "Ideology, Industrialization, and Environmental Disaster.")

European USSR has two distinct industrial orientations. One is light industrial, market-oriented production primarily focused on the Central Industrial Region of Greater Moscow and surrounding areas (Figure 10.31), where major metropolitan markets, superior transportation systems, centrality within the European portion of the country, and a readily available labor force have been the attraction and support. Leningrad is a stand-alone industrial district of the northwest, noted for labor skills since

its founder, Peter the Great, early in the 18th century brought in west European artisans to create a modern industrial base. Its industries still reflect those labor skills.

The other orientation is heavy industrial and is localized in the southern Ukrainian Donets Basin-Dnepr River district where coking coal, iron ore, fluxing materials, and ferro-alloys are found near at hand. Nearly Ruhrlike in their intensity of development and urbanization, the Donets Basin and adjacent areas were the sole source of metallurgical industry throughout late czarist times. Only the Stalinist five-year plans, with their emphasis on creation of multiple sources of supply of essential industrial goods, gave the district rivals elsewhere in the Soviet Union.

Prior to the Revolution of 1917, 90% of Russian industry was found to the west of the Volga River. The pattern of new industrial districts shown in Figure 10.31 represents the Stalinist (and later) determination to develop fully all the resources of the country, to create new

Figure 10.30

The scenic Rhine River serves as a main artery of industrial Western Europe, flowing past Europe's premier industrial complex, the Ruhr district. For western West Germany, the Rhine is the convenient connection to the North Sea and the Atlantic through the port of Rotterdam in the Netherlands. Thanks to its massive volume of river and ocean traffic, Rotterdam is Europe's leading port and the world's largest port in tonnage handled.

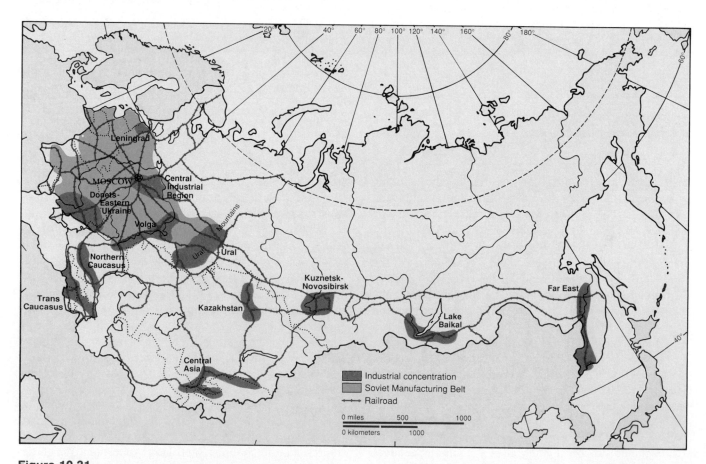

Figure 10.31

Industrial regions of the USSR. The Volga, the Central Industrial, and the Leningrad concentrations within the Soviet manufacturing belt are not dependent on local raw materials. All other USSR industrial regions have a strong orientation to materials and were developed, by plan, despite their distance from the population centers and markets of European Russia.

Ideology, Industrialization, and Environmental Disaster

Marxist theory maintains that under socialism there can be no environmental problems since, where there is no profit motive, humans and nature inherently are in harmony. That comforting philosophy does little to reduce the evident and growing environmental degradation that unrestricted and reckless industrial development has brought to Eastern Europe. Uncontrolled emissions from smelters, refineries, steel furnaces, coke ovens, chemical plants, and cities themselves have imposed enormous damages upon the forests, rivers, air, and soil of the region and upon the health and life expectancy of its populations.

Poland has identified 27 ecologically endangered areas. Five of these, containing 35% of the country's population, have been described as "ecological disasters." In them, the incidence of cancer is one-quarter to one-third higher than the national average, respiratory diseases are 50% higher, and life expectancy averages at least 2 years below the rest of the country. In Poland as a whole, air, water, and soil pollution are severe to the point that the health of at least one-third of the population is at serious risk. Dumping of untreated industrial and urban waste has so polluted the Vistula River that citizens are advised not to give babies tap water, even after boiling it. Nearly 450 cities and 3600 industrial plants are totally without facilities to treat wastewater. One authority predicts that the decline of Poland's water quality is so extreme and widespread that before the turn of the century the country's entire supply—from wells, lakes, and rivers—may be unfit for *any* use. Base metal contamination of soil and food products has markedly increased the number of mentally deficient children in Upper Silesia.

Poland, which has more detailed and restrictive regulations for environmental protection than any of the other Eastern European states, is simply the worst of a series of bad national situations. In East Germany, air pollution from power plant emissions has effectively destroyed the forests around Halle and Leipzig, where more than half of the country's power stations are concentrated. Life expectancy is reported to be considerably lower there than the national average. Hungary produces some 5 million tons of hazardous waste each year and imports for disposal thousands of additional tons from Austria, Switzerland, and West Germany—all without adequate disposal facilities. Until recently, Czechoslovakia forbade emigration from heavily industrialized northern Bohemia, but offered citizens higher wages—a form of "hazardous duty" pay—to compensate them for the health dangers of remaining in areas of extreme pollution.

In their drive to establish strong national economies and to match the level of industrial development of Western Europe, the centrally planned economies of Eastern Europe until recently ignored the environmental consequences of their programs. They are now suffering from their decisions along with areas far beyond their borders affected by air-borne pollutants. Ideological assumptions cannot alter industrial realities.

---

industrial bases at resource locations, and to strengthen the national economy through construction of multiple districts of diversified industry. The industrialization of the Urals, the Kuznetsk Basin, Kazakhstan, Central Asia, and the Soviet Far East resulted from those programs first launched in 1928.

Eastern Asia

The Eastern Asian sphere is rapidly becoming the most productive of the world's industrial regions. Japan has emerged as the overall second-ranked manufacturing country. China—building on a rich resource base, massive labor force, and nearly insatiable market demand—is industrializing rapidly and already ranks among the top ten producers of a number of major industrial commodities. Hong Kong, South Korea, and Taiwan are three of the "Four Tigers," newly industrializing Asian economies (the fourth is Singapore in Southeast Asia) that have become major presences in markets around the world.

Japanese industry was rebuilt from near-total destruction during World War II to second rank in the world—and first in some areas of electronics and other high-tech production. That recovery was accomplished largely without a domestic raw material base and primarily with the export market in mind. Dependence on imports of materials and exports of product has encouraged a coastal location for most factories and plants. The industrial core of modern Japan is the heavily urbanized belt from Tokyo to northern Kyushu (Figure 10.32).

The Tokyo megalopolis is the manufacturing heart of the country as well as its principal population concentration. Iron and steel plants, petroleum refining and petrochemicals, shipbuilding, automobiles and motorcycles, and more are part of the region's vast industrial complex, and all are based on imported raw materials. The Kobe-Osaka district, including the ancient capital of Kyoto, focuses on the Inland Sea, intensively utilized as a transport connection between the cities of the industrial core.

When the Communists assumed control of China's war-damaged economy in 1949, that country was essentially unindustrialized. Most manufacturing was small-scale production geared to local subsistence needs. Those larger textile and iron and steel plants that did exist were coastal in location, built or owned by foreigners (including the Japanese who had long occupied the northeast of the country), and oriented toward exports to the owners'

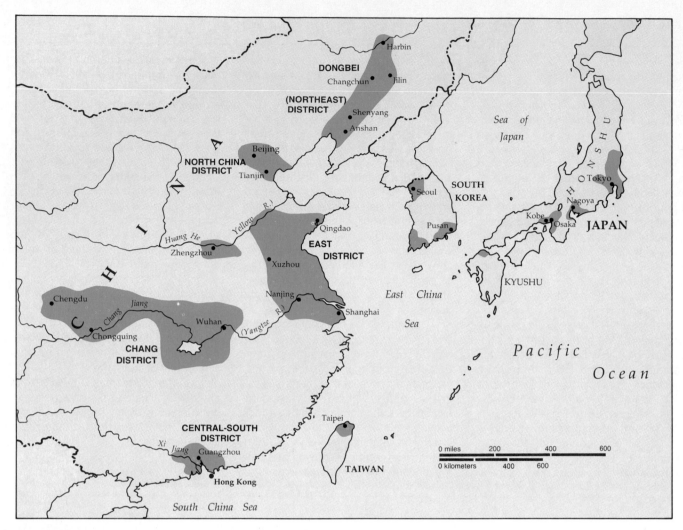

Figure 10.32
Industrial districts of eastern Asia.

home markets. A massive industrialization program initiated by the new regime—with some assistance up to the early 1960s from the Soviet Union—greatly increased the volume, diversity, and location of manufacturing in China.

Unlike Japan, China possesses a relatively rich and diversified domestic raw material base of ores and fuels. It is third in the world in coal, and has stores of petroleum, natural gas, iron ore, ferro-alloys, and nonferrous metals. The pattern of resource distribution in part accounts for the spatial pattern of industry, though coastal locations, urban agglomerations, and market orientations are equally important (Figure 10.32). Overall, the distributional pattern is one of discontinuous industrial regions extending from Harbin in the Northeast to Guangzhou (Canton) on the south coast.

The Northeast (called Dongbei by the Chinese and formerly known as Manchuria) has become a core region of Chinese industry, accounting for some 15% by value of

China's factory output and some 20% of its heavy industry. A second major industrial district—and first in total value of production with just over one-third of China's output—extends along the coast from northern Qingdao (Tsingtao) to southern Shanghai. Referred to as the East, this heavily populated coastal district includes China's biggest industrial city, Shanghai, and has access to the interior via the country's most important river, the Chang Jiang (Yangtze). Upstream on the Chang Jiang is the Chang district, a major producer of coal, iron, and steel. The fourth largest industrial production region, the Central-South, is focused on Guangzhou (Canton), where light industries including silk textiles are the rule.

Four smaller Asian economies—Hong Kong, Taiwan, South Korea, and Singapore—have become major manufacturing and trading forces in the world market. Their rise to prominence has been rapid, and their share of the industrial market in which they specialize has increased dramatically. Their successes have been largely due to an educated or trainable labor force, economic and social systems encouraging industrial enterprise, and national programs directed at capital accumulation and industrial development. Export orientation fueled the programs.

High-Tech Patterns

Major industrial districts of the world developed over time as entrepreneurs and planners established traditional secondary industries according to the pulls and orientations predicted by classical location theories. Those theories are less applicable in explaining the location of the latest generation of manufacturing activities: the high-technology, or *high-tech*, processing and production that is increasingly part of the advanced economies. For these firms, new and different patterns of locational orientation and advantage have emerged, based upon other than the traditional regional and site attractions.

High technology is more a concept than a definition. It probably is best understood as the application of intensive research and development efforts to the creation and manufacture of advanced scientific and engineering products. Professional—white collar—workers make up a large share of the total work force. The impact of high-tech industries on patterns of economic geography is expressed in at least two different ways.

First, high-tech activities are becoming major factors in employment growth in advanced economies. In the United States, for example, between 1972 and 1985, the high-tech industries added nearly 800,000 new jobs to the secondary sector of the economy, replacing thousands of other workers who lost jobs to foreign competition and changing markets. England, West Germany, Japan, and other advanced countries—though not those of Eastern Europe and the USSR—have had similar employment shifts. High tech has sustained many established industrial economies otherwise undergoing destructive change in an increasingly competitive manufacturing world (Figure 10.33).

Second, high-tech industries have tended to become regionally concentrated; they are a factor in the economic rise (and decline) of the areas in which they are (or are not) located. California, for example, has a share of U.S. high-tech employment far in excess of its share of American population. The states of the Midwest manufacturing region, however, are far below that average. And within the states or regions of high-tech concentration, specific locales have achieved prominence: "Silicon Valley" of Santa Clara County near San Francisco, North Carolina's "Research Triangle," Route 128 around Boston, or Ottawa, Canada's "Silicon Valley North," are familiar North American examples. Scotland's "Silicon Glen" and England's "Sunrise Strip" are others.

The new distributional patterns of high-tech industries suggest they respond to different localizing forces than

Figure 10.33

The Boott cotton mills at Lowell, Massachusetts, in 1852. Lowell was America's first industrial city, built in the 1820s at a water-power site on the Merrimack River as a direct copy of British cotton-mill cities. The decline of the textile industry left Lowell industrially destitute with an unemployment rate of over 12% in 1975. By the late 1980s it had transformed itself into a center of modern enterprise: 24% of its near-full employment was in high-technology industry, compared to 10% for Massachusetts as a whole. At least part of that industry is housed in the city's older vacated textile mills, showing high tech is not the exclusive property of the Sunbelt or of research parks in rural settings.

those controlling older-generation industries. Important locational considerations appear to include: (1) proximity to major universities and large pools of scientific and technical labor skills; (2) avoidance of areas with strong labor unionization; (3) locally available venture capital and entrepreneurial daring; (4) location in regions and major metropolitan areas with favorable "quality of life" reputations and an employment base sufficiently large to supply needed workers and provide job opportunities for professionally trained spouses; and (5) availability of first-quality communication and transportation facilities.

Agglomerating forces are also important. New firm formation is frequent and rapid in industries where discoveries are constant and innovation is continuous. Since many are spin-off firms founded by employees leaving established local companies, areas of established high-tech concentration tend to spawn new entrants and provide necessary labor skills. Agglomeration, therefore, is both a product and a cause of spatial associations.

Tertiary and Beyond

Primary activities, you will recall, gather, extract, or grow things. Secondary industries give form utility to the products of primary industry through manufacturing and processing efforts. *Tertiary* activities consist of those business and labor specializations that provide services to the primary and secondary sectors, the general community, and the individual. They involve services, not the production of tangible commodities. True, some services are concerned with the wholesaling or retailing of goods, providing what economists call *place utility* to items produced elsewhere. They fulfill the exchange function of advanced economies and provide the market exchanges necessary in highly interdependent societies. In commercial economies, tertiary activities also provide vitally needed information about market demand without which economically justifiable production decisions are impossible.

But most tertiary activities are concerned with personal and business services performed in shops and offices that, typically, cluster in cities. The spatial patterns of the tertiary sector are identical to the spatial distribution of *effective demand*, that is, wants made meaningful through purchasing power. Those patterns and the employment support they imply are important aspects of urban economic structure and are dealt with in Chapter 12.

Tertiary activities represent a mark of contrast between advanced and subsistence societies, and the greater their proliferation, the greater the interdependence of the society of which they are a part. Their expansion has been great in both the commercial and the planned economies. The decisive element is the development level of the society, not the form of economic administration. Within the United States, nearly three-quarters of nonfarm employment was in the mid-1980s accounted for by "services"; manufacturing accounted for only 20%. By the same date between 60% and 80% of jobs in such commercial economies as Japan, Canada, Australia, and all major West European countries and in the planned economies of the Soviet Union and Czechoslovakia were in the service sector.

The *quaternary* sector may realistically be seen as an advanced form of service, or tertiary, activity involving specialized knowledge, technical skills, communication ability, or administrative competence. These are the activities carried on in office buildings, classrooms, hospitals, doctors' offices, theaters, television stations, and the like. In advanced economies there are many individuals and groups engaged in business and institutional services and management, advertising, professional services, and the like for hire or for specific enterprises.

Quaternary-sector jobs are not spatially tied to resources, affected by the environment, or localized by market. Information, administration, and the "knowledge" activities in their broadest sense are dependent upon communication. Improvements in the technology of communication have allowed them to depart from central business districts and economic and political capitals, though they may remain close to those centers of power and population (Figure 10.34).

Figure 10.34

"Office parks," such as this complex at Tysons Corner, Virginia, outside of Washington, D.C., are increasingly familiar concentrations of quaternary workers in the outlying zones of major metropolitan centers. In the late 1980s more than 70,000 office workers were employed in this "suburban downtown."

The list of tertiary and quaternary employment is long. Its diversity and familiarity remind us of the complexity of modern life and of how far removed we are from the subsistence economies. As societies advance economically, the share of employment and national income generated by the primary, secondary, tertiary, and quaternary sectors continually changes and the spatial patterns of human activity reflect those changes. The shift is steadily away from production and processing, and toward the trade, personal, and professional services of the tertiary sector and the information and control activities of the quaternary. The transition is recognized by the now-familiar term "postindustrial."

Summary

How people earn their living and how the diversified resources of the earth are employed by different peoples and cultures are fundamental concerns of the locational tradition of geography. In seeking spatial and activity regularities we can observe that, broadly speaking, there are

three types of economic systems: *subsistence, commercial,* and *planned.* The first is concerned with production for the immediate consumption of individual producers and family members. In the second, economic decisions ideally respond to impersonal market forces and reasoned assessments of monetary gain. In the third, at least some nonmonetary social rather than personal goals influence production decisions.

We can further classify economic activities by four stages of production and the degree of specialization they represent.

Primary activities deal with food and raw material production.
Secondary production denotes processing and manufacturing.
Tertiary activities are distribution and general professional and personal services.
Quaternary activities involve administrative, informational, and technical specialities and mark highly advanced societies.

Agriculture, the most extensively practiced of the primary activities, is part of the spatial economy of both subsistence and advanced societies. In the former it is responsive to the immediate consumption needs of the producer group and reflective of the environmental conditions under which it is practiced. In the latter, agriculture reacts to consumer demand expressed through free or controlled markets. Its spatial expression reflects assessments of profitability and the dictates of social and economic planning.

Manufacturing is the dominant form of secondary activity, and is evidence of economic development beyond the subsistence level. Location theories help explain observed patterns of industrial development. Those theories are based on simplifying assumptions about fixed and variable costs of production and distribution, including costs of raw materials, power, labor, market accessibility, and transportation. *Weberian analysis* argues that least-cost locations are optimal and are strongly or exclusively influenced by transportation charges. Less rigid locational theory admits the possibility of multiple acceptable locations within a *spatial margin of profitability*. Agglomeration economies and the multiplier effect may make attractive locations not otherwise predicted for individual firms, while comparative advantage may influence production decisions of entrepreneurs. Planned economies do not necessarily ignore the locational constraints implicit in market economy decisions. They do, however, show the application of additional, frequently contradictory, bases for locational decisions uncontrolled by market forces.

A world-girdling "industrial belt" exists in which the vast majority of global secondary industrial activity occurs. The most advanced countries within that belt, however, are undergoing deindustrialization as newly industrializing countries with more favorable cost structures compete for markets. In the advanced economies, tertiary and quaternary activities become more important as secondary sector employment and share of gross national product decline. The new high-tech and postindustrial spatial patterns are not necessarily identical to those developed in response to theoretical and practical determinants of manufacturing success.

One final reminder is needed: an economy in isolation no longer exists. The world pattern of economic and cultural integration is too complete to allow totally separate national economies. Events affecting one, affect all. The potato blight in an isolated corner of Europe a century and a half ago still holds its message. Despite differences in language, culture, or ideology, we are inextricably a single people economically.

Key Words

agglomeration	least cost theory
collective farm	maximum sustainable yield
commercial economy	nomadic herding
comparative advantage	planned economy
extensive commercial	plantation
agriculture	primary activity
extensive subsistence	quaternary activity
agriculture	secondary activity
external economies	shifting cultivation
extractive industries	state farm
gathering industries	subsistence economy
Green Revolution	territorial production
intensive commercial	complex
agriculture	tertiary activity
intensive subsistence	variable costs
agriculture	von Thünen rings

For Review

1. What are the distinguishing characteristics of the economic systems labeled *subsistence, commercial,* and *planned*? Are they mutually exclusive, or can they coexist within a single political unit?

2. How is *intensive subsistence* agriculture distinguished from *extensive subsistence* cropping? Why, in your opinion, have such different land use forms developed in separate areas of the warm, moist tropics?

3. Briefly summarize the assumptions and dictates of von Thünen's agricultural model. How might the land use patterns predicted by the model be altered by an increase in the market price of a single crop? A decrease in the transportation costs of one crop but not of all crops?

4. What economic or ecological problems can you cite that do or might affect the *gathering industries* of forestry and fishing? What is *maximum sustainable yield*? Is that concept related to the problems you discerned?

5. What simplifying assumptions did Weber make in his theory of plant location? In what ways does the Weberian search for the *least cost location* differ from the recognition of the *spatial margin of profitability*?

6. How, in your opinion, do the concepts or practices of *comparative advantage* and *out-sourcing* affect the industrial structure of advanced and developing countries?

7. In what ways do the industrial development and locational principles followed in the USSR and under Marxism differ from the considerations of individual firms in free-market economies?

8. As high-tech industries and *quaternary* employment become more important in the economic structure of advanced countries, what consequences for economic geographic patterns do you anticipate? Explain.

Selected References

Amin, A., and J. B. Goddard, eds. *Technological Change, Industrial Restructuring and Regional Development.* Winchester, Mass.: Allen & Unwin, 1986.

Andreae, Bernd. *Farming, Development and Space: A World Agricultural Geography.* New York: Walter de Gruyter, 1981.

Berry, Brian J. L., Edgar C. Conkling, and D. Michael Ray. *Economic Geography.* Englewood Cliffs, N.J.: Prentice-Hall, 1987.

Breheny, Michael J., and Ronald McQuaid, eds. *The Development of High Technology Industries: An International Survey*. London: Croom Helm, 1987.

Brown, Lester R. *U.S. and Soviet Agriculture: The Shifting Balance of Power*. Worldwatch Paper 51. Washington, D.C.: Worldwatch Institute, 1982.

Castells, Manuel, ed. *High Technology, Space and Society*. Beverly Hills, Calif.: Sage, 1985.

Chandler, William U. *The Changing Role of the Market in National Economies*. Worldwatch Paper 72. Washington, D.C.: Worldwatch Institute, 1986.

Chapman, Keith, and David Walker. *Industrial Location*. 2d ed. Oxford, England: Basil Blackwell, 1990.

Clark, David. *Post-Industrial America: A Geographical Perspective*. New York and London: Methuen, 1985.

Courtenay, Percy P. *Plantation Agriculture*. 2d ed. London: Bell and Hyman, 1980.

Crosson, Pierre R., and Norman J. Rosenberg. "Strategies for Agriculture." *Scientific American* 261, no. 3 (September 1989). Special issue *Managing Planet Earth*: 128–35.

Daniels, P. W. *Service Industries: A Geographical Appraisal*. London and New York: Methuen, 1985.

Graham, Edgar, and Ingrid Floering, eds. *The Modern Plantations in the Third World*. New York: St. Martin's Press, 1984.

Greenhut, Melvin L. *Plant Location in Theory and in Practice*. Chapel Hill: University of North Carolina Press, 1956; reprint edition: Westport, Conn.: Greenwood Press, 1982.

Gregor, Howard F. *Geography of Agriculture*. Englewood Cliffs, N.J.: Prentice-Hall, 1970.

Gregory, Paul R., and R. C. Stuart. *Soviet Economic Structure and Performance*. 3d ed. New York: Harper and Row, 1986.

Hall, Peter, and Ann Markusen, eds. *Silicon Landscapes*. Boston: Allen & Unwin, 1985.

Hall, Peter, Michael Breheny, Ronald McQuaid, and Douglas Hart. *Western Sunrise: The Genesis and Growth of Britain's Major High Tech Corridor*. London: Allen & Unwin, 1987.

Harris, David R. "The Ecology of Swidden Agriculture in the Upper Orinoco Rain Forest, Venezuela." *Geographical Review* 61 (1971): 475–95.

Hartshorn, Truman A., and John W. Alexander. *Economic Geography*. 3d ed. Englewood Cliffs, N.J.: Prentice-Hall, 1988.

Hoover, Edgar M. *The Location of Economic Activity*. New York: McGraw-Hill, 1948.

International Bank for Reconstruction and Development / The World Bank. *World Development Report*. Published annually for the World Bank by Oxford University Press, New York.

Johnson, Hildegard B. "A Note on Thünen's Circles." *Annals of the Association of American Geographers* 52 (1962): 213–20.

Lardy, Nicholas R. *Agriculture in China's Modern Economic Development*. New York: Cambridge University Press, 1983.

Malecki, Edward J. "The Geography of High Technology." *Focus* 35, no. 4 (October 1985): 2–9.

———. "Industrial Location and Corporate Organization in High Technology Industries." *Economic Geography* 61, no. 4 (October 1985): 345–69.

Martin, Ron, and Bob Rowthorn, eds. *The Geography of Deindustrialization*. London: Macmillan, 1986.

Miller, E. Willard. *Manufacturing: A Study of Industrial Location*. University Park: Pennsylvania State University Press, 1977.

Morgan, W. B. *Agriculture in the Third World: A Spatial Analysis*. Boulder, Colo.: Westview Press, 1978.

Postel, Sandra, and Lori Heise. *Reforesting the Earth*. Worldwatch Paper 83. Washington, D.C.: Worldwatch Institute, 1988.

Reitsma, H. A., and J. M. G. Kleinpenning. *The Third World in Perspective*. Totowa, N. J.: Rowman and Littlefield, 1985.

Smith, David M. *Industrial Location: An Economic Geographical Analysis*. 2d ed. New York: Wiley, 1981.

Stafford, Howard A. *Principles of Industrial Facility Location*. Atlanta: Conway, 1979.

Symons, Leslie. *Agricultural Geography*. 2d rev. ed. Boulder, Colo.: Westview Press, 1979.

Taylor, Michael. "Industrial Geography." *Progress in Geography* 8, no. 2 (1984): 263–74.

Turner, B. L. II, and Stephen B. Brush, eds. *Comparative Farming Systems*. New York: Guilford Press, 1987.

Watts, H. D. *Industrial Geography*. New York: Longman Scientific & Technical/Wiley, 1987.

Webber, Michael J. *Industrial Location*. Scientific Geography Series, vol. 3. Beverly Hills, Calif.: Sage, 1984.

Weber, Alfred. *Theory of the Location of Industries*. Carl J. Friedrich, Trans. Chicago: University of Chicago Press, 1929. Reissued by Russell & Russell, New York, 1971.

Wheeler, James O., and Peter Muller. *Economic Geography*. 2d ed. New York: Wiley, 1986.

Wolf, Edward C. *Beyond the Green Revolution: New Approaches for Third World Agriculture*. Worldwatch Paper 73. Washington, D.C.: Worldwatch Institute, 1986.

Wong, Lung-Fai. *Agricultural Productivity in the Socialist Countries*. Boulder, Colo.: Westview Press, 1986.

World Resources 1990–1991. A Report by the World Resources Institute and the International Institute for Environment and Development. New York: Oxford University Press, 1990.

11

The Geography of Natural Resources

Nuclear power plant in Salem, New Jersey.

F present trends continue, the world in 2000 will be more crowded, more polluted, less stable ecologically, and more vulnerable to disruption than the world we live in now. Serious stresses involving population, resources, and environment are clearly visible ahead. Despite greater material output, the world's people will be poorer in many ways than they are today.''

The warning above is from *The Global 2000 Report to the President.* Published in 1980, the report was the result of a request made by President Jimmy Carter to 13 federal agencies, asking them to participate in an exhaustive study of likely long-term changes in the world's population, natural resources, and environment.

Alarmed by the pessimism evident in *Global 2000,* Herman Kahn, then head of the Hudson Institute, and Julian Simon, a senior fellow at the Heritage Foundation, asked a number of experts to undertake their own study of the future. Modeling their language on the earlier report, the contributors to *The Resourceful Earth* predicted:

''If present trends continue, the world in 2000 will be less crowded (though more populated), less polluted, more stable ecologically, and less vulnerable to resource-supply disruption than the world we live in now. Stresses involving population, resources, and environment will be less in the future than now. The world's people will be richer in most ways than they are today. The outlook for food and other necessities of life will be better. Life for most people on earth will be less precarious economically than it is now.''

These disparate views highlight the relationship between numbers of people, natural resources, and the environment. As Chapters 6 and 10 revealed, growing population numbers and economic development have magnified the extent and the intensity of human depletion of the treasures of the earth. Resources of land, ores, and most forms of energy are finite, but the resource demands of an expanding, economically advancing population appear to be limitless. This imbalance between resource availability and use has been a concern for over a century, at least since the time of Malthus and Darwin, but it wasn't until the 1970s that the rate of resource depletion and the environmental degradation associated with it became a major and controversial issue.

Because resources are unevenly distributed in kind, amount, and quality, and do not match uneven distributions of population and demand, a consideration of natural resources falls within the locational distributional tradition of the discipline of geography. In this chapter we survey the natural resources on which societies depend, their patterns of production and consumption, and the problem of managing those resources in light of growing demands and shrinking reserves. We do not attempt to predict which of the views expressed earlier is more likely to prove accurate, or whether some other scenario will occur, but our hope is that the chapter content will enable readers to make their own assessment.

We begin our discussion by defining some commonly employed terms.

Resource Terminology

A **resource** is a naturally occurring, exploitable material that a society perceives to be useful to its economic and material well-being. Willing, healthy, and skilled workers constitute a valuable resource, but without access to materials like fertile soil or petroleum, human resources are limited in their effectiveness. In this chapter, we devote our attention to physically occurring resources, or, as they are more commonly called, *natural* resources.

The availability of natural resources is a function of two things: the physical characteristics of the resources themselves and human economic and technological conditions. The physical processes that govern the formation, distribution, and occurrence of natural resources are determined by physical laws over which people have no direct control. We take what nature gives us. To be considered a resource, however, a given substance must be *understood* to be a resource. This is a cultural, not purely a physical, circumstance. Native Americans may have viewed the resource base of Pennsylvania as composed of forests for shelter and fuel, and as the habitat of the game animals on which they depended for food. European settlers viewed the forests as the unwanted covering of the resource that they perceived to be of value: soil for agriculture. Still later, industrialists appraised the underlying coal deposits, ignored or unrecognized as a resource by earlier occupants, as the item of value for exploitation (Figure 11.1).

Natural resources are usually recognized as falling into one of two broad classes: renewable and nonrenewable.

Renewable Resources

Renewable resources are materials that can be regenerated in nature as fast as or faster than they are exploited by society. Solar radiation, wind, water, soil, plants, and animals are among the renewable resources. The hydrologic cycle (Figure 5.4) assures that water, no matter how often used or how much abused, will return over and over to the land for further exploitation. Soils can be continuously used productively and can even be improved by proper management and fertilization practices.

If the rate of exploitation exceeds that of regeneration, however, even a renewable resource can be depleted. Groundwater extracted beyond the replacement rate in arid areas may be as permanently dissipated as if it were a nonrenewable ore. Soils can be lost by mismanagement that leads to total erosion. Forests are a renewable resource only if people are planting at least as many trees as are being cut.

Nonrenewable Resources

In their original forms, **nonrenewable resources** are generated in nature so slowly that for all practical purposes they exist in finite amounts. They include the fossil fuels (coal, crude oil, and natural gas), the nuclear fuels (uranium and thorium), and a variety of nonfuel minerals, both metallic and nonmetallic. Although the elements of which these resources are composed cannot be destroyed, they can be altered to less useful or available forms, and they are subject to depletion. The energy stored in a unit volume of the fossil fuels may have taken eons to concentrate in usable form. It can be converted to heat in an instant and be effectively lost forever.

Fortunately, many minerals can be *reused* even though they cannot be *replaced*. If they are not chemically destroyed—that is, if they retain their original chemical composition—they are potentially reusable. Aluminum, lead, zinc, and other metallic resources, plus many of the nonmetallics, such as diamonds and petroleum by-products, can be used time and time again. However, many of these materials are used in small amounts in any given manufactured object, so that recouping them is economically unfeasible. In addition, many materials are now being used in manufactured products, so that they are unavailable for recycling unless the product is destroyed. Consequently, the term *reusable resource* must be used carefully. At present, all mineral resources are being mined much faster than they are being recycled.

Resource Reserves

Some regions contain many resources, others relatively few. No industrialized country, however, has all the resources it needs to sustain itself. The United States has abundant deposits of many minerals, but it depends on other countries for such items as tin and manganese. The actual or potential scarcity of key nonrenewable resources makes it desirable to predict their availability in the future. We want to know, for example, how much petroleum remains in the earth and how long we will be able to continue using it.

Any answer will be only an estimate, and for a variety of reasons such estimates are difficult to make. Exploration has revealed the existence of certain deposits, but we have no sure way of knowing how many remain undiscovered. Further, our definition of what constitutes a usable resource depends on current *economic* and *technological* conditions. If they change—if, for example, it becomes possible to extract and process ores more efficiently—our estimate of reserves also changes. Finally, the answer depends in part on the rate at which the resource is being used, but it is impossible to predict future rates

Figure 11.1
The original hardwood forest covering these West Virginia hills was removed by settlers who saw greater resource value in the underlying soils. The soils, in their turn, were stripped away for access to the still more valuable coal deposits. Resources are as a culture perceives them, though exploitation may consume them and destroy the potential of an area for alternate uses.

of use with any certainty. The current rate could drop if a substitute for the resource in question is discovered, or increase if population growth or industrialization places greater demands on it.

A useful way of viewing reserves is illustrated by Figure 11.2. Assume that the large rectangle includes the total stock of a particular resource, all that exists of it in or on the earth. Some deposits of that resource have been discovered; they are shown in the left-hand column as "identified amounts." Deposits that have not been located are called "undiscovered amounts." Deposits that are economically recoverable with current technology are at the top of the diagram, while those labeled "subeconomic" are not attractive for any of a number of reasons (the concentration isn't rich enough, it would require expensive treatment after mining, it isn't accessible, and so on).

What is Energy?

People have built their advanced societies by using inanimate energy resources. **Energy**—the ability to do work—exists in two forms: potential and kinetic. *Potential* energy is stored energy; when released, it is in a form that can be harnessed to do work. *Kinetic* energy is the energy of motion; all moving bodies possess kinetic energy.

Assume that a reservoir contains a large amount of stored water. The water is a storehouse of *potential* energy. When the gates of the dam holding back the water are opened, water rushes out. Potential energy has become *kinetic* energy, which can be harnessed to do such work as driving a generator of electricity. No energy has been lost, it has simply been converted from one form to another.

Unfortunately, energy conversions are never complete. Not all of the potential energy of the water can be converted into electrical energy. Some potential energy is always converted to heat and dissipated to the surroundings. *Efficiency* is the measure of how well we can convert one form of energy into another without waste.

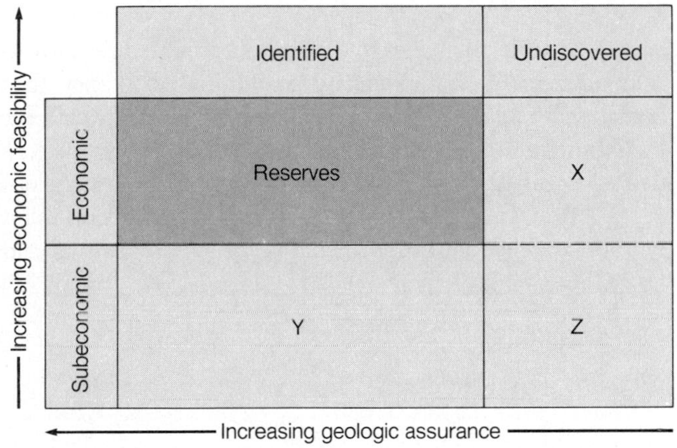

Figure 11.2
Proved reserves consist of amounts that have been identified and can be recovered at current prices and with current technology. X denotes amounts that would be attractive economically but have not yet been discovered. Identified but not economically attractive amounts are labeled Y, and Z represents undiscovered amounts that would not now be attractive even if they were discovered.

We can properly term **proved reserves**—quantities of a resource that can be extracted profitably from known deposits—only the portion of the rectangle indicated by the pink tint. These are the amounts that have been identified and that can be recovered under existing economic and operating conditions. If new deposits of the resource are discovered, the reserve category will shift to the right; improved technology or increased prices for the product can shift the reserve boundary downward. An ore that was not considered a reserve in 1950, for example, may become a reserve in 2000 if ways are found to extract it economically.

Energy Resources and Industrialization

Although people depend on a wide range of resources contained in the biosphere, energy resources are the "master"

natural resources. We use energy to make all other resources available. Without the energy resources, all other natural resources would remain in place, unable to be mined, processed, and distributed. When water becomes scarce, we use energy to pump groundwater from greater depths or to divert rivers and build aqueducts. Likewise, we increase crop yields in the face of poor soil management by investing energy in fertilizers, herbicides, farm implements, and so on. By the application of energy, the conversion of materials into commodities and the performance of services far beyond the capabilities of any single individual are made possible. Further, the application of energy can overcome deficiencies in the material world that humans exploit. High-quality iron ore may be depleted, but by massive applications of energy, the iron contained in rocks of very low iron content can be extracted and concentrated for industrial use.

Energy can be extracted in a number of different ways. Humans themselves are energy converters, acquiring their fuel from the energy contained in food. Our food is derived from the solar energy stored in plants. In fact, nearly all energy sources are really storehouses for energy originally derived from the sun. Among them are wood, water, the ocean tides, the wind, and the fossil fuels. People have harnessed each of these energy sources to a greater or lesser degree. Preagricultural societies depended chiefly on the energy stored in wild plants and animals for food, although people developed certain tools (such as spears) and customs to exploit the energy base. For example, they added to their own energy resources by using fire for heating, cooking, or clearing forestland.

Sedentary agricultural societies developed the technology to harness increasing amounts of energy. The domestication of plants and animals, the use of wind to power ships and windmills, and of water for waterwheels all expanded the energy base. For most of human history, wood was the predominant source of fuel, and even today at least half the world's people depend largely on fuelwood for cooking and heating.

However, it was the shift from renewable resources to those derived from nonrenewable minerals, chiefly fossil fuels, that sparked the Industrial Revolution, made possible the population increases discussed in Chapter 6, and gave population-supporting capacity to areas far in excess of what would be possible without inanimate energy sources. The enormous increase in individual and national wealth in industrialized countries has been built in large measure on an economic base of coal, oil, and natural gas. They are used to provide heat, generate electricity, and run engines.

Energy consumption goes hand in hand with industrial production and increases in per capita income (Figure 11.3). In general, the greater the level of energy consumption, the higher the gross national product per capita. This correlation of energy consumption with economic development points up a basic conflict between societies. Countries that can afford to consume great amounts of energy continue to expand their economies and to increase their levels of living. Countries without access to energy, or unable to afford it, see the gap between their economic prospects and those of the developed countries growing ever greater.

Nonrenewable Energy Resources

Crude oil, natural gas, and coal have formed the basis of industrialization. Figure 11.4 shows past energy consumption patterns in the United States. Burning wood supplied most energy needs until about 1890, by which time coal had risen to prominence. The proportion of energy needs satisfied by burning coal peaked about 1910; from then on, oil and natural gas were increasingly substituted for coal. The graph shows the absolute dominance of the fossil fuels as energy sources during the last 100 years. In 1986, they accounted for almost 90% of our national energy consumption.

Crude Oil

In 1988, crude oil and its by-products accounted for almost 40% of the commercial energy (excluding wood and other traditional fuels) consumed in the world. Some world regions and industrial countries have a far higher dependency. Figure 11.5 shows the main producers of crude oil.

After it is extracted from the ground, crude oil must be refined. The hydrocarbon compounds are separated and

Figure 11.3

Energy consumption rises with increasing gross national product. Because the internal combustion engine accounts for a large share of national energy consumption, this graph reflects both economic development and the roles of mass transportation, automotive efficiency, and mechanization in different national economies.

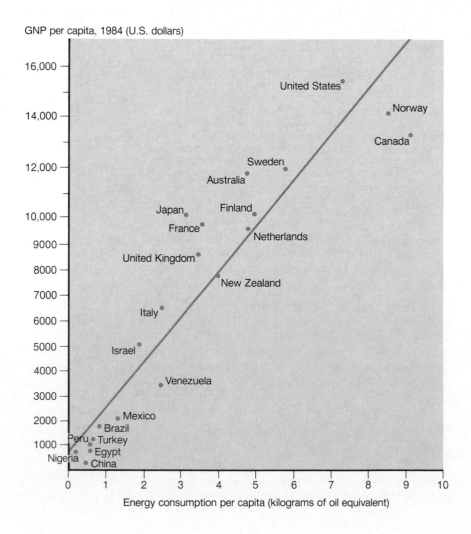

GNP per capita, 1984 (U.S. dollars)

Energy consumption per capita (kilograms of oil equivalent)

distilled into waxes and tars (for lubricants, asphalt, and many other products) and various fuels. Petroleum rose to importance because of its combustion characteristics and its adaptability as a concentrated energy source for powering moving vehicles. Although there are thousands of oil-based products, fuels, such as home heating oil, diesel and jet fuels, and gasoline are the major output of refineries. In the United States, transport fuels account for two-thirds of all oil consumption.

As Figure 11.6 shows, oil from a variety of production centers flows, primarily by water, to the industrially advanced countries. Note that the United States imports oil from a number of regions. The other major importers, Western Europe and Japan, import chiefly Middle Eastern oil.

The efficiency of pipelines, supertankers, and other modes of transport and the low cost of oil helped to create a world dependence on that fuel even though coal was still generally and cheaply available. The pattern is aptly illustrated by the American reliance on foreign oil. For many years, United States oil production had remained at about

the same level, eight to nine million barrels per day. Between 1970 and 1977, however, as domestic supplies became much more expensive to extract, consumption of oil from foreign sources increased dramatically, until almost half the oil consumed nationally was imported. The dependence of the United States and other advanced industrial economies on imported oil gave the oil-exporting countries tremendous power, reflected in the soaring price of oil in the 1970s. During that decade, oil prices rose dramatically, largely as a result of the strong market position of the Organization of Petroleum Exporting Countries (OPEC).

Among the side effects of the oil "shocks" of 1973–1974 and 1979–1980 were worldwide recessions, large net trade deficits for oil importers, a reorientation of world capital flows, and a depreciation of the U.S. dollar against many other currencies. Equally important, the soaring oil prices of the 1970s diminished total energy demand, partly because of the recession and partly because the high prices fostered conservation. Worldwide, there has been an absolute drop in the demand for oil since 1979. Industrial countries have learned to use much less oil for each unit of output. Cars, planes, and other machines are more energy efficient than they were in the 1970s, as are industries and buildings constructed in recent years.

It is particularly difficult to estimate the size of oil reserves. Not only are estimates constantly revised as oil is extracted and new reserves are located, but many governments tend to maintain some secrecy about the sizes of reserves, understating official national estimates. Nonetheless, it is clear that oil is a scarce resource and that oil reserves are very unevenly distributed among the world's countries (Figure 11.7). Some 900 billion barrels are classified as identified reserves, and another 900 billion are thought to exist in undiscovered reservoirs. If all the oil could be extracted from the earth, and if the current rate of production holds, the known reserves would last only about 40 years.

The realization that oil supplies are finite and likely to be depleted in the foreseeable future has sparked renewed interest in another fossil fuel, coal.

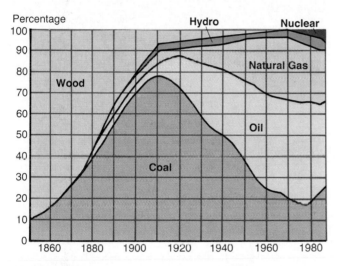

Figure 11.4
Sources of energy in the United States, 1850–1988. The fossil fuels provided 90% of the energy supply in 1988.

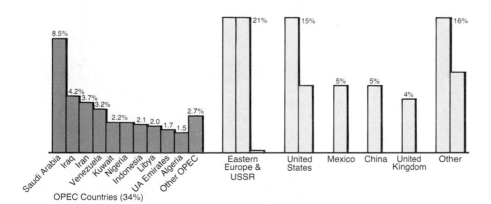

Figure 11.5

Share of total international crude-oil production, 1988. The USSR is the world's largest oil *producer*, followed by the United States. Those two countries are also the largest *consumers* of oil. Saudi Arabia and Mexico are the largest oil *exporters*.

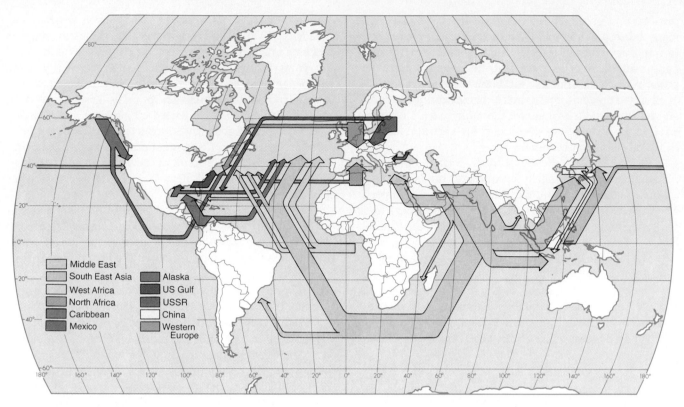

Figure 11.6

International crude-oil flow by sea, 1988. Note the dominant position of the Middle East in terms of oil exports. The arrows indicate origin and destination, not specific routes. The line widths are proportional to the volume of movement.

Legend:
- Middle East
- South East Asia
- West Africa
- North Africa
- Caribbean
- Mexico
- Alaska
- US Gulf
- USSR
- China
- Western Europe

Figure 11.7

Regional shares of proved oil reserves, in billions of barrels. Middle Eastern countries contain well over half of the proved reserves. The British Petroleum Company estimates that since 1859, almost half of all the oil in the ground has been extracted and that half of that was extracted in the 1970s. In other words, it took less than one-tenth as long to extract the second quarter as the first quarter.

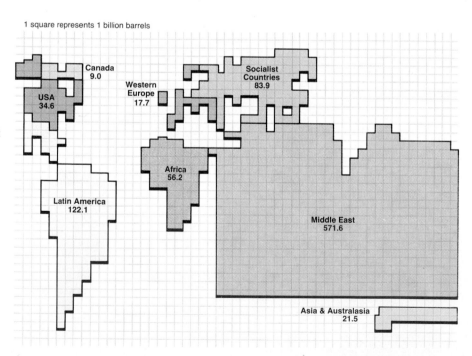

1 square represents 1 billion barrels

Canada 9.0
USA 34.6
Western Europe 17.7
Socialist Countries 83.9
Africa 56.2
Latin America 122.1
Middle East 571.6
Asia & Australasia 21.5

The Strategic Petroleum Reserve

Giant caverns created in salt beds in coastal Texas and Louisiana contain the stockpile necessary to protect the United States in an emergency: over one-half billion barrels of crude oil in the government's strategic petroleum reserve (SPR). Created under legislation passed after the Arab oil embargo of 1973–1974, the stockpile is a reserve intended to reduce the impact of any disruption in supply. Oil for the reserve comes primarily from foreign suppliers, chiefly Mexico, Canada, and Venezuela.

At current consumption levels, the stockpile is large enough to replace all oil imports for more than two months. And because the country now imports most of its oil from non-OPEC countries, the SPR could replace imports from members of that cartel for almost a year.

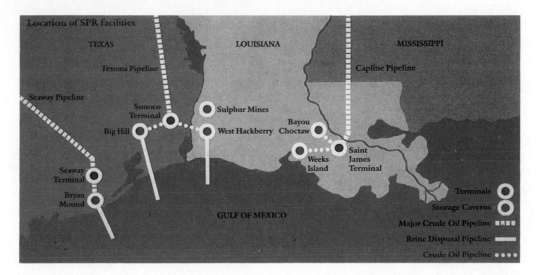

Coal

Coal was the fuel basis of the Industrial Revolution. From 1850 to 1910, the proportion of U.S. energy supplied by coal rose from 9% to almost 80%. Although the consumption of coal declined as the use of petroleum expanded, coal remained the single most important domestic energy source until 1950 (Figure 11.4).

Although coal is a nonrenewable resource, world supplies are so great that its resource life expectance may be measured in centuries, not in the decades usually cited for oil and natural gas. Of an original estimated world reserve of coal on the order of 10,000 billion (10^{13}) tons, only some 140 billion tons have so far been used. The United States alone possesses over 450 billion tons of coal considered potentially mineable on an economic basis with existing technology. At current consumption levels, these demonstrated reserves would be sufficient to meet the domestic demand for coal for another three centuries.

Worldwide, the most extensive deposits are concentrated in the industrialized middle latitudes of the Northern Hemisphere. As Figure 11.8 indicates, three countries dominate world coal production, accounting in roughly equal shares for 60% of the coal produced in the world: China, the United States, and the USSR. Note that the fourth largest producer, Poland, was responsible for

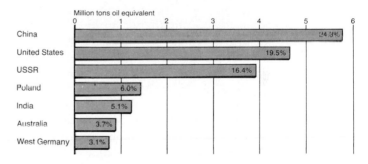

Figure 11.8
International coal production, 1988, in million tons oil equivalent.

only 6% of production in 1988, and that remaining producers each accounted for 5% or less of the total world output.

Coal is not a resource of constant quality. It ranges from lignite (barely compacted from the original peat) through bituminous coal (soft coal) to anthracite (hard coal), each *rank* reflecting the degree to which organic material has been transformed. Anthracite has a fixed carbon content of about 90% and contains very little moisture. Conversely, lignite has the highest moisture content, the lowest amount of elemental carbon, and thus the lowest heat value. About half the demonstrated reserve base in the United States is bituminous coal, concentrated primarily in the states east of the Mississippi River.

Besides rank, the *grade* of a coal, which is determined by its content of waste materials (particularly ash and sulfur), helps to determine its quality. Good-quality bituminous coals with the caloric content and the physical properties suitable for producing coke for the steel industry are decreasingly available readily and are increasing in cost. Anthracite, formerly a dominant fuel for home heating, is now much more expensive to mine and finds no ready industrial market. The Schuylkill anthracite deposits of eastern Pennsylvania are discussed as a special type of resource region in Chapter 13.

The value of a given coal deposit is determined not only by its rank and grade but also by its accessibility, which depends on the thickness, depth, and continuity of the coal seam and its inclination to the surface. Much coal can be mined relatively cheaply by surface techniques in which huge shovels strip off surface material and remove the exposed seams. Much, however, is available only by expensive and more dangerous shaft mining, as in Appalachia and most of Europe. In spite of their generally lower heating value, western United States coals are now attractive because of their low sulfur content. They do, however, require expensive transportation to markets or high-cost transmission lines if they are used to generate electricity for distant consumers (Figure 11.9).

The ecological, health, and safety problems associated with the mining and the combustion of coal must also be figured into its cost. The mutilation of the original surface and the acid contamination of lakes and streams associated with the strip mining and the burning of coal are partially controlled by environmental protection laws, but these measures add to the energy costs. Eastern U.S. coals have a relatively high sulfur content, and costly techniques for the removal of sulfur and other wastes from stack gases are now required by most industrial countries, including the United States.

The cost of moving coal influences its patterns of production and consumption. Coal is bulky and is not as easily transported as nonsolid fuels. As a rule, coal is usually consumed in the general vicinity of the mines. Indeed, the high cost of transporting coal induced the development of major heavy industrial centers directly on coalfields, for example, Pittsburgh, the Ruhr, the English Midlands, and the Donets district of the Ukraine.

Natural Gas

Coal is the most abundant fossil fuel, but natural gas has been called the nearly perfect energy source. It is a highly efficient, versatile fuel that requires little processing and is environmentally benign. Of the fossil fuels, natural gas (which is mainly methane) has the least impact on the environment. It burns cleanly. The chemical products of

Figure 11.9
Long-distance transportation adds significantly to the cost of the low-sulfur western coals because they are remote from eastern U.S. markets. To minimize these costs, unit trains carrying only coal engage in a continuous shuttle movement between western strip mines and eastern utility companies.

burned methane are carbon dioxide and water vapor, which are not pollutants, although the methane does add to the rising carbon dioxide level of the atmosphere.

As Figure 11.4 indicates, this century has seen an appreciable growth in the proportion of U.S. energy supplied by gas. In 1900, it accounted for about 3% of the national energy supply. By 1980, the figure had risen to 30%, but then declined to about 24% by 1988. The trend in the rest of the world has been in the opposite direction. Production increased significantly after the oil shock of 1973–1974 and by 1985 had more than doubled in a number of countries, including the USSR, which replaced the United States as the world's largest producer.

Most gas is used directly for industrial and residential heating. In fact, gas has overtaken both coal and oil

Coal Slurry

One way to reduce transport costs and to help unlock the large western coal reserves in the United States would be to use coal slurry pipelines. In coal slurry, coal that has been pulverized to a fine dust is mixed with fresh water to form a sludgy liquid that is pumped through pipelines and then separated at the point of discharge by power plants or other industrial users.

Mainly because of cost, coal slurry lines have not yet begun to play an important role in coal transport. The only significant slurry pipeline currently operating in the United States, the Black Mesa Pipeline, runs 273 miles (435 km) from northeastern Arizona to a power plant in southern Nevada. Environmental and agricultural groups have opposed coal slurries because of their

potential for depleting scarce supplies of fresh water in the West, but most opposition comes from the railroads, which fear loss of business to the pipelines. The railroads currently transport about two-thirds of the coal mined in the United States, and coal accounts for almost half the annual rail tonnage.

---

as a house-heating fuel, and over half the homes in the United States are now heated by gas. A portion is also used in electricity-generating plants, particularly in the Gulf Coast states, and some is chemically processed into products as diverse as motor fuels, plastics, synthetic fibers, and insecticides.

Very large natural gas fields were discovered in Texas and Louisiana as early as 1916. Later, additional large deposits were found in the Kansas-Oklahoma-New Mexico region. At that time, the south-central United States was too sparsely settled to make use of the gas, and in any case, it was oil, not gas, that was being sought. Many wells that produced only gas were capped. Gas found in conjunction with oil was vented or burned at the wellhead as an unwanted by-product of the oil industry. The situation changed only in the 1930s, when pipelines were built to link the southern gas wells with customers in Chicago, Minneapolis, and other northern cities.

Like oil, natural gas flows easily and cheaply by pipeline. Unlike oil, however, gas does not move freely in international trade. Transoceanic shipment involves costly equipment for liquefaction and for vessels that can contain the liquid under appropriate temperature conditions. **Liquefied natural gas (LNG)** is extremely hazardous because the mixture of methane and air is explosive. Although the United States has imported some LNG, chiefly from the Middle East, most gas is transferred by pipeline. In the United States, the pipeline system is over a million miles long (Figure 11.10).

Like other fossil fuels, natural gas is nonrenewable; its supply is finite. Estimates of reserves are difficult to make because they depend on what customers are willing to spend for the fuel, and estimates have risen as the price of gas has increased. Further complicating the estimate of supplies is uncertainty about potential gas resources in unusual types of geologic formations. These include tight sandstone formations, deep (below 20,000 feet, or 6000 m) geologic basins, and shale and coal beds.

Worldwide, two regions contain about two-thirds of the proved gas reserves: the USSR and the Middle East (Figure 11.11). The remaining one-third is divided roughly

equally among North America, Western Europe, Africa, Asia, and Latin America, each of which has from 5% to 7% of the total. The gas in these reserves would last about 56 more years at current production rates, but developing countries, particularly in South and Southeast Asia, may well have undiscovered deposits that could add significantly to the life expectance of world reserves if they were developed.

In the United States, the Texas-Louisiana and Kansas-Oklahoma-New Mexico regions account for about 90% of the domestic natural gas output, but there are thought to be gas deposits beneath almost all states. In addition, many offshore areas are known to contain gas. Potential Alaskan reserves are estimated to be at least twice as large as today's proved reserves in the rest of the country, containing enough gas to heat all the houses in the country for a decade. Estimates of United States reserves indicate that at the current rate of consumption, there is enough gas to last anywhere from 20 to 50 years. But if the technology necessary to produce gas from unconventional sources is developed, gas reserves may be sufficient for the next 90 or so years. Of course, these less accessible supplies will be more costly to develop and hence more expensive.

Synthetic Fuels

The rising energy prices of the 1970s stimulated interest in developing substitutes for oil and natural gas from underused resources. **Synthetic fuels** ("synfuels") are naturally occurring organic products that can be converted into synthetic petroleum and natural gas. The most promising synfuels are those produced from coal, oil shale, and tar sands.

Coal Conversion

Two broad categories of coal conversion techniques are gasification and liquefaction. **Coal gasification** is a process whereby coal is burned under high pressure in the presence of steam and either air or oxygen to produce either a high- or a low-heating-value gas. With additional processing, the product can be converted into a gas of the

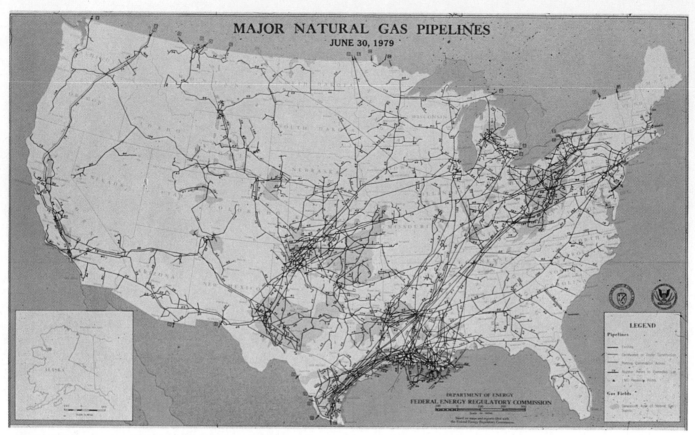

Figure 11.10
Major natural gas pipelines in the United States, 1979.

Figure 11.11
Proved natural gas reserves, 1988. The USSR and the Middle East account for two-thirds of the world's gas reserves.

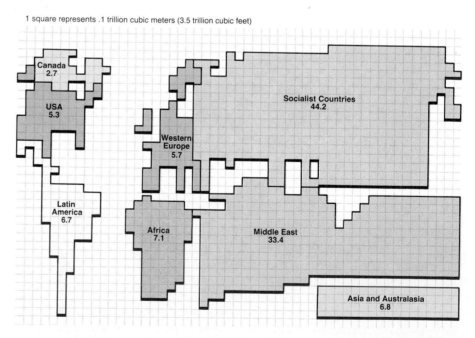

1 square represents .1 trillion cubic meters (3.5 trillion cubic feet)

Canada 2.7

USA 5.3

Western Europe 5.7

Socialist Countries 44.2

Latin America 6.7

Africa 7.1

Middle East 33.4

Asia and Australasia 6.8

quality of methane. Although gasification techniques have been employed on a small scale for several decades, the product generally has been too expensive to compete with natural gas.

Liquefaction processes are used to convert coal into liquid products comparable to petroleum. One type of liquefaction involves heating coal at relatively low temperatures and under pressure to extract the volatile (easily vaporized) matter. Volatiles are recovered as liquids and gases; the liquids are treated for the removal of organic sulfur, nitrogen, and oxygen. The resulting product, a synthetic crude oil, can be refined into gasoline and fuel oils comparable in quality to petroleum-based products. Liquefaction processes appear to be more complex and costly than gasification techniques, and the synthetic petroleum is not as clean a fuel as methane.

Producing energy from either gasified or liquefied coal is considerably more expensive than burning the coal directly and requires large capital outlays to construct the conversion facilities (Figure 11.12). Furthermore, roughly one-third of the energy in the coal is consumed in the conversion process. While synthetic gas and petroleum from coal may someday compete economically with natural gas and oil, the cost of the synfuels will be high.

Oil Shale

A similar situation colors the prospects of the extraction of oil from **oil shale**, fine-grained rock containing organic material called *kerogen*. A tremendous potential reserve of hydrocarbon energy, the rocks involved are not shales but calcium and magnesium carbonates more similar to limestone than to shale, and the hydrocarbon, kerogen, is not oil but a waxy, tarlike substance that adheres to the grains of carbonate material. The crushed rock is heated to a temperature high enough (over 900° F) to decompose the kerogen, releasing a liquid oil product, *shale oil* (Figure 11.13).

Figure 11.12
A coal-gasification plant under construction in Illinois. The plant is intended to convert Illinois high-sulfur coal to a clean, low-Btu gas.

Figure 11.13
An oil-shale retorting plant near Parachute, Colorado. Pioneers learned about "the rock that burns" from Ute Indians. Here, in roughly 1500 square miles (4000 km²) of the Piceance Basin, is one of the richest hydrocarbon deposits in the world. Retorting plants are oil-shale "ovens" that cook the rock at high temperatures to drive out the kerogen.

The richest oil shale deposits in the United States are in the Green River area of Colorado, Utah, and Wyoming. They contain astronomical potential quantities of oil, enough to supply the oil needs of the United States for another century. As oil prices eased in the 1980s, interest in synfuels waned and many projects were abandoned. Oil shale research and development efforts continue at only a few sites.

Cost and environmental damage are serious considerations in the use of oil shale. If the shale is mined by surface extraction, vast contour disruption, landscape scars, and waste heaps will result despite the best reclamation efforts. The waste heaps would pollute both air and water. Ecologically more promising are techniques for burning the rock in place in underground mines and pumping the oil to the surface. In either case, the inputs of energy, equipment, and money assure high product prices.

Tar Sands

Another potential source of petroleum liquids is **tar sand**, sandstone saturated with a viscous high-carbon petroleum called *bitumen*. Global tar-sand resources are thought to be many times larger than conventional oil resources, containing more than two trillion barrels of oil, most of it in Canada and Venezuela. Most of the tar sands in the United States are in Utah. Although there is no commercial production of tar-sand oil in the United States, oil is produced from similar deposits in Canada (Figure 11.14).

Like other synfuel technologies, producing oil from tar sands requires high capital outlays and carries substantial environmental costs. Nevertheless, because they exist in vast amounts, coal, oil shale, and tar sands could be plentiful sources of gaseous and liquid fuels. Unlike nuclear energy or most renewable energy sources (such as solar or hydroelectric power), they could provide the gasoline, jet, and other fuels on which industrialized societies depend. The lowering of oil and gas prices in the 1980s and the decline in federal government support put a temporary halt to efforts to make synfuel products economically competitive with oil imports. The prospect of depletion of oil and natural gas resources within the next century, however, indicates that at some point their prices are bound to rise. When that occurs, countries are likely to try to turn to these more unconventional sources of fuel. In the meantime, technologies employing nuclear energy as a source of nonfuel power are already developed.

Nuclear Energy

Proponents herald nuclear power as a major long-term solution to energy shortages. Assuming that the technical problems can be solved, they contend nuclear fuels could provide a virtually inexhaustible source of energy. Other commentators, pointing to the dangers inherent in any system dependent on the use of radioactive fuels, argue that nuclear power poses technological, political, social, and environmental problems for which society has no solutions.

Nuclear Fission

Basically, energy can be created from the atom in two ways: nuclear fission and nuclear fusion. The conventional form of **nuclear fission** for power production involves the controlled "splitting" of an atomic nucleus of uranium-235, the only naturally occurring fissile isotope. When

Figure 11.14

A giant bucket-wheel reclaimer about to drop freshly mined tar sands on a conveyor belt at a syncrude plant in northern Alberta, Canada. The size of the bucket wheeler is indicated by the fact that each bucket can hold six people. Production of synthetic oil from the tar sands involves four steps: removal of overburden, mining and transporting to extraction units, adding steam and hot water to separate the bitumen from the tailings residue, and refining the bitumen into coke and distillates.

U-235 atoms are split, about one-thousandth of the original mass is converted to heat. The released heat is transferred through a heat exchanger to create steam, which drives turbines to generate electricity.

A single pound of uranium-235 contains the energy equivalent of nearly 5500 barrels of oil. To tap that energy, 429 commercial nuclear reactors had been built in 26 countries around the world by 1989 (Figure 11.15). Over one-fourth of those plants are in the United States (Figure 11.16).

Some countries are much more dependent than others on nuclear power. It accounts for about 20% of the electric power generated in the United States, but two European countries (France and Belgium) receive over half of their electricity from nuclear power. Some countries have decided to reject the nuclear option altogether. Sweden plans to phase out all its reactors by 2010, and Denmark and Austria have also decided against the use of nuclear energy. Growing public opposition to nuclear power in the wake of the 1986 accident at Chernobyl resulted in the cancellation of a number of new nuclear power projects in the Soviet Union. Japan, on the other hand, which depends on nuclear power for about one-fourth of its electricity, is continuing to build new plants.

Uranium-235 is a raw material that is itself in short supply. The economically exploitable reserves are expected to be depleted by about the year 2025. The anticipated shortage of U-235 for conventional fission plants has spurred interest in the *fast breeder reactor* as a "bridge" between the fission plants of today and the nuclear fusion plants of the long-term future. Breeder reactors could stretch uranium supplies to more than 700 years.

If a core of fissionable U-235 is wrapped in a blanket of U-238 (a more abundant but nonfissionable isotope), the U-235 both generates electricity and converts its wrapping to fissionable plutonium. This is the breeder reactor, so named because it can "breed" its own fuel. Once made, plutonium can replace U-235 as the breeder's core to fuel chain reactions. In theory, a breeder could produce more plutonium fuel than it consumes.

France, Great Britain, the USSR, and Japan all operate demonstration fast breeder reactors. Funding for the first U.S. demonstration project was halted in 1982. The project had faced continuous opposition from those who argued that the breeder reactor is unnecessary, uneconomical, and unsafe.

Indeed, problems of ensuring complete safety in any fission reactor have not been solved. Although accidents at the Three Mile Island and Chernobyl nuclear power plants increased the public's awareness of safety issues connected with nuclear energy, nuclear utilities in the United States alone report about 3000 "mishaps" every year. Equally troublesome is the matter of the disposal of the great amount of radioactive waste produced by either fission method of power production. With half-lives in the hundreds and thousands of years, radioactive wastes—deadly contaminants all—represent a potential for ecological and human disaster on an unimaginable scale.

Nuclear Fusion

Nuclear fusion represents the same kind of energy generation that is carried on in the sun and other stars. The process involves forcing two atoms of deuterium to fuse into a single atom of helium, with a consequent release of large amounts of energy, about 400 times as much energy as fission of the same weight of uranium. The process is accomplished in an uncontrolled fashion in the hydrogen bomb. The problem, still not solved for commercial use despite 40 years of research, is to control fusion for a usable release of its energy.

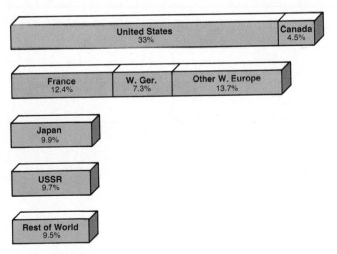

Figure 11.15
Nuclear energy consumption, 1988. Industrialized countries produce and use virtually all nuclear power. Few plants exist in developing countries.

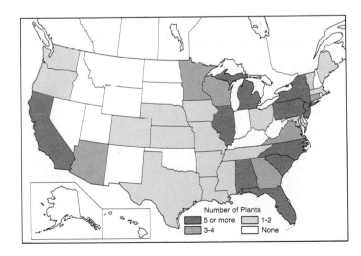

Figure 11.16
Location of nuclear power plants in the United States as of January 1, 1989. Only plants with full operating licenses are shown. One hundred and nine commercial reactors are in operation, and no new orders have been placed for over a decade. Fourteen states have no fully licensed nuclear power plants.

If the developmental problems were solved, electricity requirements would presumably be satisfied for millions of years. One cubic kilometer of ocean water, the source of deuterium atoms, contains as much potential energy as that available from the world's entire known oil reserves. Fusion also offers the advantages of the total absence of radioactive wastes and greater inherent safety in energy generation than fission plants.

Renewable Energy Resources

The problems posed by the use of nuclear energy, the threatened depletion of some finite fossil fuels, and the desire to be less dependent on foreign sources of energy have increased interest in industrialized countries in the *renewable* resources. One of the advantages of such resources is their ubiquity. Most places on earth have an abundance of sunlight, rich plant growth, strong wind, or heavy rainfall. Another advantage is ease of use. It doesn't take advanced technology to utilize many of the renewable resources, one reason for their widespread use in developing countries. The most common renewable source of energy is plant matter.

Biomass

Over half the people of the world depend on wood and other forms of **biomass**—plant material and animal waste used as fuel—for their daily energy requirements. In Burkina Faso and Malawi, biomass supplies 94% of total energy consumption; in India and Costa Rica, the figure is 42%. In contrast, energy from the conversion of wood, grasses, and other organic matter is insignificant in the developed world.

There are two major sources of biomass: (1) trees, grain and sugar crops, and oil-bearing plants like sunflowers, and (2) wastes, including crop residues, animal wastes, garbage, and human sewage. Biomass can be transformed into fuel in many ways, including direct combustion, gasification, and anaerobic digestion. Further, conversion processes can be designed to produce, in addition to electricity, solid (wood and charcoal), liquid (oils and alcohols), and gaseous (methane and hydrogen) fuels that can be easily stored and transported.

Wood

The great majority of the energy that is produced from biomass comes from *wood*. In 1850, the United States obtained about 90% of its energy needs from wood. Although wood now contributes only 3–4% to the energy mix of the country as a whole, the percentage varies by region, with wood fuel providing 16% and 13% of the energy used in Maine and Vermont, respectively. In developing countries, wood is a key source of energy, used for space heating, cooking, water heating, and lighting. This dependence on wood is leading in places to severe depletion of forest resources, a subject discussed later in this chapter.

A second biomass contribution to energy systems is *alcohol*, which can be produced from a variety of plants. Brazil, which is poorly endowed with fossil fuels, has embarked on an effort to develop its indigenous energy supplies in order to reduce the country's dependence on imported oil. By 1987, millions of its cars were running on a 20% gasoline-alcohol mixture (the alcohol is made from sugarcane and cassava), and about one-third of the country's vehicles ran on pure ethanol. Overall, alcohol met over 40% of the country's fuel needs.

Waste

Waste, including crop residues and animal and human refuse, represents the second broad category of organic fuels. Particularly in rural areas, energy can be obtained by fermenting such wastes to produce methane gas (also called *biogas*) in a process known as *anaerobic digestion*. A number of countries, including India, South Korea, and Thailand, have national biogas programs, but the largest effort to generate substantial quantities of methane gas for rural households has been undertaken in China. There, millions of backyard fermentation tanks (biodigesters) supply as many as 35 million people with fuel for cooking, lighting, and heating (Figure 11.17). The technology has been kept intentionally simple. A stone fermentation tank is fed with wastes, which can include straw and other crop residues in addition to manure. These are left to ferment under pressure, producing methane gas that is later drawn through a hose into the farm kitchen. After the gas is spent, the remaining waste is pumped out and used in the fields for fertilizer.

Figure 11.17

An anaerobic digester on a small farm in China. Animal and vegetable wastes are significant sources of fuel in countries such as India and China. The smaller unit at the left is the ''input'' tube. Wastes inserted here pass downward to an underground pit containing water and decaying organic material. The large unit in the middle is the gas-collection chamber. The copper tube leads to a community kitchen that serves ten families.

Hydroelectric Power

Biomass, particularly wood, is the most commonly used source of renewable energy. The second most common is **hydropower**, which supplies roughly one-fourth of the world's electricity (or about 7% of total commercial energy production). In the United States, where biomass does not make a major contribution to energy production, hydropower is the *largest* renewable energy resource.

Hydropower is generated when water falls from one level to another, either naturally or over a dam. The falling water can then be used to turn waterwheels, as it was in ancient Egypt, or modern turbine blades, powering a generator to produce electricity (Figure 11.18). Tied as it is to a source of water, hydroelectric power production is diffuse. In the United States, it is generated at more than 1900 sites in 47 states. Still, over 60% of the country's developed capacity is concentrated in just three areas: the Pacific states (Washington, Oregon, California), the multistate Tennessee River valley area of the Southeast, and New York. That pattern is a result both of the location of

the resource base and the role that agencies like the Tennessee Valley Authority (TVA) have played in hydropower development.

Transmitting hydroelectricity over long distances is costly, so it is generally consumed in the region where it is produced. This fact helps to account for variations in the pattern of consumption and for the energy mix used in specific regions. Thus, while hydropower provides the United States as a whole with one-tenth of its electricity, it supplies all of Idaho's needs and about 90% of the electricity used in Oregon and Washington.

Consumption patterns around the world are shown in Figure 11.19. The contribution of hydropower to a country's energy supplies varies greatly. Canada gets about 70% of its electricity from hydropower, a share roughly comparable to that found in such Western European countries as Switzerland and Austria. Almost 40 countries in Asia, Africa, and South America obtain more than half their electricity from hydropower; indeed, it supplies more than 90% of the electricity in a half-dozen of them (Bhutan, Ghana, Laos, Uganda, Zaire, and Zambia).

Figure 11.18
Power station at the Kariba Dam on the Zambezi River on the border between Zimbabwe and Zambia. Africa's hydropower potential is thought to be four or five times greater than that of the United States.

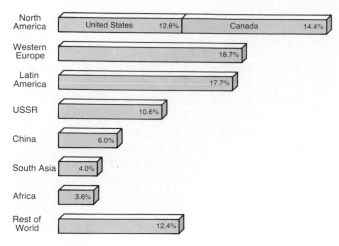

North America	United States 12.6%	Canada 14.4%
Western Europe	18.7%	
Latin America	17.7%	
USSR	10.6%	
China	6.0%	
South Asia	4.0%	
Africa	3.6%	
Rest of World	12.4%	

Figure 11.19

Consumption of hydroelectricity, 1988. In 1988, hydropower supplied the equivalent of almost 530 million tons of oil, while nuclear energy provided the equivalent of 440 million tons of oil. Note, too, that hydropower's contribution to the electricity supply is not limited to industrialized countries.

Solar Power

Each year the earth intercepts solar energy that is equivalent to many thousands of times the energy people currently use. Inexhaustible and nonpolluting, **solar energy** is the ultimate origin of most forms of utilized energy: fossil fuels and plant life, waterpower and wind power. It is, however, the direct capture of solar energy that is seen by many as the best hope of satisfying a large proportion of future energy needs with minimal environmental damage and maximum conservation of earth resources.

The technology for using solar energy for domestic uses such as water heating and space heating is well known. In the United States, both passive and active solar-heating technologies have secured a permanent foothold in the marketplace. Well over one-half million homes use solar radiation for water and space heating. Japan is reported to have some four million solar hot-water systems, and in Israel, two-thirds of all the homes heat water with solar energy.

A second type of solar-energy use involves converting concentrated solar energy into thermal energy to generate electricity. Research efforts are focusing on a variety of thermal-electric systems, including power towers, parabolic troughs, and solar ponds. Most involve the concentration of the sun's rays onto a collecting system. In a parabolic trough system, long troughs of curved mirrors, guided by computers, follow the sun, focusing solar energy onto a steel tube filled with synthetic oil (Figure 11.20). Heated to 735° F (390° C), the oil in turn heats water to produce steam to power generators. Several plants of this type are operating in the Mojave Desert, producing enough electricity for the residential needs of 270,000 people.

Figure 11.20

Parabolic trough reflectors at a solar-thermal energy plant in the Mojave Desert near Daggett, California. The facility uses sunlight to produce steam to generate electricity. Guided by computers, the parabolic reflectors follow the sun, focusing solar energy onto a steel tube filled with heat transfer fluid.

The plants just described generate electricity indirectly by first converting light to heat, but electricity can also be generated directly from solar rays by **photovoltaic (PV) cells.** In the United States, such solar-electric cells are used for a variety of specialized purposes where cost is not a constraint, such as powering spacecraft, mountaintop communications relay stations, navigation buoys and foghorns. The cells are more common in developing countries, where they power irrigation pumps, run refrigerators for remote health clinics, and charge batteries. Considerable research and development will be necessary, however, before PV power systems make a significant contribution to a country's electricity supply. At present, the costs of the energy cells are still high, and their efficiency is relatively low.

Other Renewable Energy Resources

In addition to biomass, hydropower, and solar power, there are a number of renewable sources of energy that can be exploited. Three of these are geothermal energy, the wind, and the ocean. Although none appears able to make a major contribution to the world's energy needs, each has limited, often localized, potential.

Geothermal Energy

People have always been fascinated by volcanoes, geysers, and hot springs, all of which are manifestations of **geothermal** (literally, "earth-heat") **energy.** There are several methods of deriving energy from the earth's heat as it is captured in hot water and steam trapped a mile or more beneath the earth's surface. Conventional methods of tapping geothermal energy depend on the fortuitous availability of hot-water reservoirs beneath the earth's surface. Deep wells drilled into these reservoirs use the heat energy either to generate electricity or for direct heat applications, such as heating houses or drying crops.

New Zealand, Japan, the Philippines, Mexico, and the United States are among the countries where geothermal energy is employed (Figure 11.21). Reykjavik, the capital of Iceland, is almost entirely heated by geothermal steam.

Wind Power

Although wind power was used for centuries to pump water, grind grain, and drive machinery, its contribution to the energy supply in the United States virtually disappeared over a century ago when windmills were replaced first by steam and later by the fossil fuels. Wind energy remains important in such countries as Denmark and the Netherlands. Windmills offer many advantages as sources of electrical power. They can turn turbines directly, do not use any fuel, and can be built and erected rather quickly. They need only strong, steady winds to operate, and these exist at many sites. Furthermore, wind turbine generators don't pollute the air or water and don't deplete scarce natural resources.

Eighty percent of the world's wind generation capacity is in three areas of California, where some 16,000 turbines provide about 1% of the state's electricity (Figure 11.22). California's dominance is due less to favorable wind energy conditions than to such factors as the state's commitment to developing renewable resources and the support of utility companies.

Figure 11.21
Electric generating plant powered by geothermal steam supplies electricity to Manila in the Philippines. Magma radiates heat through the rock above it, heating water in underground reservoirs. Drilled wells tap the steam and bring it to the surface, where it is piped to the power plant.

Figure 11.22
A "wind park" in the Tehachapi Mountains Wind Resource Area of California. Other wind parks are located at Altamont Pass east of San Francisco and at the San Gorgonio Pass near Palm Springs. The wind turbine generators harness wind power to produce electricity. California's wind parks produce enough electricity for more than 300,000 households.

Harnessing Tidal Power

One of the more unusual attempts to capture a renewable energy resource is the *tidal power* plant, which harnesses electricity from the ebb and flow of water as ocean tides rise and fall. The back-and-forth movement of oceans under the influence of tides represents kinetic energy. Tidal stations capture some of the kinetic energy before it is converted into heat energy and lost.

If a dam is built across an estuary or bay where the difference between high and low tides is great, electricity can be generated as water moves into and out of the bay over reversible turbines. Fewer than 50 such sites have been identified around the world and only 3 in the United States. The world's first tidal power plant, shown here, was completed in 1966 on the Rance River in France, using the tidal flows of the English Channel. Canada and the USSR also operate experimental tidal plants.

Nonfuel Mineral Resources

The mineral resources already discussed provide the energy that enables people to do their work. Equally important to our economic well-being are the *nonfuel* minerals, for they can be processed into steel, aluminum, and other metals, and into glass, cement, and other products. Our buildings, tools, and weapons are chiefly mineral in origin.

Virtually all the resources we deem essential, including metals, nonmetallic minerals, rocks, and the fuels, are contained in the earth's crust, the thin outer skin of the planet. Of all the natural elements, 12 account for over 99% of the mass of the earth's crust (Figure 11.23). They can be thought of as geologically abundant, and all others as geologically scarce.

Exploitation of a mineral resource typically involves four steps: (1) *exploration* (finding concentrated deposits of the material), (2) *extraction* (removing it from the earth), (3) *concentration* (separating the desired material from the ore), and (4) *smelting* and/or *refining* (breaking down the mineral to desired pure material). Each step requires inputs of energy and materials.

Natural processes produce minerals so slowly that they fall into the category of nonrenewable resources, existing in finite deposits. The supplies of some, however, are so abundant that a ready supply will exist far into the future. These include coal, sand and gravel, potash and magnesium. The supply of others, such as tin and mercury, is small and getting smaller as industrial societies place ever greater demands on them. Table 11.1 gives one estimate of "years remaining" for some important metals.

It should be taken as suggestive rather than definitive because mineral reserves are difficult to estimate. As we noted in the case of fossil fuels, such estimates are based on economic and technologic conditions, and one cannot predict either future prices for minerals or improvements in technology.

Although human societies began to use metals as early as 3500 B.C., world demand remained small until the Industrial Revolution. It was not until after World War II that increasing shortages and rising prices (and in the United States, increasing dependence on foreign sources) began to impress themselves on the general consciousness. Worldwide technological development has established ways of life in which minerals are the essential constituent. That industrialization has proceeded so rapidly and so cheaply is the direct result of the earlier ready availability of rich and accessible deposits of the requisite materials. Economies grew fat by skimming the cream. The question, yet unanswered, is whether the remaining supplies of scarce minerals will limit the expansion of industrialized and developing economies or whether, and how, people will find a way to cope with shortages.

The Distribution of Resources

Because the distribution of resources is the result of long-term geologic processes that concentrated certain elements into commercially exploitable deposits, it follows that the larger the country, the more likely it is to contain such deposits. And in fact, the USSR, Canada, the United

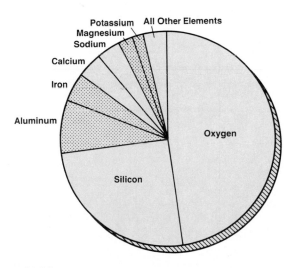

Figure 11.23

The relative abundance, by weight, of elements in the earth's crust. Only four of the economically important elements are geologically abundant, accounting for more than 1% of the total weight of the earth's crust. Fortunately, these and other commercially valuable minerals have been concentrated in specific areas within the crust. Were they uniformly disseminated throughout the crust, their exploitation would not be feasible.

Table 11.1

World Mineral Reserves

Mineral	Years Remaining
Mercury	21
Silver	22
Zinc	27
Lead	29
Copper	42
Tin	43
Manganese	111
Iron ore	195
Bauxite	257

Source: Data from *Minerals Handbook, 1984–85.* Phillip Crowson, ed. Macmillan, 1984.

Note: These figures reflect the approximate number of years the world's identified reserves of selected minerals will last, based on the expected rate of consumption between the years 1983 and 2000. Such figures are only suggestive because reserve totals and consumption rates fluctuate over time. A decline in the costs of exploitation, increases in value of the material, and/or new discoveries of deposits will extend the lifetimes shown here.

States, and Australia possess abundant and diverse mineral resources. As Figure 11.24 indicates, these are the leading mining countries. They contain roughly half of the nonfuel mineral resources and produce the bulk of the metals (e.g., iron, manganese, and nickel) and nonmetals (e.g., potash and sulfur). Great size is no guarantee of mineral resources, however. Note that other large countries, such as China and Brazil, are not as well endowed.

Many types of nonfuel minerals are concentrated in a small number of countries, and some scarce elements occur in just a few regions of the world. Thus, extensive deposits of chromium, cobalt, and diamonds are largely confined to the USSR and central-southern Africa. Some countries contain only one or two exploitable minerals—Morocco has phosphates, for example, and New Caledonia, nickel. Several countries with large populations are at a disadvantage with respect to mineral reserves. They include industrialized countries like France and Japan, which are able to import the resources, as well as developing countries like Nigeria and Bangladesh, which are less able to afford imports.

It is important to note that no country contains all the economically important mineral resources. Some, like the United States, which were bountifully supplied by nature, have spent much of their assets and now depend on foreign sources. Although the United States was virtually self-sufficient in mineral supplies in the 1940s and 1950s, it is not today. Because of its past history of use of domestic reserves and its continually expanding economy, the United States now depends on other countries for over 50% of its supply of a number of essential minerals (Figure 11.25).

The increasing costs and the declining availability of metals encourage the search for substitutes. The fact that industrial chemists and metallurgists have been so successful in the search for new materials that substitute for the traditional resources has tended to allay fears of possible resource depletion. But it must be understood that no adequate replacements have been found for some minerals, such as cobalt and chromium. Other substitutes are frequently synthetics, often employing increasingly scarce and costly hydrocarbons in their production. Many, in their use or disposal, constitute environmental hazards, and all have their own high and increasing price tags.

Copper: A Case Study

Table 11.1 indicates that the world reserves of copper will last only another 40 or so years, based on current rates of consumption and assuming that no new extractable reserves appear. Copper is a relatively scarce mineral, and

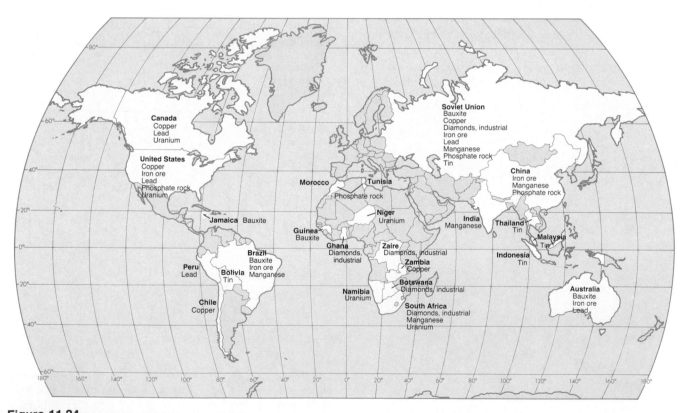

Figure 11.24

Leading producers of critical raw materials. The countries shown are not necessarily those with the largest deposits. India, for example, contains reserves of bauxite, China of tin, and South Africa of phosphates, but none is yet a major producer of those materials.

its importance to industrialized societies is evidenced by the fact that more copper is mined annually than any other nonferrous metal except aluminum. Two properties make copper desirable: it conducts both heat and electricity extremely well, and it can be hammered or drawn into thin films or wires. More than half of all copper produced is used in electrical equipment and supplies, mainly as wire. Most of the remainder is used in construction, in industrial and farm machinery, for transportation, and for plumbing pipes. In alloys with other metals, copper is used to make bronze and brass.

The leading mining regions are in western North America, western South America (Peru and Chile), the copperbelt of southern Africa (Zambia and Zaire), and in the USSR. Although the United States is a leading copper producer and has very large proved reserves, for commercial and economic reasons it still imports copper to meet part of its demands.

The threatened scarcity and the ultimate depletion of copper supplies have had several effects that suggest how societies will cope with shortages of other raw materials. First, the grade of mined ores has decreased

steadily. Those with the highest percentage of copper (2% and above) were mined early. Now, ore of 0.5% grade is the average. Thus, 1000 tons of rock must be mined and processed to yield 5 tons of copper—or, in more practical terms, 3 tons of rock are necessary to equip one automobile with the copper used in its radiator and its various electrical components. The remaining 2.985 tons is waste, generated at the mine, the concentrator, and the smelter.

Second, the recovery of copper by recycling has increased. In the United States, old scrap and new scrap from fabricating operations annually yield over a million tons of the mineral. The recycling of scrap is expected to supply as much as one-third of the copper used in the United States by the year 2000.

Developing countries that produce copper have moved to gain control over their own resources and have banded together in a cartel. Copper mines in Chile, Peru, Zambia, and Zaire, once privately owned, have all been nationalized. Those countries are all members of the Intergovernmental Council of Copper Exporting Countries (CIPEC). Should world demand for copper increase, and

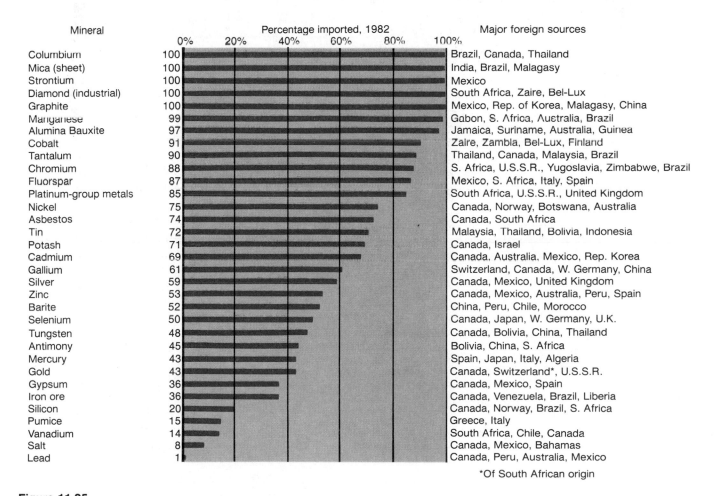

Mineral	Percentage imported, 1982	Major foreign sources
Columbium	100	Brazil, Canada, Thailand
Mica (sheet)	100	India, Brazil, Malagasy
Strontium	100	Mexico
Diamond (industrial)	100	South Africa, Zaire, Bel-Lux
Graphite	100	Mexico, Rep. of Korea, Malagasy, China
Manganese	99	Gabon, S. Africa, Australia, Brazil
Alumina Bauxite	97	Jamaica, Suriname, Australia, Guinea
Cobalt	91	Zaire, Zambia, Bel-Lux, Finland
Tantalum	90	Thailand, Canada, Malaysia, Brazil
Chromium	88	S. Africa, U.S.S.R., Yugoslavia, Zimbabwe, Brazil
Fluorspar	87	Mexico, S. Africa, Italy, Spain
Platinum-group metals	85	South Africa, U.S.S.R., United Kingdom
Nickel	75	Canada, Norway, Botswana, Australia
Asbestos	74	Canada, South Africa
Tin	72	Malaysia, Thailand, Bolivia, Indonesia
Potash	71	Canada, Israel
Cadmium	69	Canada, Australia, Mexico, Rep. Korea
Gallium	61	Switzerland, Canada, W. Germany, China
Silver	59	Canada, Mexico, United Kingdom
Zinc	53	Canada, Mexico, Australia, Peru, Spain
Barite	52	China, Peru, Chile, Morocco
Selenium	50	Canada, Japan, W. Germany, U.K.
Tungsten	48	Canada, Bolivia, China, Thailand
Antimony	45	Bolivia, China, S. Africa
Mercury	43	Spain, Japan, Italy, Algeria
Gold	43	Canada, Switzerland*, U.S.S.R.
Gypsum	36	Canada, Mexico, Spain
Iron ore	36	Canada, Venezuela, Brazil, Liberia
Silicon	20	Canada, Norway, Brazil, S. Africa
Pumice	15	Greece, Italy
Vanadium	14	South Africa, Chile, Canada
Salt	8	Canada, Mexico, Bahamas
Lead	1	Canada, Peru, Australia, Mexico

*Of South African origin

Figure 11.25
Imports supply a significant percentage of the minerals consumed in the United States. Reliance on foreign suppliers for most of the listed materials continues to increase, posing problems of national security and balance of payments.

should serious shortages appear imminent, the power of such an organization to affect supply and prices would likely grow.

Finally, price rises have spurred the search for substitutes. In many of its applications, copper is being replaced by other, less expensive materials. Aluminum is replacing copper in some electrical applications and in heat exchangers. Plastics are supplanting copper in plumbing pipes and building materials; steel can be used in shell casings and coinage.

Food Resources

People depend on the fossil fuels and nonfuel mineral resources because they provide the energy to perform work and the raw materials to construct nearly everything we use. People's own raw energy resource, however, is food, and securing an adequate supply of food is a paramount daily concern. The three most important determinants of the location of agricultural resources are sunshine, water, and soil. Plants must receive an adequate amount of solar energy and fresh water in order for the process of photosynthesis to take place. They also need soils rich in such nutrients as phosphorus, potassium, and nitrogen. As Figure 11.26 indicates, soils that are naturally fertile occupy a relatively small portion of the earth, and they are unevenly distributed among the continents.

At present, only 10% of the earth's surface is used for intensive food production. On these 3.5 billion acres (1.4 billion hectares) must be grown all the crops on which the planet depends, and rapid population growth has placed increasing pressure on them. The world's population has more than doubled since 1950; to date, the output of the world's farms has kept pace with demand. Twice as much food was produced in 1980 as in 1950, even though the amount of cultivated land per person declined during that period. This increase in agricultural efficiency occurred both in developed countries and developing countries.

In developing countries, changes in farming technology are known collectively as the Green Revolution (see Chapter 10). Wheat, corn, and rice—the three crops that account for roughly half of the world's food production by weight—have responded to applications of fertilizers and pesticides, to irrigation and farm mechanization, and to techniques of genetic engineering that produce high-yielding strains. The Green Revolution has enabled such countries as India and Thailand to join the ranks of the relatively few countries that are net exporters of grain.

If foodstuffs were evenly distributed, everyone would receive several pounds of food per day. That is, the total supply of food produced is sufficient to meet world demand. At the same time, more people than ever before are malnourished (see Figure 6.23). It is important to note that

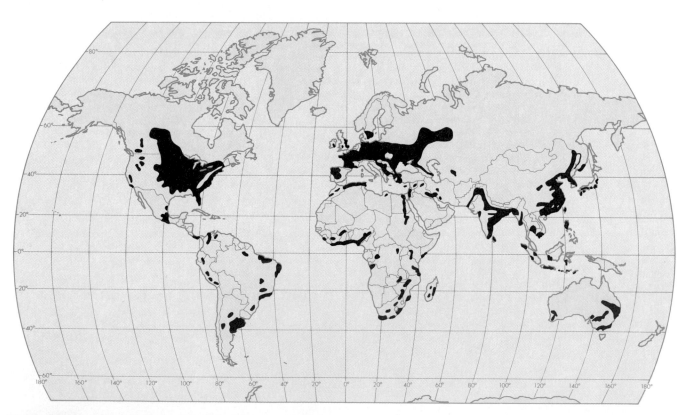

Figure 11.26

Areas with naturally fertile soils account for much of the world's grain production. They include the corn and wheat belts of North America and the Ukraine and the rice-producing regions of India and Southeast Asia.

the map conceals inequalities in food consumption *within* countries. By some estimates, one-eighth of the world's people—chiefly small children and pregnant women—suffer from clinically defined malnutrition.

This seeming contradiction between ample food supplies and widespread malnutrition results from inequalities in population growth rates, lack of access to fertile soils, local climatic catastrophes, the inadequacy of storage facilities and road systems in many countries, and problems of credits and trade imbalances that make it difficult to import food supplies. A number of countries in Latin America (e.g., Peru, Haiti), Africa (e.g., Sudan, Ethiopia), and Asia (e.g., Bangladesh, Laos) have faced serious food shortages in recent years, but on a regional basis the greatest discrepancy between population growth and food production has been in Africa (Figure 11.27).

The secondary effects of chronic or periodic shortages are of concern both to individual countries and to the international community. When malnutrition is the rule, not just starvation but low resistance to infectious diseases, high child mortality rates, mental damage, social disorder, and political unrest or upheaval are likely consequences (Figure 11.28). The interconnections of the world's peoples mean that food supply problems are not simply domestic concerns in the seemingly remote areas of their occurrence but in some form have an impact on all societies.

In the next ten years, the world's population is expected to grow from its present 5 billion to 6 billion. That is, each year, on average, another 100 million people will be added, and each year, food supplies will have to increase accordingly. Indeed, for everyone in the world to have an adequate diet, food production should actually exceed population growth in order to provide the grain reserves needed to improve diets above subsistence levels and

to compensate for variations in crop yields from year to year. In the following sections we consider methods of expanding food supplies.

Expansion of Cultivated Areas

One way to increase food production is to expand the areas being cultivated. Approximately 70% of the world's land is not suitable for intensive human use, either because it is too cold, too dry, too steep, or infertile. Essentially all activities must be concentrated on the remaining 30%,

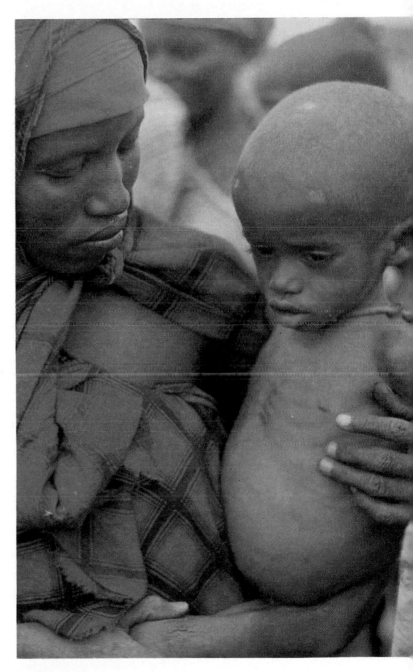

Figure 11.28

A malnourished child in Ethiopia. By some estimates, one-eighth of the world's people, chiefly small children and pregnant women, suffer from malnutrition.

Percentage

| | 0 | 1 | 2 | 3 |

All developing countries

Asia (including China)

Latin America

North Africa and Middle East

Sub-Saharan Africa

▓ Average annual growth of population 1961-1980
▓ Average annual growth of major food crop production 1961-1980

Figure 11.27

Food production and population growth in developing countries. Although food production has increased greatly since 1961, per capita production in developing countries has grown very little because population growth has canceled nearly all the gains in production. In many African countries, population growth outstrips growth in food production.

or about 11 billion acres (4.5 billion hectares). Of those 11 billion acres, only one-third (3.5 billion acres, or 1.4 billion hectares) are actually used in intensive food production. The United Nations Food and Agriculture Organization estimates that at most another 2.5 billion acres (1 billion hectares) are potentially suitable for agricultural use, depending on definitions of "suitable" and "agricultural use."

Clearly, some land is more suitable than other land for regular, sustained food production, and experts generally agree that most of the land in the world that is well suited for agriculture is already being cultivated. Few reserves of arable land remain in the temperate regions of Europe and North America, for example. Indeed, in the United States we regularly take more land *out* of production than we add. Each year more than 2 million acres (800,000 hectares), almost all of it prime agricultural land, are lost to expanding cities, highways, and other non-agricultural uses. A similar loss of cropland to urbanization occurs in Japan and many of the West European countries. Worldwide, additional millions of acres are lost annually through soil erosion and salinization and to the spread of deserts by overgrazing and deforestation.

Any attempt to expand the area of agricultural land will be made at the expense of environments that all previous human experience has judged to be marginal. These include the tropical rain forests and arid or semiarid regions where rainfall is low or unpredictable. Steppe lands, for example, normally receive an undependable rainfall of below 20 inches (50 cm) per year, sufficient for some agricultural purposes but too unreliable or insufficient for assured yields of most desirable crops. The great fluctuations in grain yields in the Soviet Union reflect in part the expansion of agriculture into dry steppe territories under the Virgin and Idle Lands Program. Instances of great crop success in years of favorable weather are offset by years of partial or total failure.

Most of the land that is deemed potentially suitable for cultivation *and* that receives enough rainfall for intensive agriculture is in the rain forests of Africa and the Amazon Basin of South America. Although food needs are great in those areas, soils (oxisols) in the tropical rain forest are delicate, low in nutrients and humus content, poor at holding water, and need time to recover their fertility after cropping.

As described in Chapter 10, farmers traditionally adapted successfully to these limitations by practicing slash-and-burn agriculture. Increasing population pressures now threaten the viability of the system, particularly in the Amazon Basin. There, some 2% of the forest is being cleared for agriculture every year, cleared areas are becoming more closely spaced, and insufficient time is being allowed for vegetation to regenerate. Within just a few years' time, nutrients are leached from the soil, soil erosion increases, and yields drop.

There is some evidence that crop yields can be sustained in tropical forests if fertilizers are used to add nutrients to the soil, and that a given plot can support three crops per year—producing more food than does slash-and-double-burn agriculture. Such continuous cultivation will become successful only if a number of conditions are met. Fertilizers will have to be available at affordable prices, soil erosion controlled, and crops rotated in order to return nutrients to the soil and maintain yields. Most important, farmers will have to be trained in the new techniques.

In summary, expanding the amount of cultivated land cannot be viewed as the solution to increasing world food supplies and may, indeed, render land that is improperly added to the agricultural base unfit for any use.

Increasing Yields

A greater potential for expanding agricultural output lies in increasing the yields on land that is already being cultivated. Thus corn yields can vary from 115 bushels per acre (7.2 tons per hectare) in commercial agricultural societies to less than 24 bushels per acre (1.5 tons per hectare) in countries relying on traditional farming techniques. Similarly, in the mid-1970s, before the impacts of the Green Revolution were fully felt, rice yields in some countries were six times greater than in others (Figure 11.29). Almost three-fourths of the increased grain production throughout the world since 1950 is thought to be due to such improved yields.

There are a number of ways yields can be increased, including:

1. multiple-cropping—growing two or three crops in rotation in one year on a single piece of land
2. use of new, high-yielding varieties of grains (mainly rice and wheat) that make more efficient use of fertilizers and, often, take less time to mature than do native strains
3. increased use of fertilizers, both organic (crop residues and animal manure) and inorganic (e.g., phosphorus, potassium, ammonia)
4. control of pests, which include thousands of species of rodents, insects, bacteria, and fungi that are estimated to destroy roughly half of the world's foodstuffs each year
5. development of irrigation systems
6. farm mechanization

As attractive as it seems, the yield-increase strategy poses a number of problems. First, the most dramatic production increments are potentially obtainable in the less developed countries, which are least able financially to make the conversion from labor-intensive to capital-intensive agriculture. High-yield agriculture is capital intensive, requiring great expenditures for equipment, fertilizers, pesticides, and energy. Requisite amounts of

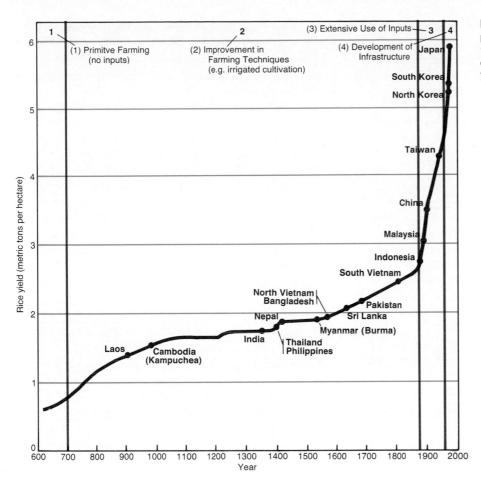

Figure 11.29
Rice yields in Asian countries in the mid-1970s. The graph shows the potential for expanding food production by increasing yields.

capital simply are not available in many developing countries. In addition, heavy farm machines are not well suited for use on small farms or in areas where holdings are fragmented.

Second, many of the proposals for yield increases are ecologically destructive. Nitrogen fertilizers consume great quantities of natural gas in their manufacture, using a scarce resource and contributing to atmospheric heat. Runoff from fields where large amounts of fertilizers and pesticides have been used pollutes surface waters. Scientists have estimated that if commercial fertilizers were used in India at the rate they are in the Netherlands, India would need nearly one-half of the present world output of fertilizer. Since it takes about 5 tons of mostly fossil fuels to produce 1 ton of fertilizer, the consequences of increasing yields by heavy fertilizer applications are serious.

Further, the Green Revolution endangers populations by encouraging dependence upon *monoculture*, total emphasis upon a single crop, which is inherently dangerous. It demands total dependence for human existence in marginally fed countries upon a single crop with a wide variety of essential inputs. Should the crop fail from disease, insect damage, or input shortage, disaster is unavoidable. Perhaps unwittingly, but wisely, subsistence agriculture is characterized by the production of a great variety of foodstuffs. Diversity is the subsistence farmer's insurance policy.

Finally, there is a point beyond which crops will not respond to higher levels of fertilizer application. During the 1950s, each additional million tons of fertilizer used annually in the United States was accompanied by a 10-million-ton rise in the grain harvest. During the early 1960s, the grain increase from that million tons of input declined to 8.2 million tons, and by the early 1970s it had declined to 5.8 million tons. Comparable arguments may be offered to demonstrate that the massive utilization of herbicides and pesticides also reaches a point of diminishing returns. Likewise, some scientists believe that a genetic ceiling exists on the increases in yields that can come from breeding improved plant strains.

Increasing Fish Consumption

The pressing nutritional need in much of the world is for protein. The amino acids contained in protein are the essential building blocks of the body. The oceans have long been viewed as a seemingly unlimited source of protein, and increasing people's consumption of fish and shellfish is a third means by which global food supplies might be augmented. Although fish and shellfish account for less than 10% of all human protein consumption worldwide, they are the major source of animal protein for about half the people in the world. Reliance on fish is greatest in the developing countries of eastern and southeastern Asia,

Africa, and parts of Latin America. Fish are also very important in the diets of some advanced economies with well-developed fishing industries—the USSR, Norway, Iceland, and Japan, for example.

Only about 10% of the annual fish supply comes from inland waters: lakes, rivers, and ponds. The other 90% of the harvest comes from the world's oceans, and most of that catch is made in coastal wetlands and relatively shallow coastal waters. Near shore, shallow embayments and marshes provide spawning grounds, and river waters supply nutrients to an environment highly productive of fish. Offshore, ocean currents and upwelling water move great amounts of nutritive salts from the ocean floor through the sunlit surface waters, nourishing *plankton*—minute plant and animal life forming the base of the marine food chain.

Commercial fishing is largely concentrated in northern waters, where warm and cold currents join and mix and where such familiar food species as herring, cod, mackerel, haddock, and flounder congregate on the broad continental shelves favorable for fish production. Two of the most heavily fished regions are the North Pacific and North Atlantic, which together supply more than half the fish catch (Figures 11.30 and 11.31).

Tropical fish species tend not to school and, because of their high oil content and unfamiliarity, to be less acceptable in the commercial market. They are, however, of great importance for local consumption. Traditional fishermen and women are estimated to number between 8 and 10 million worldwide, harvesting some 25 million tons of fish and shellfish a year—a catch usually not included in world fishery totals.

If the sea is to be a reservoir for extended exploitation, it must be more wisely handled than in the past. Between 1950 and 1970, fish supplied an expanding part of human diets as modern technology was applied to harvesting food fish. This technology included the use of sonar, helicopters, and aerial photography to locate schools of fish; used more efficient nets and tackle; and employed factory trawlers to follow the fishing fleets to prepare and freeze the catch. Annual fish harvests more than tripled during that 20-year period, rising from 23 million tons to 77 million tons.

The rapid rate of increase led to inflated projections of probable fisheries productivity and to the feeling that the resources of the oceans were inexhaustible. Quite the opposite has proved to be true. In 1970 the trend was interrupted, and the productivity of fishing areas declined as *overfishing* (catches above reproduction rates) and pollution of coastal waters seriously endangered the supplies of traditional and desired food species. Since 1976, the annual catch has never exceeded 84 million tons, and because world population has continued to grow, there has been an actual decline in the average fish catch per person.

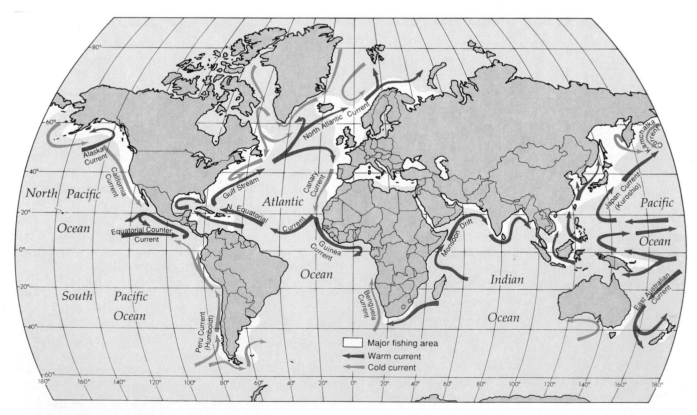

Figure 11.30
The major commercial marine fisheries of the world.

Figure 11.31
Salmon trawlers off the coast of Oregon. The increased efficiency of commercial fishing has led to serious depletion of a food source once thought to be inexhaustible.

Overfishing is partly the result of the accepted view that the world's oceans are common property, a resource open to anyone's use with no one responsible for its maintenance, protection, or improvement. The result of this "open seas" principle is but one expression of the so-called **tragedy of the commons**[1]—the economic reality that when a resource is available to all, each user thinks he or she is best served by exploiting the resource to the maximum even though this means its eventual depletion.

Despite such evidence of a decrease in productivity, some experts believe the world's fish catch could be increased to at least 100 million tons during the next decade. They point to areas that are underfished, such as the Indian

[1]The *commons* refers to undivided land available for the use of everyone. Traditionally, it meant the open land of a village that all used as pasture. The *Boston Common* originally had this meaning.

Ocean and South Atlantic. In addition, there remain vast supplies of unconventional protein. "Junk fish" can be consumed directly or used as a fish-meal protein extender of grains. An additional 10 million tons of squid might be taken annually, although few societies now use them as food. And estimates suggest that 50–100 million tons of krill—a nutritious, shrimplike crustacean—could be taken from Antarctic waters annually, though perhaps with disastrous results for the Antarctic food chain. Extensive harvesting of krill might endanger the populations that feed on them, which include whales, seals, penguins, and seabirds in addition to fish.

Another approach to increasing the fish harvest is through **aquaculture**, the breeding of fish in freshwater ponds, lakes, and canals or in fenced-off coastal bays and estuaries. Fish farming has long been practiced in Asia, where it is a major source of protein, and more recently in Africa (Figure 11.32). Fish farms in the United States

Figure 11.32
Harvesting fish at an aquaculture farm in Thailand.

produce significant amounts of catfish and crawfish, trout and oysters. While aquaculture is still small-scale and localized, it could be expanded with little difficulty and offers several advantages over conventional fishing. There is no shortage of land suitable for aquaculture, the technology is relatively simple and requires little or no fuel, farmers can choose the species they want to raise, and yields are high.

Land Resources

As we have seen, cropland occupies only 11% of the world's total land area. Other types of land resources include forests, rangelands, parks and wilderness areas. In this final section of the chapter, we examine the distribution and status of two of those resources: coastal wetlands, which comprise a minute portion of the land area, and forests, which cover about a third of the earth.

Coastal Wetlands

We noted on page 364 that the part of the sea lying above the continental shelf is the most productive of all ocean waters, supporting the major commercial marine fisheries. Because it is not very deep, this *neritic zone* is penetrated and warmed by sunlight. It also receives the nutrients flowing into oceans from streams and rivers, so that vegetation and a great variety of aquatic life can flourish. However, the neritic zone depends to a considerable extent on the continued functioning of the **estuarine zone**, the relatively narrow area of wetlands along coastlines where salt water and fresh water meet and mix (Figure 11.33).

Coastal wetlands take a variety of forms, including marshes, swamps, tidal flats, and estuaries. Extremely valuable ecological systems, they have a number of vital functions. Trapping the silt and other organic matter that rivers bring downstream, they provide shelter and food for a variety of fish and shellfish, and indeed are essential to the survival of many species by serving as their spawning grounds. Coastal wetlands are also breeding, feeding, nesting, and wintering grounds for many types of birds (Figure 11.34). Not only are these areas extraordinarily productive themselves, but they also contribute to the productivity of the neritic zone, where fish feed on the life that flows from wetlands into the sea. In addition, the wetlands absorb floodwaters, provide barriers to coastal erosion, and remove pollutants from surface waters.

In much of the world, coastal wetlands are in danger. The lands have been drained, dredged, built upon, mined for phosphate, and used as garbage dumps. They are polluted by chemicals, excess nutrients, and other water-borne wastes. Natural shorelines have been bulldozed, and artificial levees and breakwaters interfere with the flooding that nurtures wetlands with fresh infusions of sediment and water. Habitat alteration or destruction inevitably disrupts the intricate ecosystems of the wetlands. Scientists estimate that as much as half of the estuarine zone in the United States has been damaged, drastically altered, or destroyed. At present, the only way to halt this disruption seems to be to preserve the wetlands as they are, as ecological reserves. Should we fail to do so, there will be negative consequences not only for the water life they sustain but also for that in the adjoining oceans.

Forests

Coastal wetlands are only one of the renewable resources in danger of irreparable damage by human action. In many parts of the world, forests are similarly endangered.

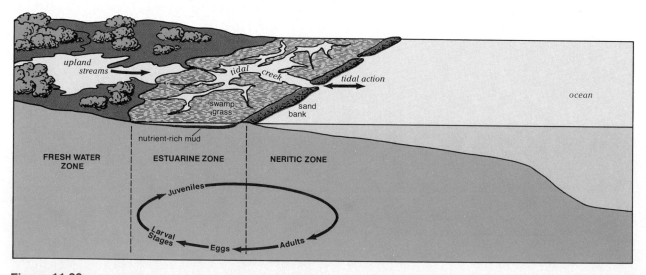

Figure 11.33

The estuarine zone. The outflow of fresh water from streams and the action of tides and wind serve to mix deep ocean waters with surface waters in estuaries, contributing to their biological productivity. The saline content of estuaries is lower than that of the open sea. Many fish and shellfish require water of low salinity at some point in their life cycles.

Figure 11.34

A salt marsh near Greenwich, Connecticut. Tidal marshlands have been subjected to dredging and filling for residential and industrial development. The loss of such areas reduces the essential habitat of waterfowl, fish, crustaceans, and mollusks. Many waterfowl breed and feed in coastal marshes and use them for rest during long migrations.

Before the rise of agriculture some 11,000 years ago, the world's forests probably covered about 45% of the earth's land area exclusive of Antarctica. Even after millennia of land clearance for agriculture and, more recently, commercial lumbering, cattle ranching, and fuelwood gathering, forests still cover about one-third of the world's land area. As an industrial raw material source, however, forests are more restricted in area. *Commercial forests* are restricted to two very large global belts (Figure 11.35). One, nearly continuous, occupies the upper-middle latitudes of the Northern Hemisphere. The second straddles the equatorial zones of South and Central America, Central Africa, and Southeast Asia.

The northern coniferous, or softwood, forest is the largest and most continuous stand. Its pine, spruce, fir, and other species are used for construction lumber and to produce pulp for paper, rayon, and other cellulose products. To its south are found the deciduous hardwoods: oak, hickory, maple, birch, and the like. These and the trees of the mixed forest lying between the hardwood and softwood belts have been much reduced in areal extent by centuries of agricultural and urban settlement and development, though they still are commercially important for hardwood applications: furniture, veneers, railroad ties, and so on. The tropical lowland hardwood forests are exploited primarily for fuelwood and charcoal, although an increasing quantity of special quality woods is cut for export as specialty lumber.

The adage about not being able to see the forest for the trees is applicable to those who view forests only for the commercial value of the trees they contain. Forests are more than trees, and timbering is only one purpose that forests serve. Chief among the other purposes are soil and watershed conservation, habitat for wildlife, and recreation. Forests also play a vital role in the global recycling of water, carbon, and oxygen.

Because forests serve a variety of purposes, the kind of management techniques employed in any one area depends on the particular use(s) to be emphasized. Thus, if the goal is to maintain a diversity of native plant species in order to provide a maximum number of ecological niches for wildlife, the forest will be managed differently than if it is designed for public recreation or the protection of watersheds. Even if the use to be emphasized is timber production, different management approaches may be taken. Logging techniques for the production of plywood or wood chips, for example, differ from those used for the production of high-quality lumber.

Commercial forests can be considered a renewable resource only if sustained-yield techniques are practiced—that is, if harvesting is balanced by new growth (see *maximum sustainable yield* in Chapter 10). Timber companies employ a number of different methods of tree harvesting and regeneration. Two quite different practices, selective cutting and clear cutting, illustrate the diversity of such approaches (Figure 11.36). *Selective*

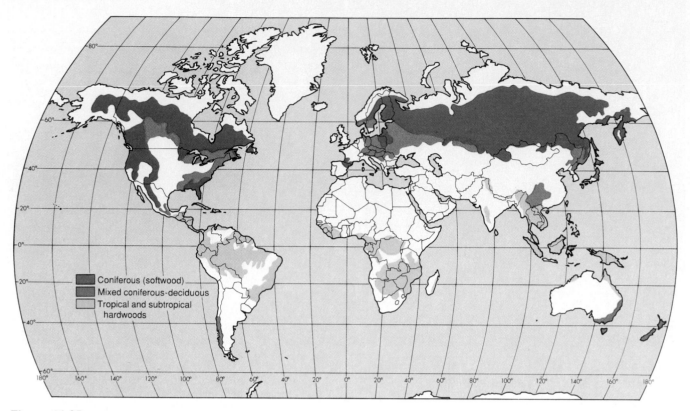

Figure 11.35

Major commercial forest regions. Much of the original forest, particularly in midlatitude regions, has been cut over. Many treed landscapes that remain do not contain commercial stands.

Significant portions of the northern forests are not readily accessible and at current prices cannot be considered commercial.

cutting is practiced in mixed-forest stands containing trees of varying ages, sizes, and species. Medium and large trees are cut either singly or in small groups, encouraging the growth of younger trees that will be harvested later. Over time, the forest will regenerate itself. With *clear cutting*, as the name implies, all the trees are removed from a given area at one time. The site is then left to regenerate naturally or it is replanted, often with fast-growing seedlings of a single species. Excessive clear cutting destroys wildlife habitats, accelerates soil erosion and water pollution, replaces a mixed forest with a wood plantation of no great genetic diversity, and reduces or destroys the recreational value of the area.

U.S. National Forests

Roughly one-third of the United States is forested, the same proportion for the world as a whole. Only some 40% of those forests provide the annual harvest of commercial timber. The remaining forests either do not contain economically valuable species, are in small, fragmented holdings, are inaccessible, or are in protected areas. Of that 40% of commercial forestland, almost half is in national forests owned by the public and managed by the U.S. Forest Service (Figure 11.37). Logging by private companies is permitted; timber companies pay for the right to cut designated amounts of timber.

By law, the national forests are to be managed under the principle of *multiple use*, which is intended to balance the needs of recreation and wildlife with those of such developmental activities as logging, mining, and drilling for oil and gas. Although no use is to be particularly favored over others, conservationists charge that the forest service increasingly supports commercial logging and that the forests are being cut at an unprecedented rate.

In 1987, 12.7 billion board feet of timber were taken from the national forests, a record that would have been surpassed the following year had not forest fires curtailed logging (Figure 11.38). Environmentalists are especially concerned that nearly half of this came from national forests in Oregon and Washington, most of it irreplaceable "old growth." These virgin forests contain trees that are among the tallest and oldest in the world, indeed, that were alive when Pilgrims set foot on Plymouth Rock.

Old-growth forests include trees of every age and size, both living and dead. Some ancient trees are immense, capable of growing 300 feet high, and they may live for more than 1000 years. They include the Douglas fir, Western red cedar, sequoia, and redwood. Tons of dead and decaying logs carpet the forest floor, where, sodden with moisture, they help control erosion and protect the forest from fire. As they decay, the logs release nutrients

(a)

(b)

Figure 11.36
(a) Selective cutting in eastern Canada. Older, mature specimens are removed at first cutting. Younger trees are left for later harvesting. **(b) Clear cutting in British Columbia.**

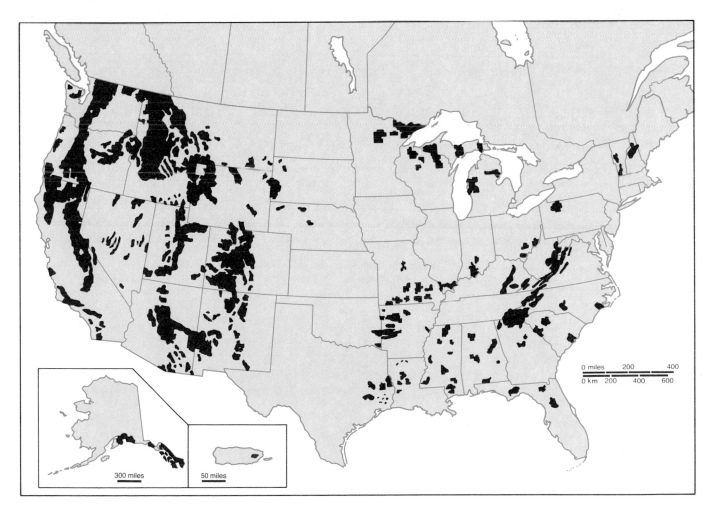

Figure 11.37
National forests of the United States. Trees cut from national forests supply about 15% of all timber harvested in the United States.

back into the soil. Such forests provide a habitat for hundreds of types of insects and animals, some of them threatened or endangered species.

The only large expanses of old-growth forest remaining in the United States are in the Pacific Northwest, most of them owned by the federal government. These last ancient forests are being logged at the rate of 60,000 acres per year. If logging continues at the present rate, they will be gone in about 20 years, ceasing to exist as they are today. Although companies plant new seedlings to replace those they cut, timber is being harvested twice as fast as new trees can replace it. Further, traditional management practices, including clear cutting, road building, and harvesting after decades, not centuries, of regeneration prevent the development of a true old-growth forest ecosystem.

It is ironic that many Americans condemn the burning of the tropical rain forests while the U.S. government not only permits the destruction of forests just as ecologically precious but also subsidizes that destruction. The federal government loses an average of $500 million per year on timber sales, because building and maintaining the logging roads costs far more than the timber companies pay for the wood. Perhaps the worst abuse has occurred in the Tongass National Forest, which covers much of Alaska's southern panhandle. The government has spent millions annually to build logging roads and promote commercial timbering and has received in return less than $100,000 from the sales. Five-hundred-year-old trees 9 feet (3 m) in diameter and over 100 feet (32 m) tall have been sold for $3 each and turned into pulp.

Tropical Forests

It is not only in the United States that government economic policies accelerate the rate of forest destruction. Much of the deforestation occurring in tropical areas also has governmental sanction. Brazil, Indonesia, and the Philippines are among the countries where governments subsidize projects aimed at converting forests to other uses, such as farming, cattle ranching, and industry. Of course, their economic policies are driven by the continual demand for resources that poverty and a growing population place on a society.

The tropical forests extend across parts of Asia, Latin America, and Africa (Figure 11.39). We noted in Chapter 5 that some 40,000 square miles (100,000 km²) are being completely cleared every year, and that almost half of their original expanse has already been either cleared or degraded. Nearly half of Asia's natural forest is gone. Seventy percent of the tropical forests of Central America and some 40% of those of South America have disappeared. Africa has lost more than half of its original

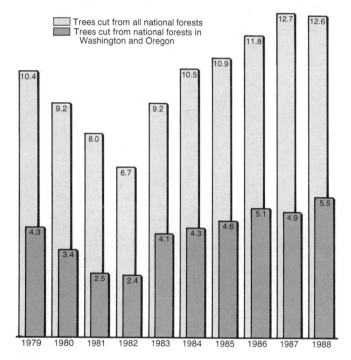

Figure 11.38

Billions of board feet cut from national forests. A board foot equals 1 foot square by 1 inch thick, and a typical 30-foot tree yields about 1000 board feet.

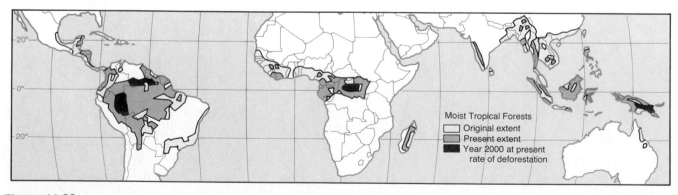

Figure 11.39

Areas of tropical rain forest. Three countries contain over half of these forests: Brazil, Zaire, and Indonesia. Every second an area of forest the size of a football field is destroyed to make way for farming, cattle ranching, commercial logging, and development projects. At current rates of deforestation, only four big blocks of tropical forest will be left in another decade.

forest, especially in the east and south. While parts of central and western Africa are still heavily forested, the United Nations estimates that at current rates of logging, Nigeria, the Ivory Coast, and other West African countries will have no forests left in another 10 years.

Why should North Americans care what happens to the tropical rain forests? Their destruction raises three principal global concerns and a host of local ones. First, all forests play a major role in maintaining the oxygen and carbon balance of the earth. People and their industries consume oxygen; vegetation both extracts the carbon from atmospheric carbon dioxide and releases oxygen back into the atmosphere. Indeed, the forests of the Amazon have been called the "lungs of the world" for their major contribution to the oxygen breathed by humankind. When the tropical forest is cleared, its role both as a carbon "sink" and oxygen-replenisher is lost.

A second global concern is the contribution of forest clearing to air pollution and climate change. Deforestation by burning releases vast quantities of carbon dioxide into the atmosphere. Brazilian scientists estimate that the thousands of fires that are set to clear the Amazon forest account for one-tenth of the global production of carbon dioxide (Figure 11.40). The destruction thus contributes to the warming of the atmosphere and, ultimately, to the

greenhouse effect. In addition, the fires generate gases (nitrogen oxides and methane) that create acid rain and contribute to the depletion of the ozone layer (see Chapter 5).

Figure 11.40

Satellite image of fires in Rondônia on a single day in 1987. The fastest and cheapest way to clear areas for farms and ranches is to burn them. About the size of Oregon, Rondônia is

the Brazilian state with the largest percentage of its forest cover destroyed by fire. Over one-tenth of the state's territory has been burned since 1976. Brazilian scientists examining satellite images counted 170,000 fires in a single year in the western Amazon. Likened by some to an environmental holocaust, the fires generate hundreds of millions of tons of gases that contribute to global warming and depletion of the earth's protective ozone layer.

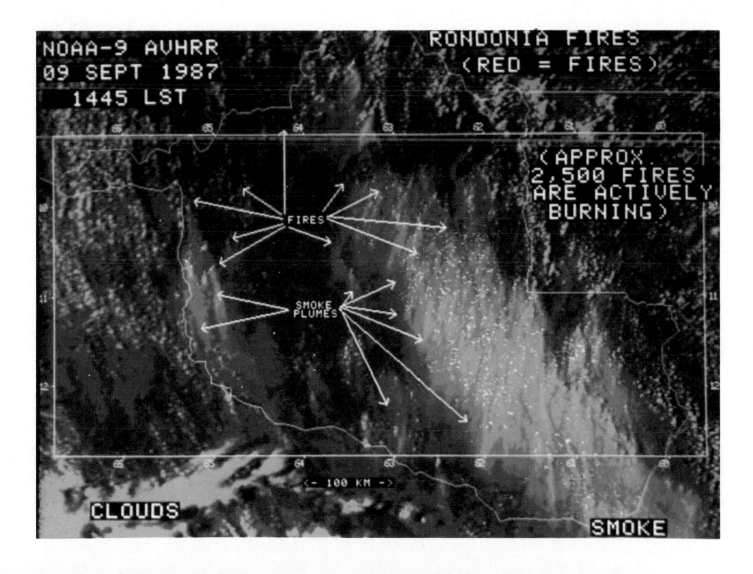

Tropical Forests and Medical Resources

Tropical forests are biological cornucopias, possessing a stunning array of plant and animal life. Costa Rica, about the size of South Carolina, contains as many bird species as all of North America, more species of insects, and nearly half the number of plant species. One stand of rain forest in Kalimantan (formerly Borneo) contains over 700 species of tree, as many as exist in North America. Forty-three species of ant inhabit a single tree species in Peru, dependent on it for food and shelter and providing in return protection from other insects.

The tropical forests yield an abundance of chemical products used to manufacture alkaloids, steroids, anesthetics, and other medicinal agents. Indeed, half of all modern drugs, including strychnine, quinine, curare, and ipecac, come from the tropical forests. A single flower, the Madagascar periwinkle, produces two drugs used to treat leukemia and Hodgkin's disease.

As significant as are these and other modern drugs derived from tropical plants, scientists believe that the medical potential of the tropical forests remains virtually untapped. They fear that deforestation will eradicate medicinal plants and traditional formulas before their uses become known, depriving humans of untold potential benefits that might never be realized. Tribal peoples make free use of plants of the rain forest for such purposes as treating stings and snakebites, relieving burns and skin fungi, reducing fevers and curing earaches. Yet botanists have only recently begun to identify tropical plants and study traditional herbal medicines to discover which of them might contain medically important compounds.

A second concern is that forest destruction will create shortages of drugs already derived from those plants. According to some experts, as many as 60,000 plants with valuable medical properties are likely to become extinct by 2050. Already endangered is reserpine, an ingredient in certain tranquilizers that is derived from *Rauwolfia serpentina*, found in India. Also threatened are cinchona, whose bark produces quinine, and foxglove varieties that are used to make the heart medications digitoxin and acetyldigitoxin.

Finally, the eradication of tropical forests is already leading to the loss of a major part of the biological diversity of the planet. The forests are one component in an intricate ecosystem that has developed over millions of years. The trees, vines, flowering plants, animals, and insects depend on one another for survival. The destruction of the habitat by clearance annually causes the extinction of thousands of plant and animal species that exist nowhere else. Of the estimated 5–30 million plant and animal species believed to exist on earth, a minimum of 40% are native to the tropical rain forest. Many of the plants have become important world staple food crops, among them rice, corn, cassava, squash, banana, pineapple, and sugarcane. Unknown additional potential food species remain as yet unexploited. In addition, the tropical forests yield an abundance of industrial products (oils, gums, latexes, and turpentines) and are the world's main storehouse of medicinal plants.

Deforestation also incurs heavy environmental, economic, and social costs on a more local basis. All forests anchor topsoil and absorb excess moisture. In a vicious cycle, forest clearance accelerates soil erosion and siltation of streams and irrigation channels, leaving the area vulnerable to flooding and drought, and leading in turn to future shortages of food and fuelwood. Within a matter of years, land that has been cleared for agriculture can become unsuitable for that use (see Figure 5.22). In the Himalayan watershed, in the Ethiopian highlands, and in numerous other places, deforestation, erosion, and rainfall runoff have aggravated floods that have killed tens of thousands and left millions homeless.

Resource Management

The destruction of the rain forests is a tragedy that yields no long-term benefits. The world is approaching the end of a period in which resources were cheap, readily available, and lavishly used. Over the centuries the earth has been viewed as an almost inexhaustible storehouse of resources for humans to exploit and, simultaneously, as a vast repository for the waste products of society. Now there is a growing realization that resources can be depleted, even renewable ones like forests, that many have life spans measured only in decades, and that the air, water, and soil—also resources—cannot absorb massive amounts of pollutants and yet retain their life-supporting abilities.

The wise management of resources entails three strategies: conservation, reuse, and substitution. By *conservation* we mean the careful use of resources so that future generations can obtain as many benefits from them as we now enjoy. It includes decreasing our consumption of resources, avoiding their wasteful use, and preserving their quality. Thus, soils can be conserved and their fertility maintained by contour plowing, crop rotation, and a variety of other practices. Properly managed, forests can be preserved even as their resources are tapped.

Opportunities to reduce the consumption of energy resources are many and varied. Nearly everything can be made more energy efficient. Motor vehicles use a significant portion of the world's oil output. Doubling their fuel efficiency by reducing vehicle weight and using more efficient engines and tires would save at least 20% of the world's total annual oil output. Industries have an enormous potential for saving energy by using more efficient equipment and processes. The Japanese steel industry, for example, uses one-third less energy to produce a ton of steel than does that industry in most other countries. Energy used for heating, cooling, and lighting homes and office buildings could be reduced by half if they were properly constructed.

The *reuse* of materials also reduces the consumption of resources. Instead of being buried in landfills, waste can be burned or decomposed and fermented to provide energy. Recycling of steel, aluminum, copper, glass, and other materials can be greatly increased, not only to recover the materials themselves but also to recoup the energy invested in their production (Figure 11.41). It has been estimated that throwing away an aluminum soft drink container wastes as much energy as filling the can half full of gasoline and pouring it on the ground.

Finally, the *substitution* of other energy sources for gas and oil, and the substitution of other materials for nonfuel minerals in short supply can be more actively pursued. If the technology to exploit them economically can be developed, coal and oil shale could supply fuel needs well into the future. In addition, the renewable energy resources, such as biomass, solar, and geothermal power, are virtually infinite in their amount and variety. While no single renewable source is likely to be as important as oil or gas, collectively they could make a significant contribution to energy needs.

Summary

Our economic and material well-being depend on our use of natural resources. Renewable resources like soil and plants are those that can be regenerated in nature as fast

Figure 11.41
Aluminum cans being recycled in Santa Monica, California.

as or faster than societies exploit them, although even renewable resources can be depleted if the rate of use exceeds that of regeneration. Nonrenewable resources—the fossil fuels and nonfuel minerals—are generated so slowly that they are thought of as existing in finite amounts. The proved reserves of a resource are the amounts that have been identified and that can be extracted profitably.

The Industrial Revolution was characterized by a shift from societal dependence on renewable resources to resources derived from nonrenewable minerals, chiefly fossil fuels. The industrially advanced countries depend for about 40% of their commercial energy on crude oil, a scarce and unevenly distributed resource whose known reserves are likely to be exhausted before the middle of the next century. Only two regions, the USSR and the Middle East, have large reserves of natural gas, another resource likely to approach depletion by the mid-21st century. Synthetic petroleum and gas can be produced from a number of sources, but at the present time all synfuels have high economic and environmental costs. Although nuclear power plants produce less than one-third of the world's electricity as a whole, the pattern is one of uneven dependence. Some countries receive over half their electricity from such plants, while others have none.

Renewable natural resources are more widely and evenly distributed than the nonrenewable ones. Wood and other forms of biomass are the primary source of energy for over half the world's people. Hydropower is a major source of electricity in many countries. Other renewable resources, including geothermal energy and wind power, make a more localized and limited contribution to energy needs.

The earth's crust contains a variety of nonfuel mineral resources from which people fashion metals, glass, stone, and other products. All are nonrenewable. Some exist in vast amounts, others in relatively small quantities; some are widely distributed, others concentrated in just a few locations.

About one-tenth of the earth's surface is used for intensive production of food. Although food production has increased as fast as world population, more people than ever before are malnourished. Methods of expanding food supplies include the expansion of cultivated areas, increasing yields on land already under cultivation, and increasing the production of fish.

Human activities have had and continue to have a severe impact on two types of land resources, coastal wetlands and forests. Both play vital ecological roles, yet the wetlands have been degraded in many parts of the world,

and forests are being destroyed faster than they can regenerate. The clearance of tropical forests is a matter of global, not just local, concern.

The growing demand for resources, induced by population increases and economic development, strains the earth's supply of raw materials. We began this chapter by presenting two significantly different views of the future. Because the shape of the future will be determined by the way the present generation of people thinks and acts, the wise and careful management of natural resources of all types is essential for all the world's economies, regardless of their stage of development.

Key Words

aquaculture	nuclear fusion
biomass	oil shale
coal gasification	photovoltaic cell
coal liquefaction	proved reserves
energy	renewable resource
estuarine zone	resource
geothermal energy	solar energy
hydropower	synthetic fuel
liquefied natural gas	tar sand
nonrenewable resource	tragedy of the commons
nuclear fission	

For Review

1. What is the basic distinction between a *renewable* and a *nonrenewable* resource? Why do estimates of *proved reserves* vary over time?

2. Why are energy resources called the "master" natural resources? What is the relationship between energy consumption and industrial production? Briefly describe historical energy consumption patterns in the United States.

3. Why has oil become the dominant form of commercial energy? Which countries are the main producers of crude oil? How long are proved reserves of oil likely to last?

4. Why has the proportion of U.S. energy supplied by coal increased since 1961? What ecological and social problems are associated with the use of coal?

5. Review some of the techniques by which *synthetic fuels* might be produced. Why are synfuels not yet economically competitive with other fossil fuels?

6. What are the different methods of generating nuclear energy? Why is there public opposition to nuclear power?

7. Which are the most widely used ways of using renewable resources to generate energy? What are the advantages of using such resources?

8. What in general are the leading mining countries? What role do developing countries play in the production of critical raw materials? How have producing countries reacted to the threatened scarcity of copper?

9. Since food resources are considered renewable, why is there concern about their exploitation? Discuss three ways of increasing food production. What problems do they pose?

10. What is the *estuarine zone*? What role does it play in the maintenance of marine life? In what ways is the ecology of the zone being assaulted?

11. What vital ecological functions do forests perform? Where are the tropical rain forests located, and what concerns are raised by their destruction?

12. Discuss three ways of reducing demands on resources.

Selected References

Blenden, John. *Mineral Resources and Their Management.* London: Longman Group, 1985.

Brown, Lester, et al. *State of the World 1989.* New York: Norton, 1989.

Cuff, David J., and William J. Young. *The United States Energy Atlas.* 2d rev. ed. New York: Macmillan, 1986.

Cutter, Susan, Hilary L. Renwick, and William H. Renwick. *Exploitation, Conservation, Preservation: A Geographic Perspective on Natural Resource Use.* Totowa, N.J.: Rowman & Allanheld, 1985.

Dasmann, R. F. *Environmental Conservation.* 5th ed. New York: Wiley, 1984.

Eckholm, Erik P. *Losing Ground: Environmental Stress and World Food Prospects.* New York: Norton, 1976.

Ehrlich, Paul R., and John P. Holdren. *The Cassandra Conference: Resources and the Human Predicament.* College Station: Texas A & M University Press, 1988.

Gates, David M. *Energy and Ecology.* Sunderland, Mass.: Sinauer Associates, 1985.

Glassner, Martin I., ed. *Global Resources: Challenges of Interdependence.* New York: Praeger, for the Foreign Policy Association, 1983.

Gradwohl, Judith, and Russell Greenberg. *Saving the Tropical Forests.* Covelo, Calif.: Island Press, 1988.

McLaren, D. J., and B. J. Skinner. *Resources and World Development.* New York: Wiley, 1987.

Miller, G. Tyler, Jr. *Resource Conservation and Management.* Belmont, Calif.: Wadsworth, 1990.

Norse, Elliott A. *Ancient Forests of the Pacific Northwest.* Covelo, Calif.: Island Press, 1989.

Owen, Oliver S. *Natural Resource Conservation.* 3d ed. New York: Macmillan, 1980.

Repetto, Robert. *World Enough and Time: Successful Strategies for Resource Management.* New Haven, Conn.: Yale University Press, 1986.

———, ed. *The Global Possible: Resources, Development, and the New Century.* New Haven, Conn.: Yale University Press, 1985.

Sawyer, Stephen W. *Renewable Energy: Progress, Prospects.* Washington, D.C.: Association of American Geographers, 1986.

Shea, Cynthia Pollock. *Renewable Energy: Today's Contribution, Tomorrow's Promise.* Worldwatch Paper 81. Washington, D.C.: Worldwatch Institute, 1988.

Stark, Linda. *State of the World 1989.* New York: Norton, 1989.

The State of the Environment 1985. Washington, D.C.: Organization for Economic Cooperation and Development, 1985.

United Nations Food and Agriculture Organization (FAO). *Land, Food and People.* Rome: FAO, 1984.

Williams, Michael, ed. *Wetlands: A Threatened Environment.* Oxford, England: Basil Blackwell, 1990.

World Resources Institute and the International Institute for Environment and Development. *World Resources 1986; World Resources 1987; World Resources 1988–90.* New York: Basic Books; *World Resources 1990–91.* New York: Oxford University Press, 1990.

C H A P T E R

12

Urban Geography

Wilshire Boulevard, Los Angeles, California.

IN the 1930s Mexico City was described as perhaps the handsomest city in North America and the most exotic capital city of the hemisphere, essentially unchanged over the years and timeless in its atmosphere. It was praised as beautifully laid out, with wide streets and avenues, still the "city of palaces" as Baron von Humboldt called it in the 19th century. The 200-foot-wide Paseo de la Reforma, often noted as "one of the most beautiful avenues in the world," was shaded by a double row of trees and lined with luxurious residences.

By the 1950s, with a population of over 2 million and an area of 20 square miles (52 km²), Mexico City was no longer unchanged. The old, rich families who formerly resided along the Paseo de la Reforma had fled from the noise and crowding. Their "palaces" were being replaced by tall blocks of apartments and hotels. Industry was expanding and multiplying, tens of thousands of peasants were flocking in from the countryside every year. By the early 1990s, with its population estimated at 20 million and its area at over 390 square miles (1000 km²), metropolitan Mexico City was adding a half-million immigrants each year as well as growing prodigiously by its own 3% birth rate.

The toll exacted by that growth is heavy. Each year the city pours more than 5 million tons of pollutants into its air, 80% coming from its estimated 3 million motor vehicles (up from 55,000 in 1950), the rest from some 35,000 industrial plants. More than 4 million people citywide have no access to tap water; in some squatter neighborhoods less than 50% do. Some 3 million residents have no access to the sewage system. Approximately one-third of all families—and they average five people—live in just a single room, and that room generally is in a hovel in one of the largest slums in the world.

The changes in Mexico City since the 1930s have been profound (Figure 12.1). Already the world's most populous metropolitan area, Greater Mexico City is expected to swell further to at least 31 million by the year 2000. Mexico City is a worst case scenario of an urban explosion that sees an increasing proportion of the world's population housed within a growing number of immense cities.

Figure 12.2 presents evidence that the growth of major metropolitan areas has been astounding in this century. There were over 280 metropolitan areas that had a population in excess of 1 million people in 1990, while at the beginning of the century there were only 13. Fifteen metropolises (Beijing, Buenos Aires, Cairo, Calcutta, London, Los Angeles, Mexico City, New York, Osaka-Kobe, Paris, Rhine-Ruhr, Rio de Janiero, São Paulo, Shanghai, Tokyo) have over 10 million people (see Figure 12.3). In 1900, there were none of that size. Of course, as we saw in Chapter 6, it follows that since the world's population has greatly increased so too would the urban component increase. But the fact remains that urbanization and metropolitanization have increased more rapidly than the growth of total population. The amount of urban growth differs from continent to continent and from country to country, but all countries have one thing in common: the proportion of their people living in cities is rising.

Table 12.1 shows world urban population by region. Note that the most industrialized parts of the world, North America and Western Europe, are the most urbanized in terms of percentage of people living in cities, while most of Asia and Africa have the lowest proportions of urban population. What is evident is that the industrialization process has been also one of urbanization. As the world continues to industrialize, especially the Third World countries, one can expect further large increases in urbanization. It is interesting to note in Table 12.1 that even though China, Southeast Asia (Vietnam, Indonesia, etc.), and South Asia (India, Pakistan, and Bangladesh) have relatively low proportions of people in urban regions, the absolute number of people in urban areas there is among the highest in the world. Given the huge populations in Asia, and the relatively heavy emphasis on agriculture (except Japan and Korea), it sometimes escapes us that there are many large cities throughout parts of the world where subsistence agriculture still engages most people.

In this chapter, our first objective is to consider the major factors responsible for the size and location of cities. The second goal is to identify the nature of land use patterns within urban areas, and third, we will attempt to differentiate cities around the world by a review of the factors that help to explain their special nature.

The Functions of Urban Areas

Except for the occasional recluse or hermit, people gather together to form couples, families, groups, organizations, towns, and so forth. This desire to be near one another, however, is more than a function of the need to socialize. Our human support systems are based on the flow of information, goods and services, and cooperation among people who are located at convenient places relative to one another. Unless individuals can produce all that they need themselves, and relatively few can, they must depend on shipments of food and supplies to their home place or convenient outlet centers. Nonsubsistence groups establish stores, places of worship, repair centers, and production sites as close to their home places as is possible and reasonable. The result is the establishment of towns. These may grow to the size of a Mexico City (about 20 million people today) or a Tokyo metropolitan area (about 18 million people).

Figure 12.1

Sprawling Mexico City. With over 15 million people, the Mexico City metropolitan area is one of the largest in the world. Ringed by mountains, the area frequently experiences temperature inversions resulting in the smog visible in this photograph.

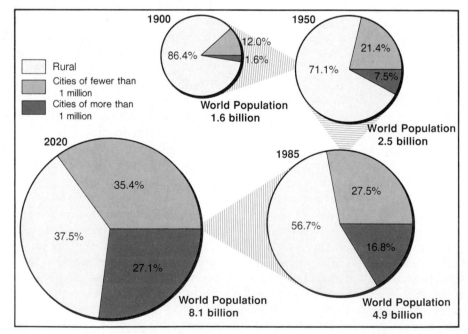

1900

12.0%
86.4%
1.6%

World Population 1.6 billion

Rural
Cities of fewer than 1 million
Cities of more than 1 million

1950

21.4%
71.1%
7.5%

World Population 2.5 billion

2020

35.4%
37.5%
27.1%

World Population 8.1 billion

1985

27.5%
56.7%
16.8%

World Population 4.9 billion

Figure 12.2
Patterns of world urbanization.

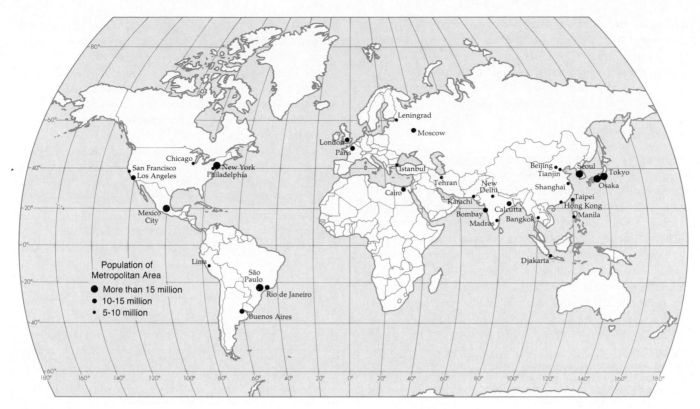

Figure 12.3
The location of metropolitan areas with over 5 million people.

Table 12.1
Estimated Urban Share of Total Population, 1950 and 1988 with Projections to 2000

Region	1950	1988	2000
North America	64%	74%	78%
Europe	56	75	79
Soviet Union	39	65	74
East Asia (except China)	43	73	79
China	12	41	50
Southeast Asia	—	25	35
South Asia	15	25	39
Latin America	41	68	77
Africa	15	30	44
Oceania	61	70	74
World	**29**	**45**	**51**

Source: Lester R. Brown and Jodi L. Jacobson, *The Future of Urbanization*. Worldwatch Paper 77 (Washington, D.C.: Worldwatch Institute, 1987), p. 8; Population Reference Bureau, *1988 World Population Data Sheets*; and composite projections.

Whether they are villages, towns, or cities, urban settlements exist for the efficient performance of functions required by the society that creates them, functions that cannot be adequately carried out in dispersed locations. They reflect the saving of time, energy, and money that the agglomeration of people and activities implies. The more accessible the producer to the consumer, the worker to the workplace, the citizen to the town hall, the worshiper to the church, or the lawyer or doctor to the client, the more efficient is the performance of their separate activities, and the more effective is the integration of urban functions.

Urban areas provide all or some of the following types of functions: retailing, wholesaling, manufacturing, business service, entertainment, religious service, political and official administration, military defensive needs, social service, public service including sanitation and police, transportation and communication service, meeting place activity, visitor service, and places for their residents to live. Because all urban functions and people cannot be located at a single point, cities themselves must take up space, and land uses and populations must have room within them. Because interconnection is essential, the nature of the transportation system will have an enormous bearing on the total number of services that can be performed and the efficiency with which the functions can be carried out (Figure 12.4). The totality of people and functions of a city constitutes a distinctive cultural landscape whose similarities and differences from place to place are the subjects for urban geographic analysis.

Some Definitions

Urban areas are not of a single type, structure, or size. What they have in common is that they are nucleated, nonagricultural settlements. At one end of the size scale, urban areas are small towns with perhaps a single main street of shops; at the opposite end, they are complex, multifunctional metropolitan areas or supercities (Figure 12.5). The word *urban* often is used in place of such terms as town, city, suburb, and metropolitan area, but it is a general term, and it is not used to specify a particular type of settlement. Although the terms designating the different types of urban settlements, such as city, are employed in common speech, they are not uniformly applied by all users. What is recognized as a city by a resident of rural Vermont or West Virginia might not at all be afforded that name and status by an inhabitant of California or New Jersey. It is necessary in this chapter to agree on the meanings of terms commonly employed but varyingly interpreted.

The words **city** and **town** denote nucleated settlements, multifunctional in character, including an established central business district and both residential and nonresidential land uses. **Towns** have smaller size and less functional complexity than cities, but still have a nuclear business concentration. **Suburb** denotes a subsidiary area, a functionally specialized segment of a large urban complex. It may be dominantly or exclusively residential, industrial, or commercial, but by the specialization of its land uses and functions, a suburb depends on urban areas

Figure 12.4

The Los Angeles freeway system. This is the world's most extensive system of high-speed highways for a metropolitan area. With continuing growth in the Los Angeles area, the system has become congested.

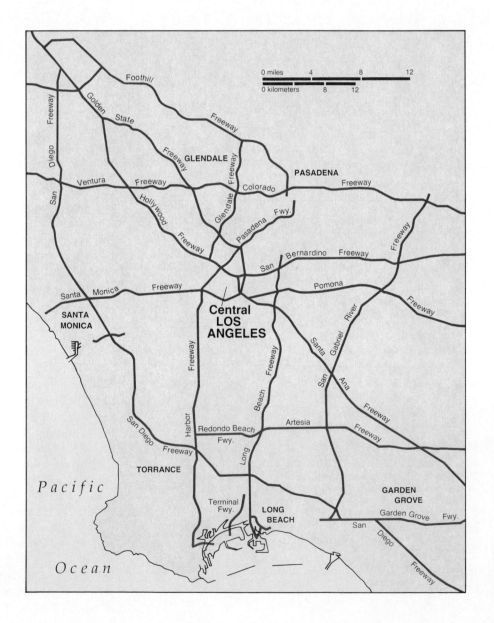

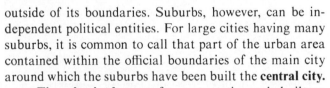

Figure 12.5
The differences in size, density, and land use complexity between New York City and a small town are immediately apparent. One is a city, one is a town, but both are urban areas.

outside of its boundaries. Suburbs, however, can be independent political entities. For large cities having many suburbs, it is common to call that part of the urban area contained within the official boundaries of the main city around which the suburbs have been built the **central city.**

The **urbanized area** refers to a continuously built-up landscape defined by building and population densities with no reference to political boundaries. It may be viewed as the physical city and may contain a central city, and many contiguous cities, towns, suburbs, and other urban tracts. A **metropolitan area,** on the other hand, refers to a large-scale functional entity, perhaps containing several urbanized areas, discontinuously built up but nonetheless operating as an integrated economic whole (Figure 12.6). (See "The Definition of 'Metropolitan' in the United States.")

The Location of Urban Settlements

Cities are functionally connected to other cities and to rural areas. In fact, the reason for the existence of a city is not only to provide services for itself, but for others outside of the city. The city is a consumer of food, a processor of materials, and an accumulator and dispenser of goods and services, but it must rely on outside areas for its supplies and as a market for its activities. In order to perform adequately the tasks that support it and to add new functions as demanded by the larger economy, the city must be ef-

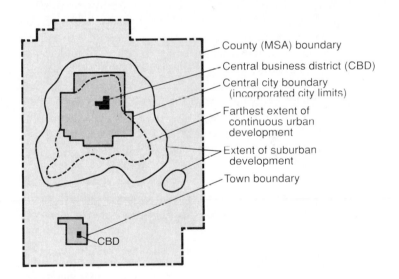

County (MSA) boundary
Central business district (CBD)
Central city boundary (incorporated city limits)
Farthest extent of continuous urban development
Extent of suburban development
Town boundary

Figure 12.6
A hypothetical spatial arrangement of urban units within a Metropolitan Statistical Area. (See the box "The Definition of 'Metropolitan' in the United States.") Sometimes official limits of the central city are very extensive and contain areas commonly thought of as suburban. Older eastern U.S. cities more often have restricted limits and contain only part of the high-density land uses associated with them.

ficiently located. That efficiency may be marked by centrality to the area served. It may derive from the physical characteristics of its site. Or placement may be related to the resources, productive regions, and transportation network of the country, so that the effective performance of a wide array of activities is possible.

Definitions of various types of urban areas must be clear if proper accounting is to be made by governmental authorities. The United States Bureau of the Census has refined and redefined the concept of "metropolitan" from time to time to summarize the realities of the changing population, physical size, and functions of urban regions.

Until 1983, the *Standard Metropolitan Statistical Area* (SMSA) was recognized. It was made up of one or more functionally integrated counties focusing upon a central city of at least 50,000 inhabitants. Now, the minimum size requirement for central cities has been dropped and central city status is determined by other qualities, such as whether a city is an employment center surrounded by bedroom community-type suburbs. Automatically, the number of central cities (and metropolitan areas) increased. The statistical structure of urban America has been altered as individual communities exercised their rights to withdraw from former metropolitan affiliations or opted to join with neighboring cities in new ones.

In the mid-1980s, old and new metropolitan areas were redefined into *Metropolitan Statistical Areas* (MSAs are economically integrated urbanized areas in one or more contiguous counties), *Primary Metropolitan Statistical Areas* (PMSAs are those counties that are part of MSAs that have less than 50% resident workers working in a different county), and *Consolidated Metropolitan Statistical Areas* (an MSA becomes a CMSA if it contains one million or more people and is composed of PMSAs). Figure 12.6 shows boundaries for a hypothetical MSA.

In discussing urban settlement location, geographers frequently differentiate between site and situation. The **site** is the exact location of the settlement and can be described in terms of latitude and longitude or in terms of the physical characteristics of the site. For example, the site of Philadelphia is an area bordering and west of the Delaware River north of the intersection with the Schuylkill River in southeast Pennsylvania (Figure 12.7). The description can be more or less exhaustive depending on the purpose it is meant to serve. In the Philadelphia case, the fact that the city is partly on the Atlantic coastal plain, partly on a piedmont (foothills), and is served by navigable rivers is important if one is interested in the development of the city during the Industrial Revolution. As Figure 12.8 suggests, water transportation was an important localizing factor when the major American cities were established.

If site suggests absolute location, **situation** indicates relative location. The relative location places a settlement in relation to the physical and human characteristics of the surrounding areas. Very often it is important to know what kinds of possibilities and activities exist in the area near a settlement, such as the distribution of raw materials, market areas, agricultural regions, mountains, and oceans. The site of central Chicago is 41°52′N, 87°40′W, on a lake plain, but more important is its situation close to the deepest penetration of the Great Lakes system into the interior of the country, astride the Great Lakes-Mississippi waterways, and near the western margin of the manufacturing belt, the northern boundary of the Corn Belt, and the southeastern reaches of a major dairy region. References to railroads, coal deposits, and ore fields would amplify its situational characteristics (Figure 12.9). From this description of Chicago's situation, implications relating to market, to raw materials, and to transportation centrality can be drawn.

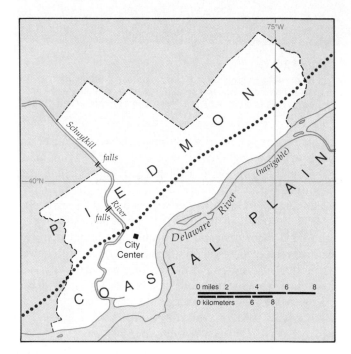

Figure 12.7
The site of Philadelphia.

The site or situation that originally gave rise to an urban unit may not long remain the essential ingredient for its growth and development. Agglomerations, originally successful for whatever reason, may by their success attract people and activities totally unrelated to the initial localizing forces. By what has been called a process of "circular and cumulative causation," a successful urban unit may acquire new populations and functions attracted by the already existing markets, labor force, and urban facilities.

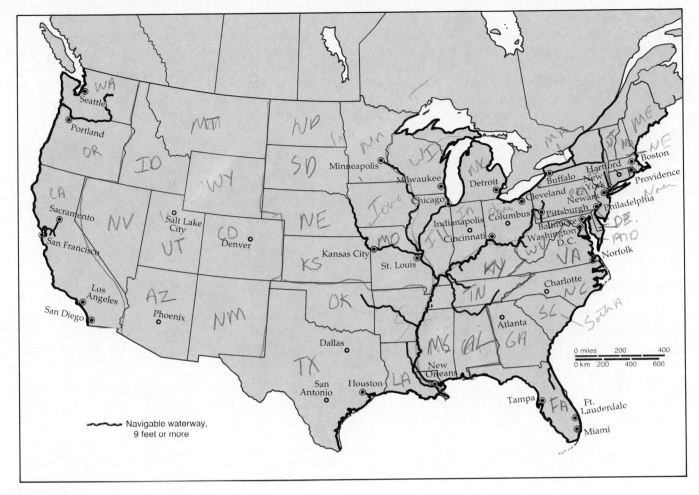

Figure 12.8

Metropolitan areas in the United States with 1 million or more residents as of July 1, 1987. Notice the association of principal cities and navigable water. Before the advent of railroads in the middle of the 19th century, all major cities were associated with waterways.

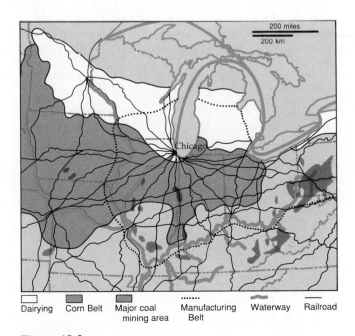

Dairying Corn Belt Major coal mining area Manufacturing Belt Waterway Railroad

Figure 12.9

The situation of Chicago helps to suggest the reasons for its functional diversity and size.

Systems of Urban Settlements

The various functions that an individual urban area performs are reflected not only in the size of that settlement but also in its location and in its relationship with other urban units in the larger system of which it is part.

The Urban Hierarchy

Perhaps the most effective way to recognize how systems of cities are organized is to consider the **urban hierarchy.** Cities can be divided into size classes on the basis of their functional complexity. One can measure the numbers and kinds of services each city or metropolitan area provides. The hierarchy is then like a pyramid; the few large and complex cities are at the top and the many smaller cities are at the bottom. There are always more smaller cities than larger ones. One can envisage, say, a 7-level hierarchy where the complexity of cities increases as one rises in the pyramid.

When a spatial dimension is added to the hierarchy as in Figure 12.10, it becomes clear that a spatial system of metropolitan centers, large cities, small cities, and towns exists. Goods, services, communication lines, and people flow up and down the hierarchy. The few high-level metropolitan areas provide specialized functions for large regions while the smaller cities have smaller regions that they serve. Note that the cities serve the area around them, but since cities of the same level provide roughly the same services, cities of the same size tend not to serve each other unless they provide some very specialized service, such as housing a political capital of a region or a major university. Thus, the cities of a given level in the hierarchy are not independent but interrelated with cities of other levels in the hierarchy. Together all cities at all levels in the hierarchy constitute an urban system.

Central Places

An effective way to realize how cities and towns are interrelated is to consider urban settlements as **central places**, that is, as centers for the distribution of economic goods

and services. In 1933, the German geographer Walter Christaller attempted to explain the size and location of settlements. He developed a framework, called **central place theory**, for understanding urban interdependence. Christaller recognized that his theory would best be developed in rather idealized circumstances. He assumed that the following propositions were true.

1. Towns that provide the surrounding countryside with such fundamental goods as groceries and clothing would develop in a plain where farmers specialized in commercial agricultural production.
2. The farm population would be dispersed in an even pattern.
3. The characteristics of the people would be uniform; that is, they would possess similar tastes, demands, and incomes.
4. Each kind of product or service available to the dispersed population would have its own threshold, or minimum number of consumers needed to support its supply. Because such goods as diamonds or fur coats are either expensive or not

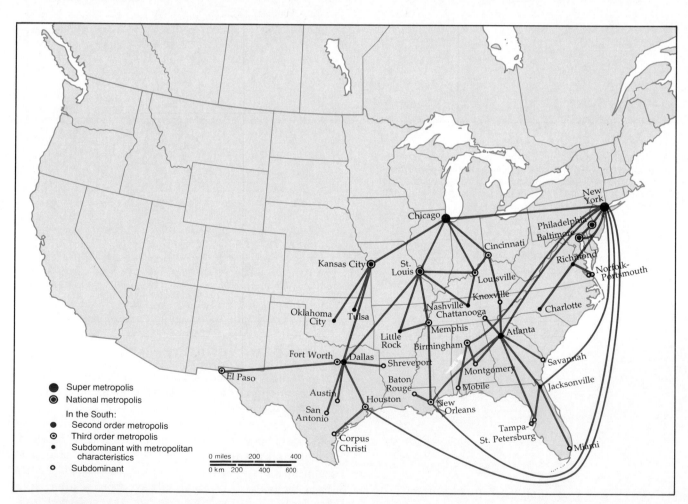

Figure 12.10
Hierarchical relationships among southern U.S. cities.

in great demand, they would have a high threshold, while a fewer number of consumers would be required to support a store selling bread.

5. Consumers would purchase goods and services from the nearest opportunity (store).

When all of the assumptions are considered simultaneously, they yield the following results.

1. The business people of the agricultural plain will divide the area into noncompeting markets where each entrepreneur has exclusive rights to the sale of a particular product.
2. A series of hexagonal market areas that cover the entire plain will emerge, as shown in Figure 12.11.
3. There will be a central place at the center of each of the hexagonal market areas.
4. The largest central places will supply all the goods and services the consumers in the area demand and can afford.
5. The size of the market area of a central place will be proportional to the number of goods and services offered from that central place.
6. Contained within or at the edge of the largest market areas are central places serving a smaller population and offering fewer goods and services.

In addition, Christaller reached two important conclusions. First, towns of the same size will be evenly spaced, and larger towns will be farther apart than smaller ones. This means that there will be many more small than large towns. In Figure 12.11 the ratio of the number of small towns to towns of the next larger size is three to one. This distinct, steplike series of towns in size classes differentiated by both size and function is called a *hierarchy of central places.*

Second, the system of towns is interdependent. If one town were eliminated, the entire system would have to readjust. Consumers need a variety of products, each of which has a different minimum number of customers required to support it. The towns containing many goods and services become regional retailing centers, and the small central places serve just the people immediately in their vicinity. The higher the threshold of a desired product, the farther, on average, the consumer must travel to purchase it.

These conclusions have been shown to be generally valid in widely differing areas within the commercial world. When varying incomes, cultures, landscapes, and transportation systems are taken into consideration, the results, although altered to some extent, hold up rather well. They are particularly applicable to agricultural areas, especially with regard to the size and spacing of cities and towns, as Figure 12.12 suggests. One has to stretch things a bit to see the model operating in highly industrialized areas, where cities are more than just retailing centers. However, if we combine a Christaller-type approach with the ideas that help us understand industrial location and transportation alignments (see Chapter 10), we have a fairly good understanding of the location of the majority of cities and towns.

The Economic Base

When one or more urban settlements within a well-linked system increases its productivity, perhaps because of an increase in demand for the special goods or services that it produces, all members of the system are likely to benefit. The concept of the **economic base** shows how settlements are affected by changes in economic conditions. In

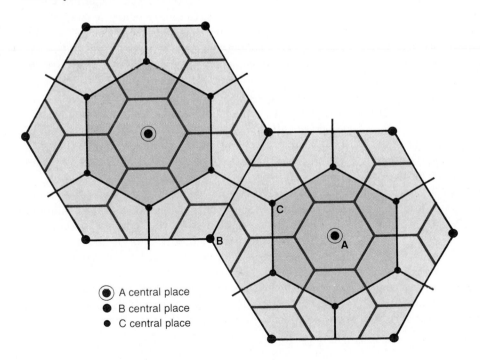

A central place
B central place
C central place

Figure 12.11
The two A central places are the largest on this diagram of one of Christaller's models. The B central places offer fewer goods and services for sale and serve only the areas of the intermediate-sized hexagons. The many C central places, which are considerably smaller and more closely spaced, serve still smaller market areas. The goods offered in the C places are also offered in the A and B places, but the latter offer considerably more and more specialized goods. Notice that the places of the same size are equally spaced.

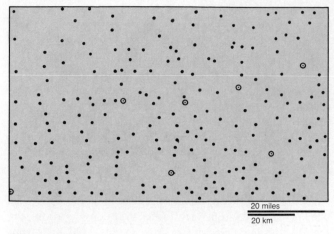

Figure 12.12

The pattern of hamlets, towns, and cities in a portion of Indiana. This map represents an area 82 by 51 miles just north of Indianapolis. Cities containing more than 10,000 people are circled. Notice that the pattern is remarkably even and includes a number of linear arrangements that correspond to highways and railroads.

the discussion that follows, we concentrate on change within one city, and from our knowledge of central place theory, we can deduce how changes in one city affect the well-being of those living in other cities.

Part of the employed population of an urban unit is engaged in the production of goods or the performance of services for areas and people outside the city itself. They are workers engaged in "export" activities, whose efforts result in money flowing into the community. Collectively, they constitute the **basic sector** of the city's total economic structure. Other workers support themselves by producing things for residents of the urban unit itself. Their efforts, necessary to the well-being and the successful operation of the city, do not generate new money for it but comprise a **service,** or **nonbasic, sector** of its economy. These people are responsible for the internal functioning of the urban unit. They are crucial to the continued operation of its stores, professional offices, city government, local transit, and school systems.

The total economic structure of a city equals the sum of its basic and nonbasic activities. In actuality, it is the rare urbanite who can be classified as belonging entirely to one sector or another. Some part of the work of most people involves financial interaction with residents of other areas. Doctors, for example, may have mainly local patients and thus are members of the nonbasic sector, but the moment they provide a service to someone from outside the community, they bring new money into the city and become part of the basic sector.

Variations in basic employment structure among urban units characterize the specific functional role played by individual cities. Most cities perform many export functions, and the larger the urban unit, the more multifunctional it becomes. Nonetheless, even in cities with a diversified economic base, one or a very small number of export activities tends to dominate the structure of the community and to identify its operational purpose within a system of cities. Figure 12.13 indicates the functional specializations of large U.S. cities.

Assuming it were possible to divide with complete accuracy the employed population of a city into totally separate basic and nonbasic components, a ratio between the two employment groups could be established. With exception for some high-income communities, this *basic/nonbasic ratio* is roughly similar for cities of similar size irrespective of their functional specializations. Further, as a city increases in size, the number of nonbasic personnel grows faster than the number of new basic workers. Thus, in cities with a population of 1 million, the ratio is about 2 nonbasic workers for every basic worker; the addition of 10 new basic employees implies the expansion of the labor force by 30 (10 basic, 20 nonbasic) and an increase in total population equal to the added workers plus their dependents. A **multiplier effect** thus exists, associated with economic growth. The term multiplier effect implies the addition of nonbasic workers and dependents to a city's total employment and population as a supplement of new basic employment; the size of the effect is determined by the city's basic/nonbasic ratio (see Figure 12.14).

The changing numerical relationships shown in Figure 12.14 are understandable when we consider how settlements add functions and grow in population. A new industry selling services to other communities requires new workers, who thus increase the basic work force. These new employees, in turn, demand certain goods and services, such as clothing, food, and medical assistance, which are provided locally. Those who perform such services must themselves have services available to them. For example, a grocery clerk must also buy groceries. The more nonbasic workers a city has, the more nonbasic workers are needed to support them, and the application of the multiplier effect becomes obvious.

We have also seen that the growth of cities may be self-generating —"circular and cumulative" in a way related not to the development of basic industry but to the attraction of what would be classified as *service* industry. Banking and legal services, a sizable market, a diversified labor force, extensive public services, and the like may generate additions to the labor force not basic by definition.

In much the same way that settlements grow in size and complexity, so do they decline. When the demand for the goods and services of an urban unit falls, obviously there is a need for fewer workers, and thus both the basic and the service components of a settlement system are affected. There is, however, a resistance to decline that impedes the process and delays its impact. Whereas cities can grow rapidly as migrants respond quickly to the need for more workers, under conditions of decline those that

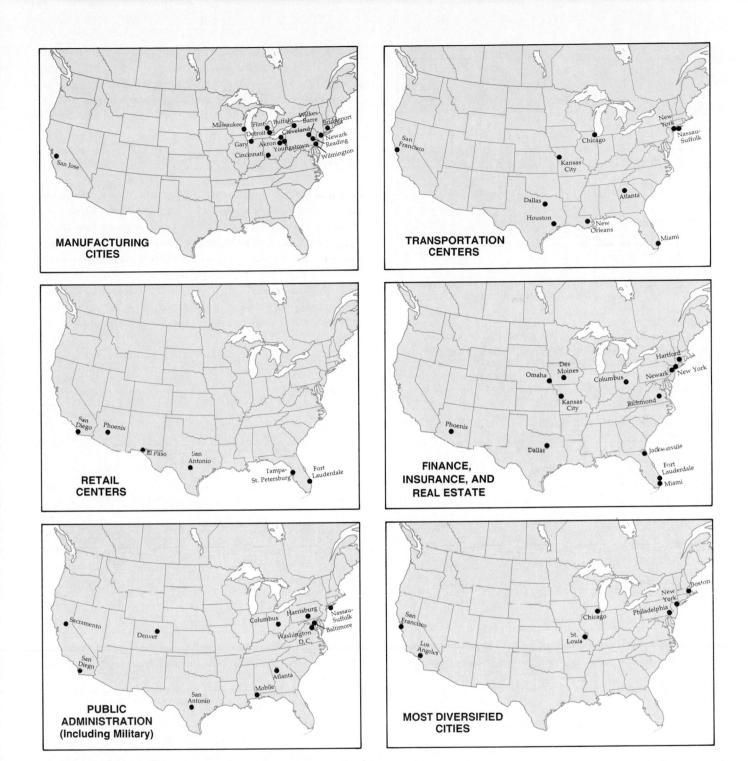

Figure 12.13

Functional specialization of U.S. metropolitan areas. Five categories of employment were selected to show patterns of specialization for some U.S. metropolitan areas. A sixth category, Most Diversified, represents those cities that have a generally balanced employment distribution. Note that the most diversified urban areas tend to be the largest.

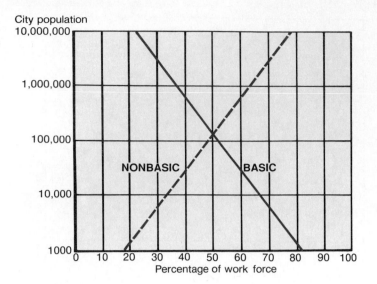

Figure 12.14

A generalized representation of the proportion of the work force engaged in basic and nonbasic activities by city size. As settlements become larger, a greater proportion of their work force is employed in nonbasic activities. Larger settlements are therefore more self-contained.

have developed roots in the community are hesitant to leave or may be financially unable to move to another locale. Figure 12.15 shows that in recent years urban areas in the South and West of the United States have been growing while decline is evident in the Northeast and the North Central regions.

Urban Influence Zones

A small city may influence a local region of, say, 25 square miles if, for example, its newspaper is delivered to that region. Beyond that area, another city may be the dominant influence. **Urban influence zones** are the areas outside of a city that are still affected by it. As the distance away from a city increases, its influence on the surrounding countryside decreases (recall the idea of distance decay discussed in Chapter 9). The sphere of influence of an urban unit is usually proportional to its size.

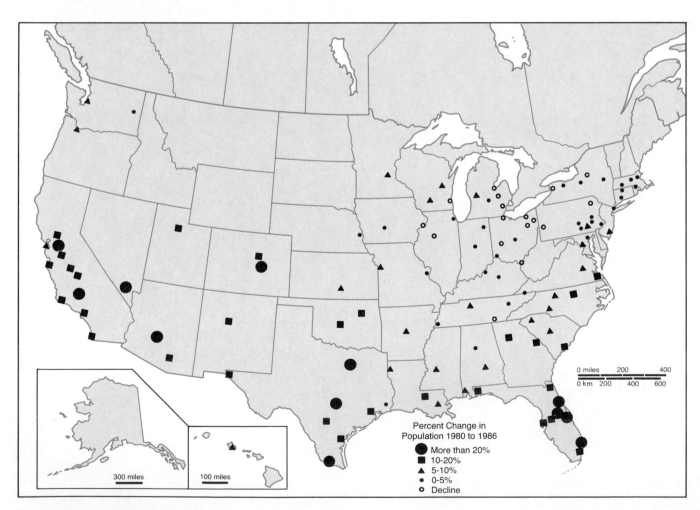

Figure 12.15

The pattern of metropolitan growth and decline in the United States, 1980–86.

A large city located 100 miles away from the small city may influence the small city and other small cities through its banking services, its TV station, and its large shopping malls. Consequently, influence zones are very much like the market areas of central place theory. There is an overlapping hierarchical arrangement, and the influence of the largest cities is felt over the widest areas (Figure 12.16).

Intricate relationships and hierarchies are common. Consider Grand Forks, North Dakota, which for local market purposes dominates the rural area immediately surrounding it. However, Grand Forks is influenced by political decisions made in the state capital, Bismarck. For a variety of cultural, commercial, and banking activities, Grand Forks is influenced by Minneapolis. A center of wheat production, Grand Forks is subordinate to the grain market in Chicago. Of course, the pervasive agricultural and other political controls exerted from Washington, D.C., on Grand Forks indicate how large and complex are the urban zones of influence.

Inside the City

An understanding of the nature of cities is incomplete without a knowledge of their internal characteristics. So far, we have explored the location, the size, and the growth and decline tendencies of cities. Now we look into the city itself in order to better understand how land uses are distributed, how social areas are formed, and how institutional controls, such as zoning regulations, affect its structure. This discussion will primarily relate to cities in the United States, although most cities of the world have been formed in a somewhat similar manner.

It is a common observation that a recurring pattern of land use arrangements and population densities exists within urban areas. There is a certain sameness to the way cities are internally organized, especially within one particular culture sphere like Anglo-America or Western Europe. The major variables responsible for shaping internal land use patterns are: accessibility, controls on the

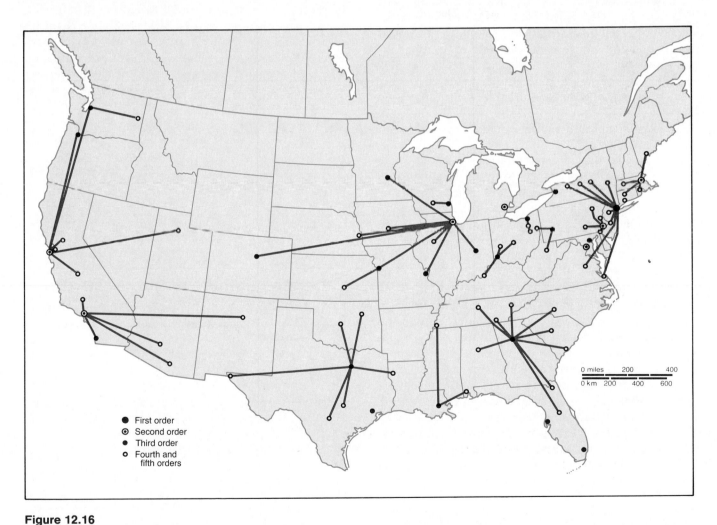

Figure 12.16

The hierarchical structure of influence zones. This map is based on Borchert's analysis of the importance of functional activities of major U.S. cities and their links of influence to cities of lesser importance. Importance is judged by the array of

manufacturing, service, and governmental activities. For map clarity, the links from New York to every second-order city are not shown.

market in land, and the transportation technologies available during the periods of urban growth. These variables will be discussed together in the following sections.

The Competitive Bidding for Land

For its effective operation, the city requires close spatial association of its functions and people. As long as these functions were few and the population small, pedestrian movement and pack-animal haulage were sufficient for the effective integration of the urban community. With the addition of large-scale manufacturing and accelerated urbanization of the economy during the 19th century, however, functions and populations—and therefore city area—grew beyond the interaction capabilities of pedestrian movement alone. Increasingly efficient, and costly, mass-transit systems were installed. Even with their introduction, however, only land within walking distance of the mass-transit routes or terminals could successfully be incorporated into the expanding urban structure.

Usable land, therefore, was a scarce commodity, and by its scarcity, it assumed high market value and demanded intensive, high-density utilization. Because of its limited supply of usable land, the industrial city of the mass-transit era was compact, was characterized by high residential and structural densities, and showed a sharp break on its margins between urban and nonurban uses. The older central cities of, particularly, the northeastern United States and southeastern Canada were of that vintage and pattern.

Within the city, parcels of land were allocated among alternate potential users on the basis of the relative ability of those users to outbid their competitors for a chosen site. There was, in gross generalization, a continuous open auction in land in which users would locate, relocate, or be displaced in accordance with "rent-paying" ability. The attractiveness of a parcel, and therefore the price that it could command, was a function of its accessibility. Ideally, the most desirable and efficient location for all the functions and the people of a city would be at the single point at which the maximum possible interchange could be achieved. Such total coalescence of activity is obviously impossible.

Because uses must therefore arrange themselves spatially, the attractiveness of a parcel is rated by its relative accessibility to all other land uses of the city. Store owners wish to locate where they can easily be reached by potential customers; factories need a convenient assembly of their workers and materials; residents desire easy connection with jobs, stores, and schools; and so forth. Within the older central city, the radiating mass-transit lines established the elements of the urban land use structure by freezing in the landscape a clear-cut pattern of differential accessibility. The convergence of that system on the city core gave that location the highest accessibility, the highest desirability, and, hence, the highest land values of

the entire built-up area. Similarly, transit junction points were accessible to larger segments of the city than locations along single traffic routes; these latter were more desirable than parcels lying between the radiating lines (Figure 12.17).

Society deems certain functions desirable without regard to their economic competitiveness. Schools, parks, and public buildings are assigned space without being participants in the auction for land. Other uses, through the process of that auction, are assigned spaces by market forces. The merchants with the highest-order goods and the largest threshold requirements bid most for, and occupy, parcels within the **central business district (CBD)**, which is localized at the convergence of mass-transit lines. The successful bidders for slightly less accessible CBD parcels are the developers of the tall office buildings of major cities, the principal hotels, and similar land uses.

Comparable, but lower-order, commercial aggregations develop at the outlying intersections (transfer points) of the mass-transit system. Industry takes control of parcels adjacent to essential cargo routes: rail lines, waterfronts, rivers, or canals. Strings of stores, light industries, and high-density apartment structures can afford and benefit from location along high-volume transit routes. The least accessible locations within the city are left for the least competitive users: low-density residences. A diagrammatic summary of this repetitive allocation of space

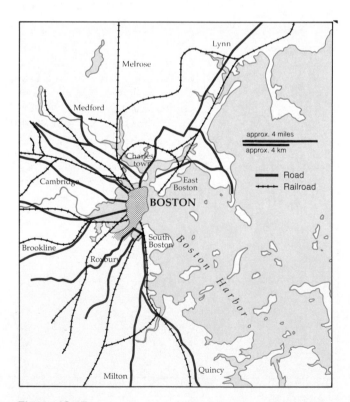

Figure 12.17

Major transit lines in Boston in 1872. Notice how the lines converge on the city center. Compare this map with the late 20th century freeway pattern of Los Angeles shown in Figure 12.4.

among competitors for urban sites is shown in Figure 12.18. Compare it to the generalized land use map of Calgary, Alberta, Canada (Figure 12.19).

Land Values and Population Density

Theoretically, the open land auction should yield two separate although related distance-decay patterns, one related to land values and the other to population density (as distance increases away from the CBD, the population density decreases). If one views the land value surface of the central city as a topographic map (Figure 12.20), with hills representing high valuations and depressions showing low prices, a series of peaks, ridges, and valleys would reflect the differentials in accessibility marked by the pattern of mass-transit lines, their intersections, and the unserved interstitial areas. Dominating these local variations, however, is an overall decline of valuations with increasing distance away from the *peak value intersection,* the most accessible and costly parcel of the central business district (Figure 12.21). As we would expect in a distance-decay pattern, the drop in valuation is precipi-

tous within a short linear distance from that point, and then the valuation declines at a lesser rate to the margins of that built-up area.

With one important variation, the population density pattern of the central city shows a comparable distance-decay arrangement, as suggested by Figure 12.22. The exception is the tendency to form a hollow *at the center,* the CBD, which represents the inability of all but the most costly apartment houses to compete for space against alternative occupants desiring supremely accessible parcels. Yet accessibility is attractive to a number of residential users and brings its penalty in high land prices. The result is the high-density residential occupancy of parcels *near to the center* of the city by those who are too poor to afford a long-distance journey to work; who are consigned by their poverty to the high-density, obsolescent

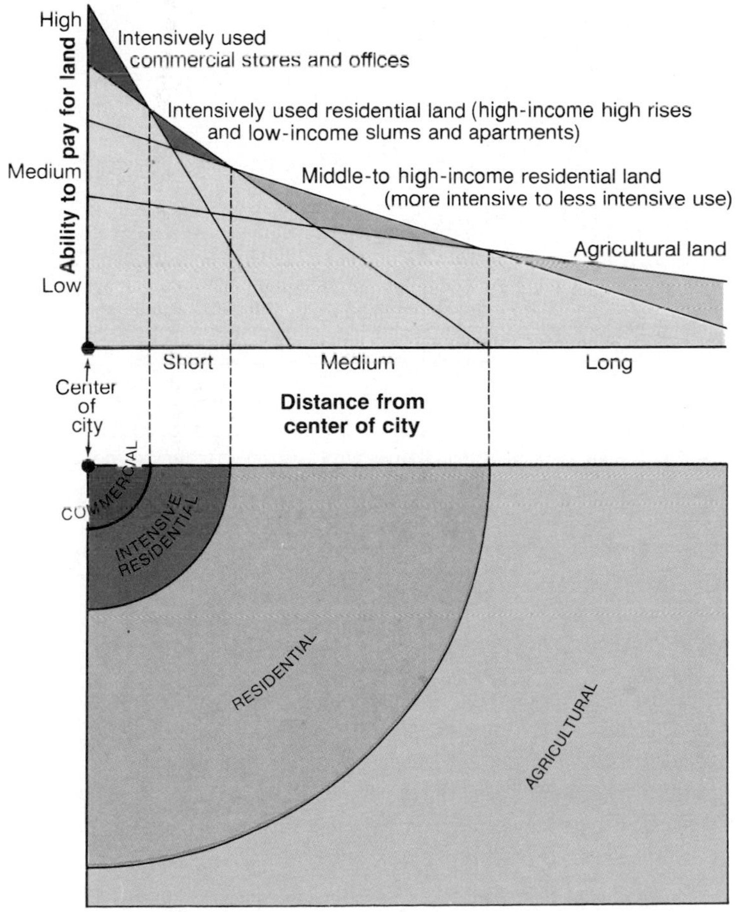

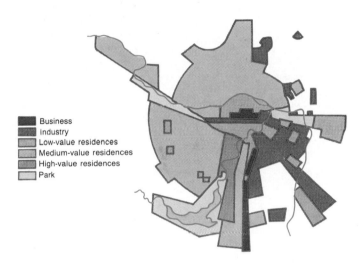

Business
Industry
Low-value residences
Medium-value residences
High-value residences
Park

Figure 12.19
The land use pattern of Calgary, Alberta, in 1961. Physical and cultural barriers and the evolution of cities over time tend to result in a sectoral pattern of similar land uses. Calgary's central business district is the focus for many of the sectors.

Figure 12.18
The location of various land uses in an idealized city where the highest bidder gets the most accessible land.

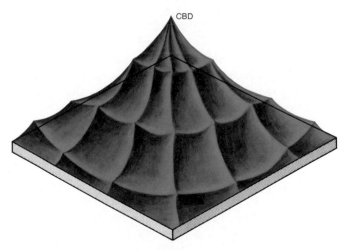

Figure 12.20
Generalized pattern of land values. The land value surface reflects demand for accessibility within a hypothetical urban area.

Urban Geography   **391**

Figure 12.21
A bird's-eye view of Toronto showing clusters of tall buildings in the central business district and at important intersections.

Figure 12.22

A generalized population density curve. As distance from the area of multistory apartment buildings increases, the population density declines.

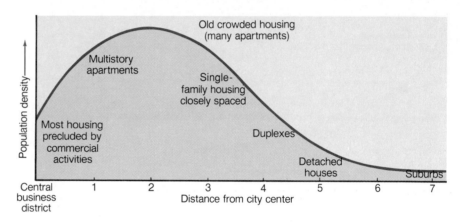

slum tenements near the heart of the inner city; or who are self-selected occupants of the high-density, high-rent apartments made necessary by the price of land. Other urbanites, if financially able, may opt to trade off higher commuting costs for lower-priced land and may reside on larger parcels away from high-accessibility, high-congestion locations. Residential density declines with increasing distance from the city center as this option is exercised.

As a city grows in population, the peak densities no longer increase, and the pattern of population distribution becomes more uniform. Secondary centers begin to compete with the CBD for customers and industry, and the residential areas become less associated with the city center and more dependent on high-speed transportation arteries. Peak densities in the inner city decline, and peripheral areas increase in population concentration.

The validity of these generalizations may be seen on Figure 12.23, a time series graph of population density

patterns for Des Moines, Iowa, over a 20-year period. The peak density was 2 miles from the CBD in 1960, but by 1980 it was at 3 miles. As the city expanded, density decreased close to the center, but beyond 3 miles from the center, population density increased.

Models of Urban Land Use Structure

Generalized models of urban growth and land use patterning have been proposed to summarize the observable results of these organizing forces and controls. The starting point of them all is the distinctive central business district possessed by every older central city. The **core** of this area is characterized by intensive land development: tall buildings, many stores and offices, and crowded streets. Just outside the core is an area of wholesaling activities, transportation terminals, warehouses, new-car dealers, furniture stores, and even light industries. Just beyond the central business district is the beginning of residential land uses.

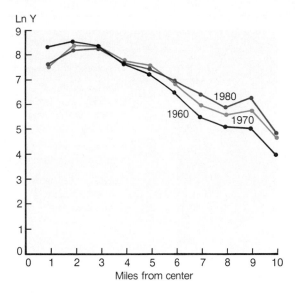

Figure 12.23
Population density-distance curves for Des Moines, Iowa, 1960–80.

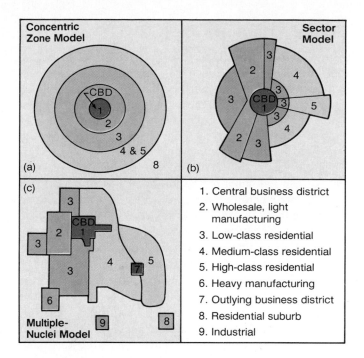

Figure 12.24
Three classic models of the internal structure of cities.

The land use models depicted in Figure 12.24 diverge in their summarization of patterns outside the CBD. The **concentric zone model** (Figure 12.24a) was developed to explain the observed sociological patterning of American cities. The model recognizes the following five concentric circles of mostly residential diversity at increasing distance in all directions from the core.

1. The high-density CBD (core) with wholesaling, warehousing, light industry, and transport depots at its margins.
2. A zone in transition marked by the deterioration of old residential structures abandoned, as the city expanded, by the former more wealthy occupants and now containing high-density, low-income slums, rooming houses, and perhaps, ethnic ghettos.
3. A zone of "independent working people's homes" occupied by industrial workers, perhaps second-generation Americans able to afford modest but older homes on small lots.
4. A zone of better residences, single-family homes or high-rent apartments occupied by those wealthy enough to exercise choice in housing location and to afford the longer, more costly journey to CBD employment.
5. A commuters' zone of low-density, isolated residential suburbs, just beginning to emerge when this model was proposed in the 1920s.

The model is dynamic; it imagines continuous expansion of inner zones at the expense of the next outer developed circles and suggests a ceaseless process of invasion, succession, and population segregation by income level.

The **sector model** (Figure 12.24b) also concerns itself with patterns of housing and wealth, but it arrives at the conclusion that high-rent residential areas dominate expansion patterns and grow outward from the center of the city along major arterials, new housing for the wealthy being added as an outward extension of existing high-rent axes as the city grows. Middle-income housing sectors lie adjacent to the high-rent areas, and low-income residents occupy the remaining sectors of growth. There tends to be a "filtering-down" process as older areas are abandoned by the outward movement of their original inhabitants, with the lowest-income populations (closest to the center of the wealthy) the dubious beneficiaries of the least desirable vacated areas. The accordance of the sector model with the actual pattern of Calgary is suggested in Figure 12.19.

These "single-node" models of growth and patterning are countered by a **multiple-nuclei model** (Figure 12.24c), which postulates that large cities develop by peripheral spread not from one but from several nodes of growth. Individual nuclei of special function—commercial, industrial, port, residential—are originally developed in response to the benefits accruing from the spatial association of like activities. Peripheral expansion of the separate nuclei eventually leads to coalescence and the juxtaposition of incompatible land uses along the lines of juncture. The urban land use pattern, therefore, is not regularly structured from a single center in a sequence of circles or a series of sectors, but based on separate expanding clusters of contrasting activities.

Social Areas of Cities

Although too simplistic to be fully satisfactory, these classical models of American city structure receive some confirmation from modern interpretations of social segregation within urban areas. The more complex cities are economically and socially, the stronger is the tendency for city residents to segregate themselves into groups based on *social status*, *family status*, and *ethnicity*. In a large metropolitan region, this territorial behavior may be a defense against the unknown or the unwanted. Most people feel more secure when they are near those with whom they can easily identify. In traditional societies, these groups are the families and tribes. In modern society, people group according to income or occupation (social status) and language or race (ethnic characteristics). Many of these groupings are fostered by the size and the value of the available housing. Land developers, especially in cities, produce homes of similar quality in specific areas. Of course, as time elapses, there is a change in the quality of houses, and new groups may replace old groups. In any case, neighborhoods of similar social characteristics evolve.

Social Status

The social status of an individual or a family is determined by income, education, occupation, and home value. In the United States, high income, a college education, a professional or managerial position, and high home value constitute high status. High home value can mean an expensive rented apartment as well as a large house with extensive grounds.

A good housing indicator of social status is persons per room. A low number of persons per room tends to indicate high status. Low status characterizes people with low-income jobs living in low-value housing. There are many levels of status, and people tend to filter out into neighborhoods where most of the heads of households are of similar rank.

In most cities, people of similar social status are grouped in sectors whose points are in the innermost urban residential areas. The pattern in Chicago is illustrated in Figure 12.25. If the number of people within a given social group increases, they tend to move away from the central city along an arterial connecting them with the old neighborhood. Major transport routes leading to the city center are the usual migration routes out from the center. Social-status patterning agrees with sector model.

Family Status

As the distance from the origin of each sector increases, the average age of the head of the household declines, or the size of the family increases, or both. Within a particular sector—say, that of high status—older people whose children do not live with them or young professionals without families tend to live close to the city center. Between these are the older families who lived at the outskirts of the city in an earlier period. The young families

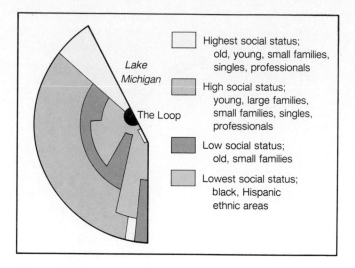

Figure 12.25

A diagrammatic representation of the major social areas of the Chicago region. The central business district is known as the "Loop."

seek space for child rearing, and older people covet more the accessibility to the cultural and business life of the city. Where inner-city life is unpleasant, there is a tendency for older people to migrate to the suburbs or to retirement communities.

Within the lower-status sectors, the same pattern tends to emerge. Transients and single people are housed in the inner city, and families, if they find it possible or desirable, live farther from the center. The arrangement that emerges is a concentric-circle patterning according to family status.

Ethnicity

For some groups ethnicity is a more important residential locational determinant than social or family status. Areas of homogeneous ethnic identification appear in the social geography of cities as separate clusters or nuclei, reminiscent of the multiple-nuclei concept of urban structure. For some ethnic groups, cultural segregation is both sought and vigorously defended, even in the face of pressures for neighborhood change exerted by potential competitors for housing space. The durability of Little Italys and Chinatowns and of Polish, Greek, Armenian, and other ethnic neighborhoods in many American cities is evidence of the persistence of self-maintained segregation.

Certain ethnic or racial groups, especially blacks, have had segregation in nuclear communities forced on them. Every city has one or more black area, which in many respects may be considered a city within a city. Figure 12.26 illustrates the concentration of blacks and Puerto Ricans in certain portions of Brooklyn, New York. The barriers to movement outside the area have always been high. Some whites have consistently blocked some blacks from gaining the social status that would allow them a greater choice in neighborhood selection. In most American cities, the poorest residents are the blacks, who are

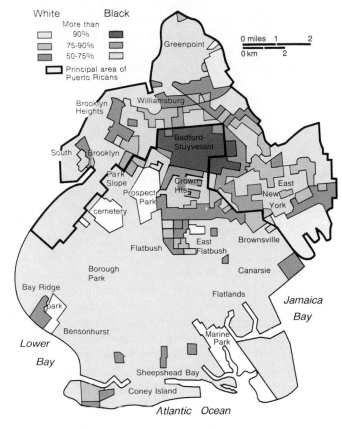

Figure 12.26

Black and Puerto Rican areas of Brooklyn, New York, 1970. The main black area is in the Bedford-Stuyvesant district, not far from Brooklyn's commercial center, which is just north of Prospect Park. In most U.S. cities, the black ghetto is in the most densely populated area close to the central business district.

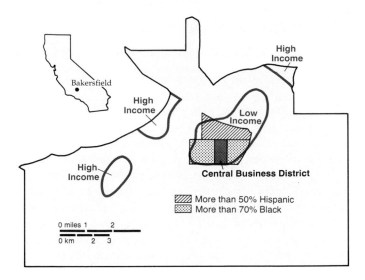

Figure 12.27

Distribution of Hispanic and black population and per capita income, Bakersfield, California, 1980.

relegated to the lowest-quality housing in the least desirable areas of the city. Similar restrictions have been placed on Hispanics and other non-English speaking minorities (Figure 12.27).

Of the three patterns, family status has undergone the most widespread change in recent years. Today, the suburbs house large numbers of singles and childless couples, and areas near the central business district have become popular for young professionals. (See "Gentrification.") Much of this is a result of changes in family structure and the advent of large numbers of new jobs for professionals in the suburbs and the central business districts but not in between.

Institutional Controls

Most governments have instituted innumerable laws to control all aspects of urban life, including the rules for using streets, the provision of sanitary services, and the use of land. In this section we touch only upon land use.

Institutional or governmental controls have strongly influenced the land use arrangements and growth patterns of most cities in the world. Cities have adopted land use plans and enacted subdivision control regulations and zoning ordinances to realize those plans. They have adopted building, health, and safety codes to assure legally acceptable urban development and maintenance. All such controls are based on broad applications of the police powers of municipalities and their rights to assure public health, safety, and well-being even when private-property rights are infringed.

These nonmarket controls on land use are designed to minimize incompatibilities (residences adjacent to heavy industry), provide for the creation in appropriate locations of public uses (the transportation system, waste disposal, government buildings, parks), and private uses (colleges, shopping centers, housing) needed for and conducive to a balanced, orderly community. In theory, such careful planning should prevent the emergence of slums, so often the result of undesirable adjacent uses, and should stabilize neighborhoods by reducing market-induced pressures for land use change.

Such controls in the United States, particularly zoning ordinances and subdivision control regulations specifying acre or larger residential building lots and large house-floor areas, have been adopted as devices to exclude from upper-income areas lower-income populations or those who would choose to build or occupy other forms of residences: apartments, special housing for the aged, and so forth. Bitter court battles have been waged, with mixed results, over "exclusionary" zoning practices that in the view of some serve to separate rather than to unify the total urban structure and to maintain or increase diseconomies of land use development. In addition, it is well known that some real estate agents "steer" people of certain racial and ethnic groups into neighborhoods that the agent thinks are appropriate.

Suburbanization in the United States

The 20 years before World War II saw the creation of a technological, physical, and institutional structure that resulted, after that war, in a sudden and massive alteration of past urban forms. The improvement of the automobile increased its reliability, range, and convenience, freeing its owner from dependence on fixed-route public transit for access to home, work, or shopping. The new transport flexibility opened up vast new acreages of nonurban land to urban development. The acceptance of a maximum 40-hour workweek guaranteed millions of Americans the time for a commuting journey not possible when workdays of 10 or more hours were common.

Finally, to stimulate the economy by the ripple effect associated with a potentially prosperous construction industry, the Federal Housing Administration was established as part of the New Deal programs under President Franklin D. Roosevelt. It guaranteed creditors the security of their mortgage loans, thus reducing down-payment requirements and lengthening mortgage repayment periods. In addition, veterans of World War II were granted generous terms on new housing.

Demands for housing, pent up by years of economic depression and wartime restrictions, were loosed in a flood after 1945, and a massive suburbanization of people and urban functions altered the existing pattern of urban America. Between 1950 and 1970, the two most prominent patterns of population growth were the metropolitanization of people and, within metropolitan areas, their

suburbanization. During the 1960s the interstate highway system was substantially completed, allowing sites 20 and 30 miles from workplaces to be within commuting distance from home places. Growth patterns for the Chicago area are shown in Figure 12.28. The high energy prices of the 1970s slowed the rush to the suburbs, but in the 1980s suburbanization again proceeded apace, although the tendency was as much for "filling-in" as it was for continued sprawl.

In the last 40 years, more and more industries of all types (first manufacturers, then service industries) moved to the suburbs. Their moves reflected the economies that began to accrue from modern single-story facilities with plenty of parking space for employees. No longer did industries need to locate near rail facilities; the freeways presented new opportunities for lower cost, more flexible truck transportation. Service industries took advantage of the large, well-educated labor force now living in the suburbs. In addition to the land that was developed around freeway intersections in the 1960s, and along major commercial routeways intersecting freeways in the 1970s, was the residential land that filled the interstitial areas between the major routeways. Now the major shopping areas of the metropolitan areas, besides the central business district, are the shopping centers at freeway intersections, the freeway frontage roads, and the major connecting highways. Table 12.2 gives strong evidence of suburbanization in the United States.

In time, in the United States, an established social and functional pattern of suburban land use emerged, giving evidence of a lower-density, more extensive repetition of the models of land use developed to describe the

Gentrification

As urban governments and state and federal authorities attempt to find ways to revitalize the ailing older American cities, a potentially significant process is occurring: **gentrification**, or the movement of middle-class people to deteriorated portions of the inner city. This movement does not yet counterbalance the exodus to the suburbs that has taken place since the end of World War II, but it marks an interesting reversal to what had seemed to be the inevitable decline of the cities. According to an estimate of the Urban Land Institute, 70% of all sizable American cities are experiencing a significant renewal of deteriorated areas. Gentrification is especially noticeable in the major cities of the North and the East, from Boston down the Atlantic Coast to Charleston, South Carolina, and Savannah, Georgia.

During the early years of suburbanization, there was also a major movement of low-income nonwhites into the central cities. That migration stream has now slowed to a trickle, and attention is centered on the movement of young, affluent, tax-paying professionals into the neighborhoods close to the city center. These formerly depressed areas are being rehabilitated and made attractive by the new residents. The prices of houses and apartments in former slums have soared. New, upscale restaurants and specialty shops open daily.

The reasons for the upsurge of interest in urban housing reflect to some extent the recent changes in American family structure and in the employment structure of central business districts. Just 20 years ago the suburbs, with their green spaces, were a pow-

erful attraction for young married couples who considered single-family houses with ample yards ideal places to raise children. Now that the proportion of single people (whether never or formerly married) and childless couples in the American population has increased, the attraction of nearby jobs and social and recreational activities has become an important residential location factor. Many of the new jobs in the central business district are designed for professionals in the fields of banking, insurance, and financial services. These jobs are replacing the manufacturing jobs of an earlier period. The gentrification process has been a positive force in the renewal of some of the depressed housing areas in neighborhoods surrounding the central business district.

structure of the central city. Multiple nuclei of specialized land uses developed, expanded, and coalesced. Sectors of high-income residential use continued their outward extension beyond the central-city limits, usurping the most scenic and the most desirable suburban areas and segregating them by price and zoning restrictions. As shown in Figure 12.29, middle-, lower-middle-, and lower-income groups found their own income-segregated portions of the fringe. Ethnic minorities were relegated to the inner city and some older industrial suburbs.

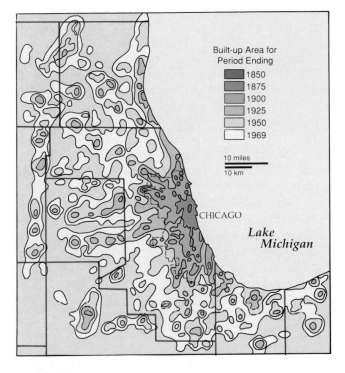

Figure 12.28

Urban sprawl. In Chicago, as in most large U.S. cities, the size of the urbanized areas has increased dramatically during the last 40 years.

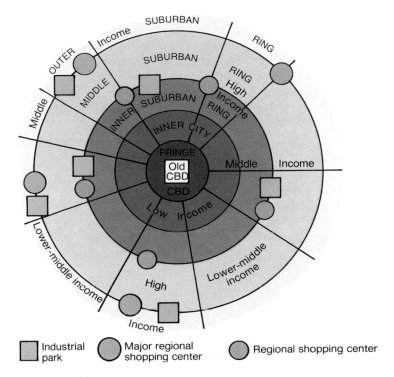

Figure 12.29

A diagrammatic representation of the present-day large U.S. city. Note that aspects of the concentric-zone, sector, and multiple-nuclei patterns are in evidence.

Table 12.2

Population Change in the Largest North and East, and South and West, Metropolitan Areas, 1980–1986

Metropolitan area	Percentage change in population, 1980 to 1986		Metropolitan area	Percentage change in population, 1980 to 1986	
	Central city	Suburbs		Central city	Suburbs
North and East			South and West		
New York	+2.7	+ 2.3	Los Angeles	+ 9.8	+ 7.2
Chicago	+0.2	+ 3.5	San Francisco	+10.3	+ 9.4
Philadelphia	−2.7	+ 2.9	Dallas	+11.0	+30.8
Detroit	−9.7	− 1.0	Houston	+ 8.4	+26.6
Boston	+2.0	+ 2.1	Miami	+ 7.8	+10.5
Washington	−1.9	+12.4	Atlanta	− 0.7	+24.9
Cleveland	−6.6	− 1.3	Seattle	− 0.2	+12.5
St. Louis	−6.0	+ 4.6	San Diego	+16.0	+20.2
Pittsburgh	−8.7	− 3.5	Tampa	+ 2.2	+21.9
Minneapolis	−3.8	+ 9.7	Denver	+ 2.6	+19.2

Note: Since 1980, in general, the large, older East and North Central central cities have lost population; in some cases, their suburbs have also lost population. In the South and West, however, the pattern is generally much like the 1950–1980 pattern: large increases are noted in the metropolitan areas, especially in the suburbs.

Source: U.S. Bureau of the Census.

With increasing suburban sprawl and rising costs implicit in the ever-greater spatial separation of the functional segments of the fringe, the limits of feasible expansion were reached, the supply of developable land was reduced (with corresponding increases in its price), and the intensity of land development grew. Changing lifestyles and cost constraints have resulted in a proliferation of suburban apartment complexes and the disappearance of open land. The maturation and the coalescence of urban land uses have resulted in the emergence of coherent metropolitan-area cities that are suburban in traditional name only, containing the business districts and the mix of land uses that the designation *city* implies. (See "The Self-Sufficient Suburbs.")

In recent years, suburban areas have expanded to the point where metropolitan areas are coalescing. Within these suburban areas new major centers of growth are appearing. Office buildings and huge shopping centers are being developed in such places as Oak Brook, Illinois (in the western Chicago suburbs), Meadowlands, New Jersey (west of New York), King of Prussia, Pennsylvania (northwest of Philadelphia), and Costa Mesa, California (south of Los Angeles) (Figure 12.30). In fact, in the 1980s more office space was created in the suburbs than in the central cities of America. The Boston to Washington corridor is now a continuously built-up region with many new centers that compete with the business districts of Boston, Providence, New York, Philadelphia, Baltimore, and Washington. One summary of the new pattern is shown in Figure 12.31. A further analysis of the northeastern U.S. urban corridor appears in "Megalopolis" in Chapter 13.

Central City Change

While the process of suburban spread continued, many central cities of the United States, especially in the Northeast and the Midwest, lost population. In the 1970s, there was a decided shift of population to what are called

Figure 12.30
Many business areas, such as Costa Mesa, south of Los Angeles, have become large centers comparable in size to the central business district of a moderate size city. These high-density, automobile-oriented, suburban regions exist outside the largest central cities.

the Sunbelt cities. This trend was partly the result of the lower living costs (especially fuel costs) in the South and West, but it is chiefly a function of manufacturers' seeking nonunionized workers or new plants unencumbered by high social and transportation costs. Perhaps as important as that trend was the movement of people out of metropolitan areas to nearby nonmetropolitan districts. In a sense, this latter trend represents a further sprawl of the metropolitan districts.

The economic base and the financial stability of the central city have been grievously damaged by the process of suburbanization. In earlier periods of growth, as new settlement areas developed beyond the political margins of the city, annexation absorbed new growth within the corporate boundaries of the expanding older city. The additional tax base and employment centers became part of the municipal whole. But in the states that recognized the right of the separate incorporation of the new growth areas, particularly in the eastern part of the United States, the

The Self-Sufficient Suburbs

The suburbanization of the American population and of the commerce and industry dependent on its purchasing power and labor has led to more than the simple physical separation of the suburbanite from the central city. It has created a metropolitan area, many of whose inhabitants have no connection with the core city, feel no ties to it, and find satisfaction of all their needs within the peripheral zone.

A *New York Times* "suburban poll," conducted in the summer of 1978, revealed a surprising lack of interest in New York City on the part of those usually assumed to be intimately involved in the economic, cultural, and social life of the "functional city," or the metropolitan area dominated by New York. The poll discovered that in 80% of suburban households, the principal wage earner did not work in New York

City. The concept of the commuter zone obviously no longer has validity if *commuting* implies, as it formerly did, a daily journey to work to the central city.

Just as employment ties have been broken, now that the majority of the suburbanites work in or near their outlying areas of residence, the cultural and service ties that were traditionally thought to bind the metropolitan areas into a single unit dominated by the core city have also been severed. The *Times* survey found that one-half of the suburbanites polled made fewer than five nonbusiness visits to New York City each year; one-quarter said that they never went there. Even the presumed concentration in the city of high-order goods and services—those with the largest service areas—did not constitute an attraction

for most suburbanites. Only 20% of the respondents thought that they would journey to New York City to see a medical specialist, and only 7% or 8% to see a lawyer or an accountant or to make a major purchase. Even needs at the upper reaches of the central-place hierarchy were satisfied within the fringe.

The suburbs, as the *Times* survey documented, have outgrown their former role as bedroom communities and have emerged as a chain of independent, multinucleated urban developments. Together, they are largely self-sufficient and divorced from the central city and have little feeling of subordination to or dependence on that city. As Brian Berry, the urban scholar, has so aptly put it, we are creating an "urban civilization without cities."

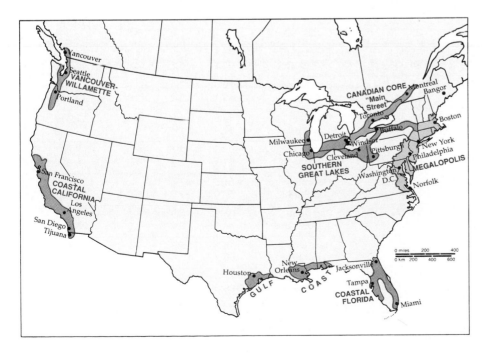

Figure 12.31
The northeastern U.S. Boston-to-Washington megalopolis is the original and largest megalopolis. The map depicts the various candidates for megalopolis status in North America.

ability of the city to continue to expand was restricted. Lower suburban residential densities made septic tanks and private wells adequate substitutes for expensive city-sponsored sewers and water mains, and lower structural densities and lower crime rates meant less felt need for high-efficiency city fire and police protection. Where possible, suburbanites opted for a separation from the central city and for aloofness from the costs, the deterioration, and the adversities associated with it. Their homes, jobs, shopping, schools, and recreation all existed outside the confines of the city from which they had divorced themselves.

The redistribution of population caused by suburbanization resulted not only in the spatial but also in the political segregation of the social groups of the metropolitan area. The upwardly mobile resident of the city—younger, wealthier, and better educated—took advantage of the automobile and the freeway to leave the central city. The poorer, older, least advantaged urbanites were left behind. As indicated in Table 12.3, the central cities and suburbs are becoming increasingly differentiated. Large

areas within the cities now contain only the poor and minority groups, a population little able to pay the rising costs of the social services that their numbers, neighborhoods, and condition require.

The services needed to support the poor include welfare payments, social workers, extra police and fire protection, health delivery systems, and subsidized housing. Central cities by themselves are unable to raise the taxes needed to support such an array and intensity of social services now that they have lost the tax bases represented by suburbanized commerce, industry, and upper-income residential uses. Lost, too, are the job opportunities formerly a part of the central-city structure. Increasingly, the poor and minorities are trapped in a central city without the possibility of nearby employment and isolated by distance, immobility, and unfamiliarity from the few remaining low-skill jobs, which are now largely in the suburbs.

The abandonment of the central city by people and functions has nearly destroyed the traditional active, open auction of urban land, which led to the replacement of

Table 12.3

Race and Income for Selected Metropolitan Areas, 1980. Central cities and their suburbs display significant differences in racial composition and household income.

Metropolitan Area	Racial Composition (%)			Median Annual Household Income	Metropolitan Area	Racial Composition (%)			Median Annual Household Income
	White	Black	Hispanic*			White	Black	Hispanic*	
North and East					*South and West*				
New York	60.7	25.2	19.9	$13,855	Los Angeles	61.2	17.0	27.5	15,746
Suburbs	89.0	7.6	0.4	23,740	Suburbs	71.9	9.6	28.9	19,410
Chicago	49.6	39.8	14.0	15,301	Houston	61.3	27.6	17.6	18,474
Suburbs	90.8	5.6	3.9	24,811	Suburbs	86.3	6.7	10.9	24,847
Philadelphia	58.2	37.8	3.7	13,169	Dallas	61.4	29.4	12.3	16,227
Suburbs	89.8	8.1	1.7	20,904	Suburbs	92.2	3.9	5.3	21,301
Detroit	34.3	63.0	0.2	13,981	San Diego	75.4	8.9	14.8	16,409
Suburbs	94.1	4.2	0.1	24,038	Suburbs	85.9	2.6	14.6	17,808
Baltimore	43.9	54.8	1.0	12,811	San Antonio	78.6	7.3	53.6	13,775
Suburbs	89.2	9.0	1.0	22,272	Suburbs	88.1	5.2	23.4	19,452
Indianapolis	77.1	21.8	0.9	17,279	Phoenix	84.0	4.9	15.1	17,419
Suburbs	98.3	1.0	0.5	26,854	Suburbs	89.4	1.4	11.2	18,085
Washington	26.8	70.3	2.7	16,211	San Francisco	58.3	12.7	12.2	15,867
Suburbs	78.5	16.7	3.1	25,728	Suburbs	81.1	6.5	10.6	22,375
Milwaukee	73.3	23.1	4.1	16,061	Memphis	51.6	47.6	—	14,040
Suburbs	98.4	0.5	1.0	23,805	Suburbs	77.8	21.0	1.0	18,753
Cleveland	53.6	43.8	2.9	12,277	San Jose	73.9	4.6	22.0	22,886
Suburbs	91.8	7.1	0.6	22,024	Suburbs	83.1	2.1	13.0	23,832
Columbus	76.2	22.1	0.8	14,834	New Orleans	46.2	55.3	—	11,834
Suburbs	97.4	1.9	0.5	20,604	Suburbs	85.4	12.6	—	19,678
Boston	69.9	22.4	6.4	12,500	Jacksonville	73.0	25.4	—	14,426
Suburbs	96.6	1.6	1.5	20,469	Suburbs	88.3	10.7	—	16,084

Source: U.S. Bureau of the Census.

*As used by the Bureau of the Census, ''Hispanic'' race has no genetic connotation.

obsolescent uses and inefficient structures in a continuing process of urban modernization (Figure 12.32). In the vacuum left by the departure of private investors, the federal government, particularly since the landmark Housing Act of 1949, has initiated urban renewal programs with or without provisions for a partnership with private housing and redevelopment investment. Under a wide array of programs instituted and funded since the late 1940s, slum areas have been cleared; public housing has been built; cleared land has been conveyed at subsidized cost to private developers for middle-income housing construction; cultural complexes and industrial parks have been created; and city centers have been reconstructed (Figure 12.33).

With the continuing erosion of the urban economic base and the disadvantageous restructuring of the central-city population base, the hard-fought governmental battle to maintain or revive the central city is frequently judged to be a losing one (Figure 12.34). In recent years, the central city has been the destination of thousands of homeless persons. Many live in public parks, in doorways, by street-level warm air exhausts of subway trains, and in subway stations. The urban economies, with their high land values, limited job opportunities for the unskilled, and inadequate resources for social services, have relegated many to a homeless existence.

Figure 12.32
A derelict slum in the South Bronx, New York City.

Figure 12.33
Some elaborate governmental projects for urban redevelopment, slum clearance, and rehousing of the poor have been failures. The demolition of the Pruitt-Igoe project in St. Louis in 1975 was a recognition, resulting from soaring vandalism and crime rates, that public high-rise developments intended to revive the central city did not always meet the housing and social needs of their inhabitants.

Figure 12.34
The Robert Taylor Homes, Chicago. Fewer than half of the 28 high-rise buildings of this massive public-housing project are shown in this photograph. The development houses more than 19,000 people in some 4200 apartments.

There are two trends, however, that do give a new flavor to the central city. One is gentrification (see box) and the other is the vitality of the CBD. New office buildings are springing up in most large central cities (Figure 12.35). These represent a shift in employment emphasis from manufacturing to more service-oriented activities, such as headquarters for industries, financial services, and business services in general. The number of jobs in the central city has not increased, but the change favors professionals over blue-collar workers and the unskilled. Many of the workers in these office buildings now live within the gentrified neighborhoods of the central city.

The Making of World Metropolitan Regions

Urban settlements have been increasing in size and number for the last 200–300 years. Until the end of World War II, one could easily distinguish a city by its distinctive central business district, the high-density housing close to the CBD, and the lower densities that ended fairly abruptly at the outer ends of the public transport system. Now that over two-thirds of the people of Western Europe, the USSR, Japan, Australia, and North America live in urban settlements (about 600–700 million people), it would seem to follow that many cities would coalesce with one another. In fact, this has been the case, and the urban region can no longer be described in simple terms. We now must talk in terms of multiple centers and greatly varying population densities not necessarily associated directly with business districts. Table 12.4 provides a list of the largest metropolitan areas of the world (see also Figure 12.3).

The coalescence of many cities into great metropolitan masses is more notable than ever before. The major **megalopolis** of the United States is the continuous urban string that stretches from north of Boston (southern New Hampshire) to south of Washington, D.C. (northern Virginia). Other North American megalopolises include:

the southern Lake Michigan regions stretching from north of Milwaukee through Chicago and across northwestern Indiana and southwest Michigan

Los Angeles, including the myriad suburbs and cities extending from Santa Barbara to San Diego and even including the one million people living in Tijuana in Mexico

Toronto and the many industrial cities in southern Ontario

Figure 12.35
An example of a dynamic central business district: the Dallas skyline with a number of new buildings under construction.

Table 12.4
Largest Urban Agglomerations in 1989

	Urban region	Country	Population
1.	Mexico City	Mexico	Over 15,000,000
2.	Tokyo-Yokohama	Japan	Over 15,000,000
3.	São Paulo	Brazil	Over 15,000,000
4.	New York-NE New Jersey	USA	Over 15,000,000
5.	Seoul	Korea	Over 15,000,000
6.	Osaka-Kobe	Japan	Over 15,000,000
7.	Calcutta	India	10–15,000,000
8.	Rio de Janiero	Brazil	10–15,000,000
9.	Bombay	India	10–15,000,000
10.	Cairo	Egypt	10–15,000,000
11.	Los Angeles-Long Beach	USA	10–15,000,000
12.	Buenos Aires	Argentina	10–15,000,000
13.	London	United Kingdom	10–15,000,000
14.	Paris	France	10–15,000,000
15.	Moscow	USSR	10–15,000,000
16.	Shanghai	China	5–10,000,000
17.	Beijing	China	5–10,000,000
18.	Djakarta	Indonesia	5–10,000,000
19.	Karachi	Pakistan	5–10,000,000
20.	Chicago	USA	5–10,000,000
21.	San Francisco Bay area	USA	5–10,000,000
22.	Tehran	Iran	5–10,000,000
23.	Delhi-New Delhi	India	5–10,000,000
24.	Manila	Philippines	5–10,000,000
25.	Philadelphia	USA	5–10,000,000
26.	Lima	Peru	5–10,000,000
27.	Bangkok	Thailand	5–10,000,000
28.	Madras	India	5–10,000,000
29.	Victoria/Kowloon	Hong Kong	5–10,000,000
30.	Istanbul	Turkey	5–10,000,000
31.	Leningrad	USSR	5–10,000,000
32.	Taipei	Taiwan	5–10,000,000
33.	Tianjin	China	5–10,000,000

Note: List of metropolitan areas prepared by Richard L. Forstall for Rand McNally and Company's *The International Atlas*, 1989. City order by A. Getis based on data from Table 6 in *Global Review of Human Settlements, Statistical Annex*, United Nations: Department of Economic and Social Affairs, 1986, pp. 77–87.

Eurasian megalopolises include:

> the north of England, including Liverpool, Manchester, Leeds, Sheffield, and the several dozen industrial towns in between
> the London region with its many cities and suburbs
> the Ruhr of West Germany, with a dozen large connected cities like Essen and Dusseldorf
> the Tokyo-Yokohama-Kawasaki region of Japan

Were it not for the mountains between Tokyo and Nagoya, and between Nagoya and Osaka-Kobe-Kyoto, there is no doubt that these huge metropolises would by now be one giant megalopolitan area.

Large metropolises are being created in developing countries as well, and wherever the automobile or the modern transport system is an integral part of the growth, the form of the metropolis takes on Western characteristics. Such places as São Paulo (Brazil) and Mexico City are two examples of extremely large urban areas with fairly modern transport systems. On the other hand, in places like Bombay (India), Shanghai (China), Lagos (Nigeria), Djakarta (Indonesia), Kinshasa (Zaire), and Cairo (Egypt), where modern roads have not yet made a significant impact and the public transport system is limited, the result has been overcrowded cities centered on a single major business district in the old tradition. In such

Population Pressure in Cairo, Egypt, and Shanghai, China

Cairo is a vast, sprawling metropolis—hot, crowded, choked by automobiles, clouded by noxious fumes from traffic and by sand that blows from the Sahara desert. The city is representative of many of the principal cities of Third World countries where population growth far outstrips economic development. A steady stream of migrants arrives in Cairo daily because it is the place where opportunities are perceived to be available, where people think that life will be better and brighter than the crowded countryside. Cairo is the symbol of modern Egypt, a place where young people are willing to undergo deprivation for the chance to "make it."

The population of the country was never more than 12 million during its 7000-year history, but now better health care has led to a big drop in infant mortality, and the result is a country of over 55 million people and an expected year 2000 population of 70 or 80 million. In the last 30 years, the population of Cairo has increased sixfold. Three decades ago, the estimated population was 2.3 million; now some 14 million reside in the greater metropolitan area. Real opportunities continue to be low relative to the size of the metropolis. The poor, of which there are millions, crowd into warrens of slums, often without water or sewage systems. Yet the city continues to grow, spreading onto valued farmland, thus decreasing the food available for the country's increasing population.

The sewer and water systems, modernized in the 1930s, were intended to serve only 2 million people. Now it is common for burst sewer mains to dump raw sewage into the city streets. When such conditions prevail in the hot summer, the chances of the spread of disease increase by many times.

It is not unusual to find people living on little boats on the river, under the arches of bridges, and in makeshift, poorly constructed buildings. On occasion, buildings cave in and many people die.

One's first impression when arriving in central Cairo is one of opulence and prosperity, a stark contrast to what lies outside the center. High-rise apartments, regional headquarters buildings of multinational corporations, and modern hotels stand amid clogged streets. This part of the city is clearly Western-oriented, but the fancy restaurants, plush apartments, and high-priced cars stand only a short distance from the slums that are home to masses of underemployed people.

societies, the impact of urbanization and the responses to it differ from the patterns and problems observable in the cities of the United States. The discussion of Cairo and Shanghai in the box indicates the results of urbanization in Third World cities.

The stories of Cairo and Shanghai can be repeated for city after city in the Third World. Bombay and Calcutta in India, Karachi in Pakistan, Bangkok in Thailand, Djakarta in Indonesia, Kinshasa in Zaire, Lagos in Nigeria, and Mexico City and São Paulo are already housing millions more than their service structures are designed to accommodate. One can foresee only worsening conditions for the future since the process of urbanization is proceeding at a faster rate than government authorities can cope with (Figure 12.36).

The developing countries, emerging from formerly dominantly subsistence economies, have experienced disproportionate population concentrations, particularly in their national and regional capitals. The term used to describe the huge, one-to-a-country (or to a large region, as is the case in India, China, and Brazil) metropolis is **primate city.** More specifically, the primate city is one that has a population much greater than twice the population of the second largest city. Greater Cairo contains 30% of Egypt's total population; 34% of all Panamanians live in Panama City; and Baghdad contains 24% of the Iraqi populace. Vast numbers of surplus, low-income rural populations have been attracted to these developed seats of wealth and political centrality in the hope of finding a job.

The metropolitan area of **Shanghai** has a population of about 12 million. The city is extremely crowded, not unlike the teeming Lower East Side of New York in the 1920s. Most people walk or bicycle to work daily. The city depends on an overworked bus system to help move millions of people to and from work daily. Work hours are staggered to alleviate the pressure on the transport system, but the streets are crowded from dawn to after dark six days a week with commuters and shoppers.

Shanghai may be the most crowded city in the world. The average population density is approximately 100,000 per square mile, about 3 times more crowded than Tokyo and 5 times more crowded than Paris. The average living space is about 6 feet by 6 feet per person. That means that 10 or 11 people reside in an average size Shanghai apartment of 350 square feet. The vast majority of people live in small 2- and 3-story houses that long ago were converted into apartments.

Crowding can be further understood by comparing the number of people per 100 square feet of living space (about the size of a typical room) by country. In the United States, the average is 0.5 persons. In France, the figure is 1.3. For Shanghai, a conservative estimate would be 3.0. Most amazing is that there are few, if any, squatter settlements. Many people live in temporary quarters, such as hallways and attics, but the use of land,

which is publicly held, is tightly controlled by a government that discourages unplanned growth.

One major government strategy to relieve the housing problem was to develop satellite towns around Shanghai. Twelve such satellites containing 300,000 jobs ring the city to a distance of 15 miles, but until recently only

45,000 people lived in them. Many were reluctant to leave cosmopolitan Shanghai for the sterile suburbs. As a result, the already overburdened bus system must deliver thousands of reverse commuters to jobs in the satellites. Commuting time for many of these workers is between 2 and 3 hours daily.

Figure 12.36
Open sewer in a slum in Lagos, Nigeria.

Figure 12.37
Modern office buildings in Lagos, Nigeria.

Although attention may be lavished on creating urban cores on the skyscraper model of Western cities (Figure 12.37), most of the new urban multitudes have little choice but to pack themselves into squatter shanty communities on the fringes of the city, isolated from the sanitary facilities, the public utilities, and the job opportunities that are found only at the center. Such impoverished squatter districts are found around most major cities in Africa, Asia, and Latin America.

Contrasts among Metropolitan Areas

Figures 12.38, 12.41, and 12.42 show typical land use patterns for three large cities in different parts of the world. The first is a diagrammatic representation of a typical European city, the second of a Latin American city, and the third an Asian city. Each reflects the way growth and development have taken place over a period of time. Notice how they differ from each other and from the American city shown in Figure 12.29.

The European City

For the European city, the institutional controls to contain cities within specified limits have had the effect of building up housing densities to heights not common in the United States. Single-family homes are more the exception than the rule. Usually a *greenbelt* rings the city as a place for relaxation and a way to protect rural areas and the environment. The historic core is now mainly a tourist and shopping district, often contained within the remains of the walls of the old city. High-incomes characterize the majority of residents of the old city. Renewal mainly takes place in the area outside of the old city where the upwardly mobile professionals are replacing the working class. Many housing districts, especially near industry, contain public housing (usually apartments) built for the industrial workers. Many parks are used for recreation and to separate incompatible land uses, such as industry and housing.

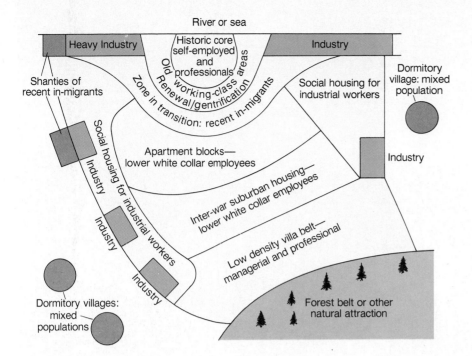

Figure 12.38
A diagrammatic representation of the
Western European city.

Labels in figure:

River or sea

Heavy Industry

Historic core self-employed and professionals

Old working-class areas

Zone in transition: recent in-migrants

Renewal/gentrification

Industry

Shanties of recent in-migrants

Social housing for industrial workers

Dormitory village: mixed population

Social housing for industrial workers

Industry

Apartment blocks— lower white collar employees

Industry

Inter-war suburban housing— lower white collar employees

Low density villa belt— managerial and professional

Industry

Dormitory villages: mixed populations

Forest belt or other natural attraction

Each European city has its own character based on a long history of change, decline, and growth. Paris is known for its boulevards; London, for its magnificent buildings, winding streets, and subway system; Rome, for its ancient ruins; Amsterdam, for its canals, and so on. But all European cities have extremely efficient transit systems. The subways run often and the commuter trains are fast. Although freeways connect the cities with each other, the favorite type of long-distance travel is by rail—auto fuel costs are higher than in the United States. Because suburban growth is restricted either by agricultural land preservation laws or by greenbelts, the cities are compact, although laws limiting tall structures have, until recently, kept the building profile low (Figure 12.39).

European cities have few slum districts; social welfare programs are more extensive than in the United States. There are ghettos, however, that contain ethnic groups who have come to European cities for low-skill and low-wage jobs. French cities have large Algerian districts, for example, while German cities have districts containing Yugoslavs and Turks.

The Latin American City

"City life" is the cultural norm in Latin America. The vast majority of the residents of Mexico, Venezuela, Brazil, Argentina, Chile, and so on live in cities, and very often in the primate city. The urbanization process is rapidly making Latin cities the largest in the world. Analysts predict that by the year 2000, six of the largest 28 cities will be in Latin America, and Mexico City and São Paulo will be the first and third largest cities, respectively.

The limited wealth of Latin cities confines most commercial activity to the central business district (CBD). The entire transportation system focuses on the down-

town, where the vast majority of jobs are found. The city centers are lively and modern with many tall office buildings, clubs, restaurants, and stores of every variety. Condominium apartments house the well-to-do who prefer living in the center because of its convenience to workplaces, theaters, museums, friends, specialty shops, and restaurants (Figure 12.40). Thousands of commuters pour into the CBD each day, some coming from the outer edge of the city (perhaps an hour or two commuting time) where the poorest people live.

The land use pattern characteristic of Latin cities is shown in Figure 12.41. There are two features of the pattern worth noting. One is the *spine*, which is a continuation of the features of the CBD outward along the main wide boulevard. Here one finds the upper-middle-class housing stock, which is again apartments and town houses. The second is the concentric rings around the center housing ever poorer people as distance increases from the center. This social patterning is just the opposite of many American cities. The slums (*barrios*, *favelas*) are on the outskirts of the city. In rapidly growing cities, like Mexico City, the barrios are found in the furthest concentric ring, which is several miles wide. Many people within these areas eke out a living by selling goods and services to other slum-dwellers.

Once Latin residents establish themselves in the city, they tend to remain at their original site, and as income permits they improve their homes. When times are good, there is a great deal of house repair and upgrading activity. Those in the city for the longest time are generally the most prosperous. As a result, the quality of housing continually improves inward toward the city center. The homes closest to the center are substantial and need little

Figure 12.39

Paris as seen from the Eiffel Tower. Even in their central areas, many European cities show a low profile. Although taller buildings—20, 30, even 50 stories tall—have become more common in major cities since World War II, they are not the universal mark of central business districts that they have become in the United States.

Figure 12.40

Buildings along the Paseo de la Reforma in Mexico City. Part of the central business district, this area contains apartment houses, theaters and nightclubs, and commercial high rises.

upgrading, but further out where slums were once in evidence one now finds modest houses with new additions being built or planned.

Unfortunately, in recent years Latin cities have been economically depressed. This has given rise to many homeless people who resort to begging or stealing. The cities are now marred by this social phenomenon. In Brazil, thousands of abandoned children form into gangs for criminal purposes. In parts of Latin America, for example, in some cities of Columbia and Peru, drug dealing has become the main economic activity.

The Asian and African City

Many of the large cities of Asia and Africa were founded and developed by European colonialists. For example, the British built Calcutta and Bombay in India and Nairobi in Africa, the French developed Ho Chi Minh City (Saigon) in Vietnam and Dakar in Senegal. The Dutch had as their main outpost Djakarta in Indonesia, and many colonial countries established Shanghai. These and many other cities in Asia and Africa have certain important similarities. They stand in marked contrast to the North American and European cities and have much in common with Latin American cities in that large areas on the periphery of the city contain slum housing. Since the transportation systems are not well developed, high-income residents usually live close to the city center, where the amenities of the city are easily available.

The main difference between Asian and African cities and their Latin American counterparts is the colonial imprint from another culture. The British usually built a fort by the water, surrounding it by a large green space to protect the fort from enemies (Figure 12.42). The city proper developed around the fort. A large section was given over to administrative headquarters of the occupying army and to trading agencies. Nearby, the central business district was established. The CBD was commercially divided into parts that served the Europeans and the local residents. A European city and an indigenous city developed side by side. In Africa, the indigenous section is typically subdivided into ethnic areas. In recent years,

with the vast majority of colonialists gone, the fort area is now a tourist center, the European sector now houses the well-to-do local administrators and business people, and many middle-class residences have appeared in outer areas away from the slums, usually where new roads have been built (Figure 12.43).

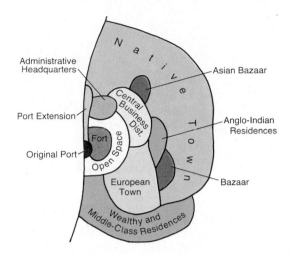

Figure 12.42
A diagrammatic representation of the colonial-based South Asian city.

Figure 12.43
Downtown Nairobi, Kenya, is a busy, modern urban core of a former colonial British city.

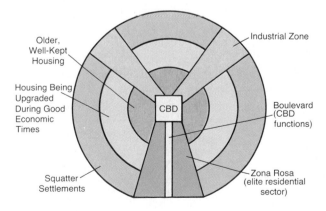

Figure 12.41
A diagrammatic representation of the Latin American city.

Figure 12.44
This scene from Bucharest, Romania, clearly shows important recurring characteristics of the socialist city: mass-transit service to boulevard-bordered "superblocks" of self-contained apartment house microdistricts with their own shopping areas, schools, and other facilities.

The Socialist City

The socialist city is somewhat different from the others in form. Typically it is completely designed by planners who have attempted to develop livable, efficient land use arrangements. Cities of the USSR and, to a certain extent, Eastern Europe are characterized by huge blocks of apartments that are as self-contained as possible (Figure 12.44). Shops, services, and some green space are available in each of the "super-blocks," which are ringed by boulevards. The center of the city is reserved for museums and public buildings, and is used for parades and other civic activities.

Summary

The city is the essential activity center in the chain of linkages that marks any society advanced beyond the level of subsistence. Although cities are among the oldest marks of civilization, only in this century have they become the home of the majority of the people in industrialized countries.

Urban development in any country is a function of a set of variables that are an outgrowth of economic and social systems and of long-run historical or cultural processes. Nevertheless, in all their incarnations, ancient and modern, recurring themes and regularities are evident among cities. There is a central-place system that accounts for the interrelatedness and size of cities when there is only an agricultural market area to serve. All cities perform services for areas beyond their boundaries. The essential elements that constitute their economies are basic and nonbasic activities. These functions generate the income necessary for cities to support themselves. No city

exists in a vacuum; each is part of a larger urban and non-urban society and economy with which it has essential reciprocal connections. Each influences cities in its region that are below it in the hierarchy of cities.

Every city has a discernible internal arrangement of land uses, social groups, and economic functions partially controlled and partially determined by the nature of the transport system. Variations in accessibility to preferred locations stimulate competition for sites among competing land uses. In the United States, ease of access to sites far from the central business district has spurred the suburbanization process and has negatively affected the central city. In European cities stringent land use regulations have brought about a compact urban form ringed by greenbelts. Many of the cities of the world, however, suffer from being overcrowded. Third World cities are currently growing faster than it is possible to provide amenities, such as decent housing, water, and sanitation services.

Key Words

basic sector	multiple-nuclei model
central business district (CBD)	multiplier effect
	nonbasic sector
central city	primate city
central place	sector model
central place theory	service sector
city	site
concentric zone model	situation
core	suburb
economic base	town
gentrification	urban hierarchy
megalopolis	urban influence zone
metropolitan area	urbanized area

For Review

1. Consider the city or town in which you live or attend school or with which you are most familiar. In a brief paragraph, discuss that community's site and situation; point out the connection, if any, between its site and situation and the basic functions that it earlier performed or now performs.

2. Describe the *multiplier effect* as it relates to the population growth of urban units.

3. What area does a *central place* serve, and what kinds of functions does it perform? If an urban system were composed solely of central places, what summary statements could you make about the spatial distribution and the urban size hierarchy of that system?

4. Is there a hierarchy of retailing activities in the community with which you are most familiar? Of how many and of what kinds of levels is that hierarchy composed? What localizing forces affect the distributional pattern of retailing within that community?

5. Briefly describe the urban land use patterns predicted by the *concentric zone,* the *sector,* and the *multiple-nuclei* models of urban development. Which one, if any, best corresponds to the growth and land use pattern of the community most familiar to you? How well does Figure 12.29 depict the land use patterns in the metropolitan area with which you are most familiar?

6. In what ways do social status, family status, and ethnicity affect the residential choices of households? What expected distributional patterns of urban social areas are associated with each? Does the social geography of your community conform to the predicted pattern?

7. How has suburbanization damaged the economic base and the financial stability of the central city?

8. Why are metropolitan areas in developing countries expected to grow larger than many Western metropolises by the year 2000? What do you expect the population density profile for Mexico City to look like in the year 2000?

9. What are *primate cities?* Why are Third World primate cities overburdened? What can be done to alleviate the difficulties?

10. What are the significant differences in the generalized pattern of land uses of North American, European, Asian and African, and socialist cities?

Selected References

Alonso, William. *Location and Land Use*. Cambridge, Mass.: Harvard University Press, 1965.

Association of American Geographers Comparative Metropolitan Analysis Project. *A Comparative Atlas of America's Great Cities: Twenty Metropolitan Regions.* Vol. 3. Minneapolis: University of Minnesota Press, 1976.

Berry, Brian J. L. *Comparative Urbanization: Divergent Paths in the Twentieth Century*. 2d rev. ed. New York: St. Martin's Press, 1981.

Berry, Brian J. L., and Frank E. Horton, eds. *Geographic Perspectives on Urban Systems*. Englewood Cliffs, N.J.: Prentice-Hall, 1970.

Borchert, John R. "America's Changing Metropolitan Regions." *Annals of the Association of American Geographers* 62 (1972): 352–73.

Bourne, Larry S., Robert Sinclair, and K. Dziewnski, eds. *Urbanization and Settlement Systems: International Perspectives*. New York: Oxford University Press, 1984.

Brunn, Stanley D., and Jack F. Williams. *Cities of the World: World Regional Development*. New York: Harper and Row, 1983.

Cadwallader, Martin. *Analytical Urban Geography*. Englewood Cliffs, N.J.: Prentice-Hall, 1985.

Christaller, Walter. *Central Places in Southern Germany*. Trans. C. W. Baskin. Englewood Cliffs, N.J.: Prentice-Hall, 1966.

Christian, Charles M., and Robert A. Harper, eds. *Modern Metropolitan Systems*. Columbus, Ohio: Charles E. Merrill, 1982.

Hall, Peter. *The World Cities*. 3d ed. New York: McGraw-Hill, 1984.

Hartshorn, Truman A. *Interpreting the City: An Urban Geography*. 2d rev. ed. New York: Wiley, 1990.

Harvey, David. *Social Justice and the City*. London: Edward Arnold, 1973.

Hughes, James, ed. *Suburbanization Dynamics and the Future of the City*. New Brunswick, N.J.: Rutgers University, Center for Urban Policy Research, 1974.

Johnston, R. J., *The American Urban System: A Geographical Perspective*. New York: St. Martin's Press, 1982.

King, Leslie J. *Central Place Theory*. Newbury Park, Calif.: Sage, 1984.

Muller, Peter O. *Contemporary Suburban America*. Englewood Cliffs, N.J.: Prentice-Hall, 1981.

Palm, Risa. *The Geography of American Cities*. New York: Oxford University Press, 1981.

Pred, Allan. *City-Systems in Advanced Economies*. New York: Wiley, 1977.

Rose, Harold. *The Black Ghetto: A Spatial Behavioral Perspective*. New York: McGraw-Hill, 1971.

Scott, Allen J. *Metropolis: From the Division of Labor to Urban Form*. Berkeley: University of California Press, 1988.

Smith, Neil, and Peter Williams, eds. *Gentrification of the City*. Winchester, Mass.: Allen & Unwin, 1986.

Ward, David. *Cities and Immigrants*. New York: Oxford University Press, 1971.

Yeates, Maurice. *The North American City*. 4th ed. New York: Harper and Row, 1990.

P A R T

4

The Area Analysis Tradition

Julius Caesar began his account of his transalpine campaigns by observing that all of Gaul was divided into three parts. With that spatial summary, he gave to every schoolchild an example of geography in action.

Caesar's report to the Romans demanded that he convey to an uninformed audience a workable mental picture of place. He was able to achieve that aim by aggregating spatial data, by selecting and emphasizing what was important to his purpose, and by submerging or ignoring what was not. He was pursuing the geographic tradition of area analysis, a tradition that is at the heart of the discipline and focuses on the recognition of spatial uniformities and the elucidation of their significance.

The tradition of area analysis is commonly associated with the term *regional geography*, the study of particular portions of the earth's surface. As did Caesar, the regional geographer attempts to view a particular area and to summarize what is spatially significant about it. One cannot, of course, possibly know everything about a region, nor would "everything" contribute to our understanding of its essential nature. Regional geographers, however, approach a preselected earth space—a continent, a country, or some other division—with the intent of making it as fully understood as possible in as many facets of its nature as they deem practicable. It is from this school of area analysis that "regional geographies" of, for example, Africa, the United States, or the Pacific Northwest emanate.

Because the scope of their inquiries is so broad, regional geographers must become thoroughly versed in all of the topical subfields of the discipline, such as those discussed in this book. Only in that way can area specialists select those phenomena that give insight into the essential unity and diversity of their region of study. In their approach, and based on their topical knowledge, regional geographers frequently seek to delimit and study regions defined by one or a limited number of criteria.

The three topical traditions already discussed have often misleadingly been contrasted to the tradition of area analysis. In actuality, practitioners of both topical and regional geography inevitably work within the tradition of area analysis. Both seek an organized view of earth space. The one asks what are the regional units that evolve from the consideration of a particular set of preselected phenomena; the other asks how, in the study of a region, its varied content may best be clarified.

Chapter 13 is devoted to the area analysis tradition. Its introductory pages explore the nature of regions and methods of regional analysis. The body of the chapter consists of a series of regional vignettes, each based on a theme introduced in one of the preceding chapters of the book. Since those chapters were topically organized, the separate studies that follow demonstrate the tradition of area analysis in the context of topical (sometimes called "systematic") geography. They are examples of the ultimate regionalizing objectives of geographers who seek an understanding of their data's spatial expression. The step from these limited topical studies to the broader, composite understandings sought by the area analyst is both short and intellectually satisfying.

Rice paddy and farmhouse near Bannomthon, Thailand.

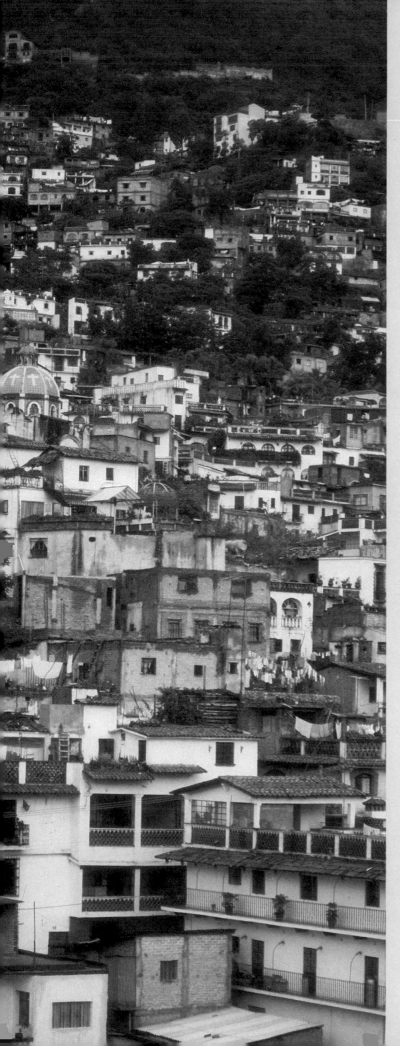

13

The Regional Concept

Hillside houses, Taxco, Mexico.

The questions geographers ask, we saw in Chapter 1 of this book, ultimately focus on matters of location and character of place. We asked how things are distributed over the surface of the earth, how physical and cultural features of areas are alike or different from place to place, how the varying content of different places came about, and what all these differences and similarities mean for people.

The Nature of Regions

In earlier chapters we examined some of the physical and cultural content of areas and some spatially important aspects of human behavior. We looked at the physical earth processes that lead to differences from place to place in the environment. We studied ways in which humans organize their actions in earth space—through political institutions, by economic systems and practices, and by cultural and social processes influencing their spatial behavior and interaction. Population and settlement patterns and areal differences in human use and misuse of earth resources all were considered as part of the mission of geography.

For each of the topics we studied, from landforms to cities, we found spatial regularities. We discovered that things are not irrationally distributed over the surface of the earth, but reflect an underlying spatial order based on understandable physical and cultural processes. We found, in short, that although no two places are *exactly* the same, it is possible and useful to recognize segments of the total world that are internally similar in some important characteristic and distinct in that feature from surrounding areas.

These regions of significant uniformity of content are the geographer's equivalent of the historian's "eras" or "ages": brief summary names for specific areas that are different in some important way from adjacent or distant territories. The **region**, then, is a device of areal generalization. It is an attempt to separate into recognizable component parts the otherwise overwhelming diversity and complexity of the earth's surface.

All of us have a general idea of the meaning of region, and all of us refer to regions in everyday speech and action. We visit "the old neighborhood" or "go downtown," we plan to vacation or retire in the "Sunbelt," or we speculate on the effects of weather conditions in the "Northeast" or the "Corn Belt" on grain surpluses or next year's food prices. In each instance we have mental images of the areas mentioned. Those images are based upon place characteristics and areal generalizations that seem useful to us and recognizable to our listeners. We have, in short, engaged in an informal place classification to pass along quite complex spatial, organizational, or content ideas. We have applied the **regional concept** to bring order to the immense diversity of the earth's surface.

What we do informally as individuals, geography attempts to do formally as a discipline—define and explain regions (Figure 13.1). The purpose is clear: to make the infinitely varying world around us understandable through spatial summaries. That world is only rarely subdivided into neat, unmistakable "packages" of uniformity. Neither the environment nor human areal actions present us with a compartmentalized order, any more than the sweep of human history has predetermined "eras," or all plant specimens come labelled in nature with species names. We all must classify to understand, and the geographer classifies in regional terms.

Regions are spatial expressions of ideas or summaries useful to the analysis of the problem at hand. Although as many possible regions exist as there are physical, cultural, or organizational attributes of area, the geographer selects for study those areal variables that contribute to the understanding of a specific topical or spatial

(a)

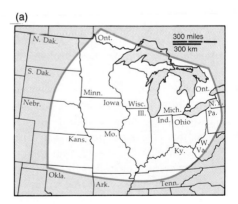

(b)

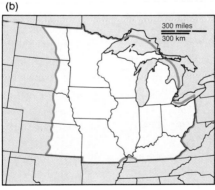

(c)

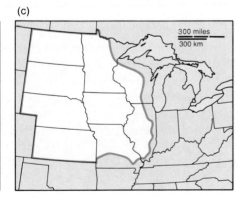

Figure 13.1

The Middle West as seen by different professional geographers. Agreement on the need to recognize spatial order and to define regional units does not imply unanimity in the selection of boundary criteria. Each of the sources concurs in the significance of the Middle West as a regional entity in the spatial structure of the United States and agrees on its core area. The sources differ, however, in their assessments of its limiting characteristics. ([a] John H. Garland, editor, *The North American Midwest*. Copyright © 1955 John Wiley & Sons, Inc., New York, NY; [b] John R. Borchert and Jane McGuigan, *Geography of the New World*. Copyright © 1961 Rand McNally, Chicago, IL; [c] Otis P. Starkey and J. Lewis Robinson, *The Anglo-American Realm*. Copyright © 1969 McGraw-Hill Publishing Company, New York, NY.)

problem. All other variables are disregarded as irrelevant. In our reference to the Corn Belt we delimit a portion of the United States showing common characteristics of farming economy and marketing practices. We dismiss—at that level of generalization—differences within the region based upon slope, soil type, state borders, or characteristics of population. The boundaries of the Corn Belt are assumed to be marked where the region's internal unifying characteristics change so materially that different agricultural economies become dominant and different regional summaries are required. It is the content of the region that suggests its definition and determines the basis of its delimitation.

Although regions may vary greatly, they all share certain common characteristics related to earth space.

1. Regions have *location*, often expressed in the regional name selected, such as the Middle West, the Near East, North Africa, and the like. This form of regional name underscores the importance of *relative location* (see Chapter 1, p. 5).
2. Regions have *spatial extent*. They are recognitions of the fact that there are differences from place to place in the physical and cultural attributes of area, and that those differences are important.
3. Regions have *boundaries* based upon the areal spread of the features selected for study. Since regions are the recognition of the features defining them, their boundaries are drawn where those features no longer occur or dominate (Figure 13.2).
4. Regions, as we saw in Chapter 1 (p. 10), may be either *formal* or *functional*.

Formal regions are areas of essential uniformity throughout in one or a limited combination of physical or cultural features. In previous chapters we encountered such formal physical regions as the humid subtropical climate zone and the Sahel region of Africa, as well as formal (homogeneous) cultural regions in which standardized characteristics of language, religion, ethnicity, or livelihood existed. The endpaper maps of topographical regions and countries show other formal regional patterns. Whatever the basis of its definition, the formal region is the largest area over which a valid generalization of attribute uniformity may be made. Whatever is stated about one part of it holds true for its remainder.

The **functional region**, in contrast, is a spatial system defined by the interactions and connections that give it a dynamic, organizational basis. Its boundaries remain constant only as long as the interchanges establishing it remain unaltered. The commuting region shown in Figure 1.9 retained its shape and size only as long as the road pattern and residential neighborhoods on which it was based remained as shown.

5. Regions are *hierarchically arranged.* Although regions vary in scale, type, and degree of

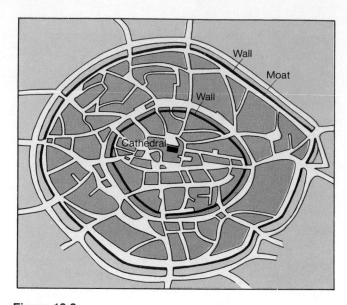

Figure 13.2
Aachen in 1649. The acceptance of regional extent implies the recognition of regional boundaries. At some defined point, *urban* is replaced by *nonurban*, the Middle West ends and the Plains begin, or the rain forest ceases and the savanna emerges. Regional boundaries are, of course, rarely as precisely and visibly marked as were the limits of the walled medieval city. Its sprawling modern counterpart may be more difficult to define, but the boundary significance of the concept of *urban* remains.

generalization, none stands alone as the ultimate key to areal understanding. Each defines only a part of spatial reality.

On a formal regional scale of size progression, the Delmarva Peninsula of the eastern United States may be seen as part of the Atlantic Coastal Plain, which is in turn a portion of the eastern North American humid continental climatic region, a hierarchy that changes the basis of regional recognition as the level and purpose of generalization alter (Figure 13.3). The central business district of Chicago is one land use complex in the functional regional hierarchy that describes the spatial influences of the city of Chicago and the metropolitan region of which it is the core. Each recognized regional entity in such progressions may stand alone and, at the same time, exist as a part of a larger, equally valid, territorial unit.

Such progressions may also reflect gradations in the intensity in spatial dominance of the phenomenon giving definition to the region. With particular attention to regions defined by the distribution of culture groups—but with application to regions delimited by other criteria—Donald Meinig has suggested the term *core* to mean a centralized zone of concentration and greatest homogeneity of regional character. Areas in which the culture (or other defining feature) is dominant, but with less intensity and totality of development, were labelled *domains*. Finally, Meinig proposed the term *sphere* to

Figure 13.3
A hierarchy of regions. One possible nesting of regions within a *regional hierarchy* defined by differing criteria. Each regional unit has internal coherence. The recognition of its constituent parts aids in understanding the composite areal unit.

recognize the zone of broadest but least-intensive expression of the regional character, where some of its defining traits are encountered but where it is no longer spatially dominant.

These generalizations about the nature of regions and the regional concept are meant to instill firmly the understanding that regions are human intellectual creations designed to serve a purpose. Regions focus our attention upon spatial uniformities, and they bring clarity to the seeming confusion of the observable physical and cultural features of the world we inhabit. Regions provide the framework for the purposeful organization of spatial data.

The Structure of This Chapter

The remainder of the chapter contains examples of how geographers have organized regionally their observations about the physical and cultural world. Each study or vignette explores a different aspect of regional reality. Each organizes its data in ways appropriate to its subject matter and objective, but in each, some or all of the common characteristics of regional delimitation and structure may be recognized. Each of the regional examples is based upon the content of one of the earlier chapters of this book, and the page numbers preceding the different selections refer to the material within those chapters most closely associated with, or explained further by, the regional case study. As an aid to visualizing the application of the regional concept, the examples are introduced by reference to the traditions of geography that unite their subject matter and approach.

Part I. Regions in the Earth-Science Tradition

The simplest of all regions to define, and generally the easiest to recognize, is the formal region based upon a single readily apparent component or characteristic. The island is land, not water, and its unmistakable boundary is naturally given where the one element passes to the other. The terminal moraine may mark the transition from the rich, black soils of recent formation to the particolored clays of earlier generation. The dense forest may break dramatically upon the glade or the open prairies. The nature of change is singular and apparent.

The physical geographer, although concerned with all the earth sciences that explain the natural environment, deals at the outset with *single factor* formal regions. Many of the earth features of concern to physical geography, of course, do not exist in simple, clearly defined units. They must be arbitrarily regionalized by the application of boundary definitions. A stated amount of received precipitation, the presence of certain important soil characteristics, the dominance of particular plant associations—all must be decided upon as regional limits, and all such limits are subject to change through time or by purpose of the regional geographer.

Landforms as Regions *(See "Stream Landscapes in Humid Areas," page 68)*

The landform region exists in a more sharply defined fashion than such transitional physical features as soil, climate, or vegetation. For these latter areas, the boundaries depend upon definitional decisions made (and defended) by the researcher. The landform region, on the other hand, arises—visibly and apparently unarguably—from nature itself, independent of human influence and unaffected by time on the human scale. Landforms constitute basic, naturally defined regions of physical geographic concern. The existence of major landform regions—mountains, lowlands, plateaus—is unquestioned in popular recognition or scientific definition. Their influence on climates, vegetational patterns, even upon the primary economies of subsistence populations has been noted in earlier portions of this book. The following discussion of a distinct landform region, describing its constitution and its relationships to other physical and cultural features of the landscape, is adapted from a classic study by Wallace W. Atwood.

The Black Hills Province.[1] The Black Hills rise abruptly from the surrounding plains (Figure 13.4). The break in topography at the margin of this province is obvious to the most casual

1. From *The Physiographic Provinces of North America* by Wallace W. Atwood. © Copyright, 1940, by Ginn and Company (Xerox Corporation). From pp. 13–15.

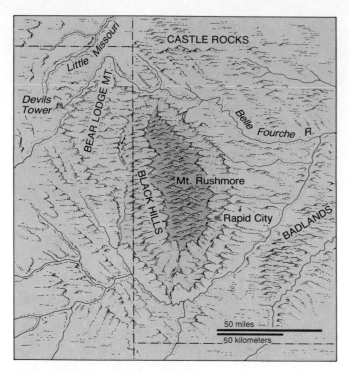

Figure 13.4
The Black Hills physiographic province.

observer who visits that part of our country. Thus the boundaries of this area, based on contrasts in topography, are readily determined.

One who looks more deeply into the study of the natural environment may recognize that in the neighboring plains the rock formations lie in a nearly horizontal position. They are sandstones, shales, conglomerates, and limestones. In the foothills those same sedimentary formations are bent upward and at places stand in a nearly vertical position. Precisely where the change in topography occurs, we find a notable change in the geologic structure and thus discover an explanation for the variation in relief.

The Black Hills are due to a distinct upwarping, or doming, of the crustal portion of the earth. Subsequent removal by stream erosion of the higher portions of that dome and the dissection of the core rocks have produced the present relief features. As erosion has proceeded, more and more of a complex series of ancient metamorphic rocks has been uncovered. Associated with the very old rocks of the core and, at places, with the sedimentary strata, there are a number of later intrusions which have cooled and formed solid rock. They have produced minor domes about the northern margin of the Black Hills.

With the elevation of this part of our country there came an increase in rainfall in the area, and with the increase in elevation and rainfall came contrasts in relief, in soils, and in vegetation.

As we pass from the neighboring plains, where the surface is monotonously level, and climb into the Black Hills area, we enter a landscape having great variety in the relief. In the foothill belt, at the southwest, south, and east, there are hogback ridges interrupted in places by water gaps, or gateways, that have been cut by streams radiating from the core of the Hills. Between the ridges there are roughly concentric valley lowlands. On the west side of the range, where the sedimentary mantle has not been removed, there is a plateau-like surface; hogback ridges are absent. Here erosion has not proceeded far enough to produce the landforms common to the east margin. In the heart of the range we find deep canyons, rugged intercanyon ridges, bold mountain forms, craggy knobs, and other picturesque features (Figure 13.5). The range has passed through several periods of mountain growth and several stages, or cycles, of erosion.

The rainfall of the Black Hills area is somewhat greater than that of the brown, seared, semiarid plains regions, and evergreen trees survive among the hills. We leave a land of sagebrush and grasses to enter one of forests. The dark-colored evergreen trees suggested to early settlers the name Black Hills. As we enter the area, we pass from a land of cattle ranches and some seminomadic shepherds to a land where forestry, mining, general farming, and recreational activities give character to the life of the people. In color and form, in topography, climate, vegetation, and economic opportunities, the Black Hills stand out conspicuously as a distinct geographic unit.

Dynamic Regions in Weather and Climate (See "Air Masses," page 96)

The unmistakable clarity and durability of the Black Hills landform region and the precision with which its boundaries may be drawn are rarely echoed in other types of formal physical regions. Most of the natural environment, despite its appearance of permanence and certainty, is dynamic in nature. Vegetations, soils, and climates change through time by natural process or by the action of humans. Boundaries shift, perhaps abruptly, as witness the recent migration southward of the Sahara. The core characteristics of whole provinces change as marshes are drained or forests are replaced by cultivated fields.

That complex of physical conditions that we recognize locally and briefly as weather and summarize as climate displays particularly clearly the temporary nature of much of the natural environment that surrounds us. Yet

Figure 13.5
The "Needles" of the Black Hills result from erosion along vertical cracks and crevices in granite.

even in the turbulent change of the atmosphere, distinct regional entities exist with definable boundaries and internally consistent horizontal and vertical properties. "Air masses" and the consequences of their encounters constitute a major portion of contemporary weather analysis and prediction. Air masses further fulfill all the criteria of *multifactor formal regions*, though their dynamic quality and their patterns of movement obviously mark them as being of a nature distinctly different from such stable physical entities as landform regions, as the following extract from *Climatology and the World's Climates* by George R. Rumney makes clear.

> *Air Masses.*[2] *An* air mass *is a portion of the atmosphere having a uniform horizontal distribution of certain physical characteristics, especially of temperature and humidity. These qualities are acquired when a mass of air stagnates or moves very slowly over a large and relatively unvaried surface of land or sea. Under these circumstances surface air gradually takes on properties of temperature and moisture approaching those of the underlying surface, and there then follows a steady, progressive transmission of properties to greater heights, resulting finally in a fairly clearly marked vertical transition of characteristics. Those parts of the earth where air masses acquire their distinguishing qualities are called* source regions.

2. G. R. Rumney, *Climatology and the World's Climates*, Macmillan, 1968.

> *The height to which an air mass is modified depends upon the length of time it remains in its source region and also upon the difference between the initial properties of the air when it first arrived and those of the underlying surface. If, for example, an invading flow of air is cooler than the surface beneath as it comes to virtual rest over a source region, it is warmed from below, and convective currents are formed, rapidly bearing aloft new characteristics of temperature and moisture to considerable heights. If, on the other hand, it is warmer than the surface of the source region, cooling of its surface layers takes place, vertical thermal currents do not develop, and the air is modified only in its lower portions. The process of modification may be accomplished in just a few days of slow horizontal drift, although it often takes longer, sometimes several weeks. Radiation, convection, turbulence, and advection are the chief means of bringing it about.*

> *The prerequisite conditions for these developments are very slowly migrating, outward spreading, and diverging air and a very extensive surface beneath that is fairly uniform in nature. Light winds and relatively high barometric pressure characteristically prevail. Hence, most masses form within the great semipermanent anticyclonic regions of the general circulation, where calms, light variable winds, and overall subsidence of the atmosphere are typical.*

Four major types of source regions are recognized: continental polar, maritime polar, continental tropical, and maritime tropical. Polar air masses are continental when they develop over land or ice surfaces in high latitudes; these are cold and dry. They are maritime when they form over the oceans in high latitudes. An air mass from these sources is cold and moist. Similarly, tropical air is continental when it originates along the Tropics of Cancer and Capricorn over northern Africa and northern Australia and is therefore warm and dry. It is maritime when it forms along the Tropics over the oceans, where it develops as a mass of warm, moist air. A single air mass usually covers thousands of square miles of the earth's surface when fully formed.

An air mass is recognizable chiefly because of the uniformity of its primary properties—temperature and humidity—and the vertical distribution of these. Secondary qualities, such as cloud types, precipitation, and visibility, are also taken into account. These qualities are retained for a remarkably long time, often for several weeks, after an air mass has traveled far from its source region, and they are thus the means of distinguishing it from other masses of air.

The principal air masses of the Americas, their source regions, and seasonal movements are shown in Figure 13.6.

Ecosystems as Regions
(See "Ecosystems," page 122)

A traditional though oversimplified definition of geography as "the study of the areal variation of the surface of the earth" suggested that the discipline centered on the classification of areas and on the subdivision of the earth into its constituent regional parts. Considerations of organization and function were secondary and even unnecessary to the implied main purpose of regional study: the definition of cores and boundaries of areas uniform in physical or cultural properties.

Newer research approaches stress the need for the study of spatial relationships from the standpoint of *systems analysis*, which emphasizes the organization, structure, and functional dynamics within an area and provides for the quantification of the linkages between the things in space. The *ecosystem*, or *biome*, introduced in Chapter 5, provides a systems-analytic concept of great flexibility. It brings together in a single framework environment, humans, and the biological realm, permitting an analysis of the relationships between these components of area. Since that relationship is structured, structure rather than spatial uniformity attracts attention and leads to new understanding of the region. The ecosystem concept, particularly, provides a point of view for investigating the complex consequences of human impact upon the natural environment, the theme of Chapter 5.

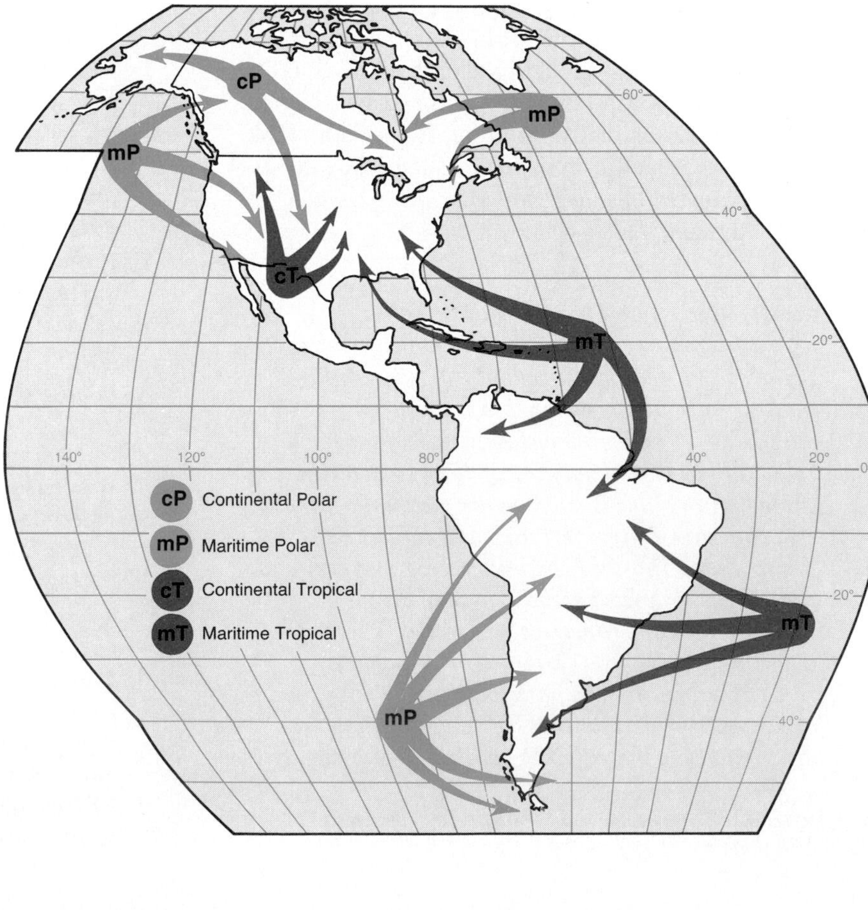

Figure 13.6
Air masses of North and South America, their source regions, and paths of movement.

cP Continental Polar
mP Maritime Polar
cT Continental Tropical
mT Maritime Tropical

Note how, in the following description drawn from an article by William J. Schneider, the ecosystem concept is employed in the recognition and analysis of regions and subregions of varying size, complexity, and nature. Introduced, too, is the concept of *ecotone,* or zone of ecological stress, in this case induced by human pressures upon the natural system.

The Everglades.[3] *The Everglades is a river. Like the Hudson or the Mississippi, it is a channel through which water drains from higher to lower ground as it moves to the sea. It extends in a broad, sweeping arc from the southern end of Lake Okeechobee in central Florida to the tidal estuaries of the Gulf Coast and Florida Bay. As much as 70 miles wide, but generally averaging 40, it is a large, shallow slough that weaves tortuously through acres of saw grass and past "islands" of trees, its waters, even in the wet season, rarely deeper than 2 feet. But, again like the Hudson or the Mississippi, the Everglades bears the significant imprint of civilization.*

Since the close of the Pleistocene, 10,000 years ago, the Everglades has been the natural drainage course for the periodically abundant overflow of Lake Okeechobee. As the lake filled during the wet summer months or as hurricane winds blew and literally scooped the water out of the lake basin, excess water spilled over the lake's southern rim. This overflow, together with rainfall collected en route, drained slowly southward between Big Cypress Swamp and the sandy flatlands to the west and the Atlantic coastal ridge to the east, sliding finally into the brackish water of the coastal marshes (Figure 13.7).

Water has always been the key factor in the life of the Everglades. Three-fourths of the annual average 55 inches (140 cm) of rainfall occurs in the wet season, June through October, when water levels rise to cover 90% of the land area of the Everglades. In normal dry seasons in the past, water covered no more than 10% of the land surface. Throughout much of the Everglades, and prior to recent engineering activities, this seasonal rain cycle caused fluctuations in water levels that averaged 3 feet. Both occasional severe flooding and prolonged drought accompanied by fire imposed periodic stress upon the ecosystem. It may be that randomly occurring ecologic trauma is vital to the character of the Everglades.

Three dominant biological communities—open water, saw grass, and woody vegetation—reflect

small, but consistent, differences in the surface elevation of peat soils that cover the Everglades (Figure 13.8). The open-water areas occur at the lower soil elevations; inundated much of the year, they contain both sparse, scattered marsh grasses and a mat of algae. The saw grass communities develop on a soil base only a few inches higher than that in the surrounding open glades. The soil base is thickest under the tree islands. The few inches' difference in soil depth apparently governs the species composition of these three communities.

Today, the Everglades is no longer precisely a natural river. Much of it has been altered by an extensive program of water management, including drainage, canalization, and the building of locks and dams. Large withdrawals of groundwater for municipal and industrial use have depleted the underlying aquifer and permitted the landward penetration of sea water through the aquifer and through the surface canals. Thousands of individual water-supply wells have been contaminated by encroaching saline water; large biotic changes have taken place in the former freshwater marshes south of Miami. Mangroves—indicators of salinity—have

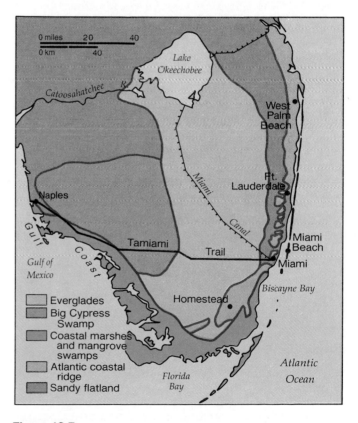

Figure 13.7

The Everglades is part of a complex of ecosystems stretching southward in Florida from Lake Okeechobee to the sea. Drainage and water-control systems have altered its natural condition.

---

3. From William J. Schneider, "Water and the Everglades." With permission from *Natural History,* November, 1966, pp. 32–40. Copyright American Museum of Natural History, 1966.

Figure 13.8
Open water, saw grasses, and tree islands, all visible in this scene, constitute separate *biomes* of the Everglades, which is home also of a teeming animal life, including the great egret shown in the foreground.

extended their habitat inland, and fires rage across areas that were formerly much wetter. The ecotone—the zone of stress between dissimilar adjacent ecosystems—is altering as a consequence of these human-induced modifications of the Everglades ecosystem.

The organization, structure, and functional dynamics of the Everglades ecosystems are thus undergoing change. The structured relationships of its components—in nature affected and formed by stress—are being subjected to distortions by humans in ways not yet fully comprehended.

Part II. Regions in the Culture–Environment Tradition

The earth-science tradition of geography imposes certain distinctive limits upon area analysis. However defined, the regions that may be drawn are based upon nature and do not result from human action. The culture–environment tradition, however, introduces to regional geography the infinite variations of human occupation and organization of space. There is a corresponding multiplication of recognized regional types and of regional boundary decisions.

Despite the differing interests of physical and cultural geographers, one element of study is common to their concerns: that of process. The "becoming" of an ecosystem, of a cultural landscape, or of the pattern of exchanges in an economic system is an important open or implied part of nearly all geographic study. Evidence of the past as an aid to understanding of the present is involved in much geographical investigation, for present-day distributional patterns or qualities of regions mark a merely temporary stage in a continuing process of change.

Population as Regional Focus
(See "World Population Distribution," page 178)

In no phase of geography are process and change more basic to regional understanding than in population studies. The human condition is dynamic and patterns of settlement are ever-changing. Although these spatial distributions are related to the ways that people utilize the physical environment in which they are located, they are also conditioned by the purposes, patterns, and solutions of those who went before. In the following extract taken from the work of Glenn Trewartha, a dean of American population geographers, notice how population regionalization—used as a focal theme—ties together a number of threads of regional description and understanding. The aspirations of colonist-conquerors, past and contemporary transportation patterns, physical geographic conditions, political separatism, and the history and practice of agriculture and rural landholdings are all introduced to give understanding to population from a regional perspective.

***Population Patterns of Latin America.**[4] A distinctive feature of the spatial arrangement of population in Latin America is its strongly nucleated character; the pattern is one of striking clusters. Most of the population clusters remain distinct and are separated from other clusters by sparsely occupied territory. Such a pattern of isolated nodes of settlement is common in many pioneer regions; indeed, it was characteristic of early settlement in both Europe and eastern North America. In those regions, as population expanded, the scantily occupied areas between individual clusters gradually filled in with settlers, and the nodes merged. But in Latin America such an evolution generally did not occur, and so the nucleated pattern persists. Expectably, the individual population clusters show considerable variations in density.*

The origin of the nucleated pattern of settlement is partly to be sought in the gold and missionary fever that imbued the Spanish colonists. Their settlements were characteristically located with some care, since only areas with precious metals for exploitation and large Indian populations to be Christianized and to provide laborers could satisfy their dual hungers. A clustered pattern was also fostered by the isolation and localism that prevailed in the separate territories and settlement areas of Latin America.

Almost invariably each of the distinct population clusters has a conspicuous urban nucleus. To an unusual degree the economic, political, and social life within a regional cluster centers on a single large primate city, which is also the focus of the local lines of transport.

The prevailing nucleated pattern of population distribution also bears a relationship to political boundaries. In some countries . . . a single population cluster represents the core area of the nation. In more instances, however, a population cluster forms the core of a major political subdivision of a nation-state, so that a country may contain more than one cluster. A consequence of this simple population distribution pattern and its relation to administrative subdivisions is that political

boundaries ordinarily fall within the sparsely occupied territory separating individual clusters. In Latin America few national or provincial boundaries pass through nodes of relatively dense settlement (Figure 13.9).

Another feature arising out of the cluster pattern of population arrangement is that the total national territory *of a country is often very different from the* effective national territory, *since the latter includes only those populated parts that contribute to the country's economic support.*

A further consequence of the nucleated pattern is found in the nature of the transport routes and systems. Overland routes between population clusters are usually poorly developed, while a more efficient network ordinarily exists within each cluster, with each such regional network joined by an overland route to the nearest port. Thus, the chief lines of transport connecting individual population clusters are often sea lanes

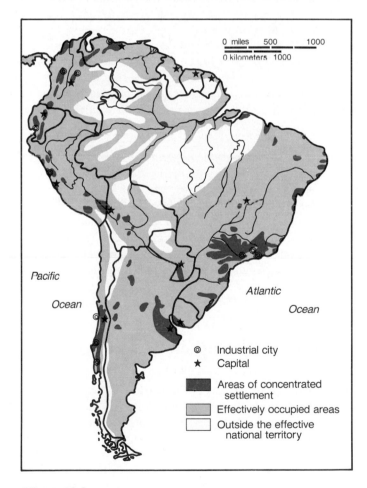

Figure 13.9
Basic settlement patterns of South America. Population clusters focused on urban cores and separated by sparsely populated rural areas were the traditional pattern in Latin American countries.

4. From Glenn T. Trewartha, *The Less Developed Realm: A Geography of Its Population.* Copyright © 1972 by John Wiley and Sons, Inc. Reprinted from pp. 43–46.

rather than land routes. Gradually, as highways are developed and improved, intercluster overland traffic tends to increase.

. . . The spectacular rates at which population numbers are currently soaring in Latin America are not matched by equivalent changes in their spatial redistributions. Any population map of Latin America reveals extensive areas of unused and underutilized land. Part of such land is highland and plagued by steep slopes, but by far the larger share of it is characterized by a moist tropical climate, either tropical wet or tropical wet and dry. Such a climatic environment, with its associated wild vegetation, soils, and drainage, admittedly presents many discouraging elements to the new settler of virgin lands. . . . [T]ropical climate alone is scarcely a sufficient explanation for the abundance of near-empty lands south of the Rio Grande. Cultural factors are involved as much as, if not more than, physical ones. One of the former is the unfortunate land-holding system that has been fastened on the continent, under which vast areas of potentially cultivable land are held out of active use by a small number of absentee landlords, who not only themselves make ineffective use of the land, but at the same time refuse to permit its cultivation by small operators. Because of the land-holding system, peasant proprietors are unable to secure their own lands, a situation that discourages new rural settlement. . . .

The changes now in progress in the spatial distributions accompanying the vast increase in numbers of people do not appear to involve any large-scale push of rural settlement into virgin territory. Only to a rather limited extent is new agricultural settlement taking place. Intercluster regions are not filling rapidly. The overwhelming tendency is for people to continue to pile up in the old centers of settlement in and around their cities, rather than to expand the frontier into new pioneer-settlement areas. . . .

Language as Region *(See "Language," page 211)*

The great culture realms of the world (outlined in Figure 7.39) are historically based composites of peoples. They are not closely identified with nation, language, religion, or technology, but with all these and more in varying combinations. Culture realms are therefore *multifactor regions* that obscure more than they clarify the distinctions between peoples that are so fundamental to the human mosaic of the earth. Basic to cultural geography is the recognition of small regions of single-factor homogeneity that give character to their areas of occurrence and that collectively provide a needed balance to the sweeping generalizations of the culture realm.

Language provides an example of such small-area variation, one partially explored in Chapter 7. The language families shown in Figure 7.18 conceal the identity of and the distinctions among the different official tongues of separate countries. These, in turn, ignore or submerge the language forms of minority populations, who may base their own sense of proud identity upon their regional linguistic separateness. In scale and recognition even below these ethnically identified regional languages are those local speech variants frequently denied status as identifiable languages and cited as proof of the ignorance and the cultural deprivation of their speakers. Yet such a limited-area, limited-population tongue contains all the elements of the classic culturally based region. Its area is defined; its boundaries are easily drawn; it represents homogeneity and majority behavior among its members; and it summarizes, by a single cultural trait, a collection of areally distinctive outlooks.

Gullah as Language.[5] Isolation is a key element in the retention or the creation of distinctive and even externally unintelligible languages. The isolation of the ancestors of the quarter-million present-day speakers of Gullah—themselves called Gullahs—was nearly complete. Held by the hundreds as slaves on the offshore islands and in the nearly equally remote low country along the southeastern United States coast from South Carolina to the Florida border (Figure 13.10), the blacks retained both the speech patterns of the African languages—Ewe, Fanti, Bambara, Twi, Wolof, Ibo, Malinke, Yoruba, Efik—native to the slave groups and over 4000 words drawn from them. Folk tales written in the Gullah creole can be heard and understood by Krio-speaking audiences in Sierra Leone today.

Forced to use English words for minimal communication with their white overseers, but modifying, distorting, and interjecting African-based substitute words into that unfamiliar language, the Gullahs kept intonations and word and idea order in their spoken common speech that made it unintelligible to white masters or to more completely integrated mainland blacks. Because the language was not understood, its speakers were considered ignorant, unable to master the niceties of English. Because ignorance was ascribed to them, the Gullahs learned to be ashamed of themselves, their culture, and their tongue, which even they themselves did not recognize as a highly structured and sophisticated separate language.

5. By Jerome Fellmann.

Figure 13.10

Gullah speakers are concentrated on the Sea Islands and the coastal mainland of South Carolina and Georgia. The isolation that promoted their linguistic distinction is now being eroded.

In common with many linguistic minorities, the Gullahs are losing their former sense of inferiority and gaining pride in their cultural heritage and in the distinctive tongue that represents it. Out of economic necessity, standard English is being taught to their schoolchildren, but an increasing scholarly and popular interest in the structure of their language and in the nature of their culture has caused Gullah to be rendered as a written language, studied as a second language, and translated into English.

In both the written and the spoken versions, Gullah betrays its African syntax patterns, particularly in its employment of terminal locator words: "Where you goin' at?" The same African origins are revealed by the absence in English translation of distinctive tenses: "I be tired" conveys the concept that "I have been tired for a period of time." Though tenses exist in the African root languages, they are noted more by inflection than by special words and structures.

"He en gut no morrataer fer mak no pie wid" may be poor English, but it is good Gullah. Its translation—"He has no more sweet potatoes for making pie"—renders it intelligible to ears attuned to English but loses the musical lilt of original speech and, more importantly, obscures the cultural identity of the speaker, a member of a regionally compact group of distinctive Americans whose territorial extent is clearly defined by its linguistic dominance.

Political Regions (See "Boundaries: The Limits of the State," page 248)

(See "Boundaries: The Limits of the State," page 248)

The most rigorously defined formal cultural region is the national state. Its boundaries are presumably carefully surveyed and are, perhaps, marked by fences and guard posts. There is no question of an arbitrarily divided transition zone or of lessening toward the borders of the basic quality of the regional core. This rigidity of a country's boundaries, its unmistakable placement in space, and the trappings—flag, anthem, government, army—that are uniquely its own give to the state an appearance of permanence and immutability not common in other, more fluid, cultural regions. But its stability is often more imagined than real. Political boundaries are not necessarily permanent. They are subject to change, sometimes violent change, as a result of internal and external pressures. The Indian subcontinent illustrates the point.

Political Regions in the Indian Subcontinent.[6] The history of the subcontinent since about 400 B.C. has been one of the alternating creation and dissolution of empires, of the extension of central control based upon the Ganges Basin, and of resistance to that centralization by the marginal territories of the peninsula. British India, created largely unintentionally by 1858, was only the last, though perhaps the most successful, attempt to bring under unified control the vast territory of incredibly complex and often implacably opposed racial, religious, and linguistic groupings.

A common desire for independence and freedom from British rule united the subcontinent's disparate populations at the end of World War II. That common desire, however, was countered by the mutual religious antipathies felt by Muslims and Hindus, each dominant in separate regions of the colony and each unwilling to be affiliated with or subordinated to the other. When the British surrendered control of the subcontinent in 1947, they recognized these apparently irreconcilable religious differences and partitioned the subcontinent into the second and seventh most populous countries on earth. The independent state of India was created out of the largely Hindu areas constituting the bulk of the former colony. Separate sovereignty was granted to most of the Muslim-majority area under the name of Pakistan. Even so, the partition left boundaries, notably in the Vale of Kashmir, dangerously undefined or in dispute.

An estimated 1 million people died in the religious riots that accompanied the partition decision. In perhaps the largest short-term mass

6. By Jerome Fellmann.

migration in history, some 10 million Hindus moved from Pakistan to India, and 7.5 million Muslims left India for Pakistan, "The Land of the Pure."

Unfortunately, the purity resided only in common religious belief, not in spatial coherence or in shared language, ethnicity, customs, food, or economy. During its 23 years of existence as originally conceived, Pakistan was a sorely divided country. The partition decision created an eastern and a western component separated by more than 1000 miles (1600 km) of foreign territory and united only by a common belief in Mohammed (Figure 13.11). West Pakistan, as large as Texas and Oklahoma combined, held 55 million largely light-skinned Punjabis with Urdu language and strong Middle Eastern cultural ties. Some 70 million Bengali-speakers, making up East Pakistan, were crammed into an Iowa-size portion of the delta of the Ganges and Brahmaputra rivers. The western segment of the country was part of the semiarid world of western Asia; the eastern portion of Pakistan was joined to humid, rice-producing Southeast Asia.

Beyond the affinity of religion, little else united the awkwardly separated country. East Pakistan felt itself exploited by a dominating western minority that sought to impose its language and its economic development, administrative objectives, and military control. Rightly or wrongly, East Pakistanis saw themselves as aggrieved and abused. They complained of a per capita income level far below that of their western compatriots, claimed discrimination in the allocation of investment capital, found disparities in the pricing of imported foods, and asserted that their exports of raw materials, particularly jute, were supporting a national economy in which they did not share proportionally. They argued that their demands for regional autonomy, voiced since nationhood, had been denied.

When, in November of 1970, East Pakistan was struck by a cyclone and tidal wave that took an estimated 500,000 lives (see p. 2), the limit of eastern patience was reached. Resentful over what they saw as a totally inadequate West Pakistani effort of aid in the natural disaster that had befallen them, the East Pakistanis were further incensed by the refusal of the central government to convene on schedule a national assembly to which they had won an absolute majority of delegates. Civil war resulted, and the separate new state of Bangladesh was created. The sequence of political change in the subcontinent is traced in Figure 13.11.

The country and nation-state so ingrained in our consciousness and so firmly defined, as displayed on the map inside the front cover of this book, is both a recent and an ephemeral creation of the cultural regional landscape. It rests upon a claim, more or less effectively enforced, of a monopoly of power and allegiance resident in a government and superior to the communal, linguistic, ethnic, or religious affiliations that preceded it or that claim loyalties overriding it. As the violent recent history of the Indian subcontinent demonstrates, nationalism may be sought, but its maintenance is not assured by the initiating motivations.

Figure 13.11

The sequence of political change on the Indian subcontinent. British India was transformed in 1947 to the countries of India and Pakistan, the latter a Muslim state with a western and an eastern component. In 1970 Pakistan was torn by civil war based on ethnic and political contrasts, and the eastern segment became the new country of Bangladesh.

Mental Regions *(See "Mental Maps," page 274)*

The regional units so far used as examples, and the methods of regionalization they demonstrate, have a concrete reality. They are formal or functional regions of specified, measurable content. They have boundaries drawn by some objective measures of change or alteration of content, and they have location upon an accurately measured global grid.

Individuals and whole cultures may operate, and operate successfully, with a much less formalized and less precise picture of the nature of the world and of the structure of its parts. The mental maps discussed in Chapter 9 represent personal views of regions and regionalization. The private world views they represent are, as we also saw, colored by the culture of which their holders are members.

Primitive societies, particularly, have distinctive world views by which they categorize what is familiar, and satisfactorily account for what is not. The Yurok Indians of the Klamath River area of northern California were no exception. Their geographic concepts were reported by T.T. Waterman, from whose paper, "Yurok Geography," the following summary is drawn.

The Yurok World View.[7] The Yurok imagines himself to be living on a flat extent of landscape, which is roughly circular and surrounded by ocean. By going far enough up the river, it is believed, "you come to salt water again." In other words, the Klamath River is considered, in a sense, to bisect the world. This whole earth mass, with its forests and mountains, its rivers and sea cliffs, is regarded as slowly rising and falling, with a gigantic but imperceptible rhythm, on the heaving, primeval flood. The vast size of the "earth" causes you not to notice this quiet heaving and settling. This earth, therefore, is not merely surrounded by the ocean but floats upon it. At about the central point of the "world" lies a place which the Yurok call qe'nek, on the southern bank of the Klamath, a few miles below the point where the Trinity River comes in from the south. In the Indian concept, this point seems to be accepted as the center of the world.

At this locality also the sky was made. Above the solid sky there is a sky-country, wo'noiyik, about the topography of which the Yurok's ideas are almost as definite as are his ideas of southern Mendocino County, for instance. Downstream from qe'nek, at a place called qe'nek-pul ("qe'nek-downstream"), is an invisible ladder leading up to the sky-country. The ladder is still

thought to be there, though no one to my knowledge has been up it recently. The sky-vault is a very definite item in the Yurok's cosmic scheme. The structure consisting of the sky dome and the flat expanse of landscape and waters that it encloses is known to the Yurok as ki-we'sona (literally "that which exists"). This sky, then, together with its flooring of landscape, constitutes "our world." I used to be puzzled at the Yurok confusing earth and sky, telling me, for example, that a certain gigantic redwood tree "held up the world." Their ideas are of course perfectly logical, for the sky is as much a part of the "world" in their sense as the ground is.

The Yurok believe that passing under the sky edge and voyaging still outward you come again to solid land. This is not our world, and mortals ordinarily do not go there; but it is good, solid land. What are breakers over here are just little ripples over there. Yonder lie several regions. To the north (in our sense) lies pu'lekūk, downstream at the north end of creation. South of pu'lekūk lies tsī'k-tsīk-ol ("money lives") where the dentalium-shell, medium of exchange, has its mythical abode. Again, to the south there is a place called kowe'tsik, the mythical home of the salmon, where also all have a "house." About due west of the mouth of the Klamath lies rkrgr', where lives the culture-hero wo'xpa-ku-mä ("across-the-ocean that widower").

Still to the south of rkrgr' there lies a broad sea, kiolaaopa'a, which is half pitch—an Algonkian myth idea, by the way. All of these solid lands just mentioned lie on the margin, the absolute rim of things. Beyond them the Yurok does not go even in imagination. In the opposite direction, he names a place pe'tskuk ("up-river-at"), which is the upper "end" of the river but still in this world. He does not seem to concern himself much with the topography there.

The Yuroks' conception of the world they live in may be summed up in the accompanying diagram (Figure 13.12).

Part III. Regions in the Locational Tradition

While location, as we have seen, is a primary attribute of all regions, regionalization in the locational tradition of geography implies far more than a named delimitation of earth space. The central concern is with the distribution of human activities and of the resources upon which those activities are based.

In this sense, world regionalization of agriculture and of the soils and climates with which it is related is within

7. *University of California Publications in American Archaeology and Ethnology,* Vol. 16, No. 5 (1920). Selection from pp. 189–93.

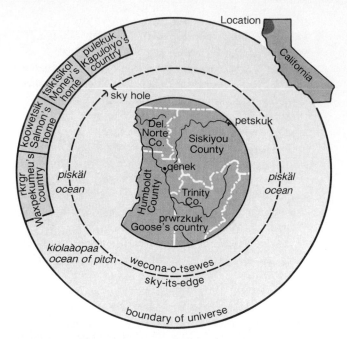

Figure 13.12
The world view of the Yurok as pieced together by T. T. Waterman during his anthropological study of the tribe. Qe'nek, in the center of the diagram, marks the center of the world in Indian belief.

the locational tradition. For practical and accepted reasons, however, such underlying physical patterns have been included under the earth-science tradition. But the point is made: the locational tradition emphasizes the "doing" in human affairs, and "doing" is not an abstract thing but an interrelation of life and the resources on which life depends.

The locational tradition, therefore, encourages the recognition and the definition of a far wider array of regional types than do the earth-science or culture–environment traditions. Any single pattern of economic activity or resource distribution invites the recognition of definable *formal regions*. The interchange of commodities, the control of urban market areas, the flows of capital, or the collection and distribution activities of ports are just a few examples of the infinite number of analytically useful *functional regions* that one may recognize.

Economic Regions *(See page 301)*

Economic regionalization is among the most frequent, familiar, and useful employments of the regional method. Through economic regions the geographer identifies activities and resources, maps the limits of their occurrence or use, and examines the interrelationships and flows that are part of the complexities of the contemporary world.

The economic region, examples of which were explored in Chapter 10, should be seen as potentially more than a device for recording what *is* in either a formal or a functional sense. It has increasingly become a device for ex-

amining what might or should be. The concept of the economic region as a tool for planning and a framework for the manipulation of the people, resources, and economic structure of a composite region first took root in the United States during the Great Depression years of the 1930s. The key element in the planning region is the public recognition of a major territorial unit in which economic change or decline is seen as the cause of a variety of interrelated problems including, for example, population out-migration, regional isolation, cultural deprivation, underdevelopment, and poverty.

Appalachia.[8] *Until the early 1960s, "Appalachia" was for most a loose reference to the complex physiographic province of the eastern United States associated with the Appalachian mountain chain. If thought of at all, it was apt to be visualized as rural, isolated, and tree-covered; as an area of coal mining, hillbillies, and folksongs (Figure 13.13).*

During the 1950s, however, the economic stagnation and the functional decline of the area became increasingly noticeable in the national context of economic growth, rising personal incomes, and growing concern with the elimination of the poverty and deprivation of every group of citizens. Less dramatically but just as decisively as the Dust Bowl or the Tennessee Valley of an earlier era, Appalachia became simultaneously a popularly recognized economic and cultural region and a governmentally determined planning region.

Evidences of poverty, underdevelopment, and social crisis were obvious to a country committed to recognize and eradicate such conditions within its own borders. By 1960, per capita income within Appalachia was $1400 when the national average was $1900. In the decade of the 1950s, mine employment fell 60% and farm jobs declined 52%; the rest of the country lost only 1% of mining jobs and 35% of agricultural employment. Rail employment fell with the drop in coal mining. Massive out-migration occurred among the young adults, with such cities as Chicago, Detroit, Dayton, Cleveland, and Gary the targets. Even with these departures, unemployment among those who remained in Appalachia averaged 50% higher than the national rate. Because of the departures, the remnant population—only 47% of whom lived, in 1960, in or near cities, as against 70% for the entire United States—was distorted in age structure. The very young and the old were disproportionally represented; the productive working-age groups were, at least temporarily, emigrants.

8. By Jerome Fellmann.

Figure 13.13

This deserted eastern Kentucky cabin is a mute reminder of the economic and social changes in Appalachia. Increases in per capita income, expansion of urban job opportunities, and improved highway networks have reduced the poverty and isolation so long associated with the region.

When these and other socioeconomic indicators were plotted by counties and by state economic areas, an elongated but regionally coherent and clearly bounded Appalachia as newly understood was revealed by maps (Figure 13.14). It extended through 13 states from Mississippi to New York, covered some 195,000 square miles (505,000 km²), and contained 18 million people, 93% of them white.

By 1963, awareness of the problems of the area at federal and state levels passed beyond recognition of a multifactor economic region to the establishment of a planning region. A joint federal-state Appalachian Regional Commission was created to develop a program designed to meet the perceived needs of the entire area. The approach chosen was one of limited investment in a restricted number of highly localized developments, with the expectation that these would spark economic growth supported by private funds.

In outline, the plan was (1) to ignore those areas of poverty and unemployment that were in isolated, inaccessible "hollows" throughout the region; (2) to designate "growth centers" where development potential was greatest and concentrate all spending for economic expansion there; the regional growth potential in targeted expenditure was deemed sufficient to overcome

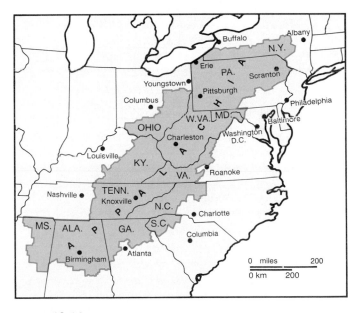

Figure 13.14

The boundary criteria of "Appalachia" as defined by the Appalachian Regional Commission were social and economic conditions colored by political considerations, not topography.

charges of aiding the prosperous and depriving the poor; and (3) to create a new network of roads so that the isolated jobless could commute to the new jobs expected to form in and near the favored growth centers. Road construction would also, of course, open inaccessible areas to tourism and strengthen the economic base of the entire planning region.

In the years since the Appalachian Regional Commission was established, and in ways not anticipated, the economic prospects of Appalachia have altered. National energy crises at least temporarily revived mining and transportation employment; new industrial jobs multiplied as manufacturing has relocated to, or has been newly developed in, the Appalachia portion of the Sunbelt. New employment opportunities have exceeded local labor pools, and out-migrants have returned home from cities outside the region, bringing again a more balanced population pyramid.

By the late 1980s, the Appalachian Regional Commission had over the years committed more than $5 billion to the area, and over $10 billion more had come from other sources. The percentage of people living in poverty within the region declined from 31% in 1960 to 14% in 1980, and per capita income, which had been 79% of the national average in 1960, rose to 85% by 1986. Population flows stabilized, with the number of people coming into Appalachia about equaling those leaving. Some 2000 miles (3200 km) of road had been laid by the end of 1986, though their construction did not give the total regional access that had been hoped. Smaller towns and their hinterlands have been by-passed and remain as isolated as ever. Some basic social services have, for the first time, reached essentially everyone in the region. Each of the 397 counties of the commission territory, for example, has been supplied with a clinic or a doctor.

Despite these evidences of progress, Appalachia still remained economically distressed in the late 1980s. The mining and manufacturing recession of the earlier years of the decade was sorely felt within a region dependent upon coal mining and with an industrial base of primary and fabricated metals, wood products, textiles, and apparel—all eroded by imports replacing domestic products. Unemployment rates continued to be between 5% and 10% above the national average, and a third of the mining and manufacturing jobs disappeared between 1979 and 1987.

Regions of Natural Resources
(See "Coal," page 345)

The unevenly distributed resources upon which people depend for existence are logical topics of interest within the locational tradition of geography. Resource regions are mapped, and raw material qualities and quantities are discussed. Areal relationships to industrial concentrations and the impacts of material extraction upon alternate uses of areas are typical interests in resource geography and in the definition of resource regions.

Those regions, however, are usually treated as if they were expressions of observable surface phenomena, as if, somehow, an oil field were as exposed and two-dimensional as a soil region or a manufacturing district. What is ignored is that most mineral resources are three-dimensional regions beneath the ground. In addition to the characteristics of an area that may form the basis for regional delimitations and descriptions of surface phenomena, regions beneath the surface add their own particularities to the problem of regional definition. They have, for example, upper and lower boundaries in addition to the circumferential bounds of surface features. They may have an internal topography divorced from the visible landscape. Subsurface relationships—for example, mineral distribution and accessibility in relation to its enclosing rock or to groundwater amounts and movement— may be critical in understanding these specialized, but real, regions. An illustration drawn from the Schuylkill field of the anthracite region of eastern Pennsylvania helps illustrate the nature of regions beneath the surface.

The Schuylkill Anthracite Region.[9] *Nothing in the wild surface terrain of the anthracite country suggested the existence of an equally rugged subterranean topography of coal beds and interstratified rock, slate, and fire clays forming a total vertical depth of 3000 feet (900 m) at greatest development. Yet the creation of the surface landscape was an essential determinant of the areal extent of the Schuylkill district, of the contortions of its bedding, and of the nature of its coal content. A county history of the area reports, "The physical features of the anthracite country are wild. Its area exhibits an extraordinary series of parallel ridges and deep valleys, like long, rolling lines of surf which break upon a flat shore." Both the surface and the subsurface topographies reflect the strong folding of strata after the coal seams were deposited; the anthracite (hard) coal resulted from metamorphic carbonization of the original bituminous beds. Subsequent river and glacial erosion removed as much as 95% of the original anthracite deposits and gave to those that remained discontinuous existence in sharply bounded fields like the Schuylkill (Figure 13.15), a discrete areal entity of 181 square miles (470 km²).*

The irregular topography of the underground Schuylkill region means that the interbedded coal seams, the most steeply inclined of all the

9. By Jerome Fellmann.

anthracite regions (Figure 13.16), outcrop visibly at the surface on hillsides and along stream valleys. The outcrops made the presence of coal known as early as 1770, but not until 1795 did Schuylkill anthracite find its first use by local blacksmiths. Reviled as "stone coal" or "black stone" that would not ignite, anthracite found no ready commercial market, although it was used in wire and rolling mills located along the Schuylkill River before 1815 and to generate steam in the same area by 1830.

The resources of the subterranean Schuylkill region affected human patterns of surface regions only after the Schuylkill Navigation Canal was completed in 1825 (Figure 13.17), providing a passage to rapidly expanding external markets for the fuel and for the output of industry newly located atop the region. Growing demand induced a boom in coal exploitation, an exhaustion of the easily available outcrop coal, and the beginning of the more arduous and dangerous underground mining.

The early methods of mining were simple: merely quarrying the coal from exposed outcrops, usually driving on a slight incline to permit natural drainage. Deep shafts were unnecessary and, indeed, not thought of, since the presence of anthracite at depth was not suspected. Later, when it was no longer possible to secure coal from a given outcrop, a small pit was sunk to a depth of 30–40 feet (9–12 m); when the coal and the water that accumulated in the pit could no longer safely be brought to the surface by windlass, the pit was abandoned and a new one was started. Shaft mining, in which a vertical opening from the surface provides penetration to one or several coal beds, eventually became a necessity; with it came awareness of the complex interrelationships between seam thickness, the nature of interstratified rock and clays, the presence of gases, and the movement of subsurface water.

The subterranean Schuylkill region has a three-dimensional pattern of use. The configuration and the variable thickness of the seams demand concentration of mining activities. Minable coal is not uniformly available along any possible vertical or horizontal cross section because of the interstratification and the extreme folding of the beds. Mining is concentrated further by the location of shafts and the construction of passages, in their turn determined by both patterns of ownership and thickness of seam. In general, no seam less than 2 feet (.6 m)

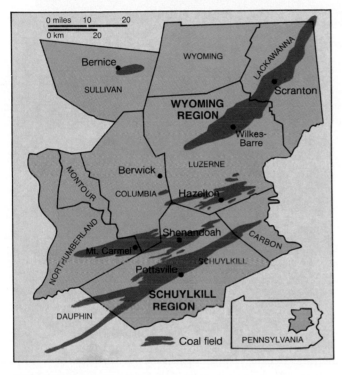

Figure 13.15
The anthracite regions of northeastern Pennsylvania are sharply defined by the geologic events that created them.

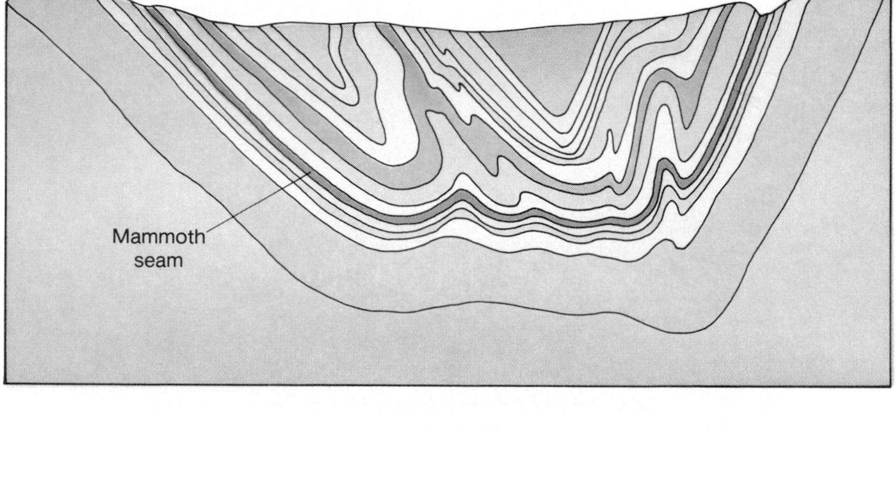

Figure 13.16
The deep folding of the Schuylkill coal seams made them costly to exploit. The Mammoth seam runs as deep as 1500 to 2000 feet (460-610 m) below the surface.

Figure 13.17
The Schuylkill Navigation Canal, like its counterpart Lehigh Canal shown here, provided an outlet to market for the anthracite region's coal resources after 1825.

is worked, and the absolute thickness—50 feet (15 m)—is found only in the Mammoth seam of the Schuylkill region.

Friable interstratified rock increases the danger of coal extraction and raises the costs of cave-in prevention. Although the Schuylkill mines are not gassy, the possibility of gas release from the collapse of coal pillars left as mine supports makes necessary systems of ventilation even more elaborate than those minimally required to provide adequate air to miners. Water is ever-present in the anthracite workings, and constant pumping or draining is necessary for mine operation. The collapse of strata underlying a river may result in sudden disastrous flooding.

The Schuylkill subterranean anthracite region presents a pattern of complexity in distribution of physical and cultural features and of interrelationships between phenomena every bit as great and inviting of geographical analysis as any purely surficial region.

Urban Regions *(See "The Making of World Metropolitan Regions," page 402)*

Urban geography represents a climax stage in the locational tradition of geography. Modern integrated, interdependent society on a world basis is urban-centered.

Cities are the indispensable functional focuses of production, exchange, and administration. They exist individually as essential elements in interlocked hierarchical systems of cities. Internally, they display complex but repetitive spatial patterns of land uses and functions.

Because of the many-sided character of urbanism, cities are particularly good subjects for regional study. They are themselves, of course, formal regions. In the aggregate, their distributions give substance to formal regions of urban concentration. Cities are also the cores of functional regions of varying types and hierarchical orders. Their internal diversity of functional, land use, and socioeconomic patterns invite regional analysis. The employment of that approach in both its formal and functional modes is clearly displayed by Jean Gottmann, who, in examining the data and landscapes of the eastern United States at midcentury, recognized and analyzed Megalopolis. The following is taken from his study.

Megalopolis.[10] The northeastern seaboard of the United States is today the site of a remarkable development—an almost continuous stretch of urban and suburban areas from southern New Hampshire to northern Virginia and from the Atlantic shore to the Appalachian foothills (Figure 13.18). The processes of urbanization,

10. From Jean Gottman, *Megalopolis: The Urbanized Northeastern Seaboard of the United States,* Copyright 1961 by the Twentieth Century Fund, New York.

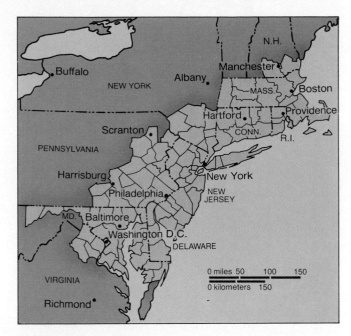

Figure 13.18
Megalopolis in 1960. The region was then composed of counties that, by U.S. census definition, were "urban" in population and economic characteristics. Much of the area is still distinctly "rural" in land use.

rooted deep in the American past, have worked steadily here, endowing the region with unique ways of life and of land use. No other section of the United States has such a large concentration of population, with such a high average density, spread over such a large area. And no other section has a comparable role within the country or a comparable importance in the world. Here has been developed a kind of supremacy, in politics, in economics, and possibly even in cultural activities, seldom before attained by an area of this size.

Great, then, is the importance of this section of the United States and of the processes now at work within it. And yet it is difficult to single this area out from the surrounding areas, for its limits cut across established historical divisions, such as New England and the Middle Atlantic states, and across political entities, since it includes some states entirely and others only partially. A special name is needed, therefore, to identify this special geographical area.

This particular type of region is new, but it is the result of age-old processes, such as the growth of cities, the division of labor within a civilized society, the development of world resources. The name applied to it should,

therefore, be new as a place name but old as a symbol of the long tradition of human aspirations and endeavor underlying the situations and problems now found here. Hence the choice of the term Megalopolis, used in this study.

As one follows the main highways or railroads between Boston and Washington, D.C., one hardly loses sight of built-up areas, tightly woven residential communities, or powerful concentrations of manufacturing plants. Flying this same route one discovers, on the other hand, that behind the ribbons of densely occupied land along the principal arteries of traffic, and in between the clusters of suburbs around the old urban centers, there still remain large areas covered with woods and brush alternating with some carefully cultivated patches of farmland (Figure 13.19). These green spaces, however, when inspected at closer range, appear stuffed with a loose but immense scattering of buildings, most of them residential but some of industrial character. That is, many of these sections that look rural actually function largely as suburbs in the orbit of some city's downtown. Even the farms, which occupy the larger tilled patches, are seldom worked by people whose only occupation and income are properly agricultural.

Thus the old distinctions between rural and urban do not apply here anymore. Even a quick look at the vast area of Megalopolis reveals a revolution in land use. Most of the people living in the so-called rural areas, and still classified as "rural population" by recent censuses, have very little, if anything, to do with agriculture. In terms of their interests and work they are what used to be classified as "city folks," but their way of life and the landscapes around their residences do not fit the old meaning of urban.

In this area, then, we must abandon the idea of the city as a tightly settled and organized unit in which people, activities, and riches are crowded into a very small area clearly separated from its nonurban surroundings. Every city in this region spreads out far and wide around its original nucleus; it grows amidst an irregularly colloidal mixture of rural and suburban landscapes; it melds on broad fronts with other mixtures, of somewhat similar though different texture, belonging to the suburban neighborhoods of other cities.

Thus an almost continuous system of deeply interwoven urban and suburban areas, with a total population of about 37 million people in 1960, has been erected along the Northeastern Atlantic seaboard. It straddles state boundaries, stretches across wide estuaries and bays, and encompasses many regional differences. In fact,

Figure 13.19
The "Pine Barrens" of New Jersey is still a largely undisturbed natural enclave in the heart of Megalopolis. Its preservation is a subject of dispute between environmentalists and developers.

the landscapes of Megalopolis offer such variety that the average observer may well doubt the unity of the region. And it may seem to him that the main urban nuclei of the seaboard are little related to one another. Six of its great cities would be great individual metropolises in their own right if they were located elsewhere. This region indeed reminds one of Aristotle's saying that cities such as Babylon had "the compass of a nation rather than a city."

The description of Megalopolis in 1960 represents, as does any regional study, a captured moment on the continuum of areal change. When that moment is lucidly described and the threads of areal character clearly delineated, the regional study serves as both a summary of the present and an augury of a future whose roots may be discerned in it. In the years since Gottmann described it, Megalopolis has continued to develop along the lines he summarized. Urbanization has continued, physically as well as functionally, to encroach upon the rural landscapes

without regard for state boundaries or even the metropolitan cores that dominated in 1960. As then, new growth centers—becoming equivalents of central cities in their own right—are linked to the expanding transportation corridors, though now the lines of importance are increasingly expressways rather than railroads.

In the south, the corridors around Washington, D.C., are the Capital Beltway in Fairfax County and an extension from it westward to Loudoun County and Dulles Airport in Virginia and, to the north into Montgomery County, Maryland, the I-270 corridor. The Virginia suburbs, focusing on Tysons Corner (Figure 10.34), have specialized in defense-related industries, but vast office building complexes and commercial centers are rapidly converting rural land to general urban uses. The Maryland suburbs specialize in health, space, and communications interests in their own new office, industrial, and commercial "parks" and complexes. Still farther to the north, in the "Princeton Corridor," a 26-mile (42 km) stretch along Route 1 from New Brunswick to Trenton, New Jersey, huge corporate parks provide office and research space to companies relocating from the New York area and to new technology firms attracted by proximity to Princeton University. Morristown, New Jersey, and White Plains in Westchester County, New York, are other similar concentrations of commercial, industrial, and office developments north and west of New York City. To the east, Stamford, Connecticut, with 150,000 daily in-commuters has become the headquarters center for major national corporations and has emerged as a truly major central city new on the scene in its present form since Megalopolis was first described. And still farther to the north, the I-495 beltway around Boston has joined Route 128 as a magnet for electronics, computer, electrical, and other technology-related companies.

Summary

The region is a mental construct, a created entity whose sole function is the purposeful organization of spatial data. The scheme of that organization, the selection of data to be analyzed, and the region resulting from these decisions are reflections of the intellectual problem posed.

This chapter has not attempted to explore all aspects of regionalism and of the regional method. It has tried, by example, only to document its basic theme: the geographer's regions are arbitrary but deliberately conceived devices for the isolation of things, patterns, interrelations, and flows that invite geographic analysis. In this sense, all geographers are regional geographers, and the regional examples of this chapter may logically complete our survey of the four traditions of geography.

Key Words

formal region	region
functional region	regional concept

For Review

1. What do geographers seek to achieve when they recognize or define regions? On what basis are regional boundaries drawn? Are regions concrete entities whose dimensions and characteristics are agreed upon by all who study the same general segment of earth space? Ask three fellow students who are not participants in this course for their definition of the "South." If their answers differed, what implicit or explicit criteria of regional delimitation were they employing?

2. What are the spatial elements or the identifying qualities shared by all regions?

3. What is the identifying characteristic of a *formal region?* How are its boundaries determined? Name three different examples of formal regions drawn from any earlier chapters of this book. How was each defined and what was the purpose of its recognition?

4. How are *functional regions* defined? What is the nature of their bounding criteria? Give three or four examples of functional regions that were defined earlier in the text.

5. The ecosystem was suggested as a viable device for regional delimitation. What regional geographic concepts are suggested in ecosystem recognition? Is an ecosystem identical to a formal region? Why or why not?

6. National, linguistic, historical, planning, and other regions have been recognized in this chapter. With what other regional entities do you have acquaintance in your daily affairs? Are fire protection districts, police or voting precincts, or zoning districts regional units identifiable with geographers' regions as discussed in this chapter or elsewhere in this book? How influenced are you in your private life by your, or others', regional delimitations?

Selected References

Freeman, T. W. "The Regional Approach." *A Hundred Years of Geography*. Chicago: Aldine, 1961.

Gottmann, Jean. *Megalopolis Revisited: 25 Years Later*. College Park: Institute for Urban Studies, University of Maryland, 1987.

McDonald, James R. "The Region: Its Conception, Design and Limitations." *Annals of the Association of American Geographers* 56 (1966): 516–28.

Meinig, Donald. "The Mormon Culture Region: Strategies and Patterns in the Geography of the American West, 1847–1964." *Annals of the Association of American Geographers* 55 (1965): 191–220.

Minshull, Roger. *Regional Geography*. Chicago: Aldine, 1967.

Murphey, Rhoads. "The Region." *The Scope of Geography*. 3d ed. New York and London: Methuen, 1982.

Whittlesey, Derwent. "The Regional Concept and the Regional Method." In *American Geography: Inventory and Prospect*. Preston James and Clarence F. Jones, eds. Syracuse, N.Y.: Syracuse University Press (for the Association of American Geographers), 1954.

1990 World Population Data

Region or Country	Population Estimate mid-1990 (millions)	Birth Rate (per 1,000 pop.)	Death Rate (per 1,000 pop.)	Natural Increase (annual, %)	Population "Doubling Time" in Years (at current rate)	Population Projected to 2000 (millions)	Population Projected to 2020 (millions)	Infant Mortality Rate[a]	Total Fertility Rate[b]	% Population Under Age 15/65+	Life Expectancy at Birth (years)	Urban Population (%)	Population with Access to Safe Water Supply (%, 1983)	Energy Consumption Per Capita, 1986 (gigajoules)/ Percent Change Since 1970	Per Capita GNP, 1988 (US$)
World	5,321	27	10	1.8	39	6,292	8,228	73	3.5	33/6	64	41	–	56/9	$ 3,470
More Developed	1,214	15	9	0.5	128	1,274	1,350	16	2.0	22/12	74	73	–	–/–	15,830
Less Developed	4,107	31	10	2.1	33	5,018	6,878	81	4.0	36/4	61	32	–	–/–	710
Less Developed (Excl. China)	2,987	35	11	2.4	29	3,738	5,382	91	4.6	40/4	59	36	51	–/–	870
Africa	661	44	15	2.9	24	884	1,481	109	6.2	45/3	52	31	41	12/44	600
Northern Africa	144	38	10	2.8	25	183	268	87	5.2	43/4	59	41	73	–/–	1,110
Algeria	25.6	40	9	3.1	22	32.7	45.6	74	6.1	46/4	60	43	90	37/243	2,450
Egypt	54.7	38	9	2.9	24	69.0	101.9	90	4.7	41/4	60	45	75	21/172	650
Libya	4.2	38	7	3.1	22	5.6	8.5	69	5.5	44/3	66	76	98	106/439	5,410
Morocco	25.6	35	10	2.6	27	31.4	43.3	82	4.8	42/4	61	43	–	9/69	750
Sudan	25.2	45	16	2.9	24	33.6	54.6	108	6.4	45/3	50	20	48	2/–46	340
Tunisia	8.1	28	7	2.0	34	10.1	13.8	59	4.1	39/4	65	53	–	21/150	1,230
Western Sahara	0.2	49	23	2.5	28	0.2	0.4	–	–	–/–	–	–	–	–/–	–
Western Africa	206	47	17	3.0	23	279	481	119	6.6	46/2	48	30	34	–/–	340
Benin	4.7	51	19	3.2	22	6.6	11.7	110	7.0	47/3	47	39	20	1/–14	340
Burkina Faso	9.1	50	18	3.2	21	12.5	23.0	126	7.2	48/4	51	8	30	1/125	230
Cape Verde	0.4	38	10	2.8	25	0.5	0.8	66	5.2	42/5	61	27	–	4/–	–
Côte d'Ivoire	12.6	51	14	3.7	19	18.5	35.4	96	7.4	49/2	53	43	20	6/5	740
Gambia	0.9	47	21	2.6	27	1.1	1.7	143	6.4	44/3	43	21	–	3/45	220
Ghana	15.0	44	13	3.1	22	20.4	33.9	86	6.3	45/3	55	32	43	3/–38	400
Guinea	7.3	47	22	2.5	28	9.2	14.4	147	6.2	43/3	42	22	17	2/–28	350
Guinea-Bissau	1.0	41	20	2.1	33	1.2	2.0	132	5.4	41/4	45	27	33	1/–40	160
Liberia	2.6	45	13	3.2	22	3.7	6.5	83	6.4	46/3	56	43	40	4/–70	450
Mali	8.1	52	22	3.0	23	10.7	19.2	117	7.2	47/3	45	18	14	1/60	230
Mauritania	2.0	46	19	2.7	25	2.7	4.5	127	6.5	44/3	46	35	–	4/–18	480
Niger	7.9	51	21	3.0	23	11.1	20.6	135	7.1	47/3	45	16	34	2/133	310
Nigeria	118.8	46	17	2.9	24	160.8	273.2	121	6.5	45/2	48	31	37	5/255	290
Senegal	7.4	46	19	2.7	26	9.7	15.2	128	6.4	44/3	46	36	44	4/8	630
Sierra Leone	4.2	48	23	2.5	28	5.4	8.9	154	6.5	44/3	41	28	23	2/–44	240
Togo	3.7	50	14	3.6	19	5.2	9.9	114	7.2	49/2	55	22	37	2/6	370

Region or Country	Population Estimate mid-1990 (millions)	Birth Rate (per 1,000 pop.)	Death Rate (per 1,000 pop.)	Natural Increase (annual, %)	Population "Doubling Time" in Years (at current rate)	Population Projected to 2000 (millions)	Population Projected to 2020 (millions)	Infant Mortality Rate[a]	Total Fertility Rate[b]	% Population Under Age 15/65+	Life Expectancy at Birth (years)	Urban Population (%)	Population with Access to Safe Water Supply (%, 1983)	Energy Consumption Per Capita, 1986 (gigajoules)/Percent Change Since 1970	Per Capita GNP, 1988 (US$)
Eastern Africa	**199**	**47**	**17**	**3.0**	**23**	**273**	**481**	**116**	**6.7**	**47/3**	**50**	**18**	**—**	**—/—**	**230**
Burundi	5.6	48	15	3.2	22	7.7	13.7	114	7.0	45/3	51	5	26	(z)/−50	230
Comoros	0.5	47	13	3.4	20	0.7	1.3	94	7.1	48/3	55	23	—	2/—	440
Djibouti	0.4	47	18	3.0	23	0.6	1.0	122	6.6	46/3	47	78	—	8/30	—
Ethiopia	51.7	44	24	2.0	34	70.8	126.0	154	6.2	46/4	41	11	—	1/33	120
Kenya	24.6	46	7	3.8	18	35.1	60.5	62	6.7	50/2	63	20	28	2/−32	360
Madagascar	12.0	46	14	3.2	22	16.6	29.6	120	6.6	45/3	54	22	23	1/−45	180
Malawi	9.2	52	18	3.4	20	11.8	22.0	130	7.7	48/3	49	14	51	1/−10	160
Mauritius	1.1	19	7	1.3	54	1.2	1.3	25.2	2.0	30/5	68	41	95	12/27	1,810
Mozambique	15.7	45	19	2.7	26	20.4	31.9	141	6.4	44/3	47	19	13	1/−76	100
Reunion	0.6	24	6	1.8	39	0.7	0.8	14	2.4	32/5	71	98	—	—/—	—/—
Rwanda	7.3	51	17	3.4	20	10.4	19.7	122	8.3	49/2	49	6	60	1/280	310
Seychelles	0.1	25	8	1.7	41	0.1	0.1	17.0	2.7	36/6	70	52	—	—/—	3,800
Somalia	8.4	51	20	3.1	23	10.4	18.7	132	6.6	47/3	45	33	31	3/175	170
Tanzania	26.0	51	14	3.7	19	36.5	68.8	106	7.1	49/2	53	19	46	1/−24	160
Uganda	18.0	52	17	3.6	20	25.1	42.2	107	7.4	49/2	49	9	16	1/−64	280
Zambia	8.1	51	14	3.8	18	11.6	22.0	80	7.2	49/2	53	45	—	8/−30	290
Zimbabwe	9.7	42	10	3.2	22	13.1	20.9	72	5.8	45/3	58	25	52	19/−22	660
Middle Africa	**68**	**45**	**16**	**3.0**	**23**	**91**	**156**	**118**	**6.1**	**45/3**	**50**	**37**	**22**	**—/—**	**420**
Angola	8.5	47	20	2.7	26	11.1	18.5	137	6.4	45/3	45	25	28	3/−35	—
Cameroon	11.1	42	16	2.6	26	14.5	23.5	125	5.8	44/3	50	42	26	9/370	1,010
Central African Republic	2.9	44	19	2.5	27	3.7	5.9	143	5.6	42/3	46	35	16	1/−26	390
Chad	5.0	44	20	2.5	28	6.2	9.4	132	5.9	43/4	46	27	26	1/12	160
Congo	2.2	44	14	3.0	23	3.0	5.0	113	6.0	45/3	53	40	29	13/210	930
Equatorial Guinea	0.4	43	17	2.6	27	0.5	0.8	120	5.5	43/4	50	60	—	2/−42	350
Gabon	1.2	39	16	2.2	31	1.6	2.6	103	5.0	33/6	52	41	—	33/139	2,970
Sao Tome and Principe	0.1	36	9	2.7	25	0.2	0.3	61.7	5.4	42/5	65	38	—	—/—	280
Zaire	36.6	47	14	3.3	21	50.3	90.0	108	6.2	46/3	53	40	19	2/2	170
Southern Africa	**45**	**36**	**9**	**2.7**	**26**	**59**	**95**	**61**	**4.7**	**40/4**	**62**	**53**	**—**	**—/—**	**2,150**
Botswana	1.2	40	11	2.9	24	1.6	2.2	64	5.3	46/3	59	22	65	—/—	1,050
Lesotho	1.8	41	12	2.8	24	2.4	3.9	100	5.8	43/4	56	17	14	—/—	410
Namibia	1.5	44	12	3.2	22	2.1	3.9	106	6.1	45/3	56	51	—	—/—	—
South Africa	39.6	35	8	2.7	26	51.5	83.5	55	4.5	40/4	63	56	—	81/11	2,290
Swaziland	0.8	46	15	3.1	22	1.1	1.8	130	6.2	47/2	50	26	—	—/—	790
Asia	**3,116**	**27**	**9**	**1.9**	**37**	**3,718**	**4,805**	**74**	**3.5**	**34/5**	**63**	**29**	**—**	**20/55**	**1,430**
Asia (Excl. China)	**1,997**	**31**	**10**	**2.1**	**33**	**2,438**	**3,308**	**88**	**4.1**	**38/4**	**60**	**33**	**50**	**—/—**	**2,140**
Western Asia	**132**	**36**	**8**	**2.8**	**24**	**175**	**283**	**71**	**5.2**	**41/4**	**64**	**58**	**70**	**—/—**	**2,860**
Bahrain	0.5	27	3	2.3	30	0.7	1.0	24	4.2	33/2	67	81	—	430/455	6,610
Cyprus	0.7	19	9	1.0	67	0.8	0.9	11	2.4	26/10	76	62	—	62/73	6,260
Gaza	0.6	50	7	4.3	16	0.8	1.4	55	7.0	50/3	65	—	—	—/—	—
Iraq	18.8	46	7	3.9	18	27.2	50.9	67	7.3	45/3	67	68	73	20/12	—
Israel	4.6	23	7	1.6	43	5.4	7.0	10.0	3.1	32/9	75	89	—	74/11	8,650
Jordan	4.1	41	6	3.5	20	5.7	9.7	54	5.9	46/2	69	64	89	30/284	1,500
Kuwait	2.1	27	2	2.5	28	2.9	4.6	15.6	3.7	37/1	73	94	89	225/59	13,680
Lebanon	3.3	28	7	2.1	33	4.1	5.8	49	3.7	40/5	68	80	92	30/50	—
Oman	1.5	46	13	3.3	21	2.1	3.8	100	7.2	46/3	55	9	—	245/5,240	5,070
Qatar	0.5	25	2	2.3	30	0.7	1.1	25	4.5	29/1	69	88	—	565/46	11,610
Saudi Arabia	15.0	42	8	3.4	20	22.0	42.2	71	7.2	45/3	63	73	93	122/800	6,170
Syria	12.6	45	7	3.8	18	18.0	32.6	48	6.8	49/4	65	50	71	35/216	1,670
Turkey	56.7	29	8	2.1	32	69.0	93.8	74	3.6	36/4	64	53	63	29/108	1,280
United Arab Emirates	1.6	23	4	1.9	36	2.0	2.6	26	4.8	31/2	71	81	93	205/185	15,720
Yemen, North	7.2	52	17	3.5	20	10.0	19.5	129	7.6	50/3	49	20	31	5/1,021	650
Yemen, South	2.6	48	14	3.4	20	3.6	6.4	110	7.0	48/3	52	40	50	20/171	430

Region or Country	Population Estimate mid-1990 (millions)	Birth Rate (per 1,000 pop.)	Death Rate (per 1,000 pop.)	Natural Increase (annual, %)	Population "Doubling Time" in Years (at current rate)	Population Projected to 2000 (millions)	Population Projected to 2020 (millions)	Infant Mortality Rate[a]	Total Fertility Rate[b]	% Population Under Age 15/65+	Life Expectancy at Birth (years)	Urban Population (%)	Population with Access to Safe Water Supply (%, 1983)	Energy Consumption Per Capita, 1986 (gigajoules)/ Percent Change Since 1970	Per Capita GNP, 1988 (US$)
Southern Asia	**1,192**	**35**	**12**	**2.3**	**30**	**1,485**	**2,065**	**101**	**4.6**	**40/3**	**57**	**26**	**50**	**–/–**	**310**
Afghanistan	15.9	48	22	2.6	27	25.4	43.0	182	7.1	46/4	41	18	10	3/110	—
Bangladesh	114.8	39	14	2.5	28	146.6	201.4	120	4.9	43/3	54	13	42	2/–	170
Bhutan	1.6	38	17	2.1	32	1.9	2.7	128	5.5	38/4	48	5	17	(z)/–	150
India	853.4	32	11	2.1	33	1,042.5	1,374.5	95	4.2	39/3	57	26	54	8/96	330
Iran	55.6	45	10	3.6	20	75.7	130.2	91	6.3	45/3	63	54	—	34/21	—
Maldives	0.2	46	9	3.7	19	0.3	0.6	76	6.6	45/2	61	26	—	–/–	410
Nepal	19.1	42	17	2.5	28	24.3	37.5	112	6.1	42/3	52	7	15	1/–	170
Pakistan	114.6	44	13	3.0	23	149.1	251.3	110	6.7	44/4	56	28	39	7/42	350
Sri Lanka	17.2	21	6	1.5	47	19.4	24.0	22.5	2.3	35/4	70	22	36	3/–16	420
Southeast Asia	**455**	**29**	**8**	**2.1**	**34**	**547**	**721**	**70**	**3.6**	**38/4**	**61**	**27**	**43**	**–/–**	**—**
Brunei	0.3	29	3	2.5	27	0.3	0.4	11	3.6	37/3	71	59	—	–/–	14,120
Cambodia	7.0	39	16	2.2	31	8.5	11.9	128	4.5	36/3	49	11	—	1/–41	—
East Timor	0.7	44	22	2.2	31	0.9	1.1	166	5.4	35/3	43	12	—	–/–	—
Indonesia	189.4	27	9	1.8	38	223.8	287.3	89	3.3	38/3	59	26	33	8/126	430
Laos	4.0	41	16	2.5	28	5.0	6.9	110	5.5	43/3	47	16	21	1/–72	180
Malaysia	17.9	30	5	2.5	28	21.5	27.3	30	3.6	38/4	68	35	80	31/88	1,870
Myanmar (Burma)	41.3	33	13	2.0	34	49.8	67.7	97	4.2	37/4	55	24	25	3/75	—
Philippines	66.1	33	7	2.6	27	82.7	117.5	48	4.3	39/3	64	42	54	7/–10	630
Singapore	2.7	20	5	1.5	47	3.0	3.4	6.9	2.0	23/6	73	100	100	122/230	9,100
Thailand	55.7	22	7	1.5	45	63.7	78.1	39	2.6	35/4	66	18	65	13/141	1,000
Viet Nam	70.2	33	8	2.5	28	88.3	119.5	50	4.2	42/4	66	20	—	4/–59	—
East Asia	**1,336**	**20**	**6**	**1.3**	**52**	**1,510**	**1,735**	**35**	**2.2**	**26/6**	**69**	**29**	**—**	**–/–**	**2,460**
China	1,119.9	21	7	1.4	49	1,280.0	1,496.3	37	2.3	27/6	68	21	—	21/105	330
Hong Kong	5.8	13	5	0.8	82	6.3	6.9	7.4	1.4	22/8	77	93	99	–/–	9,230
Japan	123.6	10	6	0.4	175	127.5	124.2	4.8	1.6	20/11	79	77	—	106/13	21,040
Korea, North	21.3	26	5	2.1	32	24.9	29.8	33	2.5	34/4	70	64	—	81/34	—
Korea, South	42.8	16	6	1.0	72	46.0	48.9	30	1.6	27/5	68	70	—	48/150	3,530
Macao	0.5	18	3	1.5	47	0.5	0.5	9	—	23/8	76	97	—	–/–	—
Mongolia	2.2	36	8	2.8	25	2.8	4.3	50	4.8	41/4	65	52	—	48/88	—
Taiwan	20.2	17	5	1.2	57	22.1	24.1	17	1.8	28/6	74	71	—	–/–	—
North America	**278**	**16**	**9**	**0.7**	**93**	**298**	**328**	**9**	**2.0**	**22/12**	**75**	**74**	**—**	**–/–**	**19,490**
Canada	26.6	14	7	0.7	96	29.3	33.1	7.3	1.7	21/11	77	77	—	285/10	16,760
United States	251.4	16	9	0.8	92	268.3	294.4	9.7	2.0	22/12	75	74	—	280/–12	19,780
Latin America	**447**	**28**	**7**	**2.1**	**33**	**535**	**705**	**54**	**3.5**	**38/5**	**67**	**69**	**70**	**–/–**	**1,930**
Central America	**118**	**32**	**6**	**2.5**	**27**	**145**	**196**	**51**	**4.1**	**42/4**	**67**	**61**	**70**	**–/–**	**1,640**
Belize	0.2	37	6	3.1	22	0.3	0.5	36	5.0	44/4	69	50	—	–/–	1,460
Costa Rica	3.0	29	4	2.5	28	3.8	5.3	17.4	3.3	36/5	76	45	93	14/27	1,760
El Salvador	5.3	35	8	2.7	26	6.5	8.8	54	4.4	45/4	62	43	51	5/–10	950
Guatemala	9.2	40	9	3.1	23	11.8	17.6	59	5.6	46/3	63	40	51	6/–11	880
Honduras	5.1	39	8	3.1	23	6.8	10.6	63	5.3	47/3	63	42	69	6/–12	850
Mexico	88.6	30	6	2.4	29	107.2	142.1	50	3.8	42/4	68	66	74	47/55	1,820
Nicaragua	3.9	42	9	3.3	21	5.1	7.7	69	5.5	47/4	62	57	53	9/–3	830
Panama	2.4	27	5	2.2	32	2.9	3.7	23	3.1	36/5	72	52	62	19/–15	2,240

Region or Country	Population Estimate mid-1990 (millions)	Birth Rate (per 1,000 pop.)	Death Rate (per 1,000 pop.)	Natural Increase (annual, %)	Population "Doubling Time" in Years (at current rate)	Population Projected to 2000 (millions)	Population Projected to 2020 (millions)	Infant Mortality Rate[a]	Total Fertility Rate[b]	% Population Under Age 15/65+	Life Expectancy at Birth (years)	Urban Population (%)	Population with Access to Safe Water Supply (%, 1983)	Energy Consumption Per Capita, 1986 (gigajoules)/ Percent Change Since 1970	Per Capita GNP, 1988 (US$)
Caribbean	**34**	**25**	**8**	**1.7**	**40**	**38**	**48**	**57**	**3.1**	**33/6**	**68**	**55**	**—**	**—/—**	**—**
Antigua and Barbuda	0.1	15	5	1.0	71	0.1	0.1	24	1.7	27/6	71	58	—	—/—	2,800
Bahamas	0.2	20	5	1.5	46	0.3	0.3	21.7	2.3	34/5	71	75	—	—/—	10,570
Barbados	0.3	16	9	0.7	100	0.3	0.3	16.2	1.8	28/11	75	32	—	39/55	5,990
Cuba	10.6	18	7	1.2	60	11.6	12.8	11.9	1.9	25/8	75	72	—	42/44	—
Dominica	0.1	26	5	2.1	33	0.1	0.1	14	2.7	34/7	75	—	—	—/—	1,650
Dominican Republic	7.2	31	7	2.5	28	8.6	11.0	65	3.8	39/3	66	52	60	14/45	680
Grenada	0.1	37	7	3.0	23	0.1	0.1	30	4.9	39/7	71	—	—	—/—	1,370
Guadeloupe	0.3	20	7	1.4	51	0.4	0.4	18.0	2.2	31/7	73	90	—	—/—	—
Haiti	6.5	35	14	2.2	32	7.8	11.4	122	5.1	40/5	53	26	33	1/−8	360
Jamaica	2.4	22	5	1.7	41	2.7	3.5	16	2.4	37/6	76	49	86	32/−8	1,080
Martinique	0.3	19	6	1.3	54	0.4	0.4	11	2.1	30/7	74	82	—	—/—	—
Netherlands Antilles	0.2	18	5	1.3	55	0.2	0.2	9	2.0	30/7	73	53	—	—/—	—
Puerto Rico	3.3	20	7	1.2	57	3.4	4.0	14.2	2.2	30/9	72	67	—	—/—	5,540
St. Kitts-Nevis	0.04	23	11	1.3	55	0.04	0.1	39.7	2.6	34/9	68	45	—	—/—	2,770
Saint Lucia	0.2	28	6	2.2	31	0.2	0.3	21.5	3.8	44/6	71	46	—	—/—	1,540
St. Vincent and the Grenadines	0.1	25	6	1.9	37	0.1	0.2	24.7	2.8	44/6	72	21	—	—/—	1,100
Trinidad and Tobago	1.3	27	7	2.0	34	1.7	2.4	13.7	3.1	34/6	70	64	99	223/71	3,350
Tropical South America	**247**	**39**	**8**	**2.1**	**33**	**298**	**395**	**59**	**3.5**	**37/4**	**65**	**71**	**72**	**—/—**	**2,020**
Bolivia	7.3	38	12	2.6	27	9.3	13.4	110	5.1	43/4	53	49	43	10/47	570
Brazil	150.4	27	8	1.9	36	179.5	233.8	63	3.3	36/4	65	74	76	22/72	2,280
Colombia	31.8	28	7	2.0	34	38.0	49.3	46	3.4	36/4	66	68	81	23/32	1,240
Ecuador	10.7	33	8	2.5	27	13.6	19.5	63	4.3	42/4	65	54	59	19/130	1,080
Guyana	0.8	25	5	1.9	36	0.8	1.1	30	2.8	37/4	67	33	80	20/−36	410
Paraguay	4.3	35	7	2.8	25	5.5	8.4	42	4.6	41/4	67	43	25	7/55	1,180
Peru	21.9	32	8	2.4	29	26.4	35.1	76	4.1	38/4	65	69	52	15/−13	1,440
Suriname	0.4	27	6	2.0	34	0.5	0.6	40	3.0	34/4	68	66	—	39/−41	2,450
Venezuela	19.6	28	5	2.3	30	24.1	33.3	33	3.5	39/4	70	83	—	99/52	3,170
Temperate South America	**49**	**21**	**8**	**1.4**	**51**	**55**	**66**	**28**	**2.8**	**33/8**	**71**	**85**	**70**	**—/—**	**2,320**
Argentina	32.3	21	9	1.3	54	36.0	43.5	32	3.0	34/9	71	85	63	51/10	2,640
Chile	13.2	22	6	1.7	41	15.3	19.0	18.5	2.5	31/6	71	84	85	27/−20	1,510
Uruguay	3.0	18	10	0.8	87	3.2	3.6	22.3	2.4	26/12	71	87	79	18/−29	2,470
Europe	**501**	**13**	**10**	**0.3**	**266**	**515**	**516**	**12**	**1.7**	**20/13**	**74**	**75**	**—**	**130/18**	**12,170**
Northern Europe	**84**	**14**	**11**	**0.2**	**286**	**86**	**88**	**9**	**1.8**	**19/15**	**75**	**85**	**—**	**—/—**	**14,300**
Denmark	5.1	12	12	0.0	(—)	5.2	4.9	7.8	1.6	18/15	75	84	—	155/−2	18,470
Finland	5.0	13	10	0.3	239	5.0	4.9	5.9	1.7	19/13	75	62	—	165/39	18,610
Iceland	0.3	19	7	1.1	61	0.3	0.3	6.2	2.3	25/10	78	89	—	151/23	20,160
Ireland	3.5	15	9	0.6	108	3.5	3.4	9.7	2.2	28/11	74	56	—	104/38	7,480
Norway	4.2	14	11	0.3	231	4.3	4.3	8.4	1.8	19/16	76	71	—	194/45	20,020
Sweden	8.5	14	11	0.2	311	8.8	9.0	5.8	2.0	18/18	77	83	—	143/−15	19,150
United Kingdom	57.4	14	12	0.2	301	59.1	60.8	9.5	1.8	19/15	75	90	—	157/11	12,800
Western Europe	**159**	**12**	**10**	**0.2**	**326**	**164**	**160**	**8**	**1.6**	**18/14**	**76**	**83**	**—**	**—/—**	**17,270**
Austria	7.6	12	11	0.1	1,155	7.7	7.6	8.1	1.4	18/15	75	55	—	118/23	15,560
Belgium	9.9	12	11	0.2	462	9.9	9.4	9.2	1.6	18/14	74	95	—	163/−1	14,550
France	56.4	14	9	0.4	157	57.9	58.7	7.5	1.8	20/14	77	73	—	114/1	16,080
Germany, West	63.2	11	11	−0.0	(—)	65.7	62.3	7.5	1.4	15/15	76	94	—	166/10	18,530
Luxembourg	0.4	12	10	0.2	346	0.4	0.4	8.7	1.4	17/13	75	78	—	333/−33	22,600
Netherlands	14.9	13	8	0.4	165	15.3	15.0	7.6	1.5	18/13	77	89	—	211/57	14,530
Switzerland	6.7	12	9	0.3	231	6.8	6.9	6.8	1.6	17/15	77	61	—	117/23	27,260

Region or Country	Population Estimate mid-1990 (millions)	Birth Rate (per 1,000 pop.)	Death Rate (per 1,000 pop.)	Natural Increase (annual, %)	Population "Doubling Time" in Years (at current rate)	Population Projected to 2000 (millions)	Population Projected to 2020 (millions)	Infant Mortality Rate[a]	Total Fertility Rate[b]	% Population Under Age 15/65+	Life Expectancy at Birth (years)	Urban Population (%)	Population with Access to Safe Water Supply (%, 1983)	Energy Consumption Per Capita, 1986 (gigajoules)/Percent Change Since 1970	Per Capita GNP, 1988 (US$)
Eastern Europe	**113**	**14**	**11**	**0.3**	**215**	**115**	**119**	**16**	**2.0**	**24/11**	**71**	**64**	**—**	**—/—**	**—**
Bulgaria	8.9	13	12	0.1	630	9.0	9.1	13.5	2.0	21/12	72	67	—	169/53	—
Czechoslovakia	15.7	14	11	0.2	289	16.3	17.0	11.9	2.1	24/11	71	75	—	183/16	—
Germany, East	16.3	13	13	0.0	6,930	15.5	15.0	8.1	1.7	19/13	73	77	—	233/29	—
Hungary	10.6	12	13	−0.2	(—)	10.6	10.4	15.8	1.8	21/13	70	60	—	109/32	2,460
Poland	37.8	16	10	0.6	122	38.9	41.7	16.2	2.1	26/10	71	61	—	138/32	1,850
Romania	23.3	16	11	0.5	141	24.5	26.0	25.6	2.3	25/9	70	54	—	131/48	—
Southern Europe	**145**	**12**	**9**	**0.3**	**250**	**150**	**149**	**14**	**1.5**	**21/12**	**75**	**68**	**—**	**—/—**	**8,650**
Albania	3.3	26	6	2.0	34	3.8	4.7	28	3.2	35/5	71	35	—	38/114	—
Greece	10.1	11	9	0.2	408	10.2	9.9	11.0	1.5	20/13	77	58	—	72/122	4,790
Italy	57.7	10	9	0.1	1,155	58.6	56.1	9.5	1.3	18/14	75	72	—	94/21	13,320
Malta	0.4	16	8	0.8	87	0.4	0.4	8.0	2.0	24/10	75	85	—	47/69	5,050
Portugal	10.4	12	10	0.2	301	10.7	10.7	14.9	1.6	22/13	74	30	—	39/76	3,670
Spain	39.4	11	8	0.3	247	40.7	40.7	9.0	1.5	22/13	77	91	—	65/50	7,740
Yugoslavia	23.8	15	9	0.6	114	25.1	26.3	24.5	2.0	23/9	71	46	—	72/74	2,680
USSR	**291**	**19**	**10**	**0.9**	**80**	**312**	**355**	**29**	**2.5**	**25/9**	**69**	**66**	**—**	**187/54**	**—**
Oceania	**27**	**20**	**8**	**1.2**	**57**	**31**	**38**	**26**	**2.6**	**27/9**	**72**	**70**	**—**	**145/35**	**9,550**
Australia	17.1	15	7	0.8	90	19.1	22.9	8.7	1.8	22/11	76	86	—	196/40	12,390
Fiji	0.8	27	6	2.2	32	0.9	1.2	21	3.3	38/3	63	39	—	11/−18	1,540
French Polynesia	0.2	31	6	2.5	27	0.2	0.3	23	3.9	37/4	68	62	—	—/—	—
New Caledonia	0.2	25	6	1.9	36	0.2	0.2	39	3.0	33/4	67	58	—	—/—	—
New Zealand	3.3	17	8	0.8	82	3.5	3.6	10.0	2.0	24/11	74	84	—	115/41	9,620
Pacific Islands[c]	0.2	36	5	3.1	22	0.2	0.3	29	5.0	47/4	71	29	—	—/—	—
Papua-New Guinea	4.0	39	12	2.7	26	5.1	7.9	59	5.7	41/2	54	13	16	9/93	770
Solomon Islands	0.3	41	5	3.5	20	0.5	0.8	40	6.3	47/3	61	9	—	7/10	430
Vanuatu	0.2	37	5	3.2	22	0.2	0.3	36	5.5	45/3	69	18	—	—/—	820
Western Samoa	0.2	34	7	2.8	25	0.2	0.3	48	4.6	40/4	66	21	—	—/—	580

[a]Infant deaths per 1000 live births
[b]Average number of children born to a woman during her lifetime
[c]Comprising the Federated States of Micronesia, Palau, and the Marshall and N. Mariana Islands
A dash (—) indicates data unavailable or inapplicable.
A (z) signifies an amount that rounds to zero.

Table modified from the *1990 World Population Data Sheet* of the Population Reference Bureau. Data for safe water supply were taken from the 1987 edition of the *World Population Data Sheet*. Data on energy consumption are from United Nation sources.

G L O S S A R Y

A

absolute location (*syn:* mathematical location)
The exact position of an object or place stated in spatial coordinates of a grid system designed for locational purposes. In geography, the reference system is the global grid of parallels of latitude north or south of the equator and of meridians of longitude east or west of a prime meridian.

accessibility
The relative ease with which a destination may be reached from other locations; the relative opportunity for spatial interaction. May be measured in geometric, social, or economic terms.

acculturation
Cultural modification or change resulting from one culture group or individual adopting traits of a more advanced or dominant society; cultural development through "borrowing."

acid rain
Precipitation that is unusually acidic; created when oxides of sulfur and nitrogen change chemically as they dissolve in water vapor in the atmosphere and return to earth as acidic rain, snow, fog, or dry particles.

activity space
The area within which people move freely on their rounds of regular activity.

adaptation
A presumed modification of heritable traits through response to environmental stimuli.

agglomeration
The spatial grouping of people or activities for mutual benefit.

agglomeration economies (*syn:* external economies)
The savings to an individual enterprise that result from spatial association with other similar economic activities.

agriculture
The science and practice of farming, including the cultivation of the soil and the rearing of livestock.

air mass
A large body of air with little horizontal variation in temperature, pressure, and humidity.

air pressure
The weight of the atmosphere as measured at a point on the earth's surface.

alluvial fan
A fan-shaped accumulation of alluvium deposited by a stream at the base of a hill or mountain.

alluvium
Sediment carried by a stream and deposited in a floodplain or delta.

anaerobic digestion
The process by which organic waste is decomposed in an oxygen-free environment to produce methane gas (biogas).

anecumene
See nonecumene.

animism
A belief that natural objects may be the abode of dead people, spirits, or gods who occasionally give the objects the appearance of life.

antecedent boundary
A boundary line established before the area in question is well populated.

aquaculture
The breeding of fish in freshwater ponds, lakes, and canals or in fenced-off coastal bays and estuaries; fish farming.

aquifer
Underground porous and permeable rock that is capable of holding groundwater, especially rock that supplies economically significant quantities of water to wells and springs.

arable land
Land that is or can be cultivated.

Arctic haze
Air pollution resulting from the transport by air currents of combustion-based pollutants to the area north of the Arctic Circle.

area analysis tradition
One of the four traditions of geography, that of regional geography.

arithmetic density
See crude density.

arroyo
A steep-sided, flat-bottomed gully, usually dry, carved out of desert land by rapidly flowing water.

artifacts
The material manifestations of culture, including tools, housing, systems of land use, clothing and the like. Elements in the technological subsystem of culture.

artificial boundary
See geometric boundary.

assimilation
The social process of merging into a composite culture, losing separate ethnic or social identity, and becoming culturally homogenized.

asthenosphere
A partially molten, plastic layer above the core and lower mantle of the earth.

atmosphere
The gaseous mass surrounding the earth.

atoll
A near-circular low coral reef formed in shallow water enclosing a central lagoon; most common in the central and western Pacific Ocean.

B

barchan
A crescent-shaped sand dune; the horns of the crescent point downwind.

basic sector
Those products of an urban unit that are exported outside the city itself, earning income for the community.

bench mark
A surveyor's mark indicating the position and elevation of some stationary object; used as a reference point in surveying and mapping.

biocide
A chemical used to kill plant and animal pests and disease organisms.

biological magnification
The accumulation of a chemical in the fatty tissue of an organism and its concentration at progressively higher levels in the food chain.

biomass
Living matter, plant and animal, in any form.

biomass fuels
Combustible and/or fermentable material of plant or animal origin, such as wood or corn cobs, that can be used as a source of energy.

biome
The total assemblage of living organisms in a single major ecological region.

biosphere (*syn:* ecosphere)
The thin film of air, water, and earth within which we live, including the atmosphere, surrounding and subsurface waters, and the upper reaches of the earth's crust.

birth rate
The ratio of the number of live births during one year to the total population, usually at the midpoint of the same year, expressed as the number of births per year per 1000 population.

boundary
A line separating one political unit from another.

boundary definition
A general agreement between two states about the allocation of territory between them.

boundary delimitation
The plotting of a boundary line on maps or aerial photographs.

boundary demarcation
The actual marking of a boundary line on the ground; the final state in boundary development.

butte
A small, flat-topped, isolated hill with steep sides, common in dry climate regions.

C

carcinogen
A substance that produces or incites cancerous growth.

carrying capacity
The numbers of any population that can be adequately supported by the available resources upon which that population subsists; for humans, the numbers supportable by the known and utilized resources—usually agricultural—of an area.

cartogram
A map that has been simplified to present a single idea in a diagrammatic way; the base is not normally true to scale.

caste
One of the hereditary social classes in Hinduism that determines one's occupation and position in society.

central business district (CBD)
The center or "downtown" of an urban unit, where retail stores, offices, and cultural activities are concentrated and where land values are high.

central city
That part of the urban area contained within the boundaries of the main city around which suburbs have developed.

central place
A nodal point for the distribution of goods and services to a surrounding hinterland population.

central place theory
A deductive theory formulated by Walter Christaller (1893–1969) to explain the size and distribution of settlements through reference to competitive supply of goods and services to dispersed rural populations.

centrifugal force
In political geography, a force that destabilizes and weakens a state.

centripetal force
In political geography, a factor that promotes unity and national identity.

CFCs
See chlorofluorocarbons.

channelization
The modification of a stream channel; specifically, the straightening of meanders or dredging of the stream channel to deepen it.

channelized migration
The tendency for migration to flow between areas that are socially and economically allied by past migration patterns, by economic trade considerations, or by some other affinity.

chemical weathering
The decomposition of earth materials due to chemical reactions that include oxidation, hydration, and carbonation.

chlorofluorocarbons (CFCs)
A family of synthetic chemicals that have significant commercial applications but whose emissions are contributing to the depletion of the ozone layer.

circumpolar vortex
High-altitude winds circling the poles from west to east.

city
A multifunctional nucleated settlement with a central business district and both residential and nonresidential land uses.

climate
A summary of weather conditions in a place or region over a period of time.

coal gasification
A process by which crushed coal is burned in the presence of steam or oxygen to produce a synthetic gas.

coal liquefaction
A process whereby coal is heated to produce a variety of liquid products that can be used as fuels.

coal slurry
A mixture of finely ground coal and water that is moved by pipeline.

cogeneration
The simultaneous use of a single fuel for the generation of electricity and low-grade central heat.

cognition
The process by which an individual gives mental meaning to information.

cohort
A population group unified by a specific common characteristic, such as age, which is treated as a statistical unit during members' lifetimes.

collective farm
In the Soviet planned economy, the term refers to the cooperative operation of an agricultural enterprise under state control of production and market, but without full status or support as a state enterprise.

commercial economy
The production of goods and services for exchange in competitive markets where price and availability are determined by supply and demand forces.

commercial energy
Commercially traded fuels, such as coal, oil, or natural gas, excluding wood, vegetable or animal wastes, or other biomass.

Common Market
See European Economic Community.

compact state
A state whose territory is nearly circular.

comparative advantage
A region's profit potential for a productive activity compared to alternate areas of production of the same good or to alternate uses of the region's resources.

concentric zone model
The idea that there is a series of circular belts of land use around the central business district, each belt containing distinct functions.

conformal map projection
One on which the shapes of small areas are accurately portrayed.

conic map projection
One based on the projection of the grid system onto a cone.

connectivity
The directness of routes linking pairs of places; all of the tangible and intangible means of connection and communication between places.

consequent boundary
A boundary line that coincides with some cultural divide, such as religion or language.

conservation
The wise use or preservation of natural resources so as to maintain supplies and qualities at levels sufficient to meet present and future needs.

contagious diffusion
The spread of a concept, practice, or article from one area to others through contact and/or the exchange of information.

continental drift
The hypothesis that an original single land mass (Pangaea) broke apart and that the continents have moved very slowly over the asthenosphere to their present locations.

contour interval
The vertical distance separating two adjacent contour lines.

contour line
A map line along which all points are of equal elevation above or below a datum plane, usually mean sea level.

convectional precipitation
Rain produced when heated, moisture-laden air rises and then cools below the dew point.

Convention on the Law of the Sea
See Law of the Sea Convention.

coral reef
A rocklike landform in shallow tropical water composed chiefly of compacted coral and other organic material.

core
In urban geography, that part of the central business district characterized by intensive land development.

core area
The nucleus of a state, containing its most developed area, greatest wealth, densest populations, and clearest national identity.

Coriolis effect
A fictitious force used to describe motion relative to a rotating earth; specifically, the force that tends to deflect a moving object or fluid to the right (clockwise) in the Northern Hemisphere and to the left (counterclockwise) in the Southern Hemisphere.

country
See state.

creole
A language developed from a pidgin to become the native tongue of a society.

critical distance
The distance beyond which cost, effort, and/or means play an overriding role in the willingness of people to travel.

crude birth rate (CBR)
See birth rate.

crude death rate (CDR)
See death rate.

crude density (*syn:* arithmetic density)
The number of people per unit area of land.

crude oil
A mixture of hydrocarbons that exists in a liquid state in underground reservoirs; petroleum as it occurs naturally, as it comes from an oil well, or after extraneous substances have been removed.

cultural convergence
The tendency for cultures to become more alike as they increasingly share technology and organizational structures in a modern world united by improved transportation and communication.

cultural divergence
The likelihood or tendency for isolated cultures to become increasingly dissimilar with the passage of time.

cultural ecology
The study of the interactions between societies and the natural environments they occupy.

cultural integration
The observation that all aspects of a culture are interconnected; no part can be altered without impact upon other culture traits.

cultural lag
The retention of established culture traits despite changing circumstances rendering them inappropriate.

cultural landscape
The natural landscape as modified by human activities and bearing the imprint of a culture group or society; the built environment.

culture
The totality of learned behaviors and attitudes transmitted within a society to succeeding generations by imitation, instruction, and example.

culture complex
An integrated assemblage of culture traits descriptive of one aspect of a society's behavior or activity.

culture–environment tradition
One of the four traditions of geography; in this text, identified with population, cultural, political, and behavioral geography.

culture hearth
A nuclear area within which an advanced and distinctive set of culture traits develops and from which there is diffusion of distinctive technologies and ways of life.

culture realm
A collective of culture regions sharing related culture systems; a major world area having sufficient distinctiveness to be perceived as set apart from other realms in its cultural characteristics and complexes.

culture region
A formal or functional region within which common cultural characteristics prevail. It may be based on single culture traits, on culture complexes, or on political, social, or economic integration.

culture system
A generalization suggesting shared, identifying traits uniting two or more culture complexes.

culture trait
A single distinguishing feature of regular occurrence within a culture, such as the use of chopsticks or the observance of a particular caste system. A single element of learned behavior.

cyclone
A type of atmospheric disturbance in which masses of air circulate rapidly about a region of low atmospheric pressure.

cyclonic precipitation
See frontal precipitation.

cylindrical projection
Any of several map projections based on the projection of the globe grid onto a cylinder.

D

database
See geographic database.

DDT
A chlorinated hydrocarbon that is among the most persistent of the biocides in general use.

death rate
A mortality index usually calculated as the number of deaths per year per 1000 population.

decomposers
Microorganisms and bacteria that feed on dead organisms, causing their chemical disintegration.

deforestation
The clearing of land through total removal of forest cover.

delta
A triangular-shaped deposit of mud, silt, or gravel created by a stream where it flows into a body of standing water.

demographic equation
A mathematical expression that summarizes the contribution of different demographic processes to the population change of a given area during a specified time period.

demographic momentum
See population momentum.

demographic transition
A model of the effect of economic development on population growth. A first stage involves both high birth and death rates; the second phase displays high birth rates and falling mortality rates and population increases. Phase three shows reduction in population growth as birth rates decline to the level of death rates. The final, fourth, stage implies again a population stable in size but larger in numbers than at the start of the transition cycle.

demography
The scientific study of population, with particular emphasis upon quantitative aspects.

density of population
See population density.

dependency ratio
The number of dependents, old or young, that each 100 persons in the productive years must support.

deposition
The process by which silt, sand, and rock particles accumulate and create landforms, such as stream deltas and talus slopes.

desertification
Extension of desertlike landscapes as a result of climatic change or human activities, such as overgrazing or deforestation, usually in semiarid regions.

developable surface
A geometric surface, such as a cylinder or cone, that may be spread out flat without tearing or stretching.

dew point
The temperature at which air becomes saturated with water vapor.

dialect
A regional or socioeconomic variation of a more widely spoken language.

diastrophism
The earth force that folds, faults, twists, and compresses rock.

dibble
Any small hand tool or stick to make a hole for planting.

diffusion
The spread or movement of a concept, practice, article, or population from a point of origin to other areas.

distance decay
The exponential decline of an activity or function with increasing distance from its point of origin.

domestication
The successful transformation of plant or animal species from a wild state to a condition of dependency upon human management usually with distinct physical change from wild forebears.

doubling time
The time period required for any beginning total, experiencing a compounding growth, to double in size.

dune
A wavelike desert landform created by wind-blown sand.

E

earthquake
Movement of the earth along a geologic fault or at some other point of weakness at or near the earth's surface.

earth-science tradition
One of the four traditions of geography, identified with physical geography in general.

ecology
The scientific study of how living creatures affect each other and what determines their distribution and abundance.

economic base
The mix of manufacturing and service activities performed by the labor force of a city to satisfy demands outside the city and in the process earning income to support the urban population.

ecosphere
See biosphere.

ecosystem
A population of organisms existing together in a particular area, together with the energy, air, water, soil, and chemicals upon which it depends.

ecumene
Permanently inhabited areas of the earth.

EEC
See European Economic Community.

electoral geography
The study of the delineation of voting districts and the spatial patterns of election results.

electromagnetic spectrum
The entire range of radiation, including the shortest as well as the longer wavelengths.

el Niño
The periodic (every 3 to 7 or 8 years) buildup of warm water along the west coast of South America; replacing the cold Humboldt current off the Peruvian coast, el Niño is associated with both a fall in plankton levels (and decreased fish supply) and with short-term, widespread weather modification.

elongated state
A state whose territory is long and narrow.

enclave
A territory that is surrounded by but is not part of a state.

energy
See also kinetic energy, potential energy. The ability to do work.

energy efficiency
The ratio of the output of useful energy from a conversion process to the total energy inputs.

environment
Surroundings; the totality of things that in any way may affect an organism, including both physical and cultural conditions; a region characterized by a certain set of physical conditions.

environmental determinism
The theory that the physical environment, particularly climate, molds human behavior.

environmental perception
The way people observe and interpret, and the ideas they have about, near or distant places.

environmental pollution
See pollution.

equal-area projection
See equivalent map projection.

equator
An imaginary line that encircles the globe halfway between the North and South poles.

equidistant map projection
One on which true distances in all directions can be measured from one or two central points.

equivalent map projection
One on which the areas of regions are represented in correct or constant proportions to earth reality; also called equal-area.

erosion
The wearing away and removal of rock and soil particles from exposed surfaces by agents such as moving water, wind, or ice.

estuarine zone
The relatively narrow area of wetlands along coastlines where salt water and fresh water mix.

estuary
The lower course or mouth of a river where tides cause fresh water and salt water from the sea to mix.

ethnicity
Social status afforded to, usually, a minority group within a national population. Recognition is based primarily upon culture traits, such as religion, distinctive customs, or native or ancestral national origin.

ethnic religion
A religion identified with a particular ethnic group and largely exclusive to it.

ethnocentrism
The belief that one's own ethnic group is superior to all others.

ethnographic boundary
See consequent boundary.

European Economic Community
An economic association established in 1957 of a number of Western European states that promotes free trade among member countries; often called the Common Market.

eutrophication
The increase of nutrients in a body of water. The nutrients stimulate the growth of algae, whose decomposition decreases the dissolved oxygen content of the water.

exclave
A portion of a state that is separated from the main territory and surrounded by another country.

exclusive economic zone (EEZ)
As proposed in the Convention on the Law of the Sea, a zone of exploitation extending 200 nautical miles seaward from a coastal state that has exclusive mineral and fishing rights over it.

expansion diffusion
Spread of ideas, behaviors, or articles from one culture to others through contact and exchange of information; the dispersion leaves the phenomenon intact or intensified in its area of origin. *See also:* relocation diffusion.

extensive agriculture
Crop or livestock system in which land quality or extent is more important than capital or labor inputs in determining output. May have either commercial or subsistence orientation.

extensive commercial agriculture
See extensive agriculture.

extensive subsistence agriculture
See extensive agriculture.

external economies
See agglomeration economies.

extractive industries
Primary activities involving the mining and quarrying of nonrenewable mineral resources.

extrusive rock
Rock solidified from molten material that has issued out onto the earth's surface.

F

false-color image
A remotely sensed image whose colors do not appear natural to the human eye.

fast breeder reactor
A nuclear reactor that uses uranium-235 to release energy from the more abundant uranium-238.

fault
A break or fracture in rock produced by stress or the movement of lithospheric plates.

fault escarpment
A steep slope formed by the vertical movement of the earth along a fault.

filtering
In urban geography, a process whereby individuals of one income group replace residents of a portion of an urban area who are of another income group.

fiord
A glacial trough whose lower end is filled with sea water.

floodplain
A valley area bordering a stream that is subject to inundation by flooding.

folding
The buckling of rock layers under pressure of moving lithospheric plates.

food chain
A sequence of organisms through which energy and materials move within an ecosystem.

footloose
A descriptive term applied to manufacturing activities for which the cost of transporting material or product is not important in determining location of production.

formal region
An earth area throughout which a single feature or limited combination of features is of such uniformity that it can serve as the basis for an areal generalization and of contrast with adjacent areas.

form utility
A value-increasing change in the form—and therefore in the utility—of a raw material or commodity.

forward-thrust capital
A capital city deliberately sited in a state's frontier zone.

fossil fuels
Hydrocarbon compounds of crude oil, natural gas, and coal that are derived from the accumulation of plant and animal remains in ancient sedimentary rocks.

fragmented state
A state whose territory contains isolated parts, separated and discontinuous.

frictional effect
In climatology, the slowing of wind movement due to the frictional drag of the earth's surface.

friction of distance
A measurement indicating the effect of distance upon the extent of interaction between two points. Generally, the greater the distance, the less the interaction or exchange or the greater the cost of achieving the exchange.

front
The line or zone of separation between two air masses of different temperatures and humidities.

frontal precipitation
Rain or snow produced when moist air of one air mass is forced to rise over the edge of another air mass.

frontier
That portion of a country adjacent to its boundaries and fronting another political unit.

frontier zone
A belt lying between two states or between settled and uninhabited or sparsely settled areas.

functional region
A region differentiated by what occurs within it rather than by a homogeneity of physical or cultural phenomena; an earth area recognized as an operational unit based upon defined organizational criteria.

G

gathering industries
Primary activities involving the subsistence or commercial harvesting of renewable natural resources from land or water; hunting, gathering, forestry, and fishing.

genetic drift
A chance modification of gene composition occurring in an isolated population and becoming accentuated through inbreeding.

gentrification
The movement into the inner portions of American cities of middle-class people who replace low-income populations and rehabilitate structures.

geodetic control data
Information specifying the horizontal and vertical positions of a place.

geographic database
In cartography, a digital record of geographic information.

geographic information system
A method of storing and manipulating geographic information in a computer; the three major components of such systems are the digital map data, the hardware used to handle those data, and the associated computer software.

geometric boundary (*syn:* artificial boundary)
A boundary without obvious physical geographic basis; often a section of a parallel of latitude or a meridian of longitude.

geomorphology
The scientific study of landform origins, characteristics, and evolutions and their processes.

geostrategy
The study of the economic, political, and military value of space.

geothermal power
Energy generated when hot water or steam is extracted from reservoirs in the earth's crust and fed to steam turbines at electric generating plants.

gerrymander
To divide an area into voting districts in such a way as to give one political party an unfair advantage in elections, to fragment voting blocks, or to achieve other nondemocratic objectives.

glacial till
Deposits of rocks, silt, and sand left by a glacier after it has receded.

glacial trough
A deep, U-shaped valley or trench formed by glacial erosion.

glacier
A huge mass of slowly moving land ice.

globe properties
Characteristics of the grid system of longitude and latitude on a globe.

gradation
The process responsible for the gradual reduction of the land surface.

grade (of coal)
A classification of coals based on their content of waste materials.

graphic scale
A graduated line included in a map legend by means of which distances on the map may be measured in terms of ground distances.

gravity model
A mathematical prediction of the interaction between two bodies as a function of their size and of the distance separating them.

gravity transfer
The downward movement of material at or near the earth's surface due to the gravitational attraction of the earth's mass.

great circle
A circle formed by the intersection of the surface of a globe with a plane passing through the center of the globe. The equator is a great circle; meridians are one half of a great circle.

greenbelt
A ring of parks, farmland, or undeveloped land around a community.

greenhouse effect
Heating of the earth's surface as shortwave solar energy passes through the atmosphere, which is transparent to it but opaque to reradiated longwave terrestrial energy. Also refers to increasing the opacity of the atmosphere through addition of increased amounts of carbon dioxide, nitrous oxides, methane, and chlorofluorocarbons.

Green Revolution
Term suggesting the great increases in food production, primarily in subtropical areas, accomplished through the introduction of very high-yielding grain crops, particularly wheat and rice.

Greenwich mean time (GMT)
Local time at the prime meridian (zero degrees longitude), which passes through the observatory at Greenwich, England.

grid system
The set of imaginary lines of latitude and longitude that intersect at right angles to form a system of reference for locating points on the earth.

gross national product (GNP)
The total value of all goods and services produced by a country per year.

groundwater
Subsurface water that accumulates below the water table in the pores and cracks of rock and soil.

H

half-life
The time required for one-half of the atomic nuclei of an isotope to decay.

hazardous waste
Discarded solid, liquid, or gaseous material that may pose a substantial threat to human health or the environment when it is improperly disposed of, stored, or transported.

heartland theory
The belief of Halford Mackinder that the interior of Eurasia provided a likely base for world conquest.

herbicide
A chemical that kills plants, especially weeds. *See also:* biocide.

hierarchical diffusion
The process by which contacts between people and the resulting diffusion of things or ideas occur first among those at the same level of a hierarchy and then among elements at a lower level of the hierarchy; e.g., small town residents acquire ideas or articles after they are common in large cities.

hierarchical migration
The tendency for individuals to move from small places to larger ones.

hierarchy of central places
The steplike series of urban units in classes differentiated by both size and function.

high-level waste
Nuclear waste with a relatively high level of radioactivity.

hinterland
The market area or region served by an urban unit.

homeostatic plateau
The equilibrium level of population that can be supported adequately by available resources. Equivalent to carrying capacity.

humid continental climate
A climate of east coast and continental interiors of midlatitudes, displaying large annual temperature ranges resulting from cold winters and hot summers. Precipitation at all seasons.

humid subtropical climate
A climate of the east coast of continents in lower-middle latitudes, characterized by hot summers with convectional precipitation and cool winters with cyclonic precipitation.

humus
Dark brown or black decomposed organic matter in soils.

hunting-gathering
An economic and social system based primarily or exclusively on the hunting of wild animals and the gathering of food, fiber, and other materials from uncultivated plants.

hurricane
A severe tropical cyclone with winds exceeding 75 mph (120 kmph) originating in the tropical region of the Atlantic Ocean, Caribbean Sea, or Gulf of Mexico.

hydroelectric power
The kinetic energy of moving water converted into electrical power by a power plant whose turbines are driven by flowing water.

hydrologic cycle
The system by which water is continuously circulated through the biosphere.

hydrosphere
All water at or near the earth's surface that is not chemically bound in rocks; includes the oceans, surface waters, groundwater, and water held in the atmosphere.

I

iconography
In political geography, the study of symbols that unite a country.

ideological subsystem
The complex of ideas, beliefs, knowledge, and means of their communication that characterize a culture.

igneous rock
Rock formed as molten earth materials cool and harden either above or below the earth's surface.

incinerator
A facility designed to burn waste.

inclination
The tilt of the earth's axis about 23½° away from the perpendicular.

Industrial Revolution
The term applied to the rapid economic and social changes in agriculture and manufacturing that followed the introduction of the factory system to the textile industry of England in the last quarter of the 18th century.

infant mortality rate
A refinement of the death rate to specify the ratio of deaths of infants age 1 year or less per 1000 live births.

infrared
Electromagnetic radiation having wavelengths greater than those of visible light.

infrastructure
The basic structure of services, installations, and facilities needed to support industrial, agricultural, and other economic development.

innovation
Introduction into an area of new ideas, practices, or objects; an alteration of custom or culture that originates within the social group itself.

insolation
The solar radiation received at the earth's surface.

intensive agriculture
The application of large amounts of capital and/or labor per unit of cultivated land to increase output. May have either commercial or subsistence orientation.

intensive commercial agriculture
See intensive agriculture.

intensive subsistence agriculture
See intensive agriculture.

interaction model
See gravity model.

International Date Line
By international agreement, the designated line where each new day begins; generally following the 180th meridian.

intrusive rock
Rock resulting from the hardening of magma beneath the earth's surface.

isochrone
A line connecting points equidistant in travel time from a common origin.

isoline
A map line connecting points of constant value, such as a contour line or an isobar.

isotropic plain
A hypothetical portion of the earth's surface where, it is assumed, the land is everywhere the same and the characteristics of the inhabitants are everywhere similar.

J

J-curve
A curve shaped like the letter J, depicting exponential or geometric growth (1, 2, 4, 8, 16. . .).

jet stream
A strong flow of rapidly moving air 30,000 to 40,000 feet (9000 to 12,000 m) high traveling from west to east in the Northern Hemisphere in an undulating pattern.

K

karst
A limestone region marked by sinkholes, caverns, and underground streams.

kerogen
A waxy, organic material occurring in oil shales that can be converted into crude oil by distillation.

kinetic energy
The energy that results from the motion of a particle or body.

L

landform region
A large section of the earth's surface characterized by a great deal of homogeneity among types of landforms.

landlocked state
A state that lacks a seacoast.

Landsat satellite
One of a series of continuously orbiting satellites that carry scanning instruments to measure reflected light in both the visible and near-infrared portions of the spectrum.

language family
A group of languages thought to have descended from a single, common ancestral tongue.

lapse rate
The rate of change of temperature with altitude in the troposphere; the average lapse rate is about 3.5°F per 1000 feet (6.4°C per 1000 m).

large-scale map
A representation of a small land area, usually with a representative fraction of 1:75,000 or less.

latitude
A measure of distance north or south of the equator, given in degrees.

lava
Molten material that has emerged onto the earth's surface.

Law of the Sea Convention
A code of sea law approved by the United Nations in 1982 that authorizes, among other provisions, territorial waters extending 12 nautical miles from shore and 200-nautical-mile-wide exclusive economic zones.

leachate
The contaminated liquid discharged from a sanitary landfill to either surface or subsurface land or water.

least cost theory (syn: Weberian analysis)
The view that the optimum location of a manufacturing establishment is at the place where the costs of transport and labor and the advantages of agglomeration or dispersion are most favorable.

levee
In agriculture, a continuous embankment surrounding areas to be flooded. See also: natural levee.

lingua franca
Any of various auxiliary languages used as common tongues among people of an area where several languages are spoken.

liquefied natural gas (LNG)
Methane gas that has been liquefied by refrigeration for storage or transportation.

lithosphere
Solid shell of rocks resting on the asthenosphere.

loam
Agriculturally productive soil containing roughly equal parts of sand, silt, and clay.

locational tradition
One of the four traditions of geography; in this text, identified with economic, resource, and urban geography.

loess
A deposit of windblown silt.

longitude
A measure of distance east or west of the prime meridian, given in degrees.

longshore current
A current that moves roughly parallel to the shore and transports the sand that forms beaches and sandspits.

low-level waste
Low-level hazardous waste, produced principally by nuclear power plants and industries.

M

magma
Underground molten material.

malnutrition
Food intake insufficient in quantity or deficient in quality to sustain life at optimal conditions of health.

Malthus
Thomas R. Malthus (1766–1834), English economist, demographer, and cleric, who suggested that unless checked by self-control, war, or natural disaster, population will inevitably increase faster than will the food supplies needed to sustain it.

map projection
A method of transferring the grid system from the earth's curved surface to the flat surface of a map.

map scale
See scale.

marine west coast climate
A regional climate found on the west coast of continents in upper midlatitudes; rainy all seasons with relatively cool summers and relatively mild winters.

maximum sustainable yield
The maximum rate at which a renewable resource can be exploited without impairing its ability to be renewed or replenished.

mechanical weathering
The physical disintegration of earth materials, commonly by frost action, root action, or the development of salt crystals.

Mediterranean climate
A climate of lower midlatitudes characterized by mild, wet winters and hot, dry, sunny summers.

megalopolis
A large, sprawled urban complex with contained open, nonurban land, created through the spread and joining of separate metropolitan areas; the name applied to the continuous functionally urban area of coastal northeastern United States from Maine to Virginia.

megawatt
A unit of power equal to 1 million watts (1000 kilowatts) of electricity.

mental map
A map drawn to represent the mental image(s) a person has of an area.

mentifacts
The central, enduring elements of a culture that express its values and beliefs, including language, religion, folklore, artistic traditions, and the like. Elements in the ideological subsystem of culture.

Mercator projection
A true conformal cylindrical projection first published in 1569, useful for navigation.

meridian
A north-south line of longitude; on the globe, all meridians are of equal length and converge at the poles.

mesa
An extensive, flat-topped elevated tableland with horizontal strata, a resistant caprock, and one or more steep sides; a large butte.

metamorphic rock
Rock transformed from igneous and sedimentary rocks by earth forces that generate heat, pressure, or chemical reaction.

metro
See unified government.

metropolitan area
A large functional entity, perhaps containing several urbanized areas, discontinuously built up but operating as a coherent economic whole.

migration
The movement of people or other organisms from one region to another.

migration field
An area that sends major migration flows to a given place or the area that receives major flows from a place.

mineral
A natural inorganic substance that has a definite chemical composition and characteristic crystal structure, hardness, and density.

ministate
An imprecise term for a state or territory small in both population and area. An informal definition accepted by the United Nations suggests a maximum of 1 million people combined with a territory of less than 270 square miles (700 km²).

monoculture
Agricultural system dominated by a single crop.

monotheism
The belief that there is only one God.

monsoon
A wind system that reverses direction seasonally, producing wet and dry seasons; used especially to describe the wind system of South, Southeast, and East Asia.

moraine
Any of several types of landforms composed of debris transported and deposited by a glacier.

mortality rate
See death rate.

mountain breeze
The downward flow of heavy, cool air at night from mountainsides to lower valley locations.

multiple-nuclei model
The idea that large cities develop by peripheral spread not from one but from several nodes of growth, and that therefore there are many origin points of the various land use types in an urban area.

multiplier effect
The expected addition of nonbasic workers and dependents to a city's total employment and population that accompanies new basic employment.

N

nation
A culturally distinctive group of people occupying a particular region and bound together by a sense of unity arising from shared ethnicity, beliefs, and customs.

nationalism
A sense of unity binding the people of a state together; devotion to the interests of a particular nation; an identification with the state and an acceptance of national goals.

nation-state
A state whose territory is identical to that occupied by a particular nation.

natural boundary
A boundary line based on recognizable physiographic features, such as mountains, rivers, or deserts.

natural gas
A mixture of hydrocarbons and small quantities of nonhydrocarbons existing in a gaseous state or in solution with crude oil in natural reservoirs.

natural hazard
A process or event in the physical environment that has consequences harmful to humans.

natural increase
The growth of a population through excess of births over deaths, excluding the effects of immigration or emigration.

natural levee
An embankment on the sides of a meandering river formed by deposition of silt during floods.

natural resource
A physically occurring item that a population perceives to be necessary and useful to its maintenance and well-being.

natural vegetation
The plant life that would exist in an area if humans did not interfere with its development.

neo-Malthusianism
The advocacy of population control programs to preserve and improve general national prosperity and well-being.

neritic zone
That relatively shallow part of the sea that lies above the continental shelf.

net migration
The difference between in-migration and out-migration of an area.

niche
The place an organism or species occupies in an ecosystem.

nomadic herding
Migratory but controlled movement of livestock solely dependent upon natural forage.

nonbasic sector
Those economic activities of an urban unit that service the resident population.

nonecumene (*syn:* anecumene)
The portion of the earth's surface that is uninhabited or only temporarily or intermittently inhabited. *See also* ecumene.

nonfuel mineral resources
Mineral used for purposes other than providing a source of energy.

nonrenewable resource
A natural resource that is not replenished or replaced by natural processes or is used at a rate that exceeds its replacement rate.

North and South poles
The end points of the axis about which the earth spins.

North Atlantic drift
The massive movement of warm water in the Atlantic Ocean from the Caribbean Sea and Gulf of Mexico in a northeasterly direction to the British Isles and the Scandinavian peninsula.

nuclear fission
The controlled splitting of an atom to release energy.

nuclear fusion
The combining of two atoms of deuterium into a single atom of helium in order to release energy.

nuclear power
Electricity generated by a power plant whose turbines are driven by steam produced by the fissioning of nuclear fuel in a reactor.

nutrient
A mineral or other element an organism requires for normal growth and development.

O

oil shale
Sedimentary rock containing solid organic material (kerogen) that can be extracted and converted into a crude oil by distillation.

organic
Derived from living organisms; plant or animal life.

Organization of Petroleum Exporting Countries (OPEC)
An international cartel composed of 13 countries that aims at pursuing common oil-marketing and pricing policies.

orographic precipitation
Rain or snow caused when warm, moisture-laden air is forced to rise over hills or mountains in its path and is thereby cooled.

orthophotomap
An aerial photograph to which a grid system and certain map symbols have been added.

out-sourcing
Producing parts or products abroad for domestic use or sale.

outwash plain
A gently sloping area in front of a glacier composed of neatly stratified glacial till carried out of the glacier by meltwater streams.

overburden
Soil and rock of little or no value that overlies a deposit of economic value, such as coal.

overpopulation
A value judgment that the resources of an area are insufficient to sustain adequately its present population numbers.

oxbow lake
A crescent-shaped lake contained in an abandoned meander of a river.

ozone
A gas molecule consisting of three atoms of oxygen (O_3) formed when diatomic oxygen (O_2) is exposed to ultraviolet radiation. As a damaging component of photochemical smog formed at the earth's surface, it is a faintly blue, poisonous agent with a pungent odor.

ozone layer
A layer of ozone in the high atmosphere that protects life on earth by absorbing ultraviolet radiation from the sun.

P

Pangaea
The name given to the supercontinent that is thought to have existed 200 million years ago.

parallel of latitude
An east-west line indicating the distance north or south of the equator.

PCBs
Polychlorinated biphenyls, compounds containing chlorine that can be biologically magnified in the food chain.

peak value intersection
The most accessible and costly parcel of land in the central business district and, therefore, in the entire urbanized area.

perforated state
A state whose territory is interrupted ("perforated") by a separate, independent state totally contained within its borders.

permafrost
Permanently frozen subsoil.

pesticide
A chemical that kills insects, rodents, fungi, weeds, and other pests. *See also* biocide.

petroleum
A general term applied to oil and oil products in all forms, such as crude oil and unfinished oils.

pH factor
The measure of the acidity/alkalinity of soil or water, on a scale of 0 to 14, increasing with increasing alkalinity.

photochemical smog
A form of polluted air produced by the interaction of hydrocarbons and oxides of nitrogen in the presence of sunlight.

photovoltaic cell
A device that converts solar energy directly into electrical energy.

physical boundary
See natural boundary.

physiological density
The number of persons per unit area of agricultural land. *See also* population density.

pidgin
An auxiliary language derived, with reduction of vocabulary and simplification of structure, from other languages. Not a native tongue, it is employed to provide a mutually intelligible vehicle for limited transactions of trade or administration.

pixel
An extremely small sensed unit of a digital image.

place utility
The perceived attractiveness of a place in its social, economic, or environmental attributes.

planar projection
Any of several map projections based on the projection of the globe grid onto a plane.

planned economy
Production of goods and services, usually consumed or distributed by a governmental agency, in quantities and at prices determined by governmental program.

plantation
A large agricultural holding, frequently foreign-owned, devoted to the production of a single export crop.

plate tectonics
The theory that the lithosphere is divided into plates that slide or drift very slowly over the asthenosphere.

playa
A temporary lake or lake bed found in a desert environment.

Pleistocene
The geological epoch dating from 2 million to about 10,000 years ago during which four stages of continental glaciation occurred.

pollution
The introduction into the biosphere of materials that, because of their quantity, chemical nature, or temperature, have a negative impact on the ecosystem or that cannot be readily disposed of by natural recycling processes.

polychlorinated biphenyls
See PCBs.

polytheism
Belief in or worship of many gods.

population density (*syn:* crude density)
A measurement of the numbers of persons per unit area of land within predetermined limits, usually political or census boundaries. *See also* physiological density.

population momentum
The tendency for population growth to continue despite stringent family planning programs because of a relatively high concentration of people in the childbearing years.

population projection
A report of future size, age, and sex composition of a population based upon assumptions applied to current data.

population pyramid
A graphic depiction of the age and sex composition of a (usually national) population.

possibilism
The philosophical viewpoint that the physical environment offers human beings a set of opportunities from which (within limits) people may choose according to their cultural needs and technological awareness.

potential energy
The energy stored in a particle or body.

precipitation
All moisture, solid and liquid, that falls to the earth's surface from the atmosphere.

predevelopment annexation
The inclusion within the central city of nonurban peripheral areas for the purpose of securing to the city itself the benefits of their eventual development.

pressure gradient force
Differences in air pressure between areas that induce air to flow from areas of high to areas of low pressure.

primary activities
Those parts of the economy involved in making natural resources available for use or further processing; includes mining, agriculture, forestry, fishing or hunting, grazing.

primate city
A country's leading city, much larger and functionally more complex than any other; usually the capital city and a center of wealth and power.

prime meridian
An imaginary line passing through the Royal Observatory at Greenwich, England, serving by agreement as the zero degree line of longitude.

private plot
In the Soviet Union, a small garden parcel allotted to collective farmers and urban workers.

projection
See map projection.

prorupt state
A state of basically compact form that has one or more narrow extensions of territory.

proved reserves
That portion of a natural resource that has been identified and can be extracted profitably with current technology.

psychological distance
The way an individual perceives distance.

pull factor
A characteristic of a region that acts as an attractive force, drawing migrants from other regions.

push factor
A characteristic of a region that contributes to the dissatisfaction of residents.

Q

quaternary activity
That employment concerned with research, with the gathering or disseminating of information, and with administration, including administration of the other economic activity levels.

R

race
A subset of human population whose members share certain distinctive, inherited biological characteristics.

radar
A device for detecting distant objects by analysis of very high frequency radio waves beamed at and reflected from their surfaces.

rank (of coal)
A classification of coals based on their age and energy content; those of higher rank are more mature and richer in energy.

rate
The frequency of occurrence of an event during a specified time period.

rate of natural increase
Birth rate minus the death rate, suggesting the annual rate of population growth without considering net migration.

recycling
The reuse of disposed materials after they have passed through some form of treatment (e.g., melting down glass bottles to produce new bottles).

reflection
The process of returning to outer space some of the earth's received insolation.

region
In geography, the term applied to an earth area that displays a distinctive grouping of physical or cultural phenomena or is functionally united as a single organizational unit.

regional autonomy
A measure of self-governance for a subdivision of a country.

regional concept
The view that physical and cultural phenomena on the surface of the earth are rationally arranged by complex but comprehensible spatial processes.

regionalism
In political geography, minority-group identification with a particular region of a state rather than with the state as a whole.

relative humidity
A measure of the relative dampness of the atmosphere; the ratio between the amount of water vapor in the air and the maximum amount that it could hold at the same temperature, given in percent.

relative location
The position of a place or activity in relation to other places or activities.

relict boundary
A former boundary line that is still discernible and marked by some cultural landscape feature.

relocation diffusion
The transfer of ideas, behaviors, or articles from one place to another through the migration of those possessing the feature transported; also, spatial relocation in which a phenomenon leaves an area of origin as it is transported to a new location. *See also* expansion diffusion.

remote sensing
Any of several techniques of obtaining images of an area without having the sensor in direct physical contact with it, as by air photography or satellite sensors.

renewable resource
A naturally occurring material that is potentially inexhaustible either because it flows continuously (as solar radiation or wind) or is renewed within a short period of time (as biomass). *See also* sustained yield.

replacement level
The number of children per family just sufficient to keep total population constant. Depending on mortality conditions, replacement level is usually calculated at between 2.1 and 2.5 children.

representative fraction (RF)
The scale of a map expressed as a ratio of a unit of distance on the map to distance measured in the same unit on the ground, e.g., 1:250,000.

reradiation
A process by which the earth returns solar energy to space; some of the shortwave solar energy that is absorbed into the land and water is returned to the atmosphere in the form of longwave terrestrial radiation.

resource
See natural resource.

return migration
The stream of migrants who subsequently decide to return to their point of origin.

rhumb line
A line of constant compass bearing; it cuts all meridians at the same angle.

Richter scale
A logarithmic scale used to express the magnitude of an earthquake.

rimland theory
The belief of Nicholas Spykman that domination of the coastal fringes of Eurasia would provide a base for world conquest.

S

Sahel
The semiarid zone between the Sahara desert and the savanna area to the south in West Africa; district of recurring drought, famine, and environmental degradation.

salinization
The process by which soil becomes saturated with salt, rendering the land unsuitable for agriculture; occurs when land that has poor drainage is improperly irrigated.

sandbar
An offshore shoal of sand created by the backwash of waves.

sanitary landfill
Disposal of solid wastes by spreading them in layers covered with enough soil or ashes to control odors, rats, and flies.

savanna
A tropical grassland characterized by widely dispersed trees and experiencing pronounced yearly wet and dry seasons.

scale
In cartography, the ratio between length or size of an area on a map and the actual length or size of that same area on the earth's surface; map scale may be represented verbally, graphically, or as a fraction. In more general terms, scale refers to the size of the area studied, from local to global.

S-curve
The horizontal bending, or leveling, of an exponential J-curve.

secondary activities
Those parts of the economy involved in the processing of raw materials derived from primary activities; includes manufacturing, construction, power generation.

sector model
The idea that wedge-shaped sectors of different land uses radiate outward from the central business district.

sedimentary rock
Rock formed from particles of gravel, sand, silt, and clay that were eroded from already existing rocks.

seismic waves
Vibrations within the earth set off by earthquakes.

service sector
See nonbasic sector.

shaded relief
A method of representing the three-dimensional quality of an area by use of continuous graded tone to simulate the appearance of sunlight and shadows.

shale oil
The crude oil resulting from the distillation of kerogen in oil shales.

shamanism
A form of tribal religion based on belief in a hidden world of gods, ancestral spirits, and demons responsive only to a shaman, or interceding priest.

shifting cultivation (*syn:* slash and burn agriculture, swidden agriculture)
Crop production of forest clearings kept in cultivation until their quickly declining fertility is lost. Cleared plots are then abandoned and new sites are prepared.

sinkhole
A deep surface depression formed when ground collapses into a subterranean cavern.

site
The place where something is located; the immediate surroundings.

situation
The location of something in relation to the physical and human characteristics of a larger region.

slash and burn agriculture
See shifting cultivation.

small-scale map
A representation of a large land area on which small features (highways, buildings) cannot be shown true to scale.

sociofacts
The institutions and links between individuals and groups that unite a culture, including family structure and political, educational, and religious institutions. Components of the sociological subsystem of culture.

sociological subsystem
The totality of expected and accepted patterns of interpersonal relations common to a culture or subculture.

soil
The thin layer of fine material that rests on bedrock and is capable of supporting plant life.

soil depletion
The loss of some or all of the vital nutrients from soil.

soil erosion
See erosion.

soil horizon
A layer of soil distinguished from other soil zones by color, texture, and other characteristics resulting from soil-forming processes.

solar energy
Radiation from the sun, which is transformed into heat primarily at the earth's surface and secondarily in the atmosphere.

solar power
The radiant energy generated by the sun; sun's energy captured and directly converted for human use. *See also* photovoltaic cell.

solid waste
Unwanted materials generated in production or consumption processes that are solid rather than liquid or gaseous in form.

source region
In climatology, a large area of uniform surface and relatively consistent temperatures where an air mass forms.

southern oscillation
The atmospheric conditions occurring periodically near Australia that create the el Niño condition off the coast of South America.

spatial diffusion
The outward spread of a substance, concept, or population from its point of origin.

spatial distribution
The arrangement of things on the earth's surface.

spatial interaction
The movement (e.g., of people, goods, information) between different places.

spatial margin of profitability
The set of points delimiting the area within which a firm's profitable operation is possible.

spatial search
The process by which individuals evaluate the alternative locations to which they might move.

spine
In urban geography, a continuation of the features of the central business district outward along the main wide boulevard characteristic of Latin American cities.

spring wheat
Wheat sown in spring for ripening during the summer or autumn.

standard language
A language substantially uniform with respect to spelling, grammar, pronunciation, and vocabulary and representing the approved community norm of the tongue.

standard parallel
The tangent circle, usually a parallel of latitude, in a conic projection; along the standard line the scale is as stated on the map.

state (*syn:* country).
An independent political unit occupying a defined, permanently populated territory and having full sovereign control over its internal and foreign affairs.

state farm
Under Soviet and other planned economies, a government agricultural enterprise operated with paid employees.

steppe
Name applied to treeless midlatitude grasslands.

strategic petroleum reserve (SPR)
Petroleum stocks maintained by the federal government for use during periods of major supply interruption.

stratosphere
The layer of the atmosphere that lies above the troposphere and extends outward to about 35 miles (56 km).

stream load
Eroded material carried by a stream in one of three ways, depending on the size and composition of the particles: (1) in dissolved form, (2) suspended by water, (3) rolled along the stream bed.

subduction
The process by which one lithospheric plate is forced down into the asthenosphere as a result of collision with another plate.

subsequent boundary
A boundary line that is established after the area in question has been settled, and that considers the cultural characteristics of the bounded area.

subsidence
Settling or sinking of a portion of the land surface, sometimes as a result of the extraction of fluids such as oil or water from underground deposits.

subsistence agriculture
Any of several farm economies in which most crops are grown for food, nearly exclusively for local consumption.

subsistence economy
A system in which goods and services are created for the use of producers or their immediate families. Market exchanges are limited and of minor importance.

substitution principle
In industry, the tendency to substitute one factor of production for another in order to achieve optimum plant location and profitability.

suburb
A functionally specialized segment of a large urban complex located outside the boundaries of the central city.

superimposed boundary
A boundary line placed over, and ignoring, an existing cultural pattern.

sustained yield
The practice of balancing harvesting with growth of new stocks so as to avoid depletion of the resource and ensure a perpetual supply.

swidden agriculture
See shifting cultivation.

syncretism
The development of a new form of, for example, religion or music, through the fusion of distinctive parental elements.

synthetic fuel (synfuel)
Crude oil and natural gas substitutes that can be synthesized from a variety of nonoil and nongas sources, including coal, tar sands, and wood.

systems analysis
An approach to the study of large systems through (1) segregation of the entire system into its component parts, (2) investigation of the interactions between system elements, and (3) study of inputs, outputs, flows, interactions, and boundaries within the system.

T

talus slope
A landform composed of rock particles that have accumulated at the base of a cliff, hill, or mountain.

tar sands
Sand and sandstone impregnated with heavy oil.

technological subsystem
The complex of material objects together with the techniques of their use by means of which people carry out their productive activities.

technology
An integrated system of knowledge and skills developed within a culture to carry out successfully purposeful and productive tasks.

tectonic forces
Processes that shape and reshape the earth's crust, the two main types being diastrophic and volcanic.

temperature inversion
The condition caused by rapid reradiation in which air at lower altitudes is cooler than air aloft.

territoriality
Persistent attachment of most animals to a specific area; the behavior associated with the defense of the home territory.

territorial production complex
In Soviet economic planning, a design for large regional industrial, mining, and agricultural development leading to regional self-sufficiency and the creation of specialized production for a larger national market.

tertiary activities
Those parts of the economy that fulfill the exchange function and that provide market availability of commodities; includes wholesale and retail trade and associated transportation, government, and information services.

thermal pollution
The introduction of heated water into the environment, with consequent adverse effects on plants and animals.

thermal scanner
A remote sensing device that detects the energy (heat) radiated by objects on earth.

Third World
Those nations that, in contrast to the industrialized countries, typically have low per capita incomes, high population growth rates, and the majority of their labor forces engaged in agriculture.

threshold
In economic geography, the minimum market needed to support the supply of a product or service.

tidal power
The kinetic energy of ocean tides as they rise and fall harnessed to generate electricity as the water flows over reversible turbines.

topographic map
One that portrays the surface features of a relatively small area, often in great detail.

toponymy
The place-names of a region or, especially, the study of place-names.

tornado
A small, violent storm characterized by a funnel-shaped cloud of whirling winds that can form beneath a cumulonimbus cloud in proximity to a cold front and that moves at speeds up to 300 mph (480 kmph).

total fertility rate
The average number of children that would be born to each woman if during her childbearing years she bore children at the current year's rate for women that age.

town
A nucleated settlement that contains a central business district but that is smaller and less functionally complex than a city.

traditional religion
See tribal religion.

tragedy of the commons
The observation that in the absence of collective control over the use of a resource available to all, it is to the advantage of all users to maximize their separate shares even though their collective pressures may diminish total yield or destroy the resource altogether.

transform fault
A break in rocks that occurs when one lithospheric plate slips past another in a horizontal motion.

tribal religion
An ethnic religion specific to a small, localized, preindustrial culture group.

tropical rain forest
Tree cover composed of tall, high-crowned evergreen deciduous species, associated with the continuously wet tropical lowlands.

tropical rain forest climate
The continuously warm, frost-free climate of tropical (and equatorial) lowlands, with abundant moisture year-round.

troposphere
The atmospheric layer closest to the earth, extending outward about 7 to 8 miles (11 to 13 km) at the poles to about 16 miles (26 km) at the equator.

truck farming
The production of fruits and vegetables for market.

tsunami
A seismic sea wave generated by an earthquake or volcanic eruption.

tundra
The treeless area lying between the tree line of arctic regions and the permanently ice-covered zone.

typhoon
Name given to hurricanes occurring in the western Pacific Ocean region.

U

ubiquitous industry
A market-oriented industry whose establishments are distributed in direct proportion to the distribution of population (market).

underpopulation
A value statement reflecting the view that an area has too few people in relation to its resources and population-supporting capacity.

unified government (*syn:* Unigov, Metro)
Any of several devices for federating or consolidating governments within a metropolitan region.

Unigov
See unified government.

unitary state
A state in which the central government dictates the degree of local or regional autonomy and the nature of local governmental units; a country with few cultural conflicts and a strong sense of national identity.

United Nations Convention on the Law of the Sea (UNCLOS)
See Law of the Sea Convention.

universalizing religion
A religion that claims global truth and applicability and seeks the conversion of all humankind.

urban hierarchy
A ranking of cities based on their size and functional complexity.

urban influence zone
An area outside of a city that is nevertheless affected by the city.

urbanization
Transformation of a population from rural to urban status; the process of city formation and expansion.

urbanized area
A continuous built-up landscape defined by building and population densities with no reference to the political boundaries of the city; may contain a central city and many contiguous cities, towns, suburbs, and unincorporated areas.

V

valley breeze
The flow of air up mountain slopes during the day.

variable costs
In economic geography, costs of production inputs that display place to place differences.

verbal scale
A statement of the relationship between units of measure on a map and distance on the ground, as "one inch represents one mile."

vernacular
(1) The nonstandard indigenous language or dialect of a locality; (2) of or related to indigenous arts and architecture, such as a vernacular house; (3) of or related to the perceptions and understandings of the general population, such as a vernacular region.

volcanism
The earth force that transports heated material to or toward the surface of the earth.

von Thünen model
Model developed by Johann H. von Thünen (1783–1850) to explain the forces that control the prices of agricultural commodities and how those variable prices affect patterns of agricultural land utilization.

von Thünen rings
The concentric zonal pattern of agricultural land use around a single market center proposed in the von Thünen model.

W

warping
The bowing of a large region of the earth's surface due to the movement of continents or the melting of continental glaciers.

wash
A dry, braided channel in the desert that remains after the rush of rainfall runoff water.

water table
The upper limit of the saturated zone and therefore of groundwater.

weather
The state of the atmosphere at a given time and place.

weathering
Mechanical and chemical processes that fragment and decompose rock materials.

Weberian analysis
See least cost theory.

Weber model
Analytical model devised by Alfred Weber (1868-1958) to explain the principles governing the optimum location of industrial establishments.

wetland
A lowland area that is saturated with moisture, such as a marsh or tidal flat.

wind power
The kinetic energy of wind converted into mechanical energy by wind turbines that drive generators to produce electricity.

winter wheat
Wheat sown in fall for ripening the following spring or summer.

Z

zero population growth (ZPG)
A term suggesting a population in equilibrium, with birth rates and death rates nearly identical.

zoning
Designating by ordinance areas in a municipality for particular types of land use.

C R E D I T S

Photo Credits

Part Openers and Table of Contents

Part 1, p. iii: © J. Robert Stottlemeyer/Biological Photo Service; **Part 2,** p. iv: © Robert Frerck/Odyssey Productions; **Part 3,** p. v: © B. Templeman/Valan Photos; **Part 4,** p. vi: © Cameramann International

Chapter 1

Opener: © Robert Frerck/Odyssey Productions; **1.1:** NOAA; **1.5:** © Walter Frerck/Odyssey Productions; **1.6:** NASA; **1.7:** U.S. Geological Survey; **1.9:** Chicago Area Transportation Study, "Final Report," vol. I, p. 44, 1959

Chapter 2

Opener: © Maptec International LTD/Science Photo Library/Photo Researchers, Inc.; **p. 36:** © Elizabeth Hovinen; **2.27:** NASA; **2.28:** Exxon Corporation and American Petroleum Institute; **2.30:** Courtesy of Intergraph Corporation

Chapter 3

Opener: © Robert Frerck/Odyssey Productions; **p. 52:** NASA; **p. 74:** L. A. Yehle/USGS; **3.10b:** © B. F. Molnia/Terraphotographics/BPS; **3.13b:** © N. Myers/Bruce Coleman, Inc.; **3.14a:** © Kevin Shafer/Tom Stack and Associates; **3.15a:** Wide World Photos; **3.15b:** M. Celebi/USGS; **3.17a:** Dr. Mullineux/USGS; **3.17b:** Austin Post/USGS; **3.18a:** © B. F. Molnia/Terraphotographics/BPS; **3.20a:** © W. D. Kleck/Terraphotographics/BPS; **3.20b:** © John Cunningham/Visuals Unlimited; **3.21:** © William E. Ferguson; **3.22:** NASA; **3.23a:** © John Cunningham/Visuals Unlimited; **3.23b:** © B. F. Molnia/Terraphotographics/BPS; **3.24:** © W. D. Kleck/Terraphotographics/BPS; **3.25:** © Arthur Getis; **3.27b:** NASA; **3.30:** © Dr. Carla Montgomery; **3.31:** © Steve McCutcheon/Alaska Photo; **3.32b:** © Dr. Carla Montgomery; **3.34:** © Carl May/Terraphotographics/BPS; **3.35:** © Arthur Getis

Chapter 4

Opener: © Carl Purcell/Photo Researchers, Inc.; **4.11b:** © David Baird/Tom Stack and Associates; **4.19a:** © Stephen Trimble; **4.19b:** © William E. Ferguson; **4.19c:** © Steve McCutcheon/Alaska Photo; **4.27:** © H. Oscar/Visuals Unlimited; **4.32:** © Lawrence Naylor/Photo Researchers, Inc.; **4.33a:** © Aubrey Lang/Valan Photos; **4.33b:** © Arthur Getis; **4.36:** © Ann Duncan/Tom Stack and Associates; **4.39:** © William E. Ferguson; **4.44:** © John Cunningham/Visuals Unlimited; **4.47, 4.50:** © William E. Ferguson

Chapter 5

Opener: © J. A. Wilkinson/Valan Photos; **5.1:** © Joe Sohm/The Image Works; **p. 136:** © Robert Eckert/EKM-Nepenthe; **p. 147, top:** © Gene Alexander/Soil Conservation Service, USDA; **bottom:** © Robert Frerck/Odyssey Productions; **5.8:** © Cameramann International; **5.9:** © William E. Ferguson; **5.12:** © Stephen Trimble; **5.15:** © Will McIntyre/Photo Researchers, Inc.; **5.18:** NASA; **5.20a:** Courtesy of Kennecott; **5.20b:** © Stephen Trimble; **5.21:** © Lawrence E. Stager/Anthro-Photo; **5.22:** Courtesy of the California Department of Water Resources; **5.23:** © Kirschenbaum/Stock Boston; **5.24:** © F. Botts/United Nations; **5.27:** Australian Information Service; **5.28:** © Mark Boulvon/Photo Researchers, Inc.; **5.34:** © ISP/PSI Fotofile

Chapter 6

Opener: © Cameramann International; **6.2:** © Bob Daemmrich/Stock Boston; **6.11:** © Patrick Ward/Stock Boston; **6.17:** Historical Pictures Service, Inc.; **6.18:** © William E. Ferguson; **6.22:** © Robert Frerck/Odyssey Productions; **6.26:** © Mark Antman/The Image Works; **6.27:** United Nations; **6.29:** © Ben Mathes; **6.30:** © Christine Osborne/Valan Photos

Chapter 7

Opener: © Eugene Gordon; **7.1a:** © Robert Frerck/Odyssey; **7.1b:** © Robert Eckert/EKM-Nepenthe; **p. 198:** © Joe Sohm/The Image Works; **7.2:** © J. Popp/Anthro-Photo; **7.3:** © Bob Daemmrich/The Image Works; **7.4a:** © Robert Frerck/Odyssey Productions; **7.4b:** © R. Aguirre and G. Switkes, AMAZONIA/Terraphotographics/BPS; **7.8a:** © Warren Garst/Tom Stack and Associates; **7.8b:** © Robert Frerck/Odyssey Productions; **7.8c:** © Cary Wolinsky/Stock Boston; **7.8d:** © Harriet Gans/The Image Works; **7.9:** © Aubrey Lang/Valan Photos; **7.10a:** © Mark Antman/The Image Works; **7.10b:** © Marc Bernsau/The Image Works; **7.10c:** © Owen Franken/Stock Boston; **7.14a:** © Robert Frerck/Odyssey Productions; **7.14b:** © William H. Allen, Jr.; **7.15:** © David Austen/Stock Boston; **7.16:** © Jean Fellmann; **7.17:** © Mark Antman/The Image Works; **7.24:** © Charles Gateman/The Image Works; **7.30a:** © David Burney; **7.30b:** © Robert Frerck/Odyssey; **7.32:** © T. Joyce/Valan Photos; **7.33:** © Fred Bruemmer/Valan Photos; **7.35, 7.36:** © Wolfgang Kaehler

Chapter 8

Opener: © Bob Daemmrich/The Image Works; **8.1:** © Joe Sohm/The Image Works; **8.8:** Australian Information Service; **8.10:** © AP/Wide World Photos; **8.16:** © Michael Siluk; **8.22:** AP/Wide World Photos; **8.28:** © Cameramann International

Chapter 9

Opener: © Genzo Sugino/Tom Stack; **9.1:** © Bob Daemmrich/The Image Works; **9.6:** AP/Wide World; **9.15:** © Robert Frerck/Odyssey Productions; **9.20:** © J. K. Isaac/United Nations

Chapter 10

Opener: © Robert E. Ford/Terraphotographics/BPS; **10.1:** © John Maher/EKM-Nepenthe; **10.3:** © Pierre R. Chabot/Valan Photos; **10.5:** © Kurt Thorson/EKM-Nepenthe; **10.7:** © Philip Jon Bailey/Stock Boston; **10.8:** © V. Crowl/Visuals Unlimited; **10.10:** © Jean Fellmann; **10.15:** © Brian Brake/Photo Researchers, Inc.; **10.16:** © Jonathan Wright/Bruce Coleman, Inc.; **10.18:** © Thomas Kitchin/Valan Photos; **10.19:** © Cameramann International; **10.20:** United Nations; **10.25:** © Cameramann International; **10.26:** Courtesy of John Deere; **10.30:** © Aubrey Diem/Valan Photos; **10.33:** Culver Pictures, Inc.; **10.34:** Courtesy Fairfax County Economic Development Authority

Chapter 11

Opener: © David M. Doody/Tom Stack and Associates; **11.1:** © Grant Heilman/Grant Heilman; **p. 356:** © American Petroleum Institute; **11.9:** © Visuals Unlimited; **11.12:** © ISP/PSI Fotofile; **11.13:** © American Petroleum Institute, Tosco Corporation; **11.14:** © American Petroleum Institute, Gulf Oil Corporation; **11.17:** © Harold W. Scott; **11.18:** © Christine Osborne/Valan Photos; **11.20:** © Cameramann International; **11.21:** © American Petroleum Institute and Union Oil Company of California; **11.22:** Courtesy Southern California Edison Company and the American Petroleum Institute; **11.28:** © Christine Osborne/Valan Photos; **11.31:** © John Maher/Stock Boston; **11.32:** © Cameramann International; **11.34:** © John Bova/Photo Researchers, Inc.; **11.36a:** © R. Moller/Valan Photos; **11.36b:** © Halle Flygare Photos/Valan Photos; **11.39b:** Goddard Space Flight Center/NASA; **11.41:** © Joe Sohm/The Image Works

Chapter 12

Opener: © Louis Goldman/Photo Researchers, Inc.; **p. 405**: © Arthur Getis; **12.5a**: © B. F. Molnia/ Terraphotographics/BPS; **12.5b**: © Carl May/ Biological Photo Service; **12.21**: © John Jakle; **12.30**: © Ellis Herwig/Stock Boston; **12.32**: © Eugene Gordon; **12.33**: © The Pulitzer Publishing Company/St. Louis Post-Dispatch; **12.34**: Courtesy of The Chicago Housing Authority; **12.35**: © David Woo/Stock Boston; **12.36**: © Robin L. Ferguson/ William E. Ferguson; **12.37**: © Robert Frerck/ Odyssey Productions; **12.39**: © IPA/The Image Works; **12.40, 12.43**: © Carl Purcell; **12.44**: © Aubrey Diem/Valan Photos

Chapter 13

Opener: © Robert E. Ford/Terraphotographics/ BPS; **13.5**: © B. F. Molnia/Terraphotographics/ BPS; **13.8**: © William E. Ferguson; **13.13**: © J. Y. Rabeuf/The Image Works; **13.17, 13.19**: © Elizabeth L. Hovinen

Illustration Credits

Chapter 1

Fig. 1.8 Source: Redrawn by permission from Howard J. Nelson and William A. V. Clark, *The Los Angeles Metropolitan Experience*, p. 49. Association of American Geographers, 1976. **Fig. 1.12** Source: Reproduced by permission from the *Annals of the Association of American Geographers*, John R. Borchert, Vol. 62, p. 358, Association of American Geographers, 1972. **Fig. 1.10** Source: From Thomas J. Mason, et al., *An Atlas of Mortality from Selected Diseases*. NIH Publication #81-2397, May 1981. National Institutes of Health, Bethesda, Maryland.

Chapter 2

Fig. 2.9b Source: Copyright 1977, Brooks-Roberts; with permission. **Fig. 2.11** Copyright The University of Chicago Department of Geography. **Fig. 2.12a** Source: Copyright—Florence Thierfeldt, Milwaukee, Wisconsin. **Fig. 2.14** Source: U.S. Geological Survey. **Fig. 2.15** Source: U.S. Geological Survey. **Fig. 2.18a** From George O. Carney, "From Down Home to Uptown" in Journal of Geography 76: 107, 1977. Copyright © 1977 National Council for Geographic Education, Indiana, PA. Reprinted by permission. **Fig. 2.18b** From Daniel B. Arreola, "Urban Mexican Americans" in Focus. Vol. 34, No. 3, January/ February, 1984. Copyright © 1984 American Geographical Society. **Fig. 2.21** Source: U.S. Census Bureau. **Fig. 2.22** Source: Reprinted with permission from *The Economist*, London. The *Economist* The Economist Building, 25 St. James' Street, London SW1A 1HG ENGLAND. **Fig. 2.23** From Gunther Glebe and John O'Loughlin, editors, Foreign Minorities in Continental European Cities, Vol. 84 of Erkundliches Wissen. Copyright © 1987. Franz Steiner Verlag, Weisbaden, West Germany. Reprinted by permission. **Fig. 2.24** Source: U.S. Geological Survey. **Fig. 2.31** From Stephen C. Guptill and Lowell E. Starr, "Making Maps with Computers" in American Scientist, Vol. 76, March-April, 1988 Sigma XI, the Scientific Research.

Chapter 3

Fig. 3.1 Source: After U.S. Geological Survey publication, "Geologic Time." **Fig. 3.3** Source: Exxon Corporation, 1983. **Fig. 3.4** Source: American Petroleum Institute. **Fig. 3.6** Source: From Barazangi and Dorman, *Bulletin of Seismological Society of America,* 1969. Reprinted by permission. **Fig. 3.12b** Copyright © 1986 by the New York Times Company. **Fig. 3.19** From Carla W. Montgomery, *Physical Geology,* 2d edition. Copyright © 1990 Wm. C. Brown Publishers, Dubuque, Iowa. All rights reserved. Reprinted by permission. **Fig. 3.36** From Charles B. Hunt, Geology of Soils: Their Evolution, Classification, and Uses. Copyright © W.H. Freeman and Company.

Chapter 4

Fig. 4.23 From T. McKnight, Physical Geography: A Landscape Approach, 2d ed. Copyright © 1987 Prentice-Hall, Inc., Englewood Cliffs, NJ. Reprinted by permission. **Fig. 4.26** Source: After U.S. Navy Oceanographic Office. **Illustration, Page 103** From T. L. McKnight, Physical Geography: A Landscape Appreciation, 2nd ed. Copyright © 1987 Prentice-Hall, Inc., Englewood Cliffs, NJ. Reprinted by permission.

Chapter 5

Fig. 5.6 Source: Redrawn from "Channelization: Shortcut to Nowhere" by Raymond V. Corning, in *Virginia Wildlife,* February 1975. Copyright *Virginia Wildlife.* Used with permission. **Table 5.1** Reproduced from *Environmental Issues* by Lawrence G. Hines, by permission of W. W. Norton & Co., Inc., copyright 1973 by W. W. Norton & Co., Inc. **Fig. 5.10a,b** Sources: National Air Pollution Control Administration; U.S. Department of Health and Human Services. **Fig. 5.16** Source: Environmental Protection Agency. **Fig. 5.25** Sources: Based upon H. E. Dregne, *Desertification of Arid Lands,* figure 1.2. Copyright 1983 Harwood Academic Publishers; and *A World Map of Desertification.* Unescol FAO. **Fig. 5.26** Source: Based on data from 1934 Reconnaissance Erosion Survey of the United States and other soil conservation surveys by the Soil Conservation Service—U.S. Soil Conservation Service. **Fig. 5.30** Charles J. Krebs, "Ecology," *The Economist,* April 26, 1986, p. 93. *The Economist* The Economist Building 25 St. James' Street London SW1A 1HG ENGLAND. **Fig. 5.35** Artwork by Richard Farrell. Copyright © 1983, Richard Farrell.

Chapter 6

Fig. 6.3 Source: Data from Population Reference Bureau. **Fig. 6.4** Source: Data from United Nations and U.S. Bureau of the Census. **Fig. 6.5** Source: Data from Population Reference Bureau. **Fig. 6.6** Source: Data from Population Reference Bureau. **Fig. 6.7** Source: Data from the U.S. Bureau of the Census and Population Reference Bureau. **Fig. 6.8** Sources: Data from the United Nations, *Demographic Yearbook*; Population Reference Bureau; and the censuses of Mexico, the United States, and West Germany. **Fig. 6.9** Sources: Carl Haub and Mary Mederios Kent, *1988 World Population Data Sheet* (Washington D.C.: Population Reference Bureau, 1988). **Fig. 6.10** Source: Data from Population Reference Bureau. **Fig. 6.12** Sources: U.S. Bureau of the Census and "The Population Future of West Germany: Three Possibilities," *Population Today,* Vol. 13, No. 10 (Washington, D.C.: Population Reference Bureau, October 1985) p. 5. **Fig. 6.13** Source: Data from Population Reference Bureau. **Fig. 6.23** Sources: Data from World Bank and FAO. **Fig. 6.24** Source: World Bank, *Population Growth and Policies in sub-Saharan Africa* (A World Bank Policy Study), 1986. Redrawn with permission. **Fig. 6.25** Source: Data from Population Reference Bureau. **Table 6.1** Sources: Thomas W. Merrick, with PRB staff, "World Population in Transition," *Population Bulletin* Vol 41, no. 2 (Washington, D.C.: Population Reference Bureau, 1986); and *1989 World Population Data Sheet,* Population Reference Bureau. **Table 6.5** Source: The World Bank, *World Development Report 1984.* **Table 6.6** Brinley Thomas, *Migration and Economic Growth* (2nd ed.) Cambridge: Cambridge University Press, 1973. p. 131. **Table 6.7** K. C. Zachariah and Julien Conde, *Migration in West Africa: The Demographic Aspects,* Table 30, p. 46. World Bank Research Publication Service. **Table 6.8** Sources: *1989 World Population Data Sheet* (Population Reference Bureau) and data derived from UN Food and Agricultural Organization (FAO) and United States Department of State.

Chapter 7

Fig. 7.5 Source: Data from World Bank. **Fig. 7.22** Redrawn from *A Word Geography of the Eastern United States* by Hans Kurath. Copyright © 1949 by The University of Michigan. By permission of the University of Michigan Press. **Fig. 7.29** Redrawn from Isma'il al-Farugi and David E. Soper (eds.), *Historical Atlas of the Religions of the World.* New York: Macmillan, 1974, p. 235. **Fig. 7.38** Redrawn from St. Clair Drake and Horace R. Cayton, *Black Metropolis.* New York: Harper & Row, 1962. **Table 7.2** Sources: Based upon *Britannica Book of the Year* and other sources; 1989 population data from Population Reference Bureau, Inc.

Chapter 8

Fig. 8.4 From *World Regional Geography: A Question of Place* by Paul Ward English, with James Andrew Miller. **Fig. 8.11** Ernst Griffin, "Latin America" in D. Gordon Bennett *Tension Areas of the World.* Champaign, IL, Park Press, 1982. **Fig. 8.17** Reproduced by permission from The Professional Geographer of the Association of American Geographers, Vol. 39, 1987, p. 16, Fig. 1, p. 21, Fig. 2, V. L. Mote. **Fig. 8.18** Source: Redrawn from *Oxford Atlas for New Zealand,* 1960. Reproduced by permission of Oxford University Press. **Fig. 8.29** Source: From Edward F. Bergman, *Modern Political Geography,* Copyright 1975 Wm. C. Brown Publishers, Dubuque, Iowa. All Rights Reserved. Redrawn by permission of the author.

Chapter 9

Fig. 9.3 Redrawn with permission from Herbert A. Whitney, "Preferred Locations in North America: Canadians Clues and Conjectures," in *Journal of Geography,* 83, No. 5, 1984. Copyright © 1984 National Council for Geographic Education, Indiana, PA. **Fig. 9.4** From Los Angeles City Planning Department, The Visual Environment of Los Angeles, 1971. Used by permission. **Fig. 9.9** Source: Data from Chicago Area Transportation Study, *1970 Travel Characteristics.* **Fig. 9.8** Source: Redrawn with permission from Robert A. Murdie, "Cultural Differences in Consumer Travel" in *Economic Geography,* Vol 41, No. 3, July 1965. Copyright 1965 *Economic Geography.* **Fig. 9.10** Source: Metropolitan Toronto and Region Transportation Study, by Maurice Yeates. The Queen's Printer, Toronto, 1966, fig. 42. Used with permission. **Fig. 9.12** Reproduced by permision from Resource Publications for College Geography, Spatial Diffusion, Peter R. Gould, Page 4, Association of American Geographers, 1969. **Fig. 9.14** Source: Redrawn with permission from Paul F. Mattingly, "Disadoption of Milk Cows by Farmers in Illinois: 1930-1978" *The Professional Geographer,* Vol. 36, No. 1, p. 45. Association of American Geographers, Washington, D.C., 1984.

Fig. 9.17 Source: Redrawn with permission from Yoshio Sugiura, "Diffusion of Rotary Clubs in Japan, 1920–1940: A Case of Non-Profit Motivated Innovation Diffusion Under A Decentralized Decision-Making Structure" in *Economic Geography,* Vol. 62, No. 2, 1986. Copyright 1986 *Economic Geography.* Fig. 9.18 Source: Redrawn with permission from Yoshio Sugiura, "Diffusion of Rotary Clubs in Japan, 1920–1940: A Case of Non-Profit Motivated Innovation Diffusion Under A Decentralized Decision-Making Structure" in *Economic Geography,* Vol. 62, No. 2, 1986. Copyright 1986 *Economic Geography.* Fig. 9.22 Source: Redrawn with permission from C. C. Roseman, *Association of American Geographers Resource Paper* 77.2, p. 5. Association of American Geographers, Washington, D.C., 1977. Table 9.1 Source: Ian Burton and Robert Kates, "The Perception of Natural Hazards in Resource Management." Reprinted with permission from *3 Nat. Res. J.* 435 (1964), published by the University of New Mexico School of Law, Albuquerque, N.M. Fig. 9.23 Data from T. Hägerstrand, *Innovation Diffusion as a Spatial Process.* Copyright © 1967 The University of Chicago Press, Chicago, IL. Fig. 9.25 Source: Redrawn with permission from C. C. Roseman, "Channelization of Migration Flows From the Rural South to the Industrial Midwest," *Proceedings of the Association of American Geographer,* Vol. 3, p. 142. Association of American Geographers, Washington, D.C., 1971. Fig. 9.24 Source: J. A. Jakle, S. Brunn, and C. C.Roseman, *Human Spatial Behavior: A Social Geography* (1976), reissued 1985 by Waveland Press, Inc., P.O. Box 400, Prospect Heights, Illinois 60070, p. 167. Reprinted with permission. Fig 9.26 Redrawn by permission from Annals of the Association of American Geographers, Vol. 76, 1986, J. O. Huff. Part 3 Opener Source: From *Partick Geddes: Spokesman for Man and the Environment,* ed, by Marshall Stalley. New Brunswick, N.J.: Rutgers University Press, 1972, p. 123.

Chapter 10

Fig. 10.4 Source: Copyright permission: Hammond Incorporated. Maplewood, New Jersey 07040. Fig. 10.9 Source: Redrawn from Robert E. Huke, "The Green Revolution," *Journal of Geography,* Vol. 84, No. 6, 1985, Fig. 1, p. 248. With permission of the National Council for Geographic Education. Fig. 10.11b Modified from Bernd Andreae, Farming, Development and Space, (Berlin: De Gruyter, 1981), Figure 63. Fig. 10.13 Sources: U.S. Bureau of Agricultural Economics; Agriculture Canada; Mexico. Secretaria de Agricultura y Recursos

Hidraulicos. Fig. 10.23 Redrawn from Truman A. Harshorn and John W. Alexander, Economic Geography, 3rd ed. Englewood Cliffs, NJ: Prentice-Hall, 1988, Fig. 13–2, P. 198. Fig. 10.27 Redrawn from: Charles Whynne-Hammond, *Elements of Human Geography* (2nd ed.). London: George Allen and Unwin, 1985. Figure 9.7, p. 124.

Chapter 11

Fig. 11.2 Adapted from Macmillan Publishing Company, a Division of Macmillan, Inc., from the United States Energy Atlas, 2nd edition, by David J. Cuff and William J. Young. Copyright © 1985 by Macmillan Publishing Company. Fig. 11.3 Source: From World Bank. *World Development Report, 1986,* published by Oxford University Press. Used with permission. Fig. 11.5 Source: Data from British Petroleum Company, *BP Statistical Review of World Energy,* 1989. Fig. 11.6 Source: British Petroleum Company, *BP Statistical Review of World Energy,* 1989. Fig. 11.7 British Petroleum Co. BP Statistical Review of World Energy, 1989, Page 3. Fig. 11.10 Sources: Courtesy of U.S. Department of Energy and American Petroleum Institute. Fig. 11.11 British Petroleum Co. BP Statistical Review of World Energy, 1989, Page 3. Fig. 11.16 Source: U.S. Council for Energy Awareness. Table 11.1 Source: Data from *Minerals Handbook, 1984–85.* Phillip Crowson, ed. Macmillan, London, 1984. Fig. 11.25 Source: Data from U.S. Bureau of Mines, *Mineral Commodity Summaries,* 1983. Fig. 11.27 J. W. Mallor and R. H. Adams, Jr. "Feeding the Underdeveloped World," Chemical and Engineering News, 62(17), 1984. Pg. 32–39. Fig. 11.38 Source: U.S. Forest Service.

Chapter 12

Fig. 12.3 Source: List of metropolitan areas prepared by Richard L. Forstall for Rand McNally and Company's *The International Atlas,* 1989. Used by permission. Fig. 12.2 Source: Redrawn with permission from Elaine M. Murphy, *World Population Toward the Next Century,* Population Reference Bureau, Inc., May, 1985. Fig 12.10 Source: Redrawn by permission from R. B. Vance and S. Smith, "Metropolitan Dominance and Integration" in *The Urban South,* R. B. Vance and N. S. Demcruth, Eds. Copyright 1954 University of North Carolina Press, Chapel Hill, North Carolina. Fig. 12.11 Source: From A. Getis and J. Getis, "Christaller's Central Place Theory," *Journal of Geography,* 1966. With permission of the National Council for Geographic Education. Fig. 12.15

Source: Data from U.S. Bureau of the Census. Fig. 12.19 Source: Redrawn with permission from P. J. Smith, "Calgary: A Study in Urban Patterns" in *Economic Geography,* Vol. 38, 1962. Copyright 1962 *Economic Geography.* Fig. 12.20 Redrawn from B. J. L. Berry, Commercial Structure and Commercial Blight. Research Paper 85, Department of Geography. Research Series, The University of Chicago, 1963. Fig. 12.23 Redrawn from Arthur Getis, "Urban Population Spacing Analysis" in Urban Geography, Vol. 6, 1985. Copyright © 1985 V. H. Winston and Son, Inc. Siver Srings, MD. Fig. 12.24 Redrawn with permission from "The Nature of Cities" by Chauncy D. Harris and Edward L. Ullman in Volume CCXLII of the Annals of the American Academy of Political and Social Science. Copyright © 1945 The American Academy of Political and Social Science, Philadelphia, PA. Fig. 12.25 Source: Redrawn with permission from Philip Rees, "The Factorial Ecology of Metropolitan Chicago" M.A. Thesis, University of Chicago, 1968. Fig 12.26 Source: Redrawn from E. F. Bergman and T. W. Pohl, *A Geography of the New York Metropolitan Region.* Fig 12.28 Source: Redrawn with permission from B. J. L. Berry, Chicago: *Transformation of an Urban System.* Fig. 12.29 Figure from The North American City by Maurice Yeates and Barry Garner. Copyright © 1980 by Maurice Yeates and Barry J. Garner. Reprinted by permission of Harper & Row, Publishers, Inc. Table 12.3 Source: U.S. Bureau of the Census. Fig. 12.38 Paul White, *The West Europe City: A Social Geography.* Copyright © 1984, Longman Group U K Limited, Essex England. Reprinted by permission. Fig. 12.41 Redrawn with permission from Ernst Griffin and Larry Ford, "A Model of Latin American City Structure" in *Geographical Review,* 70, p. 406, 1980. Copyright © 1980 The American Geographical Society, New York, NY. Fig. 12.42 Figure from Cities of the World. World Regional Development, by Stanley D. Brunn and Jack F. Williams. Copyright © 1983 by Harper & Row, Publishers, Inc. Reprinted by permission.

Chapter 13

Fig. 13.9 From Preston E. James. Introduction to Latin America. Copyright © 1964 Odyssey Press. Macmillan Publishing Co., Inc. Fig 13.12 Source: Redrawn with permission from T. T. Waterman, "Yorok Geography," *University of California Publications in American Archaeology and Ethnology,* Vol. 16, 1920.

INDEX

Outer Seven, 262
Out-sourcing, 324, 325
Outwash plain, 72
Overfishing, 363–65
Overland Trail, 273
Overpopulation, 182–84
Overturned fold, *56*
Oxbow-shaped lake, 69
Oxidation, 65
Oxisols, 106, 362
Ozone, 136, 137, 138

P

Pacific Northwest, 51
Pacific Ocean, 50–51
    winds, 93
Pakistan, 145, 179, 242, 244, 249, 253, *261*
    birth/death rate, 173–74, 185
    political regions, 425
    population, 377
Palisades, 64
Panama City, 405
Pancasila, 220
Pangaea, 47
Parachute, Colorado, *349*
Parallel, 20, 22
    of latitude, 24
Paris, 247, 377, *408*
Pattern, map, 31–35
Pattison, William D., 10
PCBs. *See* Polychlorinated biphenyls (PCBs)
Pennsylvania, *56*, 269
Perception
    environment, 274–78
    mental maps and, 274–75, *276, 277*
    of natural hazards, 275–78
Perestroika, 304, 316, 325
Permafrost, 74
Persian Empire, 3
Peru, 93
Pest control, 149, 151
Pesticide resistance, 151
Peters, Arno, 23
Petrochemical dump, *6*
Petroleum, 54
pH factor, 134
Philippines, 50, 51, 185, 244, 307, 320
    agriculture, 315
    energy, 355
    forest, 370
Phosphorus cycle, 123, *124*
Photography, aerial, 35–36
Photovoltaic (PV) cells, 355
Pidgins, 217
Pine Barrens, *434*
Pipelines, 74, 347, *348*
Pits, 142
Pittsburgh, 346
Pixels, 40
Place
    interactions, 7
    names, 219
    physical and cultural attributes of, 6–7
    similarity, regional, 9–10
    utility, 290–91, 334
Plain, outwash, 78, 79, 109

Plane
    of the ecliptic, *83*
    table, *322*
Plantation, 315
Plants
    human impact on, 146–51
    location problem, *322*
    pollution and, 149–51
Plate tectonics, 46, 47
    African, *49*
    lithospheric, *49, 50*
    North American, 47, *49, 50, 59*
    Pacific, 47, *49, 50*
Playas, 69
Pleistocene epoch, 70, 72
Plymouth Plantation, 240
Points, quantity at, 31–33
Poisoning, 149–51
Poland, 175, 331, 345
Polar easterlies, 91
Polar high, 91
Political alliances, 264–65
Political geography, 239–71
Political systems
    international, 261–65
    local and regional, 265–70
    national, 240–60
Politics
    economic activity and, 301
    fragmentation, 267–70
    international systems of, 261–65
    regional organization and, 265–70
    religion and, 220, 255–56
Pollution
    agricultural, 128
    air and climate, 132–40
    chemical, 130
    control, 131–32
    environmental, 127
    industrial, 128, 130, 133
    mining, 128, 130
    ozone, 136–37
    plant and animal, 146–51
    sewage, 131
    solids and, 151–56
    thermal, 130
    urban drainage, 130–31
    water, 127–28, 130–31
Polychlorinated biphenyls (PCBs), 130, *150*
Pond, *71*
Population
    birth rate, 163–64, *176*
    cohort, 163
    controls, 188–90
    data, 184–87
    death rates, 166–67, *176*
    demographic equation, 177–78
    demographic transition model, 172–76
    density, 133, 179–84, 391–92
    dependency ratio, 169
    dilemma in Europe, 175
    distribution, *33, 34, 35,* 178–79
    doubling times, 170–72
    earth, 161
    explosion, 161, *176*

fertility rates, 164–65
futures, *170*
geography, 160–93
growth, 161–62, *178*
homeostatic plateau, 189
immigration and, 177–78
land value and, 391–92
Latin America, 423–24
measures, 163
metropolitan area, 403, 404, 405
momentum, 190, 191
natural increase, 170, *171*
nucleated pattern, 423
overpopulation, 182–84
predictions, 187
pressure, 404, 405
projections, 184, 187–88
prospects, 190–91
pyramids, 167–70
rates, 163
regions, 422–24
relocations, 177
replacement level, 175, 191
Third World, 174–76
urban, 377
urbanization and, 184
Western Europe, 173
world distribution, 178
world growth, 361
Population Reference Bureau, Inc., 161, 185
Portsmouth, England, 18
Portugal, 262
Portuguese, 218
Positional disputes, 250–52
Potato famine, 301
Potomac River, 131
Power
    geothermal, 355
    hydroelectric, 353, *354*
    nuclear, 153, 351
    solar, 354–55
    tidal, 356
    wind, 355, *356*
Prairie, 103
Precipitation, 82, 94, 100
    acid, 131, 133, 134–35, 136
    convectional, 96
    cyclonic, 96
    frontal, 96
    hydrologic cycle, 125
    orographic, 96, *97*
    types, 96–98
    variability, 118
Pressure
    barometric, 35
    belts, *92*
    equatorial low, 90
    gradient, 89, 90
    subtropical high, 91, 109
Primary activities, 302, 305–20
Primary Metropolitan Statistical Area, 382
Primary products, 320
Primate city, 247, 405
Prime meridian, 17, 18, 19
Procession, sun, 84
Profit, *323*

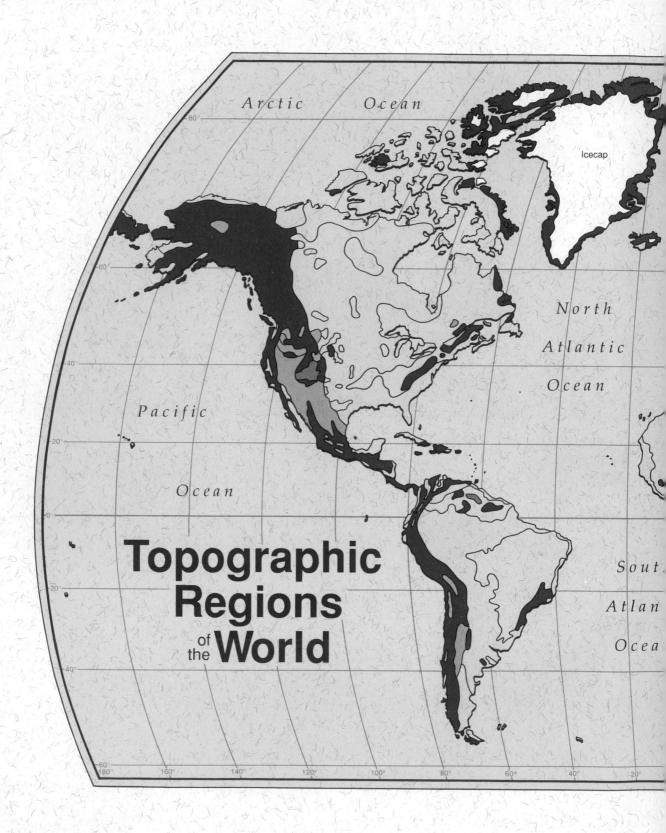

**Topographic Regions
of the World**

Arctic Ocean

North Atlantic Ocean

Pacific Ocean

Icecap

Sout Atlan Ocea